MW01011689

GUIDELINES FOR HAZARD EVALUATION PROCEDURES

Third Edition

Center for Chemical Process Safety
New York, New York

An AIChE Industry Technology Alliance

A Joint Publication of the Center for Chemical Process Safety of the American Institute of Chemical Engineers and John Wiley & Sons, Inc.

Published by John Wiley & Sons, Inc., Hoboken, New Jersey.
Published simultaneously in Canada.

For general information on our other products and services or for technical support, please contact our Customer Care Department within the United States at (800) 762-2974, outside the United States at (317) 572-3993 or fax (317) 572-4002.

Wiley also publishes its books in a variety of electronic formats. Some content that appears in print may not be available in electronic format. For information about Wiley products, visit our web site at www.wiley.com.

Library of Congress Cataloging-in-Publication Data is available.

ISBN 978-0-471-97815-2

Printed in the United States of America.

In Honor of Tom Carmody

This third edition of CCPS' *Guidelines for Hazard Evaluation Procedures* is dedicated to Tom Carmody. Tom served as the first Director of CCPS from its first year in 1985 until 1993. He made extensive use of his leadership skills from Union Carbide in establishing the basis for the organization—how it would function, the products it would develop, the acquisition of sponsors, and the development of relationships with various national and international organizations interested in process safety. Although he had not been personally involved in process safety when he took over the position, he learned rapidly, making the best use of the technical experts from the various sponsor organizations. He was the right person at the right time to develop and grow CCPS. It is only fitting that this third edition be dedicated to Tom, since the first edition of *Guidelines for Hazard Evaluation Procedures* was CCPS' very first publication and the first fruits of his leadership.

Tom and his wife Jill reside in Amelia Island, Florida.

Contents

Part I – Hazard Evaluation Procedures

Part II - Worked Examples and Appendices

Acknowledgments

The Center for Chemical Process Safety (CCPS) thanks all of the members of the HEP3 (Hazard Evaluation Procedures, 3rd Edition) Subcommittee of CCPS' Technical Steering Committee for providing input, reviews, technical guidance and encouragement to the project team throughout the preparation of this book. CCPS also expresses appreciation to the members of the Technical Steering Committee for their advice and support.

The CCPS staff liaison for this project was Bob Ormsby, who also coordinated meetings and facilitated subcommittee reviews and communications. The subcommittee had the following members, whose significant efforts and contributions are gratefully acknowledged:

Jonathan Babcock
Eli Lilly and Company

Kumar Bhimavarapu
FM Global

Christine E. Browning
Eastman Chemical Company

Paul Butler
Buckman Laboratories

Ken Harrington, HEP3 Subcommittee Chair
Chevron/Phillips

Wayne Jamison
Intel

Jim Johnston
Wyeth

Bob Lenahan
Bayer BMS

Donald Lorenzo
ABS Consulting

Narayanan Sankaran
UOP

John C. Stoney
BP

Angela Summers
SIS-TECH Solutions, LP

Tim Wagner
The Dow Chemical Company

Joe Wilson
Syngenta

Unwin Company (Columbus, Ohio) prepared this Third Edition of the *Guidelines for Hazard Evaluation Procedures,* building on the previous work of Battelle Memorial Institute (First Edition) and JBF Associates, Inc. (Second Edition). Robert W. Johnson was Unwin Company's lead author and project manager for the Third Edition. John F. Murphy was a principal author, and Steven W. Rudy, John E. Corn, and Bryan T. Haywood authored and reviewed particular sections within their areas of expertise. William G. Bridges and Revonda Tew of Process Improvement Institute, Inc. (Knoxville, Tennessee) contributed the new section on hazard evaluation of procedure-based operations.

CCPS and the Unwin Company project team also gratefully acknowledge the valuable suggestions and feedback submitted by the following persons who provided peer review comments on the final draft manuscript.

Jeffrey Castillo	Monsanto Company
Carol Garland	Eastman Chemical Company
Richard C. Griffin	Chevron Phillips Chemical Company LP
Kevin L. Klein	Solutia, Inc.
Mark M. Moderski	Lummus
Adrian L. Sepeda	CCPS Emeritus
Martin Sich	
Steve Sigmon	Honeywell Specialty Materials
Robert J. Stack	The Dow Chemical Company

In addition, comments on specific sections were provided by Paul Delanoy, Gregory Schultz and David Wechsler of The Dow Chemical Company.

List of Tables

Table

Table

Table

Table

List of Figures

Figure

Abbreviations and Acronyms

ACC	American Chemistry Council
ACGIH	American Conference of Government and Industrial Hygienists
AEGL	Acute Exposure Guideline Level
AIChE	American Institute of Chemical Engineers
AIHA	American Industrial Hygiene Association
ALARP	As low as reasonably practicable
ANSI	American National Standards Institute
API	American Petroleum Institute
ARC®	Accelerating Rate Calorimeter; accelerating rate calorimetry
ASME	American Society of Mechanical Engineers
ASSE	American Society of Safety Engineers
BLEVE	Boiling liquid expanding vapor explosion
BPCS	Basic process control system
CCA	Cause-Consequence Analysis
CCF	Common cause failure
CCPS	AIChE Center for Chemical Process Safety
CEI	Chemical Exposure Index
CHAZOP	Chemistry HAZOP *or* Computer HAZOP
CPI	Chemical process industry
CPQRA	Chemical Process Quantitative Risk Analysis
CSB	U.S. Chemical Safety and Hazard Investigation Board
DAP	Diammonium phosphate
DIERS	AIChE Design Institute for Emergency Relief Systems
DIPPR	AIChE Design Institute for Physical Property Data
EHS	Environmental, health and safety
EPA	U.S. Environmental Protection Agency
ERPG	Emergency Response Planning Guideline
ETA	Event Tree Analysis
F&EI	Fire and Explosion Index
FMEA	Failure Modes and Effects Analysis
FMECA	Failure Modes, Effects, and Criticality Analysis
FTA	Fault Tree Analysis
HAZOP	Hazard and Operability Study [*or* Analysis]
HE	Hazard evaluation
HEP	Hazard evaluation procedures
HEP3	*Guidelines for Hazard Evaluation Procedures, 3rd Edition*
HRA	Human Reliability Analysis

IChemE	Institution of Chemical Engineers (United Kingdom)
ICI	Imperial Chemical Industries
IEC	International Electrotechnical Commission
ISA	The Instrumentation, Systems, and Automation Society
IDLH	Immediately dangerous to life and health
IPL	Independent protection layer
LC_{LO}	Lethal concentration low
LD_{50}	Lethal dose, 50% mortality
LEL	Lower explosive limit
LFL	Lower flammable limit
LOPA	Layer of Protection Analysis
MCS	Minimal cut set
MSDS	Material safety data sheet
MORT	Management Oversight and Risk Tree
NFPA	National Fire Protection Association
OSHA	U.S. Occupational Safety and Health Administration
PEL	Permissible exposure limit
PFD	Process flow diagram *or* Probability of failure on demand
P&ID	Piping and instrumentation diagram
PHA	Process hazard analysis[1]
PreHA	Preliminary Hazard Analysis[1]
PSF	Performance shaping factor
PSM	Process safety management
R&D	Research and development
SCBA	Self-contained breathing apparatus
SHI	Substance Hazard Index
SIF	Safety instrumented function
SIL	Safety integrity level
SIS	Safety instrumented system
SOP	Standard operating procedure
STEL	Short term exposure limit; 15 min time-weighted-average maximum concentration
TLV®	Threshold Limit Value; occupational exposure limit recommended by ACGIH
UEL	Upper explosive limit
UFL	Upper flammable limit
VPP	[OSHA] Voluntary Protection Program
VSP2™	Vent Sizing Package, Version 2
WI	What-If [Analysis]
WI/CL	What-If/Checklist [Analysis]

---

[1] The first and second editions of these *Guidelines* used the abbreviation "PHA" for Preliminary Hazard Analysis; however, use of this abbreviation has been changed to PreHA to avoid confusion with the now more common term Process Hazard Analysis which is associated with the acronym PHA.

Glossary

See Part I, Sections 1.3 (Anatomy of an Incident) and 1.4 (The Role of Safeguards) to understand how some of the Glossary terms fit together in the context of hazard evaluation procedures.

Abnormal situation**:* A disturbance in an industrial process with which the basic process control system of the process cannot cope. In the context of hazard evaluation procedures, synonymous with ***deviation.

***Acute hazard**:* The potential for injury or damage to occur as a result of an instantaneous or short duration exposure to the effects of an incident.

***Administrative control**:* A procedural requirement for directing and/or checking engineered systems or human performance associated with plant operations.

***ALARP**:* As low as reasonably practicable; the concept that efforts to reduce risk should be continued until the incremental sacrifice (in terms of cost, time, effort, or other expenditure of resources) is grossly disproportionate to the incremental risk reduction achieved. The term *as low as reasonably achievable* (ALARA) is often used synonymously.

***Audit (process safety audit)**:* An inspection of a plant or process unit, drawings, procedures, emergency plans, and/or management systems, etc., usually by an independent, impartial team. (See "Safety Review" for contrast.)

***Autoignition temperature**:* The lowest temperature at which a fuel/oxidant mixture will spontaneously ignite under specified test conditions.

***Basic event**:* An event in a fault tree that represents the lowest level of resolution in the model such that no further development is necessary (e.g., equipment item failure, human failure, or external event).

***Basic process control system (BPCS)**:* A system that responds to input signals from the process and its associated equipment, other programmable systems, and/or from an operator, and generates output signals causing the process and its associated equipment to operate in the desired manner and within normal production limits.

***Branch point**:* A node with two paths in an event tree or cause-consequence diagram. One path represents success of a safeguard and the other path represents failure of the safeguard.

Cause**:* In the context of hazard evaluation procedures, an ***initiating cause.

***Cause-Consequence Analysis**:* A method for illustrating the possible outcomes arising from the logical combination of selected input events or states. A combination of fault tree and event tree models.

Checklist (traditional): A detailed list of desired system attributes or steps for a system or operator to perform. Usually written from experience and used to assess the acceptability or status of the system or operation compared to established norms.

Chronic hazard: The potential for injury or damage to occur as a result of prolonged exposure to an undesirable condition.

Common cause failure: The occurrence of two or more failures that result from a single event or circumstance.

Consequence: Result of a specific event. In the context of qualitative hazard evaluation procedures, the *consequences* are the effects following from the initiating cause, with the consequence description taken through to the loss event and sometimes to the loss event impacts. In the context of quantitative risk analyses, the *consequence* refers to the physical effects of the loss event usually involving a fire, explosion, or release of toxic or corrosive material.

Consequence analysis: The analysis of the effects of incident outcome cases independent of frequency or probability.

CPQRA: The abbreviation for Chemical Process Quantitative Risk Analysis. The process of hazard identification, followed by numerical evaluation of incident consequences and frequencies, and their combination into an overall measure of risk when applied to the chemical process industry. Ordinarily applied to episodic events. Related to Probabilistic Risk Assessment (PRA) used in the nuclear industry.

Deviation: A process condition outside of established design limits, safe operating limits, or standard operating procedures.

Dow Chemical Exposure Index (CEI): A method, developed by The Dow Chemical Company, used to identify and rank the relative acute health hazards associated with potential chemical releases. The CEI is calculated from five factors: a measure of toxicity; the quantity of volatile material available for a release; the distance to each area of concern; the molecular weight of the material being evaluated; and process variables that can affect the conditions of a release such as temperature, pressure, and reactivity.

Dow Fire and Explosion Index (F&EI): A method, developed by The Dow Chemical Company, for ranking the relative potential fire and explosion effect radius and property damage / business interruption impacts associated with a process. Analysts calculate various hazard and exposure factors using material characteristics and process data.

Emergency response planning guidelines (ERPG): A system of guidelines for airborne concentrations of toxic materials prepared by the AIHA. For example, ERPG-2 is the maximum airborne concentration below which it is believed nearly all individuals could be exposed for up to one hour without experiencing or developing irreversible or other serious health effects or symptoms that could impair an individual's ability to take protective action.

Engineered control: A specific hardware or software system designed to maintain a process within safe operating limits, to safely shut it down in the event of a process upset, or to reduce human exposure to the effects of an upset.

Episodic event: An unplanned event of limited duration, usually associated with an incident.

Episodic release: A release of limited duration, usually associated with an incident.

Error-likely situation: A work situation in which the performance-shaping factors are not compatible with the capabilities, limitations, or needs of the worker. In such situations, workers are much more likely to make errors, particularly under stressful conditions.

Event: An occurrence involving the process caused by equipment performance or human action or by an occurrence external to the process.

Event sequence: See ***Incident sequence***.

Event tree: A logic model that graphically portrays the combinations of events and circumstances in an incident sequence.

External event: Event external to the system caused by (1) a natural hazard — earthquake, flood, tornado, extreme temperature, lightning, etc., or (2) a human-induced event — aircraft crash, missile, nearby industrial activity, fire, sabotage, etc.

Failure: Cessation of equipment to operate as specified.

Failure mode: A symptom or condition by which a failure is observed. A failure mode might be identified as loss of function; premature function (function without demand); an out-of-tolerance condition; or a simple physical characteristic such as a leak observed during inspection.

Failure Modes and Effects Analysis (FMEA): A systematic, tabular method for evaluating and documenting the effects of known types of component failures.

Failure Modes, Effects, and Criticality Analysis (FMECA): A variation of FMEA that includes a quantitative estimate of the severity of consequence of a failure mode.

Fault event: A failure event in a fault tree that requires further development.

Fault tree: A logic model that graphically portrays the combinations of failures that can lead to a specific main failure or incident of interest (Top event).

Frequency: Number of occurrences of an event per unit time (e.g., 1 event in 1000 yr $= 1 \times 10^{-3}$ events/yr).

Hazard: A physical or chemical condition that has the potential for causing harm to people, property, or the environment.

Hazard analysis: See ***Hazard evaluation.***

Hazard and Operability (HAZOP) Study: A scenario-based hazard evaluation procedure in which a team uses a series of guide words to identify possible deviations from the intended design or operation of a process, then examines the potential consequences of the deviations and the adequacy of existing safeguards.

Hazard checklist: An experience-based list of hazards, potential incident situations, or other process safety concerns used to stimulate the identification of hazardous situations for a process or operation.

Hazard evaluation: Identification of individual hazards of a system, determination of the mechanisms by which they could give rise to undesired events, and evaluation of the consequences of these events on health (including public health), environment, and property. Uses qualitative techniques to pinpoint weaknesses in the design and operation of facilities that could lead to incidents.

Hazard identification: The pinpointing of material, system, process, and plant characteristics that can produce undesirable consequences through the occurrence of an incident.

Hazardous event: See ***Loss event.***

Human error: Any human action (or lack thereof) that exceeds some limit of acceptability (i.e., an out-of-tolerance action) where the limits of human performance are defined by the system. Includes actions by designers, operators, or managers that may contribute to or result in incidents.

Human factors: A discipline concerned with designing machines, operations, and work environments to match human capabilities, limitations, and needs.

Human Reliability Analysis (HRA): A method used to evaluate whether necessary human actions, tasks, or jobs will be completed successfully within a required time period. In these *Guidelines*, HRA is used strictly in a qualitative context. HRA is also used to determine the probability that no extraneous human actions detrimental to the system will be performed.

HRA event tree: A graphical model of sequential events in which the tree limbs designate human actions and other events as well as different conditions or influences upon these events.

Impact: A measure of the ultimate loss and harm of a loss event. *Impact* may be expressed in terms of numbers of injuries and/or fatalities, extent of environmental damage, and/or magnitude of losses such as property damage, material loss, lost production, market share loss, and recovery costs.

Incident: An unplanned event or sequence of events that either resulted in or had the potential to result in adverse impacts.

Incident sequence: A series of events composed of an initiating cause and intermediate events leading to an undesirable outcome.

Initiating cause: In the context of hazard evaluation procedures, the operational error, mechanical failure, or external event or agency that is the first event in an incident sequence and marks the transition from a normal situation to an abnormal situation. Synonymous with ***initiating event.***

Initiating event: See ***Initiating cause.***

Intermediate event: An event that occurs after the initiating cause and before the loss event in an incident sequence.

Layer of protection: A physical entity supported by a management system that is capable of preventing an initiating cause from propagating to a specific loss event or impact.

Layer of Protection Analysis (LOPA): An approach that analyzes one incident scenario (cause-consequence pair) at a time, using predefined values for the initiating cause frequency, independent protection layer failure probabilities, and consequence severity, in order to compare an order-of-magnitude scenario risk estimate to tolerable risk goals for determining where additional risk reduction or more detailed analysis is needed. Scenarios are identified elsewhere, typically using a scenario-based hazard evaluation procedure such as a HAZOP Study.

Likelihood: A measure of the expected probability or frequency of occurrence of an event.

Loss event: Point of time in an abnormal situation when an irreversible physical event occurs that has the potential for loss and harm impacts. Examples include release of a hazardous material, ignition of flammable vapors or ignitable dust cloud, and overpressurization rupture of a tank or vessel. An incident might involve more than one loss event, such as a flammable liquid spill (first loss event) followed by ignition of a flash fire and pool fire (second loss event) that heats up an adjacent vessel and its contents to the point of rupture (third loss event). Generally synonymous with ***hazardous event.***

Minimal cut set: A combination of failures and conditions necessary and sufficient to cause the occurrence of the Top event in a fault tree.

Mitigate: Reduce the impact of a loss event.

Mitigative safeguard: A safeguard that is designed to reduce loss event impact.

Operator: An individual responsible for monitoring, controlling, and performing tasks as necessary to accomplish the productive activities of a system. Often used in a generic sense to include people who perform all kinds of tasks (e.g., reading, calibration, maintenance).

Passive equipment: Hardware that is not physically actuated in order to perform its function, such as secondary containment or a blast wall.

Performance shaping factor (PSF): Any factor that influences human performance. PSFs include factors intrinsic to an individual (personality, skill, etc.) and factors in the work situation (task demands, plant policies, hardware design, training, etc.).

Process safety management: A program or activity involving the application of management principles and analytical techniques to ensure the safety of process facilities. Sometimes called ***process hazard management.***

Preventive safeguard: A safeguard that forestalls the occurrence of a particular loss event, given that an initiating cause has occurred; i.e., a safeguard that intervenes between an initiating cause and a loss event in an incident sequence. (Note that *containment and control measures* are also preventive in the sense of preventing initiating causes from occurring; however, the term *preventive safeguard* in the context of hazard evaluation procedures is used with the specific meaning given here.)

Quantitative risk analysis: The systematic development of numerical estimates of the expected frequency and severity of potential incidents associated with a facility or operation based on engineering evaluation and mathematical techniques.

Rare event: An event or incident whose expected frequency is very small. The event is not statistically expected to occur during the normal life of a facility or operation.

Recovery factors: Feedback factors that limit or prevent the undesirable consequences of a human error.

Risk: The combination of the expected frequency (events/year) and severity (effects/event) of a single incident or a group of incidents.

Risk assessment: The process by which the results of a risk analysis (i.e., risk estimates) are used to make decisions, either through relative ranking of risk reduction strategies or through comparison with risk targets.

Risk management: The systematic application of management policies, procedures, and practices to the tasks of analyzing, assessing, and controlling risk in order to protect employees, the general public, the environment, and company assets.

Risk measures: Ways of combining and expressing information on likelihood with the magnitude of loss or injury (e.g., risk indexes, individual risk measures, and societal risk measures).

Safeguard: Any device, system, or action that would likely interrupt the chain of events following an initiating cause or that would mitigate loss event impacts. See ***Preventive safeguard; Mitigative safeguard.***

Safety Review: An inspection of a plant or process unit, drawings, procedures, emergency plans, and/or management systems, etc., usually by a team and usually problem-solving in nature. (See "Audit" for contrast.)

Safety system: Equipment and/or procedures designed to limit or terminate an incident sequence, thus mitigating the incident and its consequences.

Scenario: An unplanned event or incident sequence that results in a loss event and its associated impacts, including the success or failure of safeguards involved in the incident sequence.

Scribe/recorder: A hazard evaluation team member who is responsible for capturing the significant results of discussions that occur during a hazard evaluation team meeting.

Source term: For a hazardous material and/or energy release to the surroundings associated with a loss event, the release parameters (magnitude, rate, duration, orientation, temperature, etc.) that are the initial conditions for determining the consequences of the loss event. For vapor dispersion modeling, it is the estimation, based on the release specification, of the actual cloud conditions of temperature, aerosol content, density, size, velocity and mass to be input into the dispersion model.

Task analysis: A human error analysis method that requires breaking down a procedure or overall task into unit tasks and combining this information in the form of event trees. It involves determining the detailed performance required of people and equipment and determining the effects of environmental conditions, malfunctions, and other unexpected events on both.

Top event: The loss event or other undesired event at the "top" of a fault tree that is traced downward to more basic failures using Boolean logic gates to determine its possible causes.

Two Guide Word Analysis: A procedure-based hazard evaluation technique, similar to a HAZOP Study, in which the adequacy of existing safeguards is evaluated by asking what would happen if each step in a procedure was (1) skipped or (2) performed incorrectly.

Undeveloped event: An event in a fault tree that is not developed because it is of no significance, because more detailed information is unavailable, or because its frequency or probability can be estimated without determining its basic events.

What-If Analysis: A scenario-based hazard evaluation procedure using a brainstorming approach in which typically a team that includes one or more persons familiar with the subject process asks questions or voices concerns about what could go wrong, what consequences could ensue, and whether the existing safeguards are adequate.

What-If/Checklist Analysis: A What-If Analysis that uses some form of checklist or other listing of broad categories of concern to structure the what-if questioning.

Worst case: A conservative (high) estimate of the consequences of the most severe incident identified.

Worst credible case: The most severe incident considered plausible or reasonably believable.

Part I

Hazard Evaluation Procedures

Preface

The American Institute of Chemical Engineers (AIChE) has been closely involved with process safety and loss control issues in the chemical and allied industries for more than four decades. Through its strong ties with process designers, constructors, operators, safety professionals, and members of academia, AIChE has enhanced communication and fostered continuous improvement of the industry's high safety standards. AIChE publications and symposia have become information resources for those devoted to understanding the causes of incidents and discovering better means of preventing their occurrence and mitigating their consequences.

The Center for Chemical Process Safety (CCPS) was established in 1985 by AIChE to develop and disseminate technical information for use in the prevention of major chemical incidents. CCPS is supported by nearly 100 sponsoring companies in the chemical process industry (CPI) and allied industries; these companies provide the necessary funding and professional experience for its technical subcommittees.

CCPS' first project was the preparation of *Guidelines for Hazard Evaluation Procedures*. The goal of that groundbreaking project was:

> ***"...to produce a useful and comprehensive text prepared to foster continued personal, professional, and technical development of engineers in the areas of chemical plant safety, and to upgrade safety performance of the industry... The document will be updated periodically, and will serve as a basis for additional related topics such as risk management."***

CCPS achieved its stated goal with the publication of the *Guidelines* in 1985, and has since continued to foster the development of process safety professionals in all industries. For example, CCPS has developed 85 Guideline and Concept Books and has sponsored 23 international meetings since its inception. Planning and work on many other projects are also underway. This activity has occurred in the midst of many other changes and events that over the past years have fostered an unprecedented interest in hazard evaluation:

- A number of incidents have occurred, even though many companies are seeking continuous improvement of process safety and have embraced the ideal of striving for "zero incidents." Industry is learning from these incidents, and this hard-earned experience is an important additional source of information for process safety professionals in their quest to prevent major chemical incidents in the future.
- Both private and public organizations, including government agencies, have become more concerned with ensuring the safety of industrial operations. This is exemplified by the formation and activities of the U.S. Chemical Safety and Hazard Investigation Board (Chemical Safety Board, or CSB), which has made several recommendations related to hazard evaluations.

- Many organizations—including companies, industrial groups, and others concerned with the safe handling of hazardous materials—have made clear and definite commitments to the management of process safety. In 1989, CCPS published *Guidelines for the Technical Management of Chemical Process Safety,* followed in 2007 by *Guidelines for Risk Based Process Safety.* These publications outline strategies for companies to consider when designing management systems for use in preventing major chemical incidents. Other organizations have followed suit by proposing their own approaches for process safety management (PSM). In all of these PSM models, the use of hazard evaluation techniques plays a central role in helping to manage the risk of facilities and operations.

- Many laws and regulations now place demands on organizations that handle hazardous materials. These include U.S. federal and state legislative initiatives, as well as international requirements such as the European Union's Seveso II Directive. In 1992, the U.S. Occupational Safety and Health Administration (OSHA) promulgated a standard for *Process Safety Management of Highly Hazardous Chemicals* (29 CFR 1910.119). The U.S. Congress also amended the Clean Air Act by adding chemical incident prevention provisions that included broad-based process safety requirements for companies that use hazardous chemicals (U.S. Environmental Protection Agency's Risk Management Program Rule, 40 CFR Part 68). These laws and regulations require facility owners and operators to employ hazard evaluation methods such as those recognized by CCPS. These requirements have sparked an increasing demand for practitioners who are qualified to use these methods.

- International standards related to instrumented protective systems, notably IEC 61511 and its U.S. implementation (ANSI/ISA-84.00.01, IEC 61511 Mod), reference the use of scenario-based hazard evaluation procedures as part of the process of specifying required safety integrity levels for safety instrumented systems.

Because of the experience gained in the use of hazard evaluation techniques since 1985, and the increased impetus for companies to become involved in performing these studies, CCPS decided to revise the original *Guidelines for Hazard Evaluation Procedures*. Thus, as promised in CCPS' original project mission statement, a significantly updated and expanded version was produced in 1992—*Guidelines for Hazard Evaluation Procedures, Second Edition with Worked Examples*. Recognition of further changes in the field of hazard evaluation and refinement in various methodologies led CCPS' Technical Steering Committee to conclude a Third Edition was warranted and a project was initiated. This project has now been completed and the *Guidelines for Hazard Evaluation Procedures, Third Edition* is the result. Besides considerable updating of terminology, especially as it relates to the elements of an incident scenario, the major changes from the Second Edition include the following.

- A new section on inherent safety reviews has been added, and the hazard evaluation method descriptions have been expanded to indicate how inherent safety concepts can be considered.

- The hazard evaluation methodologies have been reorganized into scenario-based and non-scenario-based methods, with the recognition that scenario-based methods can be used in conjunction with aids such as risk matrices to determine the adequacy of safeguards and the priority to be placed on follow-up actions.

- Qualitative and order-of-magnitude quantitative scenario risk estimation approaches are presented in a new chapter. These approaches are now in common use for determining the adequacy of safeguards.
- A new section summarizing Layer of Protection Analysis (LOPA) has been added, and descriptions are given of how LOPA has been combined with hazard evaluation techniques.
- Use of the cause-by-cause approach to documenting HAZOP Studies has been emphasized, to lessen the likelihood of overestimating scenario risks or crediting safeguards that do not apply to particular initiating cause / loss event combinations.
- Other new sections have been added on evaluating procedure-based operations, evaluating the hazards of programmable systems, and addressing issues related to facility siting. New text on addressing human factors has been added to consideration of the Human Reliability Analysis technique.
- An even greater emphasis has been placed on process life cycle considerations as they relate to hazard evaluations, including hazard reviews for management of change, and a new section discusses integrating hazard evaluations with other considerations such as reliability and security.
- Additional checklists and forms have been included in the book chapters and in Appendix A.

Part I — Hazard Evaluation Procedures of these Guidelines describes methods used to identify and assess the significance of hazardous situations found in process operations or activities involving hazardous materials. However, these approaches are not limited in their application to the chemical manufacturing industry; they are also appropriate for use in any industry where activities create situations that have the potential to harm workers or the public; damage equipment or facilities; or threaten the environment through hazardous material releases, fires, or explosions.

Part I contains an overview for management and nine chapters. Appendices are located at the end of the book. The following list describes the organization of ***Part I***.

Management Overview

- Summarizes the use of hazard evaluation techniques as an integral part of a process safety management program
- Describes how these techniques can be used throughout the life of a process to support many PSM activities
- Lets managers know what they can realistically expect from a hazard evaluation and discusses important limitations found in the most commonly used techniques

Chapter 1 — Introduction to the Guidelines

- Describes how hazard evaluation techniques fit into an overall PSM program
- Relates the use of hazard evaluation techniques to risk management strategies
- Introduces terminology used for evaluating process hazards in the context of a typical incident sequence of events

- Introduces the role of safeguards in preventing and protecting against process upsets and mitigating the impacts of loss events
- Shows how hazard evaluation techniques can be used throughout the lifetime of a process or operation
- Outlines important theoretical and practical limitations of hazard evaluation techniques and summarizes what practitioners and management can reasonably expect from the use of these approaches

Chapter 2 — Preparation for Hazard Evaluations

- Describes the infrastructure needed to support a hazard evaluation program
- Gives examples of appropriate statements of scope for hazard evaluations
- Outlines the skills and information needed to perform these studies
- Addresses schedule and logistical considerations associated with the efficient execution of hazard evaluations

Chapter 3 — Hazard Identification Methods

- Discusses the importance of identifying hazards and the contemporary approaches used in hazard identification
- Illustrates the use of experience in analyzing material properties and process conditions for hazards
- Presents several structured approaches for hazard identification, with examples
- Describes the types of results that can be expected from hazard identification techniques, which can be used in subsequent hazard evaluation efforts

Chapter 4 — Non-Scenario-Based Hazard Evaluation Procedures

- Explains the difference between scenario-based and non-scenario based hazard evaluations
- Provides the following information for each of four non-scenario-based hazard evaluation techniques: purpose, description, types of results, resource requirements and analysis procedure
- Illustrates each method with a brief example

Chapter 5 — Scenario-Based Hazard Evaluation Procedures

- Provides the following information for each of eight hazard evaluation methods that are capable of being used to generate incident scenarios and evaluate scenario-based risks: purpose, description, types of results, resource requirements and analysis procedure
- Illustrates each method with a brief example

Chapter 6 — Selection of Hazard Evaluation Techniques

- Discusses factors that can influence selection of an appropriate hazard evaluation technique
- Lists selection criteria and provides a flowchart of questions to help choose an appropriate method for a particular application

Chapter 7 — Risk-Based Determination of the Adequacy of Safeguards

- Gives guidelines for when it is appropriate to perform a more detailed evaluation of scenario risks
- Introduces the basic concepts of estimating loss event impacts, initiating cause frequency, and safeguard effectiveness
- Gives examples of how these scenario risk estimates can be compared to risk criteria for determining the adequacy of safeguards
- Introduces Layer of Protection Analysis (LOPA) as a technique to evaluate scenarios on an order-of-magnitude basis

Chapter 8 — Analysis Follow-Up Considerations

- Discusses the importance of prioritizing the results and properly documenting a hazard evaluation
- Gives general guidelines for communicating these results to managers so they can make appropriate risk management decisions
- Presents strategies for tracking the changes made as a result of a hazard evaluation

Chapter 9 — Extensions and Special Applications

- Gives further information on special related topics including human factors; facility siting; and evaluating hazards of procedure-based operations, programmable control systems, and reactive chemical systems
- Discusses the combining of tools such as HAZOP with LOPA

Appendices. Located at the end of ***Part II — Worked Examples***, the Appendices provide:

- Example checklists and forms to help analysts perform various hazard evaluations
- A legend of symbols and abbreviations used in drawings in Part II
- A list of commercially available software aids for performing hazard evaluations
- A chemical compatibility chart to aid in identifying hazards
- A listing of organizations offering process safety enhancement resources.

The *Guidelines for Hazard Evaluation Procedures* contain information useful to both the inexperienced analyst and the accomplished practitioner. Chapters 1 through 3 are important for both the beginner and experienced hazard analyst. The experienced analyst may wish to scan the ideas on selecting an appropriate hazard evaluation method (Chapter 6); after that, to proceed directly to the appropriate sections in Chapters 4 and 5, which give the detailed steps for performing the chosen technique, and/or to Chapter 9, which gives information on special applications. Chapters 7 and 8 advise all analysts—regardless of their hazard evaluation experience—of ways to prioritize, document, and communicate the results of the hazard evaluations. The Overview figure on the next page shows how these chapters are interrelated.

Part II — Worked Examples for Hazard Evaluation Procedures, the companion to the *Guidelines,* provides the novice hazard analyst with realistic examples in which various hazard evaluation techniques are used throughout the life of a process. Experienced hazard analysts that are selected to provide in-house training will find *Part II* extremely helpful as they develop training programs. Moreover, even the experienced practitioner should find the *Worked Examples* helpful when designing and executing corporate PSM programs.

As was true for the original *Guidelines for Hazard Evaluation Procedures* and the *Second Edition with Worked Examples,* these *Guidelines* do not contain a complete program for managing the risk of chemical operations, nor do they give specific advice on how to establish a hazard analysis program for a facility or an organization. However, they do provide some of the insights that should be considered when making risk management decisions and designing risk management programs. Furthermore, they describe what users can reasonably expect from their performance of high quality hazard evaluations.

These *Guidelines* cannot replace hazard evaluation experience. This book should be used as an aid for the initial training of hazard analysts and as reference material for experienced practitioners. Only through frequent use will beginners become skilled in hazard evaluation techniques and be able to perform efficient hazard evaluations. Using these *Guidelines* within the framework of a complete PSM program will help organizations continually improve the safety of their facilities and operations.

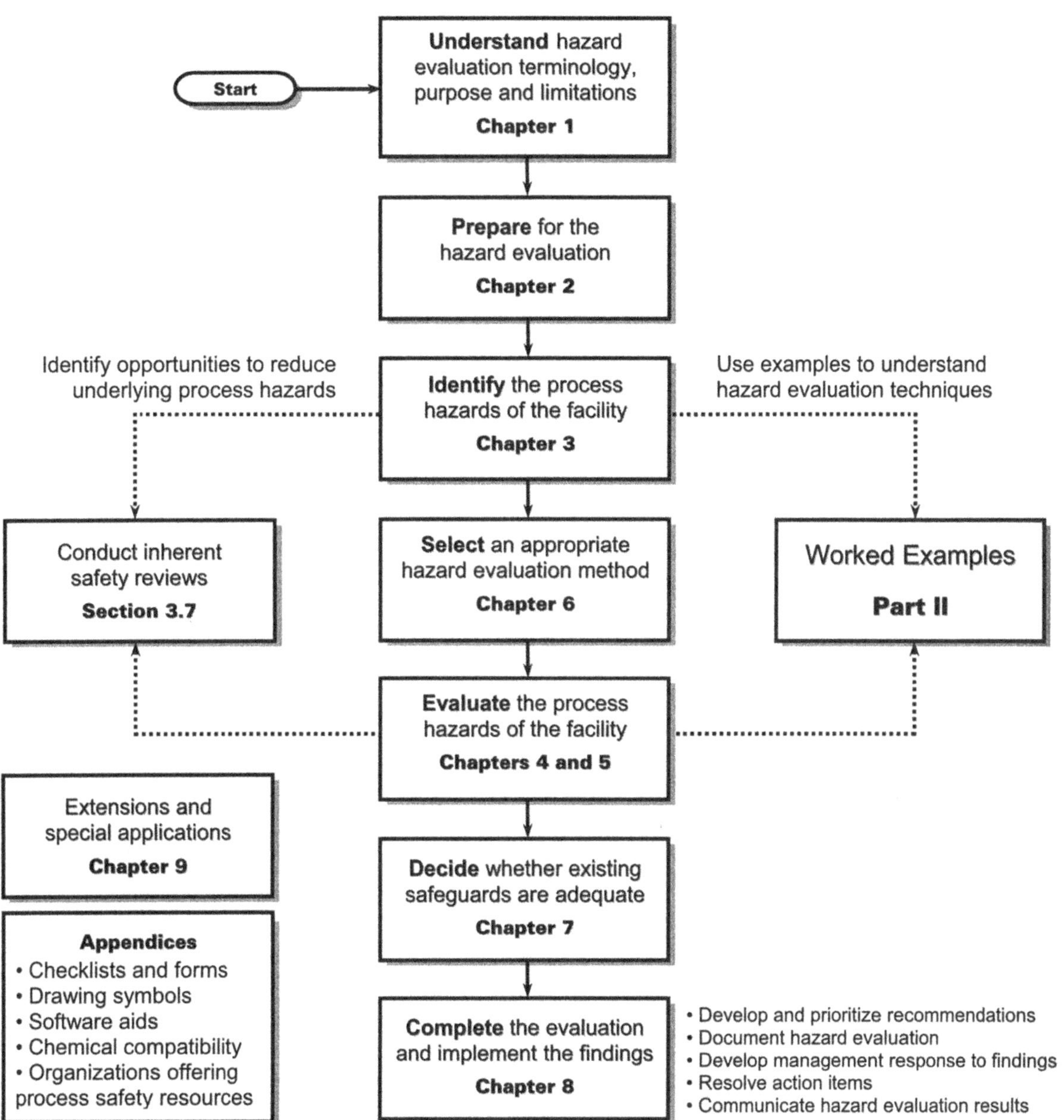

Overview Interrelation of book chapters

Management Overview

A hazard evaluation is an organized effort to identify and analyze the significance of hazardous situations associated with a process or activity. Specifically, hazard evaluations are used to pinpoint weaknesses in the design and operation of facilities that could lead to chemical releases, fires, or explosions. These studies provide organizations with information to aid in making decisions for improving safety and managing the risk of operations. Hazard evaluations usually focus on process safety issues, like the acute effects of unplanned chemical releases on plant personnel or the public. These studies complement more traditional industrial health and safety activities in which protection against slips or falls, use of personal protective equipment, and monitoring for employee exposures to industrial chemicals are considered. Although primarily directed at providing safety-related information, many hazard evaluation techniques can also be used to investigate operability, economic, and environmental concerns.

Hazard evaluation is the cornerstone of an organization's overall process safety management (PSM) program. Although hazard evaluations typically involve the use of qualitative techniques to analyze potential equipment failures and human errors that can lead to incidents, the studies can also highlight gaps in the management systems of a process safety program. In addition, individual hazard evaluation techniques can be used as a part of many other PSM program elements. For example, hazard evaluation techniques can be used (1) to investigate the possible causes of an incident that has occurred; (2) as part of a facility's management of change program; and (3) to identify critical safety equipment for special maintenance, testing, or inspection as part of a facility's mechanical integrity program.

Hazard evaluations should be performed throughout the life of a process as an integral part of an organization's PSM program. These studies can be performed to help manage the risk of a process from the earliest stages of research and development (R&D); in detailed design and construction; periodically throughout the operating lifetime; and continuing until the process is decommissioned and dismantled. By using this "life cycle" approach in concert with other PSM activities, hazard evaluations can efficiently reveal deficiencies in design and operation before a unit is sited, built, or operated, thus making the most effective use of resources devoted to ensuring the safe and productive life of a facility.

Part I — Hazard Evaluation Procedures contains a brief overview of the purpose, benefits, costs, and limitations of various hazard evaluation techniques for those with a need for basic information. It also contains "how to" details on preparing for hazard evaluations, techniques for identifying hazards, strategies for selecting appropriate hazard evaluation techniques, procedures for using hazard evaluation methods, and advice on documenting and using the results of a study. Part I contains specific steps for performing a hazard evaluation using the following techniques:

- Preliminary Hazard Analysis
- Safety Review
- Relative Ranking
- Checklist Analysis
- What-If Analysis
- What-If/Checklist Analysis
- Hazard and Operability Study
- Failure Modes and Effects Analysis
- Fault Tree Analysis
- Event Tree Analysis
- Cause-Consequence Analysis

These techniques represent the approaches for hazard evaluation most often used in the chemical process industry (CPI). For completeness, other less commonly used techniques are also briefly reviewed. The advice contained in the *Guidelines for Hazard Evaluation Procedures* is based on the experience of process safety professionals with many years of practice in applying hazard evaluation techniques in the CPI and allied industries.

Part II — Worked Examples is included for those who wish to become more experienced in the use of hazard evaluation technology and for those responsible for training analysts to use these methods. With the guidance provided in Parts I and II, analysts should be able to understand the basics of hazard evaluation and begin performing hazard evaluations of simple processes using the less complicated hazard evaluation techniques. With practice using the techniques described in the *Hazard Evaluation Procedures* and the *Worked Examples*, and with experience gained from participating in actual studies, a hazard analyst should be able to scope, organize, lead, and document hazard evaluations of most types of processes and operations with a minimum of outside assistance.

The benefits of a hazard evaluation program can be substantial, although none of these effects can easily be measured over a short period of time. These benefits can include:

- Fewer incidents over the life of a process
- Reduced consequences of incidents that do occur
- Improved emergency response
- Improved training and understanding of the process
- More efficient and productive operations
- Improved regulatory and community relations

However, these benefits cannot be realized without a significant investment. Depending upon the size and complexity of a process or operation, a hazard evaluation can require from several hours to many months to complete. Moreover, the documentation, training, and staff resources required to support a hazard evaluation program over the life of a facility can be extensive. Because of the large resource commitments necessary to maintain a vigorous hazard evaluation program, it is important that an organization have strategies in place to use properly trained and skilled people for performing this type of work. It is also extremely important that the appropriate hazard evaluation techniques are selected for each process or operation to ensure that effort is not wasted by over-studying a problem with a more detailed approach than is necessary.

Users and reviewers of hazard evaluations need to be aware that even in an efficient and high quality hazard evaluation program there are a number of limitations:

1. Analysts can never be certain they have identified all hazards, potential incident situations, causes, and effects.

2. Most of the time, the results and benefits of performing hazard evaluations cannot be directly verified. The savings from incidents that are prevented cannot be readily estimated.

3. Hazard evaluations are based on existing knowledge of a process or operation. If the process chemistry is not well known, if the relevant drawings or procedures are not accurate, or if the process knowledge available from a study team does not reflect the way the system is actually operated, then the results of a hazard evaluation may be invalid. This could lead managers to make poor risk management decisions.

4. Hazard evaluations are very dependent on the subjective judgment, assumptions, and experience of the analysts. The same process, when analyzed by different teams of competent analysts, may yield somewhat different results.

Performing high quality hazard evaluations throughout the lifetime of a process cannot guarantee that incidents will not occur. However, when used as part of an effective process safety management program, hazard evaluation techniques can provide valuable input to managers who are deciding whether or how to reduce the risk of chemical operations. With programs such as these in place, organizations will be well positioned to strive for continual improvement in process safety.

Understanding hazards and risks is one of the four pillars upon which risk-based process safety is established. Another of the pillars is committing to process safety. To commit to process safety, facilities need to focus on:

- Developing and sustaining a culture that embraces process safety
- Identifying, understanding, and complying with codes, standards, regulations, and laws
- Establishing and continually enhancing organizational competence
- Soliciting input from and consulting with all stakeholders, including employees, contractors, and neighbors.

Each of the above areas of focus is essential to conducting effective hazard evaluations. In addition, since the management systems for each of these elements should be based on an organization's current understanding of the risk associated with the process with which the workers will interact, it can be seen that understanding hazards and risks is an important part of a commitment to process safety.

1

Introduction to the Guidelines

A *hazard* is a physical or chemical condition that has the potential for causing harm to people, property, or the environment. A *hazard evaluation* is an organized effort to identify and analyze the significance of hazardous situations associated with a process or activity. Specifically, hazard evaluations are used to pinpoint weaknesses in the design and operation of facilities that could lead to hazardous material releases, fires, or explosions. These studies provide organizations with information to help them improve the safety and manage the risk of their operations.

Hazard evaluations usually focus on process safety issues, like the acute effects of unplanned chemical releases on plant personnel or the public. These studies complement more traditional industrial health and safety activities, such as protection against slips or falls, use of personal protective equipment, monitoring for employee exposure to industrial chemicals, and so forth. Many hazard evaluation techniques can also be used to help satisfy related needs (e.g., operability, economic, and environmental concerns). Although hazard evaluations typically analyze potential equipment failures and human errors that can lead to incidents, the studies can also highlight gaps in the management systems of an organization's process safety program. For example, a hazard evaluation of an existing process may reveal gaps in the facility's management of change program or deficiencies in its maintenance practices.

From its inception, the Center for Chemical Process Safety (CCPS) has recognized the importance of hazard evaluations; in fact, the first book in CCPS' series of guidelines dealt with hazard evaluation procedures.[1] Because of the ongoing and increased emphasis on performing hazard evaluations, CCPS commissioned the development of the *Guidelines for Hazard Evaluation Procedures, Third Edition.* The purpose of *Part I — Hazard Evaluation Procedures* is to provide users with a basic understanding of the concepts of hazard evaluation, as well as information about specific techniques so they will be able to perform high quality hazard evaluations within a reasonable amount of time. Several chapters on new topics, including preparing for studies, identifying hazards, and following up after completed analyses are included in the *Guidelines.*

In addition, because of the ongoing need to train a large number of competent hazard evaluation practitioners, this document includes the companion, *Part II — Worked Examples.* The *Worked Examples* give detailed illustrations of how the various hazard evaluation techniques can be used throughout the lifetime of a process as a part of a company's process safety management (PSM) program. People responsible for hazard evaluation training in their organizations will find both the *Hazard Evaluation Procedures* and the *Worked Examples* to be valuable resources.

The remainder of the Introduction explains some basic terminology and concepts of hazard evaluation and its relationship to risk management. It outlines various incident prevention and risk management strategies and discusses how hazard evaluation can provide important information to organizations who are striving for incident-free operation. This section also discusses how hazard evaluations can be performed throughout the life of a process as part of a PSM program. Finally, some limitations that should influence the interpretation and use of hazard evaluation results are presented.

1.1 Background

Formal hazard evaluations have been performed in the chemical process industry (CPI) for more than thirty years. Other less systematic reviews have been performed for even longer. Over the years, hazard evaluations have been called by different names. At one time or another, all of the terms listed in Table 1.1 have been used as synonyms for hazard evaluation, with some of the terms having different shades of meaning depending on the context and usage.

An important prerequisite or starting point for performing a hazard evaluation is the identification of process hazards, since hazards that are not identified cannot be further studied. Chapter 3 describes frequently used hazard identification methods and discusses their use in hazard evaluation efforts. An efficient and systematic hazard evaluation, preceded by a thorough hazard identification effort, can increase managers' confidence in their ability to manage risk at their facilities.

Hazard evaluations usually focus on the potential causes and consequences of episodic events, such as fires, explosions, and unplanned releases of hazardous materials, instead of the potential effects of conditions that may routinely exist such as a pollutant emitted from a registered emission point. Also, hazard evaluations usually do not consider situations involving occupational health and safety issues, although any new issues identified during the course of a hazard evaluation are not ignored and are generally forwarded to the appropriate responsible person. Historically, these issues have been handled by good engineering design and operating practices. In contrast, hazard evaluations also focus on ways that equipment failures, software problems, human errors, and external factors (e.g., weather) can lead to fires, explosions, and releases of toxic material or energy.

Hazard evaluations can occasionally be performed by a single person, depending upon the specific need for the analysis, the technique selected, the perceived hazard of the situation being analyzed, and the resources available. However, ***most high-quality hazard evaluations require the combined efforts of a multidisciplinary team.*** The hazard evaluation team uses the combined experience and judgment of its members along with available data to determine whether the identified problems are serious enough to warrant change. If so, they may recommend a particular solution or suggest that further studies be performed. Sometimes a hazard evaluation cannot give decision makers all the information they need, so more detailed methods may need to be used such as Layer of Protection Analysis (LOPA) or Chemical Process Quantitative Risk Analysis (CPQRA).

The purpose of these *Guidelines* is to provide practitioners and potential users of the results of hazard evaluations with information about identifying hazards, selecting a hazard evaluation technique appropriate for a particular need, using a particular method, and following up on the results. This document is designed to be useful to the veteran hazard analyst as well as the novice. It also provides some guidance to those faced with using, reviewing, or critiquing the results of hazard evaluations so they will know what to reasonably expect from them. Special emphasis is placed on the theoretical and practical limitations of the various hazard evaluation techniques presented.

Table 1.1 Hazard evaluation synonyms

▪ Process hazard(s) analysis	▪ Predictive hazard evaluation	▪ Hazard and risk analysis
▪ Process hazard(s) review	▪ Hazard assessment	▪ Hazard identification and risk analysis
▪ Process safety review	▪ Process risk survey	
▪ Process risk review	▪ Hazard study	

1.2 Relationship of Hazard Evaluation to Risk Management Strategies

Over the past few years, remarkable progress has been made toward institutionalizing formal process safety management (PSM) programs within chemical process industry companies. This crescendo of activity was sparked by a variety of factors including (1) the occurrence of major industrial incidents, (2) aggressive legislative and regulatory process safety initiatives reflecting a reduced public risk tolerance, and (3) the evolution and publication of model PSM programs by several industrial organizations.[2-9] Perhaps even more significant was the increased awareness and the enlightened self-interest of companies that realized, in the long run, operating a safer plant leads to more profitable business performance and better relationships with communities and regulatory agencies.

In 1989, CCPS published its *Guidelines for Technical Management of Chemical Process Safety*, which outlined a twelve-element strategy for organizations to consider when adopting management systems to ensure process safety in their facilities.[10] This strategy has been more recently updated and expanded to reflect an emphasis on risk-based process safety, as reflected in the twenty elements listed in Table 1.2.[16] Two of the elements in this table address the identification of hazards, assessment of risk, and selection of risk control alternatives throughout the operating lifetime of a facility. Other elements such as management of change, incident investigation, and asset integrity and reliability can also involve the use of hazard evaluation techniques.

Implementing a PSM program can help an organization manage the risk of a facility throughout its lifetime. Managers must, at various times, be able to develop and improve their understanding of the things that contribute to the risk of the facility's operation.[11-13] Developing this understanding of risk requires addressing three specific questions (also shown in Figure 1.1):

- ***What can go wrong?***
- ***What is the potential impact*** *(i.e., how severe are the potential loss event consequences)?*
- ***How likely is the loss event to occur?***

Table 1.2 CCPS elements of risk-based process safety

Commit to process safety	Manage risk	Learn from experience
▪ Process safety culture	▪ Operating procedures	▪ Incident investigation
▪ Compliance with standards	▪ Safe work practices	▪ Measurement and metrics
▪ Process safety competency	▪ Asset integrity and reliability	▪ Auditing
▪ Workforce involvement	▪ Contractor management	▪ Management review and continuous improvement
▪ Stakeholder outreach	▪ Training and performance assurance	
Understand hazards and risk	▪ Management of change	
▪ Process knowledge management	▪ Operational readiness	
▪ Hazard identification and risk analysis	▪ Conduct of operations	
	▪ Emergency management	

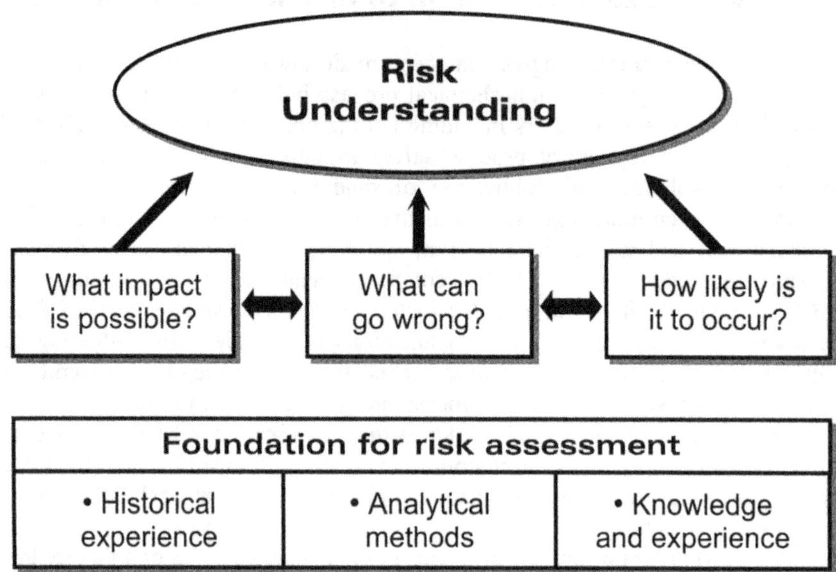

Figure 1.1 Aspects of understanding risk

The effort needed to develop this understanding of risk will depend upon (1) how much information the organization possesses concerning potential incidents and (2) the specific circumstance that defines the organization's need for better risk information. In any case, managers should first use their experience and knowledge to understand the risk their organizations face in operating a facility. If the organization has a great deal of pertinent, closely related experience with the subject process or operation, then little formal analysis may be needed. In these situations, experienced-based hazard evaluation tools (e.g., checklists) are commonly used to manage risk.

On the other hand, if there is not a relevant or adequate experience base, an organization may have to rely on analytical techniques for developing "answers" to the three risk questions to satisfactorily meet the organization's risk management needs. In these situations, organizations typically use predictive hazard evaluation techniques to creatively evaluate the significance of potential incidents.

Using hazard evaluation techniques is one way to increase a company's understanding of the risk associated with a planned or existing process or activity so that appropriate risk management decisions can be made.

1.3 Anatomy of a Process Incident

One definition of process safety is the sustained absence of process incidents at a facility. To prevent these process incidents, one must understand how they can occur. Using hazard evaluation methods can help organizations better understand the risks associated with a process and how to reduce the frequency and severity of potential incidents. Section 1.2 showed how hazard evaluation procedures fit into an overall strategy for risk management. The purpose of Sections 1.3 and 1.4 is to discuss some of the salient features of process incidents by presenting the "anatomy" of typical process incidents.[14]

A process hazard represents a threat to people, property and the environment. Examples of process hazards are given in Table 1.3. Process hazards are always present whenever hazardous materials and hazardous process conditions are present. Under normal conditions, these hazards are all contained and controlled.

An ***incident*** is defined as an unplanned event or sequence of events that either resulted in or had the potential to result in adverse impacts. Thus, an ***incident sequence*** is a series of events that can transform the threat posed by a process hazard into an actual occurrence.

Table 1.3 Elements of process incidents

Process hazards	Initiating causes	Incident outcomes
Significant inventories of:	Containment failures	Loss events
Flammable materials	Pipes, ducts, tanks, vessels, containers, flexible hoses, sight glasses, gaskets/seals	Discharges or releases
Combustible materials		Fires
Unstable materials		Pool fires
Corrosive materials		Jet fires
Asphyxiants	Equipment malfunctions	Flash fires
Shock-sensitive materials	Pumps, compressors, agitators, valves, instruments, sensors, control failures	Fireballs
Highly reactive materials		Explosions
Toxic materials		Confined explosions
Inert gases	Spurious trips, vents, reliefs	Unconfined vapor cloud explosions
Combustible dusts	Loss of utilities	
Pyrophoric materials		Vessel rupture explosions
	Electricity, nitrogen, water, refrigeration, air, heat transfer fluids, steam, ventilation	
Physical conditions		BLEVEs
High temperatures		Dust explosions
Cryogenic temperatures		Detonations
High pressures	Human errors	Condensed-phase detonations
Vacuum	Operations	
Pressure cycling	Maintenance	Impacts
Temperature cycling	External events	Toxic, corrosive, thermal, overpressure, missile, and other effects on:
Vibration/liquid hammering	Vehicle impact	
Ionizing radiation	Extreme weather conditions	Community
High voltage/current	Earthquake	Workforce
Mass storage	Nearby incident impacts	Environment
Material movement	Vandalism/sabotage	Company assets
Liquefied gases		Production

The first event in an incident sequence is called the ***initiating cause,*** also termed the ***initiating event*** or, in the context of most hazard evaluation procedures, just the ***cause.*** The types of events that can initiate incident sequences are generally equipment or software failures, human errors, and external events. Table 1.3 gives some examples.

The initiating cause can be understood by considering the anatomy of an incident from an operations perspective, as presented in Figure 1.2. In the **Normal** operations mode, all process hazards are contained and controlled, and the facility is operating within established limits and according to established operating procedures. The operational goals during normal operation can be summarized as optimizing production and keeping the facility within the bounds of the normal operating procedures and limits. Key systems involved in keeping the facility operating normally include the primary containment system typically consisting of piping and vessels, the basic process control system (BPCS) including sensors and final control elements, functional process equipment such as pumps and distillation columns, and the execution of operational tasks according to established operating procedures. These key systems are supported by activities such as inspections, functional testing, preventive maintenance, operator training, management of change, and facility access control.

An ***initiating cause*** has as its result a shift from a **Normal** to an **Abnormal** operations mode, as soon as the operation departs from its established operating procedures or safe operating limits. In the context of hazard evaluation procedures, this abnormal mode is termed a ***deviation***. For example, loss of cooling water supply to an exothermic reaction system can be an initiating cause for a runaway reaction incident sequence. As soon as the cooling water supply (pressure and/or flow rate) drops below the minimum established limit, it can be considered an initiating cause, and the plant is in an "abnormal situation." The plant operational goal changes when an abnormal situation is detected. Instead of the goal of keeping the plant operating within normal limits, the operational goal becomes returning the plant to normal operation if possible; and, if this is not possible, bringing it to a safe state such as shutting down the unit before a loss event can occur.

If the situation in this example is allowed to continue uncorrected, a runaway reaction may result, with possible outcomes of an emergency relief discharge to the atmosphere (if the system is so configured) or a vessel rupture due to overpressurization. At this point, the operating mode transitions from an abnormal situation—which may be able to be corrected and brought back under control—to an **Emergency** situation. (The term "emergency" in this context refers to the emergency operations mode after a loss event occurs. Emergency procedures may actually be activated even before the relief discharge or vessel rupture event.) The operational goal again changes in an emergency situation, with the objective now being to minimize injuries and losses (*mitigate* the loss event impacts).

Key Concept: the <u>Loss Event</u>

In the anatomy of an incident, the beginning of an **Emergency** situation is termed the ***<u>loss event</u>*** (Figure 1.3), since some degree of loss or harm is likely to ensue once a loss event has occurred. The loss event is the point of time in an incident sequence when an ***<u>irreversible physical event</u>*** occurs that has the potential for loss and harm impacts. Examples include opening of a non-reclosing emergency relief device such as a rupture disk, release of a hazardous material to the environment, ignition of flammable vapors or an ignitable dust cloud, and overpressurization rupture of a tank or vessel. Other examples are given in Table 1.3. Note that an incident might involve more than one loss event, such as a flammable liquid spill (first loss event) followed by ignition of a flash fire and pool fire (second loss event) that heats up an adjacent vessel and its contents to the point of rupture (third loss event).

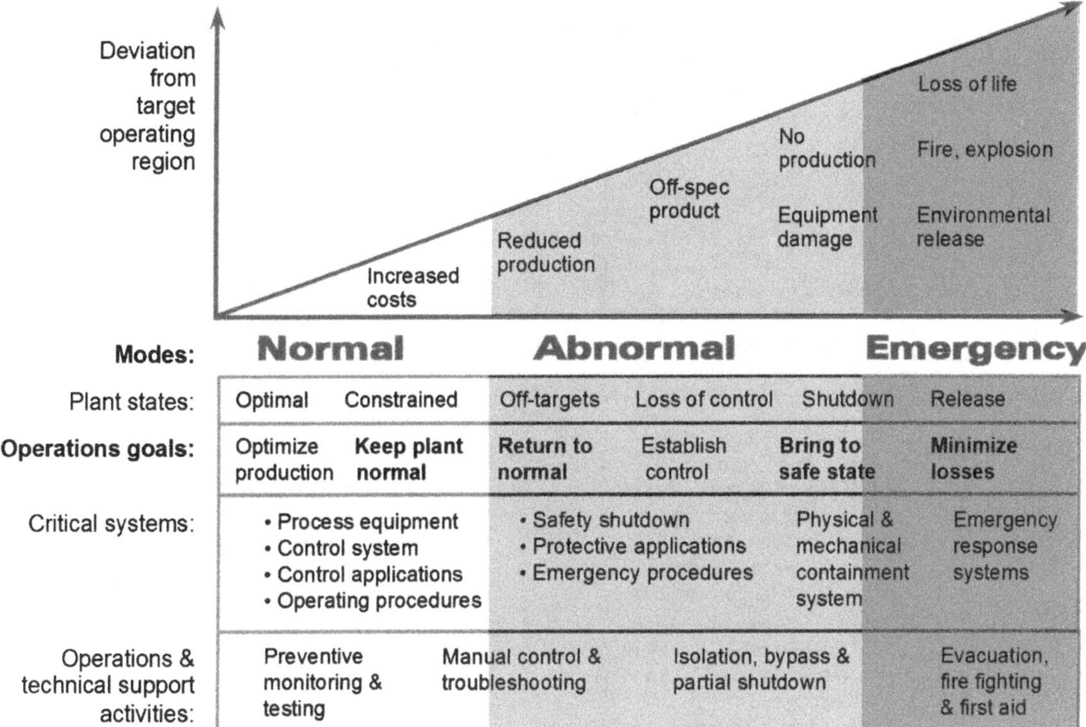

Figure 1.2 Anatomy of a catastrophic incident, from Reference 17

(Note: This Figure is included only to help understand initiating causes and loss events in relation to Normal, Abnormal, and Emergency operational modes and highlighted key operational goals)

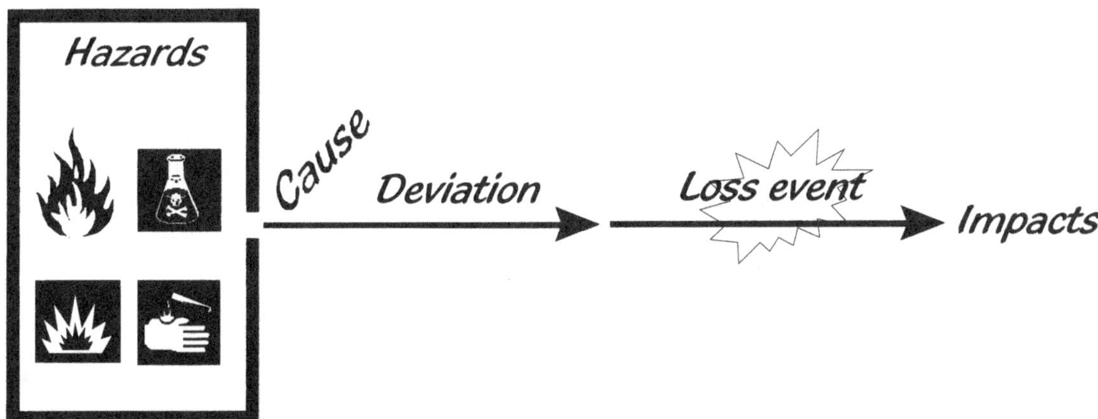

Figure 1.3 Basic incident sequence without safeguards

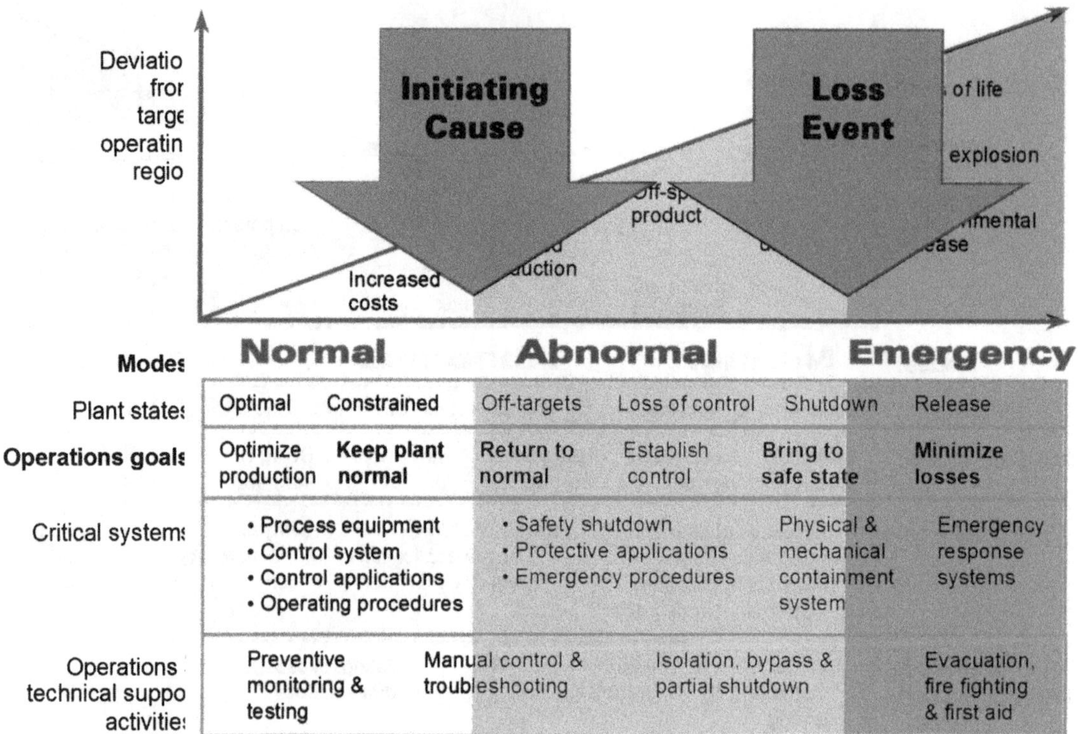

Figure 1.4 Identifying the initiating cause and the loss event in an incident scenario

Figure 1.4 might be helpful in identifying the initiating cause and loss event in an incident sequence. The initiating cause is at the transition from the **Normal** to the **Abnormal** mode of operation, and the loss event is at the transition from the **Abnormal** to the **Emergency** mode of operation.

The ***initiating cause*** may proceed directly to the ***loss event*** if there are no intervening safeguards or if the initiating cause is so severe that the design basis for the safeguards is violated. An example would be sufficient vehicle movement to cause mechanical failure of a simple unloading hose while transferring a hazardous material. As soon as the vehicle-movement initiating cause occurs, the irreversible physical event (unloading hose failure with release of hazardous material to the surroundings) would be realized. More often, there is a series of intermediate events that link an initiating cause to the loss event, due to the presence of preventive safeguards as described in Section 1.4.

The severity of consequences of the loss event is termed the ***impact*** (see Figure 1.3). The impact is a measure of the ultimate loss and harm of a loss event. It may be expressed in terms of numbers of injuries and/or fatalities, extent of environmental damage, and/or magnitude of losses such as property damage, material loss, lost production, market share loss, and recovery costs.

The full description of a possible incident sequence is a ***scenario***. A scenario is an unplanned event or incident sequence that results in a loss event and its associated impacts, including the success or failure of safeguards involved in the incident sequence (see Section 1.4 regarding the role of safeguards). Thus, each scenario starts with an initiating cause as previously described, and terminates with one or more incident outcomes. The outcomes may involve various physical or chemical phenomena, which can be evaluated using ***consequence analysis*** methodologies, to determine the loss event impacts.

Hazard evaluation methods can help users understand the significance of potential incident sequences associated with a process or activity. This understanding leads to identification of ways to reduce the frequency and severity of potential incidents, thus improving the safety of process operations.

1.4 The Role of Safeguards

In the context of hazard evaluation procedures, any device, system or action that would likely interrupt the chain of events following an initiating cause is known as a ***safeguard***.[18] Different safeguards can have very different functions, depending on where in an incident sequence they are intended to act to reduce risks, as illustrated in an event-tree format in Figure 1.5.

One way of characterizing safeguards that is useful in hazard evaluations is to view the safeguards in relation to the loss event. A ***preventive safeguard*** intervenes after an initiating cause occurs and prevents the loss event from ensuing. A ***mitigative safeguard*** acts after the loss event has occurred and reduces the loss event impacts. Thus, preventive safeguards affect the likelihood of occurrence of the loss event, whereas mitigative safeguards lessen the severity of consequences of the loss event. As will be discussed later, more than one loss event is possible for a given initiating cause, depending on the success or failure of safeguards. Figure 1.6, which is a "bow-tie" diagram as further described in Section 5.7, provides another illustration of how preventive and mitigative safeguards relate to hazards, initiating causes, loss events, and impacts.

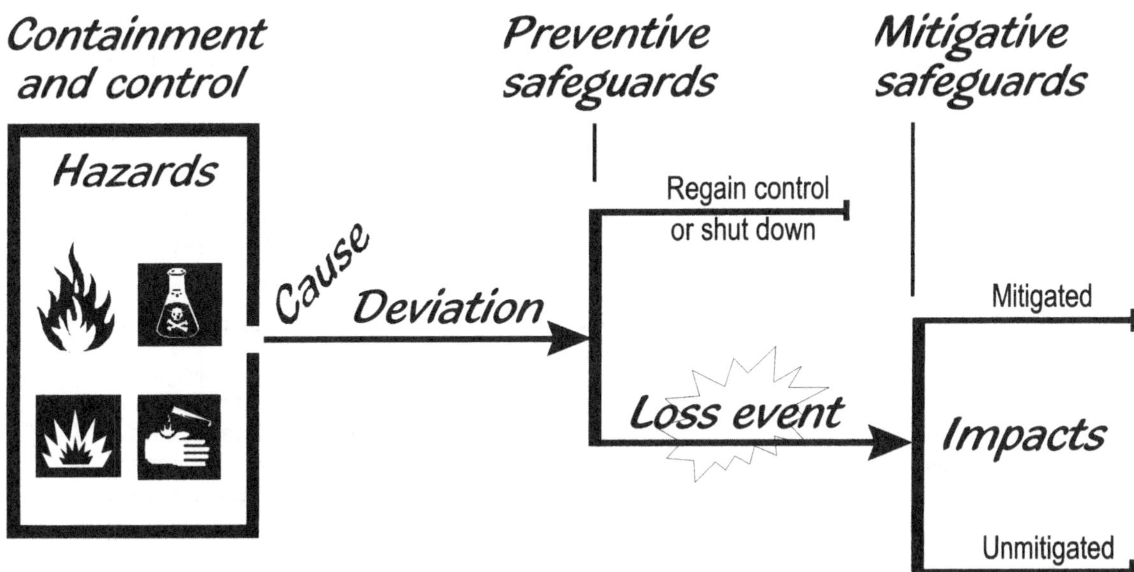

Figure 1.5 Preventive and mitigative safeguards function after an initiating cause has occurred

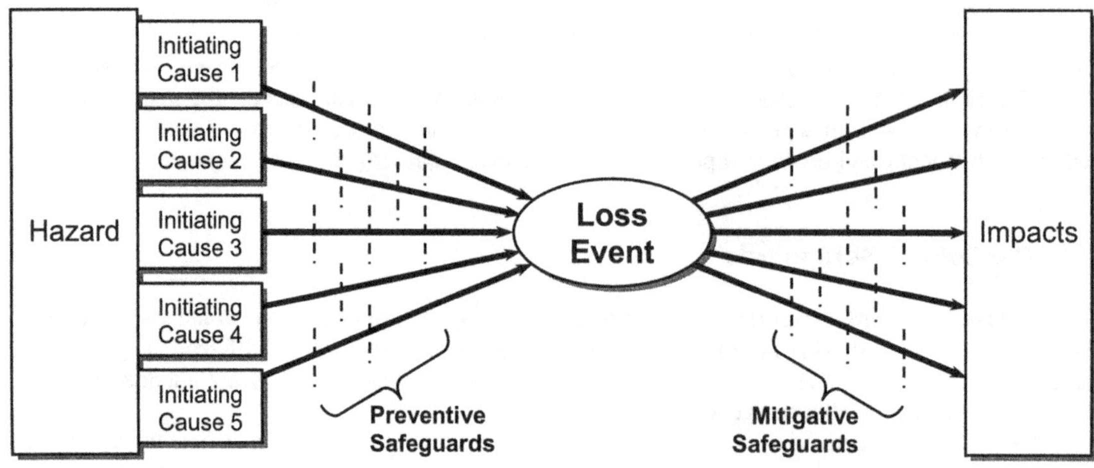

Figure 1.6 Generic "bow-tie" diagram showing relation of safeguards to loss event

Contain and Control

Although not considered to be safeguards as defined above, the containment and control of process hazards serve critical functions in avoiding or reducing the likelihood of initiating causes and ensuing incident scenarios. Note that, in this context, "containment" refers to the primary containment system consisting of piping, vessels and other process equipment designed to keep hazardous materials and energies contained within the process. Secondary containment systems such as diked areas and berms are mitigative safeguards.

Typical ***contain and control*** measures include:

- Proper design and installation of the primary containment system, along with inspections, testing, and maintenance to ensure the ongoing mechanical integrity of the primary containment system
- Guards and barriers to reduce the likelihood of an external force such as maintenance activities or vehicular traffic impacting process piping or equipment
- Basic process control system (BPCS) design, installation, management, and maintenance to ensure successful control system response to anticipated changes and trends such as variations in feed compositions, fluctuations in utility parameters such as steam pressure and cooling water temperature, ambient condition changes, gradual heat exchanger fouling, etc.
- Operator training to reduce the likelihood of a procedure being improperly performed

- Segregation, dedicated equipment, and other provisions to reduce the likelihood of incompatible materials coming into contact with each other
- Management of change with respect to materials, equipment, procedures, personnel, and technology.

The objectives of ***contain and control*** are to keep process material confined within its primary containment system and to keep the process within safe design and operating limits, thus avoiding abnormal situations and loss of containment events that could lead to loss, damage and injury impacts. Containment and control measures, such as those listed above, affect the frequency of initiating causes.

It should be noted that many practitioners consider containment and control measures to also be "safeguards." However, they do not meet the definition of a safeguard given earlier as "any device, system, or action that would likely interrupt the chain of events following an initiating cause." Most of these measures apply not only to individual scenarios but to the entire process or facility, so the repeated listing of measures such as "Operator training" and "Mechanical integrity program" in the Safeguards column on hazard evaluation worksheets only makes it more difficult for the review team to assess the overall effectiveness of the preventive and mitigative safeguards in interrupting the chain of events following the initiating cause. If the desire is to give credit for having these general measures in place, they can be listed in a separate "Primary Containment and Control of Process Hazards" or similar section in a hazard evaluation report, rather than be included throughout the hazard evaluation worksheets.

Preventive Safeguards

Preventive safeguards intervene after an initiating cause has occurred and process conditions are abnormal or out of control. They act to regain control or achieve a safe state when an abnormal process condition is detected, thus interrupting the propagation of the incident sequence and avoiding the loss event (irreversible physical event with potential for loss and harm impacts, such as a hazardous material release, fire, or explosion). Preventive safeguards do not affect the likelihood of initiating causes, but do affect the probability that a loss event will result, given that an initiating cause occurs. Preventive safeguards thus affect the overall scenario frequency. Typical preventive safeguards include:

- Operator response to bring an upset condition back within safe operating limits
- Operator response to a safety alarm or upset condition to manually shut down the process before a loss event can occur
- Instrumented protective system designed and implemented to automatically bring the system to a safe state upon detection of a specified abnormal condition
- Ignition source control implemented to reduce the probability of ignition given the presence of an ignitable mixture, thus preventing the loss event of a fire, dust explosion, confined vapor explosion or vapor cloud explosion
- Emergency relief system acting to relieve vessel overpressurization and prevent the loss event of a bursting vessel explosion
- Other last-resort preventive safety systems such as manual dump or quench systems.

The objective of preventive safeguards is to avoid a loss event or a more severe loss event, given the occurrence of an initiating cause. An example of avoiding a more severe loss event is if mechanical failure of a piping system immediately results in loss of containment of a flammable liquid (which is both an initiating cause and a loss event, since no preventive safeguards intervene), ignition source control can avoid a different, more severe loss event of a fire or vapor cloud explosion.

Preventive safeguards should be considered as systems that must be designed, maintained, inspected, tested, and operated to ensure they are effective against particular incident scenarios. For safety instrumented systems, this is termed the safety integrity level (SIL). CCPS[19] provides guidance on the life cycle management of instrumentation and control systems to achieve a specified level of integrity.

Both qualitative and quantitative methodologies can be used to identify and classify safeguards. Layer of Protection Analysis (LOPA), an order-of-magnitude method that builds on traditional hazard evaluation results to determine the required integrity of safeguards, is summarized in Section 7.6.

The following example illustrates how operator response to a safety alarm can be considered as a preventive safeguard "system" having several essential parts. Figure 1.7 shows the example reaction process used to illustrate Fault Tree Analysis in Section 5.5. The process consists of a reactor for a highly unstable process that is sensitive to small increases in temperature. It is equipped with a deluge for emergency cooling to protect against an uncontrolled reaction. To prevent a runaway reaction during an increase in temperature, the inlet flow of process material to the reactor must be stopped or the deluge must be activated. The reactor temperature is monitored by a sensor (T1) that automatically activates the deluge by opening the deluge water supply valve when a temperature rise is detected. At the same time, sensor T1 sounds an alarm in the control room to alert the operator of the temperature rise. When the alarm sounds, standard operating procedure calls for the operator to push the inlet valve close button to shut the inlet valve and stop inlet flow to the reactor and to push the deluge open button in the control room if the deluge is not activated by sensor T1. If the inlet valve closes or the deluge is activated, system damage due to an uncontrolled reaction is averted. (Note that the example process is described in this manner for illustrative purposes only; this would not likely be the best way to arrange a reactive process of this nature.)

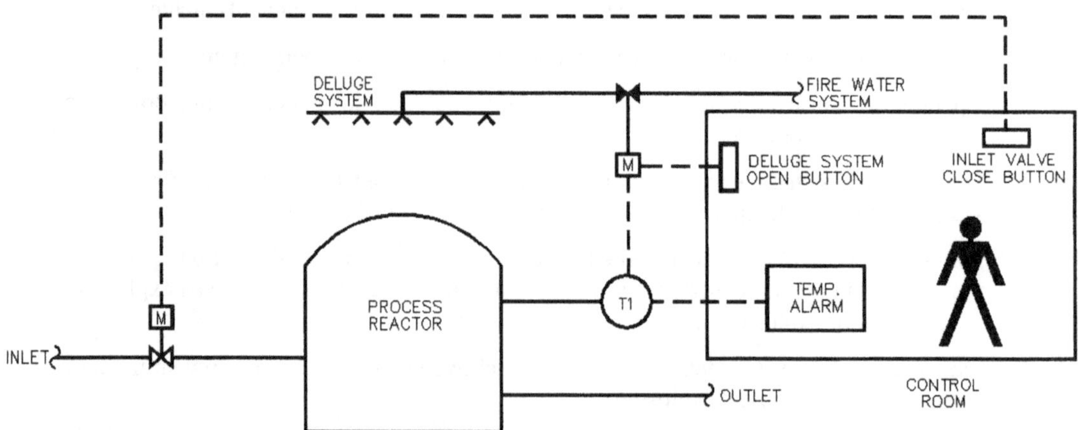

Figure 1.7 Emergency cooling system schematic

The operator response preventive safeguard system would require all of the following to occur in order to successfully protect against the consequence of concern:

1. The temperature sensor is at the right location and responds with inconsequential time delay, giving a correct output signal corresponding to the increase in reactor temperature.

2. A relay or other device successfully operates at the proper safety limit setting to send a signal to the alarm module.

3. The high temperature alarm functions to annunciate the proper audible and/or visual warning in the control room.

4. The operator is present in the control room at the time the alarm sounds.

5. The ambient noise level and distractions are sufficiently minimal such that the operator is alerted by the alarm signal.

6. The operator decides to respond to the alarm and not just acknowledge it.

7. The operator makes the correct diagnosis as to the meaning of the alarm based on the operator's training, experience, and preconceptions of the state of the process.

8. The operator responds to the alarm in time to avert the loss event.

9. The operator actuates the correct push buttons to stop the inlet flow and/or activate the deluge.

10. The inlet flow is stopped in time by successful functioning of the inlet valve close button and the inlet valve; or, the deluge is activated in time by successful functioning of the deluge push button, deluge valve, and deluge piping and nozzles, and an adequate supply of fire water is available.

It should be noted that, for this example, the operator response to the alarm to actuate the deluge system is not independent of the automatic deluge system, since they share a common temperature sensor and a common final control element (deluge water supply valve). Likewise, the operator-actuated inlet flow isolation system and the deluge system are not independent of each other, since they share a common temperature sensor. Thus, a hazard evaluation team would need to evaluate the effectiveness of the overtemperature safeguards by examining both the operator responses and the automatic safety systems together rather than as independent protective systems. This assessment of the independence of preventive safeguards is an important part of a hazard evaluation, regardless of whether the evaluation is performed using a qualitative or a quantitative technique.

Mitigative Safeguards

A *mitigative safeguard* acts to reduce the severity of consequences of a loss event; i.e., the sum total of safety, business, community, and environmental impacts resulting from a fire, explosion, toxic release, or other irreversible physical event. Typical mitigative safeguards include:

- Reclosing emergency relief devices such as safety relief valves, acting to reduce the duration of a hazardous material release loss event if the emergency relief discharges to the atmosphere
- Secondary containment (e.g., double-walled system, secondary enclosure)
- Explosion blast and missile containment structures / barricades
- Fire/release detection and warning systems
- Automatic or remotely actuated isolation valves
- Fire extinguishers, sprinkler systems, and fire water monitors
- Deluge, foam, and vapor mitigation systems
- Fire-resistant supports and structural steel
- Storage tank thermal insulation
- Blast-resistant construction of occupied buildings
- Loss-event-specific personal protective equipment (e.g., splash goggles, flame-retardant clothing, escape respirators)
- Emergency response and emergency management planning.

The objective of mitigative safeguards is to detect and respond to emergency situations in such a way as to reduce the impacts of loss events as compared to the unmitigated impacts without the safeguards.

When performing detailed, scenario-based hazard evaluations, a useful distinction can be drawn between those mitigative safeguards designed to act after the loss event occurs and affect the *source term* (i.e., the release parameters of magnitude, rate, duration, orientation, temperature, etc. that are the initial conditions for determining the consequences of the loss event) and those mitigative safeguards designed to reduce the impacts of the released material or energy on people, property and the environment. Examples of the first category of mitigative safeguards (which could be called *source-mitigative safeguards*) include excess flow valves, dry-break connections on unloading hoses, automatic release detection and isolation systems, and engineered vapor release mitigation systems such as deluges and water curtains. Examples of the second category of mitigative safeguards (which could be called *receptor-mitigative safeguards* and are sometimes termed *response* rather than *mitigation*) include the buffer distance to surrounding populations, occupied building blast resistance, fire-resistant construction, specialized personal protective equipment evacuation or shelter-in-place procedures, and other emergency response actions including firefighting. Section 7.2 includes a discussion of how these different types of safeguards are evaluated when assessing scenario risk.

1.5 Hazard Evaluation Throughout a Plant Lifetime

Many organizations have published model programs for process safety management (PSM). All of these PSM approaches embrace a consistent theme: *Hazard evaluations should be performed throughout the life of a facility.* As an integral part of its PSM program, an organization can use the results of hazard evaluations to help manage the risk of each phase of process activity. Hazard evaluations can be done efficiently from the earliest stages of R&D, in detailed design and construction, during commissioning and start-up, periodically throughout the operating lifetime, and until the process is decommissioned and dismantled.[10,15] A more complete discussion of hazard evaluation at different plant life cycle stages, including as part of managing changes, can be found in Chapter 6, Sections 6.4 and 6.6. Two aspects of hazard evaluation throughout a plant lifetime warrant particular emphasis:

- Using this life cycle approach in association with other PSM activities can efficiently reveal deficiencies in design and operation before a unit is sited, built or operated, thus making the most effective use of resources devoted to ensuring the safe and productive life of a facility.
- Regardless of the technique used for conducting hazard evaluations throughout the operating lifetime of a facility, each study, along with its documented information and assumptions, should be updated or revalidated on a periodic basis.

An important part of performing hazard evaluations throughout a plant lifetime is knowing which technique is the best one for the study. Chapter 5 discusses many factors that influence this decision and provides the logic behind choosing an appropriate technique. One of the most important factors that influences which hazard evaluation technique an analyst chooses is how much information is available to perform the work. Some hazard evaluation methods may be inappropriate or impossible to perform at a particular life cycle stage because of inadequate process information.

1.6 Hazard Evaluation and Regulations

Although most companies in the chemical processing industries conduct hazard evaluations voluntarily because they believe they are necessary to control hazards and to manage risk at an acceptable level, many companies in the world also conduct hazard evaluations because they are required by regulation. For example, in the United States, the U.S. OSHA regulation 29 CFR 1910.119 (Process Safety Management Standard) requires a hazard evaluation to be performed for covered processes in these elements:

- The Process Hazard Analysis element requires a hazard evaluation that meets certain criteria every five years.
- The Management of Change element includes a requirement to assure the safety and health impact is addressed prior to any change being made.

The techniques discussed in this book can be used in part to fulfill the hazard evaluation requirements for this and other international regulations. When hazard evaluations are required by regulation, specific documentation of the study and the follow-up of recommendations may be mandatory so that regulators can be shown that proper hazard evaluations have been completed. Table 1.4 gives a partial list of international regulations requiring some form of hazard identification and evaluation. Many other countries have hazard evaluation requirements that are based on U.S. or European requirements. This list is not comprehensive; companies need to contact government authorities having jurisdiction to determine what regulations apply to them and what hazard evaluation procedures can be used.

1.7 Limitations of Hazard Evaluation

Hazard evaluation, whether one uses experience-based or predictive methods, is subject to a number of theoretical and practical limitations.[11,14] Managers should realize that the quality of any risk management decisions they base on hazard evaluation results will be directly related to their appreciation of the limitations of such studies. Table 1.5 lists five limitations of hazard evaluations discussed in this section. Some of these may be relatively unimportant for a specific study, depending upon its objectives, while others may be minimized through care in execution and by limiting expectations about the applicability of the results. However, both practitioners and users of these studies must respect these limitations when chartering, executing, and using the results of a hazard evaluation.

Table 1.4 Governmental regulations related to identifying and evaluating process hazards

Country or region	Regulation
Australia	National Standard for Control of Major Hazard Facilities [NOHSC:1014 (1996)]
European Union	Seveso II Directive 2003/105/EC
	ATEX 137 Workplace Directive 1999/92/EC
Mexico	NOM-028-STPS-2004, Occupational organization – Safety in the Processes of Chemical Substances
Singapore	National Environment Agency (one-time QRA Report for new chemical plants)
South Korea	Industrial Safety and Health Act — Article 20, Preparation of Safety and Health Management Regulations
United Arab Emirates	Federal Law No 8 of 1980 Regulations of Labour Relations
	Federal Law No 24 of 1999 for the protection and development of the environment
United Kingdom	U.K. Health & Safety Executive, Control of Major Hazards (COMAH) regulations
United States	29 CFR 1910.119, U.S. Occupational Safety and Health Administration (OSHA) Process Safety Management of Highly Hazardous Chemicals
	40 CFR 68, U.S. Environmental Protection Agency (EPA) Risk Management Program for Chemical Accident Release Prevention

Table 1.5 Classical limitations of hazard evaluations

Issue	Description
Completeness	There can never be a guarantee that all incident situations, causes, and effects have been considered
Reproducibility	Various aspects of hazard evaluations are sensitive to analyst assumptions; different experts, using identical information, may generate different results when analyzing the same problem
Inscrutability	The inherent nature of some hazard evaluation techniques makes the results difficult to understand and use
Relevance of experience	A hazard evaluation team may not have an appropriate base of experience from which to assess the significance of potential incidents
Subjectivity	Hazard analysts must use their judgment when extrapolating from their experience to determine whether a problem is important

Completeness

The issue of completeness affects a hazard evaluation in two ways. First, it arises in the hazard identification step — an analyst can never be certain that all hazardous conditions or potential incident scenarios have been identified. Second, for those hazards that have been identified, a hazard analyst can never guarantee that all possible causes and effects of potential incidents have been considered. It is impossible for a hazard analyst to identify and assess the significance of all possible things that can go wrong — even for a very limited, well-defined set of circumstances. But one can reasonably expect trained and experienced practitioners, using systematic hazard evaluation techniques and relevant experience, to identify the most important incidents, causes, and effects.

Moreover, a hazard evaluation is a "snapshot in time" evaluation of a process. Any changes in design, procedures, operation or maintenance (however small) may have a significant impact on the safety of the facility.

Reproducibility

Probably the least appreciated limitation of hazard evaluation techniques is that the results of a hazard evaluation, because of their highly subjective nature, are difficult to duplicate by independent experts. Even with the variety of experience-based and predictive methods available for use in a hazard evaluation, the performance of a high quality hazard evaluation is still largely dependent on good judgment. The subtle assumptions that hazard analysts and process experts necessarily make while performing hazard evaluations can often be the driving force behind the results. Analysts should always highlight their known assumptions when documenting their work so future users can identify the places where additional research is necessary for better hazard information or data. As organizations gain experience in using these approaches, they will appreciate that the assumptions made during a study are as important as any of the results.

Inscrutability

Hazard evaluations can generate hundreds of pages of tables, minutes of review team meetings, models such as fault trees and event trees, and other information. Attempting to assimilate all of the details of a hazard evaluation, depending upon the method chosen and the size of the problem, can be an overwhelming task. Combined with hazard analysts' tendencies to use copious amounts of jargon, reviewers can find themselves wondering what to do with all this information. Fortunately, not all hazard evaluations result in this much paperwork; instead, effective hazard evaluation analysts produce a summary of potential improvements or areas for additional study that management should consider pursuing to improve the safety of the process. These lists, by themselves, are usually straightforward; however, depending upon the technique used to perform the hazard evaluation, the underlying technical bases of the problems and the potential effectiveness of the solutions may be difficult to understand.

Relevance of Experience Base

Some hazard evaluation methods depend solely on the analysts' experience with similar operations (e.g., the Checklist Analysis technique). Other hazard evaluation techniques involve analysts predicting the causes and effects of potential incidents based on their creativity and judgment. All of the techniques hope to capitalize, to some extent, on an organization's experience with a hazardous process. In cases where the experience base is limited, not very relevant, or nonexistent, hazard analysts should use more predictive, systematic techniques such as HAZOP Study or Fault Tree Analysis. Even then, users of the results of these studies must be cautious, since the foundation of knowledge on which the study is based may not justify the use of a sophisticated hazard evaluation technique. Use of a detailed analysis technique does not guarantee better risk understanding; the relevance of the experience base that underlies the analysis is more important than the use of a particular hazard evaluation method.

Reliance on Subjective Judgment

Hazard evaluations use qualitative techniques to determine the significance of potential incident situations. Inherently, the conclusions of such a study are based on the collective knowledge and experience of the hazard evaluation team. Because many of the events considered by the team may never have happened before, the team must use their creativity and judgment to decide whether the potential causes and effects of the incident pose a significant risk. The subjective nature of these deliberations may trouble some people who use the results of these studies, as this subjectivity can create a lack of confidence in the results. Some people incorrectly believe that if an analyst uses quantitative methods to express the significance of a problem, then the limitation of subjectivity will simply fade away. However, this is not the case. Although quantifying the risk parameters can reduce some of the subjectivity in estimating likelihoods and impact, the apparent numerical precision of a chemical process quantitative risk analysis can mask (1) a great deal of the judgment that influenced the identification of incident scenarios and the selection of incident models and (2) large uncertainties associated with the data used to estimate risk. In fact, the quality of hazard evaluation and CPQRA studies alike depend upon having performed an exhaustive search for what can go wrong. In the end, the user must have confidence in both the team and the technique selected for performing the hazard evaluation.

The limitations discussed above should not be reasons for rejecting the use of hazard evaluation techniques. Learning from experience alone may be adequate when the consequences of an incident are small. However, the consequences of potential incidents are not always small, and gaining an empirical perspective of risk through experiencing high-consequence incidents is not acceptable. Hazard evaluation techniques can help analysts find ways to reduce both the frequency and severity of life-threatening loss events. In this way, hazard evaluation techniques can form the basis for a sound and cost-effective risk management program.

Chapter 1 References

1. Center for Chemical Process Safety, *Guidelines for Hazard Evaluation Procedures*, American Institute of Chemical Engineers, New York, 1985.
2. Recommended Practice RP-750, "Management of Process Hazards," American Petroleum Institute, Washington, DC, January 1990, reaffirmed May 1995.
3. Organization Resources Counselors, Inc., *Process Hazards Management of Substances with Catastrophic Potential*, Process Hazard Management Task Force, Washington, DC, 1988.
4. *Responsible Care Management System® Guidance and Interpretations (RCMS 102)*, American Chemistry Council, Arlington, Virginia, 2004.
5. 29 CFR 1910.119, *Process Safety Management of Highly Hazardous Chemicals*, Occupational Safety and Health Administration, Washington, DC, 1992.
6. 40 CFR Part 355, "Extremely Hazardous Substances List," U.S. Environmental Protection Agency, Washington, DC, 1987.
7. N.J.A.C. 7:31, Toxic Catastrophe Prevention Act of 1987, New Jersey Department of Environmental Protection, Trenton, 1987.
8. State of California Health and Safety Code, *Risk Management and Prevention Program*, Chapter 8.95 of Division 20, Sacramento, 1987.
9. Extremely Hazardous Substances Risk Management Act of 1989, Delaware Department of Natural Resources and Environmental Control, Wilmington, DE, 1989.
10. Center for Chemical Process Safety, *Guidelines for Technical Management of Chemical Process Safety*, American Institute of Chemical Engineers, New York, 1989.
11. Arendt, J. S. et al., *Evaluating Process Safety in the Chemical Industry—A Manager's Guide to Quantitative Risk Assessment*, American Chemistry Council, Arlington, Virginia, 1989.
12. V. L. Grose, *Managing Risk—Systematic Loss Prevention for Executives*, Prentice Hall, Englewood Cliffs, New Jersey, 1987.
13. G. L. Head, *The Risk Management Process*, Risk and Insurance Management Society, Inc., New York, 1978.
14. Center for Chemical Process Safety, *Guidelines for Chemical Process Quantitative Risk Analysis, 2nd Ed.*, American Institute of Chemical Engineers, New York, 1999.
15. J. Stephenson, *System Safety 2000—A Practical Guide for Planning, Managing, and Conducting System Safety Programs*, ISBN 0-0442-23840-1, Van Nostrand Reinhold, New York, 1991.
16. Center for Chemical Process Safety, *Guidelines for Risk Based Process Safety*, American Institute of Chemical Engineers, New York, 2007.
17. Abnormal Situation Management® Consortium, "Anatomy of a Catastrophic Incident," www.asmconsortium.com. Figure 1.3 is based on original concepts created by K. Emigholz of ExxonMobil, I. Nimmo of UCDS LLC and J. Errington of NOVA Chemicals. Earlier version published in I. Nimmo, "Adequately Address Abnormal Operations," *Chemical Engineering Progress 91*(9), 36-45, September 1995.
18. Center for Chemical Process Safety, *Layer of Protection Analysis: Simplified Process Risk Assessment*, ISBN 0-8169-0811-7, American Institute of Chemical Engineers, New York, 2001.
19. Center for Chemical Process Safety, *Guidelines for Safe and Reliable Instrumented Protective Systems*, ISBN 978-0-471-97940-1, American Institute of Chemical Engineers, New York, 2007.

2

Preparation for Hazard Evaluations

Successful hazard evaluations are the product of concerted efforts throughout an organization. This chapter discusses the first phase of a successful hazard evaluation: ***preparation***. The success of all subsequent studies depends upon adequate preparation.

2.1 Infrastructure

To realize the full benefits of a hazard evaluation program, management must foster a corporate culture and build an infrastructure that will support hazard evaluation teams as they perform and implement the results of hazard evaluations. These are often done under the auspices of an organization's process safety management (PSM) program.

The foundation of an effective PSM program is a clear commitment to process safety that is articulated in a corporate safety policy. This policy must empower managers to commit organizational resources to PSM activities. But the policy statement alone is not sufficient; managers must actively support the employees directly involved with the PSM program.

In addition to this fundamental requirement, there are three key commitments that must be made in order to build the necessary infrastructure for a cost-effective hazard evaluation program. These are: a commitment to organize and maintain process knowledge and information, a commitment to assign and support personnel, and a commitment to act on the results of hazard evaluations in a timely manner.

The first commitment can be accomplished by developing a system that will allow the organization to develop and document knowledge about its processes and equipment. For example, essential drawings and operating procedures should be kept up-to-date and internal company standards, guidelines, and checklists should be periodically reviewed. This may be time consuming for companies that have out-of-date process documentation. Nevertheless, the quality of any hazard evaluation depends directly on this information, so it is essential that the process documentation be accurate. Furthermore, an organization should have a practical management of change program to ensure that all modifications are properly reviewed and drawings and procedures are kept up-to-date.

The second commitment requires that management throughout the organization provide knowledgeable and competent personnel to participate in hazard evaluations and that they recognize the importance of the participants' contributions (by writing letters, supporting promotions, etc.). Some participants may be vendors, contractors, or consultants, but company personnel (e.g., top operators, unit supervisors, maintenance planners) must also participate, because only they know how the facility is truly operated and maintained, and because they will be the ones who must act on the results of any hazard evaluation. Some companies may also decide to use dedicated corporate or business division staff that will support facility hazard evaluation programs by training and advising local facility personnel.

The third commitment involves creating a system that not only documents and maintains the results of hazard evaluations, but also records management's response to these studies and ensures timely resolution of action items. This requires a person or group to establish consistent documentation guidelines, keep a permanent file of the hazard evaluation reports, and track the status of recommendations made during the evaluations. It is vital that this person or group document the resolutions (i.e., acceptance, rejection, substitution, or modification) of all recommendations and include them in the file along with the original report. Management must reinforce their commitment to process safety by allocating resources to implement reasonable recommendations for risk reduction. They must also enforce an audit system to ensure that all approved corrective actions are implemented in a timely manner.

2.2 Analysis Objectives

An organization's leadership must clearly define the objectives for any hazard evaluation. This input is essential if the analysis is to be performed efficiently. Analysts who lack clear guidance may waste time examining parts of the process or situations that may be of relatively minor concern to the organization.

The appropriate objective for a hazard evaluation depends upon several factors, including the life cycle phase the project is in when the hazard evaluation is performed. Obviously, as a project evolves, the types of hazardous situations investigated change from general questions about basic process chemistry to more detailed questions about equipment and procedures. Table 2.1 lists some typical hazard evaluation objectives at different stages of a process lifetime.

Although management input is necessary to define the objectives of the study, that input must not be so detailed or restrictive that it stifles the hazard evaluation team. For example, a team may be directed to investigate potential explosion hazards associated with a dust collector. If, in the course of the review, the team recognizes that the procedure for removing collected particulates could expose the operator to hazardous materials, they should have the latitude to document this unexpected finding and call it to management's attention. Similarly, analysts investigating the hazards associated with a piping modification might discover a potential level control problem in one of the vessels connected to the piping. Even though the piping modification neither created nor exacerbated the level control problem, the analysts should communicate their findings for further review.

In general, hazard evaluation teams must have the freedom to exercise good judgment in their investigations of potential hazards. However, hazard analysts should not allow their attention to be diverted from the main purpose of the study. Periodically throughout a study, a team should assess whether the path they are taking will ultimately satisfy the original goal set for the hazard evaluation.

Deadlines are an integral part of any discussion of the analysis objectives. (Deadlines are defined as real, nonnegotiable dates by which analysis results must be available to satisfy the defined need.) Deadlines place genuine limits on the time available to collect and/or develop information and to perform the analysis. If the desired objectives cannot be accomplished within the available time using the most appropriate hazard evaluation technique, then management must provide additional analysis resources or, perhaps, revise the analysis objectives.

Table 2.1 Typical hazard evaluation objectives at different stages of a process lifetime

Process phase	Example objectives
Research and development	Identify chemical reactions or interactions that could cause runaway reactions, fires, explosions, or toxic gas releases
	Identify process safety data needs
Conceptual design	Identify opportunities for inherent safety
	Compare the hazards of potential sites
	Provide input to facility layout and buffer zones
Pilot plant	Identify ways for hazardous materials to be released to the environment
	Identify ways to deactivate the catalyst
	Identify potentially hazardous operator interfaces
	Identify ways to minimize hazardous wastes
Detailed engineering	Identify ways for a flammable mixture to form inside process equipment
	Identify how a reportable spill might occur
	Identify which process control malfunctions will cause runaway reactions
	Identify ways to reduce hazardous material inventories
	Evaluate whether designed safeguards are adequate to control process risks to tolerable, required or ALARP level
	Identify safety-critical equipment that must be regularly tested, inspected, or maintained
Construction and start-up	Identify error-likely situations in the start-up and operating procedures
	Verify that all issues from previous hazard evaluations were resolved satisfactorily and that no new issues were introduced
	Identify hazards that adjacent units may create for construction and maintenance workers
	Identify hazards associated with vessel cleaning procedures
	Identify any discrepancies between as-built equipment and the design drawings
Routine operation	Identify employee hazards associated with the operating procedures
	Identify ways an overpressure transient might occur
	Update previous hazard evaluation to account for operational experience
	Identify hazards associated with out-of-service equipment
Process modification or plant expansion	Identify whether changing the feedstock composition will create any new hazards or worsen any existing ones
	Identify hazards associated with new equipment
Decommissioning	Identify how demolition work might affect adjacent units
	Identify any fire, explosion, or toxic hazards associated with the residues left in the unit after shutdown

2.3 Developing the Review Scope and Boundaries

This may seem obvious to the hazard evaluation team, but unless the scope of the study and boundaries of the process to be studied are clearly agreed to by the team before the study begins, much time can be wasted evaluating parts of the process not requiring an evaluation, or worse yet, important parts of the process can be neglected. The boundaries must be clearly identified on the piping and instrumentation diagrams (P&IDs) by designating a specific process connection that can be identified in the field such as a piping flange or valve. The boundary of the evaluation might also be designated as a certain geographical part of the plant such as a building or part or a building, or columns in a process area that can be clearly marked on a plot plan.

Before the process boundaries of the study can be determined, the study scope must be established. First, if the hazard evaluation is being done to comply with regulatory requirements, the process covered by regulation must clearly be identified. For example, the U.S. OSHA Process Safety Management Standard requires a hazard evaluation on covered processes. In this case, a covered process is one that contains the listed material at its threshold quantity or greater, including connected equipment and adjacent processes that could affect the covered process.

The team must then decide other scoping questions such as:

- Should the study include utilities and environmental control equipment that connect to the process being evaluated?
- Should the hazard evaluation include adjacent processes not covered by the regulations?

Another scope-related decision should be made concerning what consequences are to be addressed in the study. Most hazard evaluations will, of course, address safety-related consequences. A severity threshold might be determined, such as the impact of a recordable or lost-workday injury, or perhaps permanent injuries or fatalities if the focus of the study is to be on major hazards. Depending on the objectives of the study, the boundaries of the study may be drawn to also include non-injury community impacts, property damage, business interruption, and/or environmental impacts, with a threshold severity also drawn for each additional type of impact.

After the review scope decisions are made, the boundaries can be clearly described by designating them in tabular form, marked on P&IDs, and/or marked on a plot plan before the hazard evaluation begins so that the study can be done completely and efficiently.

2.4 Information Requirements

Hazard evaluations can be performed using whatever process information is available, including preliminary flowsheet sketches; a comprehensive set of chemical process data; piping, instrumentation, and control drawings; data sheets; and procedures. Obviously, the more information and knowledge one has about a process, the more thorough and valuable the hazard evaluation can be (Table 2.2). In fact, the unavailability of detailed process P&IDs and operating procedures may preclude a hazard evaluation team from using some detailed techniques such as HAZOP Studies or Fault Tree Analysis. Ultimately, the quality of any hazard evaluation depends directly on the quality of the information available to the analyst(s). Thus, as stated previously, a system for collecting and documenting process knowledge is a key part of the infrastructure needed to support a hazard evaluation program.

Realistically, the information available to perform hazard evaluations varies over the lifetime of a process (Figure 2.1).[1] During the early stages, hazard evaluation teams may only have access to basic chemical data, such as boiling points, vapor pressures, flammability limits, toxicity data, regulatory limits, etc. By the time a process reaches the detailed design phase, however, most of the information listed in Table 2.2 should be available and used in the evaluation.

It is helpful to bring to the hazard evaluation the consensus standards that have been used to design and build the process (e.g., National Fire Protection Association codes and standards, American Petroleum Institute recommended practices, and American Society of Mechanical Engineers codes). This information is useful for most hazard evaluations, not only those required by regulation.

Other information, such as plot plans and incident reports, provides the reviewers with more perspective on potential hazards. Information from sister plants or industry groups such as the Chlorine Institute, the Vinyl Institute or the American Petroleum Institute is also very useful. A well-organized team also brings invaluable information to the hazard evaluation from the experience of the individual team members. This information, in addition to the information used in earlier stages, forms a substantial basis for reviewing the design and for reviewing any subsequent changes to the design. However, hazard evaluations should never be considered a substitute for an organization's customary design review activities. Further, hazard evaluations are best used to supplement project design reviews, and are most effective when a third-party creatively studies the design in an attempt to pinpoint its weaknesses.

At the end of a project's life, information is needed on issues such as process residues; potential reactions of process and equipment materials with air, water, or rust; exposure and release limits for any hazardous wastes; and potential impacts to adjacent units. Process flows, controls, operating procedures, etc., may no longer be relevant.

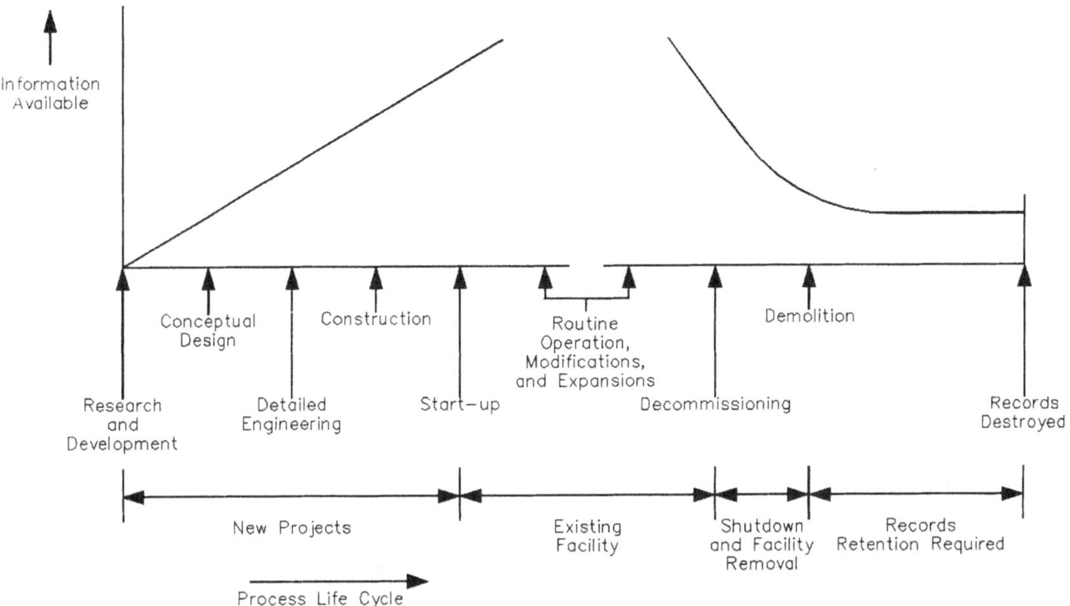

Figure 2.1 Information available for hazard review (adapted from Reference 1, Figure 5.2)

Table 2.2 Examples of information used to perform a hazard evaluation

Management system information

❑ Operating procedures (with critical operating parameters)

❑ Maintenance procedures

❑ Emergency response plan and procedures

❑ Incident reports

❑ Previous safety studies

❑ Internal standards and checklists

❑ Corporate safety policies

❑ Relevant industry experience

❑ Regulatory limits and/or permit limits

❑ Applicable codes and standards

❑ Variances

Process technology information

❑ Chemical reaction equations and stoichiometry for primary and important secondary or side reactions

❑ Type and nature of catalysts used

❑ Kinetic data for important process reactions, including the order, rate constants, approach to equilibrium, etc.

❑ Kinetic data for undesirable reactions, such as decompositions and autopolymerizations

❑ Process limits stated in terms of pressure, temperature, concentration, feed-to-catalyst ratio, etc., along with a description of the consequences of operating beyond these limits

❑ Process flow diagrams and a description of the process steps or unit operations involved, starting with raw material storage and feed preparation and ending with product recovery and storage

❑ Design energy and mass balances

❑ Major material inventories

❑ Description of general control philosophy (i.e., identifying the primary control variables and the reasons for their selection)

❑ Discussion of special design considerations that are required because of the unique hazards or properties of the chemicals involved

Chemical hazard information

❑ Reactive chemical data on all streams, including in-process chemicals

❑ Safety, health, and environmental data for raw materials, intermediates, products, by-products, and wastes

Equipment information

❑ Area electrical classification drawings

❑ Building and equipment layouts

❑ Electrical classifications of equipment

❑ Piping and instrumentation drawings

❑ Mechanical equipment data sheets

❑ Equipment catalogs

❑ Vendor drawings; operation & maintenance manuals

❑ Valve and instrumentation data sheets

❑ Piping specifications

❑ Utility specifications

❑ Test and inspection reports

❑ Electrical one-line drawings

❑ Instrument loop drawings and logic diagrams

❑ Control system and alarm description

❑ Computer control system hardware and software design

❑ Relief system design basis

❑ Ventilation system design basis

❑ Safety system(s) design basis

❑ Fire protection system(s) design basis

❑ Fire protection system(s) design basis

Site information

❑ Plot plans

❑ Meteorological data

❑ Population distribution data

❑ Site hydrology data

2.5 Use of Software Programs

Early hazard evaluations were documented on paper or by the use of overhead transparencies. With the close fit between the elements of incident scenarios and the fields of a database, most hazard evaluations are amenable to being documented using computer-based systems. Likewise, most hazard evaluation teams now gravitate toward using computers and some sort of software to facilitate team-based studies. For any but the smallest analysis scopes, computer software can greatly enhance the efficiency of analysis facilitation and documentation.

Two configurations are common when using hazard evaluation software with team reviews. One approach is for one person (often called the *scribe*) to document the hazard evaluation on a computer but with the results not displayed to anyone but the scribe. The other, more common approach is to have the results continuously presented to all team members, such as by use of an LCD projector or large display. The latter approach can offer several benefits:

- The team discussions can more easily be kept on track, since the team members can see exactly what checklist item, scenario, etc. is being discussed.
- Key information such as the design intent or the wording of a procedural step can be displayed simultaneously with the hazard evaluation.
- All team members can see exactly what is being documented, so what is captured by the software can be truly considered as the consensus of the team.
- Documentation errors can be caught by team members and immediately corrected.
- Previously completed work can be displayed if it has a bearing on the current discussion.

Some hazard evaluation teams have found it helpful to have two projection devices available, with one displaying the hazard evaluation and the other capable of bringing up pertinent data, drawings, reports, or other information that may be on a local computer or a company intranet. Chemical properties, regulatory requirements and guidance, web searches, etc. can also be displayed using an internet browser. It should be noted, however, that there can be some disadvantages to using projectors or displays. For example, team members can get bogged down with perfecting the language rather than capturing the information, or the room lighting may need to be dimmed if the projection system is not bright enough.

When considering whether to use a software package to facilitate team-driven hazard evaluations, or which software to use, there are several factors to consider that pertain to the capabilities of the software, as well as how it makes use of available hardware, particularly projection devices. Similar considerations also apply to selecting software for following up hazard evaluation results.

- Importing and exporting data. If the organization wishes to make use of the data outside of the internal capabilities of the software (e.g., the analysis itself and action item tracking), then the selected software should either (a) have sophisticated enough export features to deliver the data in a usable form such as a common spreadsheet or database file, or (b) be built on an open database format and not be encrypted. Likewise, if a previous hazard evaluation is being updated but different software is being considered, the compatibility of the previous study files for importing into the new software should be verified. If the software cannot support the level of data integration desired, then redundant data entry will be required.

- Display limitations. Some software shows a useful, readable display at screen resolutions found on the average workstation, but may not be legible at the resolution of the available projection device. This issue is less commonly encountered, given advances in projection equipment. Also, using a database or spreadsheet program not specifically set up for a given hazard evaluation technique may result in needing to constantly scroll across fields or columns.

- User interface. More difficult to judge is the user interface by which the facilitator will focus the attention of the analysis team on the sequence of tasks involved in performing a hazard evaluation and record the results of the evaluation, and by which those assigned follow-up responsibilities will know the scope of their responsibilities and the associated schedule, and those overseeing, administering, and/or auditing follow-up status will be able to initialize, track and audit the status of follow-up issues. Considerations include ease of finding previously documented information, sophistication of querying and reporting features, and extent of ability to customize such as to change risk tolerance boundaries or follow-up report formats.

- Types of hazard studies supported. If the company wishes to make use of a variety of hazard evaluation techniques, then the selected software needs to have the flexibility to support them.

See Appendix D for a list of commercially available software programs for facilitating and documenting hazard evaluations.

2.6 Personnel and Skills

The composition of the team performing a hazard evaluation is essential to the success of the study.[6] The level and types of skills that personnel must have to participate in hazard evaluations depend upon several factors, including the type and the complexity of process or operation analyzed, the hazard evaluation technique selected, and the objective of the analysis. Most process safety managers recognize that the lead hazard analyst should have had training with or experience using the selected hazard evaluation technique, should have good interpersonal and leadership skills, and should have enough technical expertise to understand the process or operation being analyzed. Other participants in a study provide the detailed knowledge (e.g., operating and maintenance practices) or special skills such as control system design that are necessary for a thorough hazard evaluation. However, managers may sometimes inadvertently undermine a hazard evaluation effort unaware of the range of skills needed in a hazard evaluation and the resulting number of participants on a team. It is the duty of the lead analyst to work with management to clearly articulate the study scope and identify necessary resources to achieve the study objectives. It is essential that the scope be well defined before resources are assigned and committed.

All hazard evaluation techniques are more thorough and effective when performed by a team of reviewers. Depending upon the technique chosen and the complexity or size of the subject process, a team may be as few as two or as many as seven or eight people. The optimum size of a hazard evaluation team varies according to the needs of a study, but it is important to avoid having too large a team, which inevitably makes it harder for the leader to keep the meeting on track.

In some specialized cases, it is possible, but not recommended, that a team of one (the analyst) could accomplish a hazard evaluation. For example, an analyst performing a traditional checklist evaluation on a process unit may simply interview one or two people and visually inspect the process equipment during a one- or two-day visit to a facility. Or a single analyst may apply hazard evaluation techniques for evaluating work orders, procedure change requests, and batch recipe modifications as a part of a facility's management of change program. However, the assertion stated in the previous paragraph remains true, that all hazard evaluation techniques are more thorough and effective when performed by a team of reviewers. In the specific case of a HAZOP Study, the person leading the analysis is required by the definition of the technique to work with a team.

There are three basic roles filled by participants that typically make up a hazard evaluation team: (1) leader, (2) scribe (which is a role that can be filled by any member of the hazard evaluation review team, including the leader), and (3) contributor(s) that may be considered expert(s) in one or more areas. Some desirable attributes of the team are the independence of the leader and diversity of knowledge brought to the table by the team. Additional attributes for each role are discussed in the following paragraphs.

The ***team leader (facilitator)*** provides direction for the analysis, organizes and executes analysis activities, and facilitates any team meetings that may be held as part of the study. Effective hazard evaluation leaders have strong interpersonal skills and an in-depth understanding of the scope and objectives of the study. This person may also be the main contact with management personnel. Normally, the hazard evaluation team leader is the member most experienced in the application of the selected hazard evaluation technique. In most cases, the success of the study depends directly upon the skill of the leader. In addition, the most effective facilitators have some independence and separation from the design team; i.e., it is desirable to have the leader be someone other than the person who designed the facility. The viable option of using a third party or outside consultant as team leader/ facilitator would provide that desirable separation from the design team. This would be an effective (maybe the only) option for companies not doing frequent hazard evaluations. However, the preparation time may be extended due to the time needed to learn the process and understand some of the corporate culture and requirements.

The ***scribe*** is the individual designated to formally document the discussions that take place during studies using meeting-oriented techniques such as HAZOP Study. Typically, if the facilitator is not also serving as the study scribe, then the scribe might be someone who is not as experienced as the team leader in the use of the chosen hazard evaluation method but who has had some basic hazard evaluation experience. If using a computer-based means of documenting the hazard evaluation, the scribe will also need to be proficient in the use of both the hardware and the software that is employed. It is helpful if the scribe has good language and organizational skills, since he or she will be deluged with information to sort out in the meeting. Sometimes, organizations assign the duties of scribe to relatively junior personnel; however, many have found this to be ineffective. Scribes with more process experience can better sort out the meeting discussions that should be documented from issues that should not.

The remainder of the team is composed of *contributors* or *experts* in various aspects of the design and operation of the process being evaluated. It is possible that some core team members will have key experience but may not be considered an 'expert'. These other team members are chosen for their specific knowledge about some aspect of the process being examined, such as the process chemistry, the equipment design, the operating procedures, the control strategy, or maintenance practices. Some organizations have found it effective to routinely include line managers on an otherwise complete hazard evaluation team. Having line managers participate helps support the safety consensus developed by the team concerning potential safety improvements.

As shown in Table 2.3, a variety of people may be asked to be part of a hazard evaluation team. The exact team composition will depend upon the type of process being analyzed, what stage the process is in, what the objectives of the evaluation are, what hazard evaluation technique is selected, and what resources are available.

Occasionally, a single individual may satisfy more than one of the basic hazard evaluation team roles. For example, the leader may be a metallurgical engineer who has expert knowledge concerning the inspection practices of the subject process. Or, one person may be the "design expert" as well as the "safety expert" for the process. However, managers assembling hazard evaluation teams should not allow people to "wear too many hats." For example, as mentioned in an earlier paragraph, it would be unrealistic to expect any expert whose full-time input is needed to efficiently perform the duties of the scribe. Experience has shown that scribes playing multiple roles on a team sometimes fail to document all the important results because they are too busy presenting their own perspectives on a hazardous situation of interest to be able to effectively participate as a scribe (or vice versa). The number and type of team members that can efficiently play combined roles is in large measure a function of the complexity of the process and the experience of the team member.

Table 2.3 Candidates for membership on a hazard evaluation team

Chemist	Mechanic/pipefitter/electrician
Civil engineer	Mechanical engineer
Construction representative	Metallurgist
Corporate process safety technology leader	Operations supervisor
Electrical engineer	Operator/technician
Environmental engineer	Outside consultant
Expert from another plant	Process engineer
Fire protection engineer	Process control programmer
Hazard evaluation expert/leader	Project engineer
Human factors specialist	R&D engineer
Industrial hygienist	Safety engineer
Inspection engineer/technician	Shift foreman
Instrument engineer/technician	Toxicologist
Interpreter	Transportation specialist
Maintenance supervisor	Vendor representative
Maintenance planner	

In general, the team should have members with practical experience in operations, maintenance, and engineering. On an as-needed basis, it should also should have people with special knowledge of process chemistry, inspections, instrumentation, environmental regulations, and corporate and industry safety standards. For reviews of existing units (or new units that are similar to existing units), it is highly desirable to involve frontline personnel, as they usually have the most accurate knowledge of the process equipment and procedures used during day-to-day operation. They are also highly motivated to identify and eliminate hazards. Recently retired employees, if available, may also be excellent team members because of their knowledge of the unit and its history. Some team members may not need to participate in the entire review; some may only attend part of the meetings and others may simply be on-call to help resolve specific issues. The team should have a variety of experience and expertise, it should be willing to consider the potential for incidents, and it should be capable of objectively analyzing any identified hazards.

2.7 Schedule and Execution

Once the scope and objectives of a hazard evaluation have been established, the hazard evaluation team participants have been selected, and pertinent information has been gathered, the team leader can schedule the review meetings and tours (if they are studying an existing process) necessary to perform the evaluation. Hazard evaluations conducted by team reviews are generally held as face-to-face meetings with all team members present in person. Although various kinds of meetings can be held with one or all members participating remotely by teleconference, videoconference, internet connection, or other means, this will hinder the efficiencies and synergistic benefits of in-person meetings, and make the team leader's job more challenging as the meeting facilitator. Some companies insist that all of the primary hazard evaluation meetings be conducted in person. However, it may be possible to conduct follow-up meetings with one or more persons participating from a remote location.

The review of the process should take place expeditiously, with a minimum of interruptions between team meetings, but the desire for immediate results should not dictate the schedule. Other factors must be considered such as the availability of qualified personnel, the need to maintain production at the facility and the need to maintain the mental alertness of the team. To avoid fatigue, meetings should normally not exceed four to six hours per day; however, a particular situation such as a tight engineering project schedule or the participation of key personnel from out of town might necessitate longer days. The need to adequately prepare for the hazard evaluation must also be considered. This is particularly important for team-oriented studies because some team members often need as much time to adequately prepare for the study as they do to participate in the evaluation.

Much of the burden for orchestrating hazard evaluations will fall on the team leader, who must interact with management to define the scope and objectives, propose an appropriate technique, and choose suitable team members who are (or will be) adequately trained to participate in the review. The leader must then assign information-gathering responsibilities and negotiate a schedule for performing the evaluation. Finally, the leader must assemble the necessary checklists, worksheets, and other materials to be used in the study.

The leader must also determine where to hold review meetings. A basic requirement is that the meeting room be large enough to accommodate the participants, that the room be appropriately equipped with marker boards, projectors, chart pads, computers, etc., that the seating arrangements are comfortable, and that there are adequate wall and/or table space to spread out the necessary documents (e.g., large drawings). If the review concerns existing equipment, it may be desirable to conduct the meetings on-site or in the immediate vicinity of the complex to accommodate tours and inspections.

Likewise, if all the information resources the team will need are on-site, the review should be held there. Off-site locations are favored if (1) the review team members will be interrupted frequently or called out of an on-site meeting or (2) if there is no on-site location that is large enough or can be reserved long enough to accommodate the review team. Either on-site or off-site locations are acceptable as long as the team can work without interruption and can have access to the information it needs.

Once the meeting attendees, location and schedule are established, the leader must make the additional arrangements necessary for a successful meeting. This might include organizing a plant tour, arranging plant passes and transportation, and providing appropriate personal protective equipment and safety training. The leader may also need to arrange temporary office space for any team members visiting the site. To improve team efficiency, the leader can ensure there are refreshments readily available at break times and consider having meals delivered if the meetings will extend into the afternoon or evening.

During meetings, the leader should promote the participation of all team members and ensure that the chosen hazard evaluation technique is properly applied.[3] The team may need to select a different technique if it becomes apparent that the originally chosen technique cannot be applied due to limitations in the available information, personnel, or time. The leader must stress that the team's objective is to identify and evaluate hazards and to make the unit safer and more productive, not to criticize individuals or to design solutions to identified problem areas. The leader must work to keep the meeting on track and to build team consensus. Good leadership skills, such as those listed in Table 2.4, are essential for success.[4,5] Ultimately, the leader must present the team's findings to management, coordinate the production of a formal report, be prepared to interpret the team's suggestions, and represent the team in any subsequent reviews of their findings.

Table 2.4 Important team leader responsibilities

- Ensure that the proper hazard evaluation method is selected and correctly applied
- Organize the team review and negotiate for resources
- Communicate with personnel at all levels in the organization
- Motivate a group to achieve a common goal
- Work with a wide range of personalities (possibly including individuals who are highly defensive and argumentative, quick and direct, or rambling and talkative)
- Interpret engineering drawings and understand process operations
- Ask questions and probe for further explanations without making team members defensive
- Maintain objectivity, honesty, and ethical conduct, reporting all significant findings regardless of the potential discomfort to the leader, team or management
- Encourage, direct, and focus group discussions
- Judge the relative importance of issues and help the team drop those not worth pursuing
- Summarize issues, negotiate a compromise, and forge a consensus
- Appreciate different points of view and empathize with team members
- Remain impartial and maintain the team's respect
- Manage the pace of team discussions and tactfully maintain the meeting schedule
- Sense team fatigue, boredom, unsuitability, etc., and implement corrective action
- Keep the team working together
- Suspend discussion of issues that cannot be resolved by the team
- Fulfill team members' psychological needs without letting any one ego, including the leader's, dominate the team

2.8 Initial Team Review Meeting

A complete, team-based review of a facility's process hazards and risks may involve several topics in addition to the use of one or more hazard evaluation techniques as presented in Chapters 4 and 5. These additional elements are often covered in an initial team review meeting, which may last anywhere from a half day for a small system to two or three days for a large process. The initial team meeting may start off with introductions if everyone does not already know every team member and their backgrounds and intended contributions. Guidelines for other items generally covered in an initial team review meeting are given in this section.

Orientation

The team leader or other designated person usually gives an up-front orientation so that all team members have a proper understanding of what the review is to accomplish and how it is to be achieved. This orientation generally includes the specific objectives, scope, and boundaries of the study (Sections 2.2 and 2.3). Site, company and regulatory requirements that pertain to the review should be clearly communicated, as well as the roles and responsibilities of each team member. Some or all of the participants may need to be given a training session on the specific hazard evaluation procedure(s) to be employed, including terminology (Sections 1.3 and 1.4) and any scenario risk evaluation procedures to be used (Chapter 7). The orientation should cover the proposed schedule but also clearly communicate that a hazard evaluation is a detailed process that will require patience and determination on the part of each participant (i.e., the schedule should be determined by the hazard evaluation process, and not vice versa).

Process Overview

The next step in preparing to participate in a hazard evaluation is typically to have a person with detailed knowledge of the subject process give an overview of the process. This may include his or her perceptions of the major hazards.

The process flow can be described using either detailed block flow diagrams or process flow diagrams or, for relatively simple processes, the piping and instrument diagrams (P&IDs). The overview can begin with the input of each chemical and utility, and proceed in the same direction as the process flow. For processes that are large or very complex, this overview should be broken down into "systems" so that enough detail can be discussed. A "system" is a sub-part of the entire process or could be a utility associated with the process, such as cooling water system, scrubbers, dust collectors, etc.

Facility Walkdown

If evaluating an operating facility, after the process overview discussion, the entire hazard evaluation team is often required to participate in a detailed walkdown of the process to be studied. This may also include all the utility systems associated with the process. Many times the best guides for this walkdown are a senior operator and the process engineer in charge of the process. Together they can provide a clear picture of both how the process is designed and intended to be operated as well as how the process was actually built and is run on a day-to-day basis. The team should, of course, abide by all safety requirements for entering the process area while conducting this walkdown. Team members should not be rushed during the walkdown, and should be encouraged to ask questions about what they are observing, hearing, and perhaps smelling. If the process is in the design stages and not yet in operation, a sister facility may be able to be visited to give the review team a sense for what the new process will look like. Some companies use a walk-through checklist as an aid for helping remember to examine key process safety aspects, such as relief discharge locations and human factor considerations.

Documentation of Attendance

Attendance of participants should be taken and documented for each session. Many times hazard evaluations are broken down in manageable time segments over weeks and months. When conducted in this manner, maintaining a consistent team can often be challenging. The company, team, or facilitator should set a minimum number of attendees that must be present before the hazard evaluation can progress. Attendance documentation should be sufficiently detailed to show who was present for each node studied. Each participant's name, position, and qualifications (which may include degree of experience or expertise, training, or short biography of related experience) and association with the process should be included in this documentation. See Section 2.6 for further details regarding team composition.

Review of Process Safety Documentation

Along with the process overview and facility walkdown, a review of the process safety documentation should be conducted before starting the formal hazard evaluation. This includes the documented chemical, technology, and equipment data associated with the process and its connected utilities, and the operating procedures used to operate the process. Other information such as vendor information, literature review, and broader company experience might also be communicated. At this time, the process engineer(s) should inform and educate the hazard evaluation team as to the standards and codes to which the process is built and maintained (i.e., API, ASME, NFPA, etc.) and the electrical classifications of the process. The participants in the hazard evaluation should have a clear understanding of the intent and hazards of the process being studied and how the process is supposed to be operated and maintained.

Previous Incidents Review

Before the hazard evaluation actually begins, the participants should review previous incidents involving the process to be studied. This review of incidents should include all types of incidents involved in the process, including personnel injuries and near misses, for at least the previous five years. The incidents should be reviewed before starting the actual hazard evaluation and should be studied in more detail when the team is studying the segment of the process where the incident actually took place. The team should not forget about those incidents that have occurred outside their facility but in processes very similar to theirs. These incidents could have occurred at a sister plant within their company or within their industry. If industry trade associations or safety organizations have issued warnings and safety alerts regarding the process or chemicals involved in the process, these are also valuable resources to be reviewed to assure the learnings from previous incidents can be applied to the hazard evaluation.

Hazard Identification and Inherent Safety Review

The initial team meeting may also include use of one or more approaches to systematically identify and document the underlying hazards of the process, then review opportunities to eliminate or reduce the hazards and thereby make the facility inherently safer. Detailed guidance on these topics is given in Chapter 3.

Global Facility Siting and Human Factors Considerations

If human factors and/or facility siting are considered globally; i.e., for the entire process being studied, then supplemental reviews covering these topics might be included as part of the initial team review meeting. An alternative often employed is to consider one or both of these topics at the final team review meeting, after the hazard evaluation is completed using one or more of the techniques in Chapters 4 and 5. Human factors and facility siting are defined and discussed in Sections 9.5 and 9.6, respectively.

Chapter 2 References

1. Center for Chemical Process Safety, *Guidelines for Technical Management of Chemical Process Safety,* AIChE, New York, 1989.
2. 29 CFR 1910.119, *Process Safety Management of Highly Hazardous Chemicals*, OSHA, Washington, DC, 1992.
3. *Discussion Leading*, Du Pont Company, Wilmington, DE, 1974.
4. R. D. Gamache and K. W. Eastman, "Running a Creative Meeting," *Chemical Engineering*, November 1990.
5. R. D. Gamache and R. L. Kuhn, *The Creativity Infusion*, Harper & Row, New York, 1989.
6. W. L. Frank, J. E. Giffin, and D. C. Hendershot, “Team Makeup...An Essential Element of a Successful Process Hazard Analysis,” *Proceedings of the International Process Safety Management Conference and Workshop,* American Institute of Chemical Engineers, New York, 1993, 129-146.

3

Hazard Identification Methods

Two separate and distinct steps are involved in a thorough hazard evaluation: first, identifying process hazards; second, evaluating whether existing or proposed safeguards are adequate to control those hazards. This chapter describes methods used to accomplish the first step of hazard identification. Chapters 4 and 5 will then describe methods to evaluate the adequacy of safeguards.

Hazards are associated with every activity, but analysts cannot begin to evaluate the hazards until they know what they are. As defined previously, a *hazard* is a physical or chemical characteristic that has the potential for causing harm to people, property, or the environment. Thus, hazard identification involves two key tasks: (1) identification of specific undesirable consequences and (2) identification of material, system, process, and plant characteristics that could produce those consequences.

The first task is relatively easy, but it is essential because it defines the scope of the second task. Undesirable consequences can be broadly classified as human impacts, environmental impacts, or economic impacts. Within these broad classifications, there may be specific consequence categories as illustrated in Figure 3.1. Each of these categories can be further subdivided by the type of damage they can cause (e.g., toxic exposure, thermal exposure, overpressure, mechanical force, radiation, electrical shock). Thus, the more precisely the consequences of concern are defined, the easier it will be to identify hazards. For example, there may be many hazards with potential human impacts but only two that could result in serious off-site injuries.

Once the consequences of concern are defined, the analyst can identify those system, process, and plant characteristics that could be a hazard of interest. It is essential that the hazard identification technique be thorough enough to identify all the important hazards. However, if the approach does not provide some discrimination between more important and less important hazards, then subsequent hazard evaluation attempts will be overwhelmed by the sheer number of potential hazards to be examined. Common methods for hazard identification include analyzing process material properties and process conditions, reviewing organization and industry process experience, developing interaction matrixes, and applying hazard evaluation techniques.

3.1 Analyzing Material Properties and Process Conditions

Every proposed or existing process is based on a certain body of knowledge. An important part of this process knowledge is data on all of the chemicals used or produced in the process, including chemical intermediates that can be isolated. This information is the foundation of all hazard identification efforts.

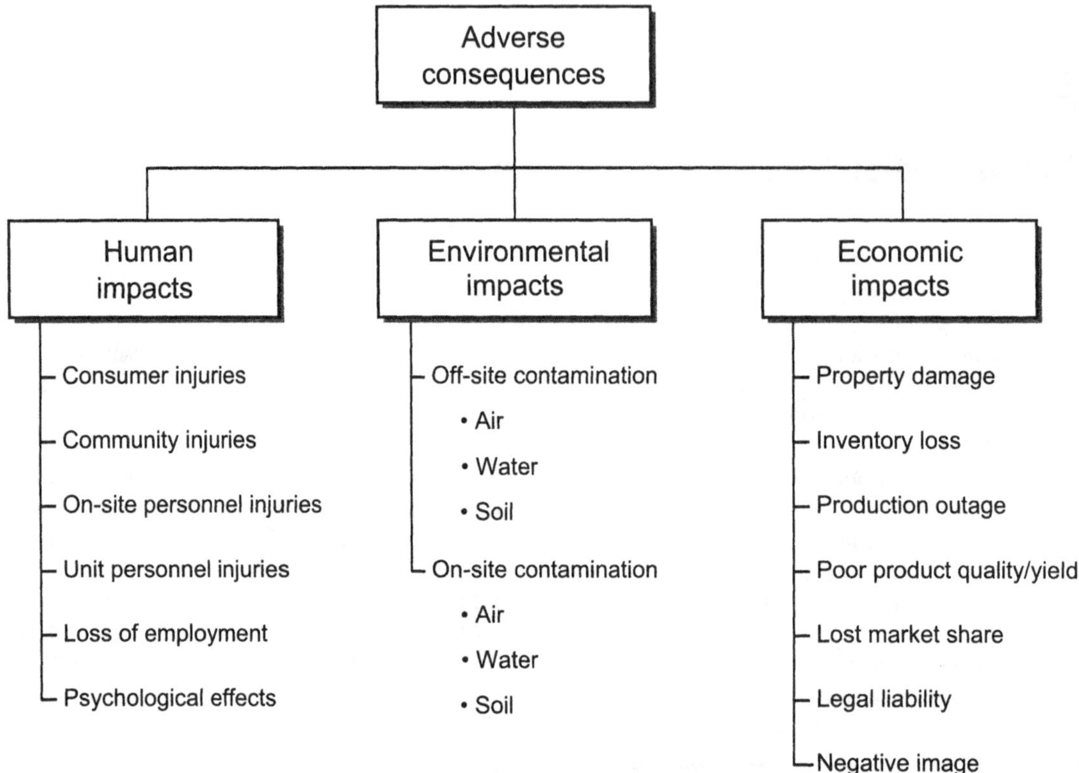

Figure 3.1 Adverse consequences associated with process hazards

Typical material properties that are useful in hazard identification are listed in Table 3.1. In addition to the information an organization has developed about a specific material, there are many other public information resources (such as those listed in Appendix F and in the reference section of this chapter) that can provide information about material properties.[1-4] Some of the best resources are the chemical manufacturers and/or suppliers; they can provide product literature, access to their chemical experts, and material safety data sheets (MSDSs). In addition, there may be a specific industrial group or association, like the Chlorine Institute, that will provide detailed information about safely handling specific types of chemicals. Even if there is not a specific industrial group, there may be information available from professional and industrial organizations such as the American Institute of Chemical Engineers, the American Petroleum Institute, or the American Chemistry Council.

There are many published sources of chemical data. Sax's *Dangerous Properties of Industrial Materials* is one frequently used reference, as are the databases maintained by the Chemical Abstracts Service and the AIChE Design Institute for Physical Property Data (DIPPR).[5-9] Government agencies and funded organizations like the U.S. Coast Guard, the Environmental Protection Agency, the Federal Emergency Management Agency, and the World Bank have also published chemical data.[10,11] Specific legal limits applicable to certain chemicals are included in federal, state, and local legislation and regulations.[12] An initial hazard identification can be performed by simply comparing the material properties available from these diverse resources to the consequences of concern. For example, if an analyst is concerned about the consequences of a fire, he or she can identify which process materials are flammable or combustible. The analyst could then classify all of those materials as fire hazards and perform more detailed hazard evaluations.

Process conditions also create hazards or exacerbate the hazards associated with the materials in a process. For example, water is not classified as an explosion hazard based on its material properties alone. However, if a process is operated at a temperature and pressure that exceeds water's boiling point, then a rapid introduction of water presents the potential for a steam explosion. Similarly, a heavy hydrocarbon may be difficult to ignite at ambient conditions, but if the process is operated above the hydrocarbon's flash point, a spill of the material may ignite. Therefore, it is not sufficient to consider only the material properties when identifying hazards; the process conditions must also be considered.

Table 3.1 Common material property data for hazard identification

Acute toxicity
- inhalation (e.g., LC_{LO})
- oral (e.g., LD_{50})
- dermal

Chronic toxicity
- inhalation
- oral
- dermal

Carcinogenicity

Mutagenicity

Teratogenicity

Exposure limits
- TLV®
- PEL
- STEL®
- IDLH
- ERPG
- AEGL

Biodegradability

Aquatic toxicity

Persistence in the environment

Odor threshold

Physical properties
- freezing point
- coefficient of expansion
- boiling point
- solubility
- vapor pressure

Physical properties (cont'd)
- density or specific volume
- corrosivity/erosivity
- heat capacity
- specific heats

Reactivity
- process materials
- desired reaction(s)
- side reaction(s)
- decomposition reaction(s)
- undesired reaction(s)
- kinetics
- materials of construction
- raw material impurities
- contaminants (air, water, rust, lubricants, etc.)
- decomposition products
- incompatible chemicals
- pyrophoric materials

Stability
- shock
- temperature
- light
- polymerization

Flammability/Explosibility
- LEL/LFL
- UEL/UFL
- minimum oxygen concentration for combustion
- dust explosion parameters
- minimum ignition energy
- flash point
- autoignition temperature
- energy production

Considering the process conditions may also enable an analyst to eliminate some materials from further evaluation as significant hazards. For example, a material may have a flash point greater than 400 °C. If the material is only present at ambient temperature and atmospheric pressure, then it may not be considered a significant fire hazard that warrants further evaluation. However, when identifying hazards, it is important to consider both normal and abnormal process conditions. Consider the following three cases:

- A pyrophoric material is normally processed with an inert gas blanket. The material warrants further evaluation as a fire hazard because there are many potential abnormal events that could expose the material to air.

- A combustible liquid is processed at high pressure. The material warrants further evaluation as a fire hazard because it could create a flammable mist if unintentionally sprayed into the air.

- A monomer is normally processed at relatively low temperatures and pressures. The material warrants further evaluation as an explosion hazard because it could undergo uncontrolled polymerization if a high temperature upset occurred.

These examples show how consideration of material properties and process conditions must be combined to identify process hazards. This approach is relatively quick and easy, and it can be applied to both new and existing processes.

3.2 Using Experience

Whenever possible, a company should use its own experience to supplement the hazard identification process.[7,13] Problems that have occurred demonstrate where hazards exist. However, basing hazard identification solely upon a company's (or even the industry's) experience is never fully satisfactory because many hazards may be overlooked. Good experience may only demonstrate that the hazards have been adequately controlled, not that hazards do not exist. Assuming that something cannot happen simply because it has not happened is a very poor approach to hazard identification.

The proper use of experience helps build a base of knowledge about the process that can be used in hazard identification activities. Analysts can always use knowledge of basic chemistry as the starting point. Then laboratory experiments may reveal the basic physical properties of a compound, its toxic effects, and its reaction kinetics. Computer software, such as ASTM's CHETAH™ program, can be used to predict the heats of reactions as well as the stability of new compounds.[14] Pilot plant experience may reveal unexpected byproducts of the reaction, show that the process conditions must be changed to achieve optimum performance, and corroborate speculation on the effects of typical process contaminants.[15] Even decommissioning a unit can add important process experience because it may reveal conditions that were not apparent (or not accessible) in the system during normal operation or unit shutdowns (e.g., evidence of incipient metal fatigue in a process application previously unknown to be vulnerable to cyclic fatigue failure). Table 3.2 lists several classes of chemical compounds that, based on industry experience, would warrant investigation as hazards.

If the hazards of a well-established process are being investigated, then the analyst can review the operating experience of similar full-scale units. Where have releases occurred? Why have emergency shutdowns occurred? What caused unscheduled outages? The answers to these and similar questions may point out hazards that are not obvious from an abstract review of material properties and process conditions.

If the organization's experience has been documented, it should be used just like any other source of data for hazard identification. If the experience has not been recorded, then it may be necessary to assemble a team of knowledgeable personnel to participate in the hazard identification process. It is usually more efficient if other hazard identification activities precede this team's review. Then the team can simply confirm that their experience matches, contradicts, or does not address the information gathered from other sources and they can point out any additional hazards that they have observed in the existing system(s). Even if the team members do not have specific experience with the chemical(s) of interest, they may have experience with similar materials that represent similar hazards.

3.3 Developing Interaction Matrices

The interaction matrix technique, a simple tool for identifying interactions among specific parameters (including materials, energy sources, environmental conditions, etc.), is a structured approach to hazard identification. As a practical matter, the technique is usually limited to two parameters (as illustrated in Figure 3.2) because the number of potential interactions increases as more and more simultaneous interactions are considered. However, there is nothing to prevent the analyst from adding a third parameter, such as stable mixtures of chemicals, to show higher-order interactions. For example, in Figure 3.2, Mixture 1 might be a mixture of Chemicals C and D, and the table would show the interaction of Mixture CD with Chemical A, Mixture CD with Contaminant 1, etc. Or the analyst could build an n-dimensional matrix if he or she believes many higher-order interactions are potentially significant and has the resources to investigate them in detail.

Table 3.2 Examples of hazardous chemical compounds

Acids	Ethers
Aldehydes	Halogens
Alkaline metals	Hydrocarbons
Alkyl metals	Hydroxides
Amines	Isocyanates
Ammonia and ammonium compounds	Mercaptans
Azo and diazo compounds and hydrazines	Nitro compounds — organic
Carbonyls	Organophosphates
Chlorates and perchlorates	Peroxides and hydroperoxides
Cyanides	Phenols and cresols
Epoxides	Silanes and chlorosilanes

To construct an interaction matrix, include all the materials of interest on each axis. This might include not only process materials, stable intermediates, materials of construction, utility streams, and other chemicals handled in adjacent processes, but also chemicals used by either site or contractor personnel during clean-out, maintenance and turnaround activities.

Normally, the matrix will exhibit bilateral symmetry, so it is only necessary to complete half the matrix (the unshaded area in the matrix portion of Figure 3.2) because Chemical A interacting with Chemical B is generally the same as Chemical B interacting with Chemical A. However, Chemical A interacting with itself may be a unique, potentially important reaction such as by polymerization. If the order of mixing is important, make sure to put this information in a table note.

The matrix parameters should not be limited only to chemicals, since other parameters could also introduce hazards to people or the environment. For example, a compound might decompose explosively if exposed to temperatures greater than 100 °C. Table 3.3 lists other parameters that may reveal potential undesirable consequences in an interaction matrix. Normally, it is adequate to list such additional parameters on only one axis of the matrix, because analysts are interested only in the parameter's interaction with the process materials, not with other parameters.

When constructing an interaction matrix, it is important to define what process conditions are being considered. Sometimes it is necessary to construct several interaction matrixes to account for both normal and abnormal process conditions. If only one matrix is constructed, the analyst should at least note the potential for hazardous interactions under other process conditions.

Once the matrix is constructed, the analyst should examine the potential consequences associated with each interaction represented in the matrix (each row-column intersection). If the consequences of the interactions are unknown, additional research or experimentation may be necessary. The type and severity of known consequences can be noted in the appropriate cell(s) of the matrix or in footnotes. (A single interaction may produce more than one type of consequence.) The results of the interaction matrix can then be compared to the consequences of concern to identify potential hazards that warrant further evaluation. Examples of interaction matrixes are included in Appendix A and in the references at the end of this chapter.[4,9,11,16]

Table 3.3 Other parameters commonly used in an interaction matrix

- Process conditions such as temperature, pressure, or static charge
- Environmental conditions such as temperature, humidity, and dust
- Materials of construction such as carbon steel, stainless steel, and asbestos gaskets
- Concentration of material in diluent and the particular diluent used
- Common contaminants such as air, water, rust, salt, and lubricants
- Contamination from other materials handled in the same process equipment or area
- Order of mixing of the interacting materials; ratio of materials combined
- Human health effects including short-term and long-term exposure limits
- Environmental effects including odor thresholds and aquatic toxicity limits
- Legal limits for inventory, spills, or waste disposal

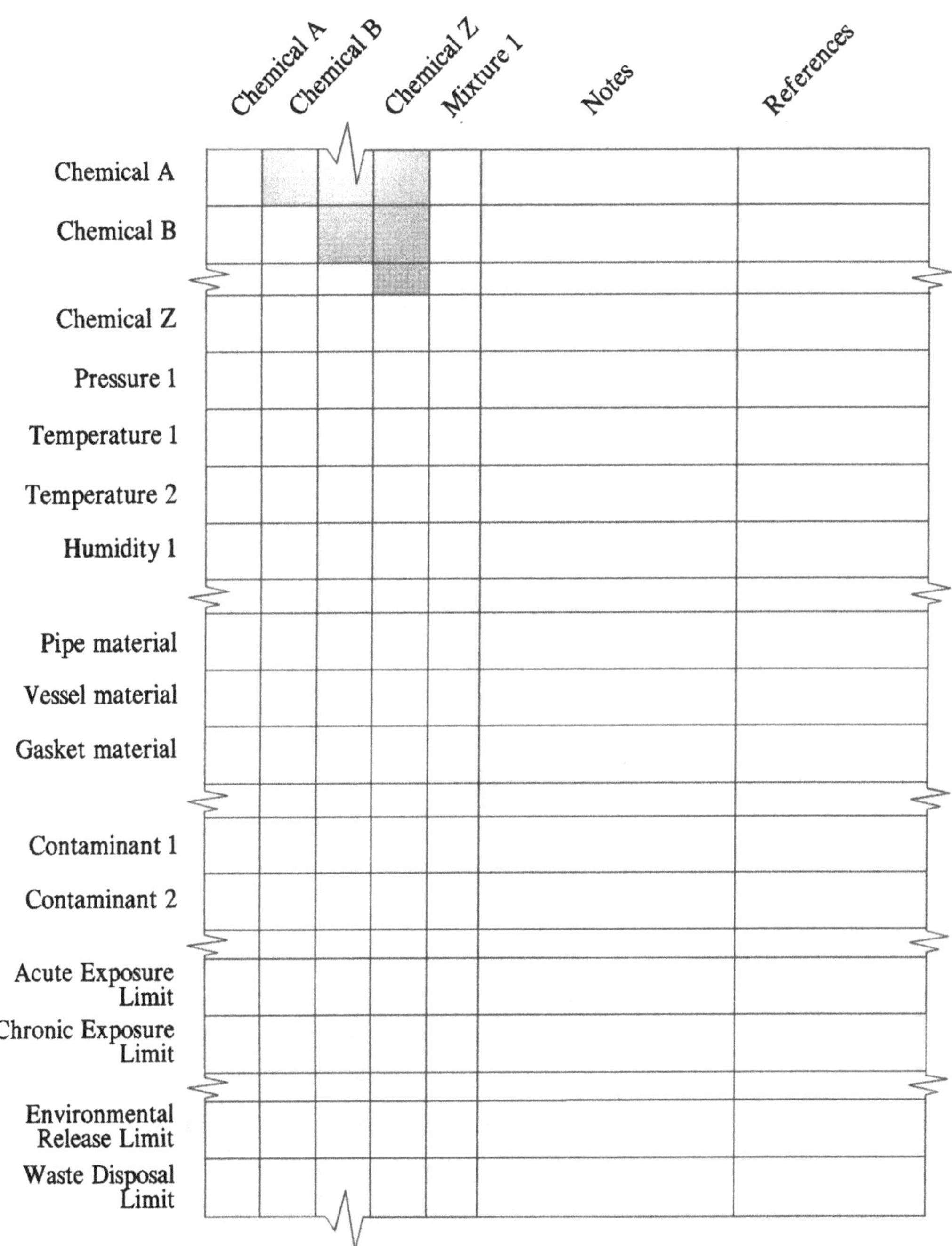

Figure 3.2 Typical interaction matrix

Interaction matrixes are most commonly used to identify chemical incompatibility. ASTM International has developed a "Standard Guide for the Preparation of a Binary Chemical Compatibility Chart" (ASTM E 2012) that also includes an example compatibility chart.[24] This and similar methods are discussed in the CCPS Concept Book *Essential Practices for Managing Chemical Reactivity Hazards*.[26] The U. S. National Oceanic and Atmospheric Administration has developed a Chemical Reactivity Worksheet, with over 6000 chemicals in its database.[27] The program predicts the results of binary mixtures by reactive group combinations and indicates possible hazardous interactions. It also sets up a compatibility chart and indicates possible consequences of the interaction, such as "Heat generation by chemical reaction, may cause pressurization." Figure 3.3 shows a Chemical Reactivity Worksheet compatibility display.

3.4 Hazard Identification Results

Usually, hazard identification efforts result in simple lists of materials or conditions that could result in hazardous situations, such as those in Table 3.4. An analyst can use these results to define the appropriate scope and select the appropriate technique for performing a hazard evaluation. A hazard identification effort may also involve a concurrent inherent safety review (see Section 3.7), by documenting the feasibility of reducing or eliminating each of the identified hazards. A worked example showing hazard identification results can be found in Part II, Chapter 12.

In general, the scope and complexity of subsequent hazard evaluations will be directly proportional to the number and type of hazards identified and the depth to which they are understood. If the extent of some hazards is unknown (e.g., if it is not known where there may be a potential for an incompatible heat transfer fluid to leak into the process), additional research or testing may be required before the hazard evaluation can proceed.

Table 3.4 Typical hazard identification results

- List of flammable/combustible materials
- List of toxic/corrosive materials and by-products
- List of energetic materials and explosives
- List of explosible dusts
- List of hazardous reactions; chemical interaction matrix
- Fundamental hazard properties e.g. flash point, toxic endpoint
- Others e.g. simple asphyxiants, oxidizers, etc.
- Total quantities of each hazardous material
- List of chemicals and quantities that would be reportable if released to the environment
- List of physical hazards (e.g., pressure, temperature, mechanical energy, radiation, electrical energy, etc.) associated with a system
- List of contaminants and process conditions that lead to a runaway reaction

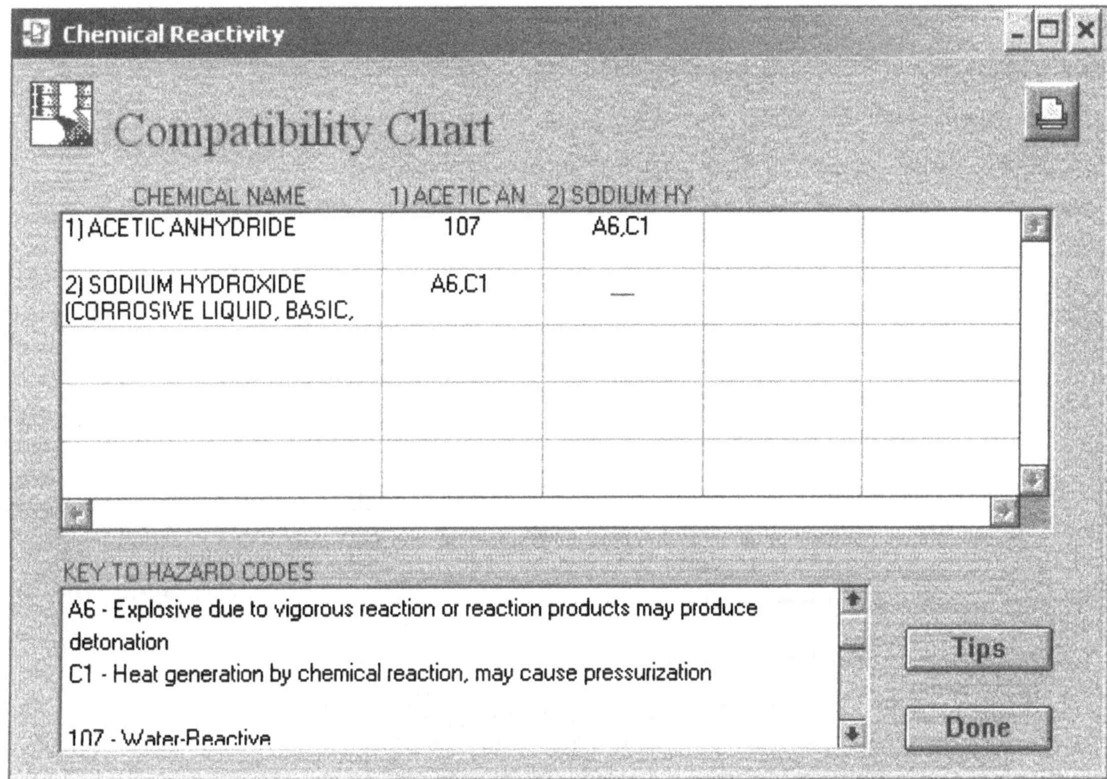

Figure 3.3 NOAA Worksheet compatibility chart display

3.5 Using Hazard Evaluation Techniques to Identify Hazards

Most of the hazard evaluation techniques discussed in Chapters 4 and 5 are not well-adapted for hazard identification purposes. However, one of these techniques, Checklist Analysis, can effectively be used for hazard identification.

The Checklist Analysis technique, as described in Section 4.4, provides a list of questions and issues that should be resolved for each chemical in the process and for the process as a whole. Table 3.5 lists typical checklist items used for hazard identification. The drawback of this technique is that no checklist can anticipate every potentially hazardous situation. The most thorough checklists also tend to be the longest, which makes them very tedious to complete. Nevertheless, checklists are appealing because they can be customized for a particular process or company and because they help ensure that analysts will review processes consistently. As long as the analysts are given the freedom to add their own insights to the review, checklists can be a very powerful hazard identification tool.

Table 3.5 Examples of checklist questions used in hazard identification

- ❑ Is the flash point of the material below 100 °F?
- ❑ Is the material shock sensitive?
- ❑ Does the material polymerize? If so, what accelerates polymerization?
- ❑ Does the material react with water?
- ❑ What material spills are reportable?
- ❑ Is the material toxic if inhaled?
- ❑ Does the process operate above any material's autoignition temperature?
- ❑ Is the vapor space of any vessel in the flammable range?

Other hazard evaluation procedures, such as the What-If Analysis and HAZOP Study techniques described in Chapter 5, allow experts to use their experience more creatively. Each technique offers a different way for questions to be focused. Both methods, however, challenge the review team to create and answer a series of questions, thereby revealing the potential for undesirable consequences. Hazard identification using these methods is only indirect, by uncovering loss events that imply underlying hazards that may not have been previously recognized. For example, a What-If Analysis might identify a scenario whereby a previously unanticipated shock-sensitive material could be formed.

Because the What-If Analysis and HAZOP Study techniques are creative processes, they are more likely to uncover unique or unexpected hazards in processes. However, unless the team leader is highly skilled and the team members have appropriate experience, important hazards may be overlooked. Thus, many companies combine Checklist Analysis and brainstorming approaches to take advantage of the rigor and consistency of the Checklist Analysis while retaining the flexibility and creativity of the brainstorming techniques.

In general, it is not efficient to use these techniques for the sole purpose of hazard identification when there is adequate information for the team to perform both hazard identification and evaluation. However, when information is limited, as may be the case for newly developed processes or conceptual designs of new units, hazard evaluation techniques can be used to effectively identify hazards as long as the hazard analysts limit their study to a fairly general level of detail. The What-If/Checklist Analysis technique is a widely used combination for hazard identification and evaluation and is described in Section 5.2; however, checklists can be combined with any of the methods covered in the *Guidelines*.

3.6 Initial Assessment of Worst-Case Consequences

After specific process hazards are identified and before hazard evaluation techniques are employed, some companies find it beneficial to estimate the worst-case consequences for the identified hazards, or at least for what are judged to be the most significant identified hazards. This can serve one or more purposes:

- It highlights what consequences are possible if the hazards are not contained and controlled

- It can be used to help decide an appropriate analysis methodology (e.g., a more rigorous analysis method may be used if the potential consequences are more severe)
- It can aid in the later hazard evaluation by helping the review team determine the severity of consequences of scenarios involving the hazardous material or energy
- It can focus the review team's efforts on the parts of the process that involve the greater hazards.

This assessment of the worst-case consequences can take two basic forms. For each hazard of concern, the consequences can be expressed as (1) an effect radius or distance and/or (2) an estimated impact in terms of e.g. number of persons potentially affected or maximum probable property damage. For example, the presence of a propane storage tank on-site poses multiple hazards, ranging from the flammability, liquefied gas, and simple asphyxiant hazards of the propane to the physical hazards of vapor stored under pressure and the mass of the system. An assessment of the worst-case consequences might focus on the flammability and liquefied gas hazards and, through experience, pass over the simple asphyxiant hazard as having less severity potential. The assessment may then calculate an effect radius for a worst-case scenario of tank rupture and propane ignition, and may also go on to determine what the impacts may be on surrounding populations, buildings, and equipment.

Various methods, from simplified to fully quantitative, can be used to determine the worst-case consequences. For example, the Hazard Assessment methods included in the guidance documents for complying with the EPA Risk Management Program (RMP) Rule, 40 CFR Part 68, can be used to determine a worst-case effect distance for toxic releases, fires, and vapor cloud explosions involving the "extremely hazardous substances" covered by the RMP Rule. A computerized version of the RMP calculation methods (RMP*Comp) is available from the U.S. EPA.[28] The Dow Fire/Explosion Index, described below, is another commonly employed method. Various CCPS books provide approaches for more detailed calculations; *Guidelines for Consequence Analysis of Chemical Releases* [21] is a good starting point.

The Dow Fire/Explosion Index and Chemical Exposure Index

The Dow Fire and Explosion Index (F&EI)[22] has been used for many years for hazard identification and hazard evaluation and is published by AIChE. The Dow F&EI is used to identify and evaluate the existence and significance of fire and explosion hazards by dividing the process facility into pertinent process units. The pertinent process units are assigned factors based on material, physical and chemical characteristics; process conditions; plant layout and other process attributes. The F&EI is calculated from the various factors. The index then can be used as a measure of the process hazard magnitude by predicting a hazard radius within which process and equipment damage can be expected if a fire or explosion occurs. A maximum probable property damage (MPPD) is also calculated, using the replacement cost of equipment within the hazard radius. Experienced analysts can use the index to evaluate the need for additional safety improvements such as fire protection, equipment spacing, and instrumented protective systems.

The Dow Chemical Exposure Index (CEI)[23] is used in similar fashion to identify and rank the relative acute health hazards associated with potential chemical releases. It is also published by AIChE. The CEI is calculated from five factors: a measure of toxicity; the quantity of volatile material available for a release; the distance to each area of concern; the molecular weight of the material being evaluated; and process variables that can affect the conditions of a release such as temperature, pressure, and reactivity. It should be noted that if dusting or misting is possible, the CEI may underestimate the hazard.

Both indexes are screening tools that can be used to determine the need for more sophisticated hazard evaluation techniques, consequence modeling, and quantitative risk assessment. They can also be used to screen specific incident scenarios for evaluation by Layer of Protection Analysis (Section 7.6).

3.7 Hazard Reduction Approaches and Inherent Safety Reviews

The emphasis in environmental protection has moved over the years from environmental restoration to waste management, pollution prevention, and more recently to inherently cleaner and greener processes. Similarly, the emphasis in process safety has moved from disaster recovery to incident mitigation and prevention, and more recently to inherently safer processes. Inherently safer principles are thus aligned with the philosophy of "green chemistry," which is the design of chemical products and processes in such a way as to reduce or eliminate the use and generation of hazardous substances, in order to minimize or prevent the creation of pollution and hazardous waste. Inherent safety embraces the concepts of reducing or eliminating the hazards in chemical processing. Identifying and discussing opportunities to make a facility inherently safer has become a more integral part of hazard evaluations.

Introduction to Hazard Reduction Approaches and Inherently Safer Systems

Inherent safety is one of the strategies that can be used to reduce the risk of a major process safety incident. Rather than applying controls and safeguards, inherent safety fundamentally involves the elimination or reduction of the underlying chemical or physical hazard. Inherent safety principles can also be used to make safeguards more robust and reliable, although not reducing or eliminating the hazard.

Reduction of major process safety incident risks can be accomplished either by application of inherent safety strategies to reduce process hazards or by more effective containment and control of the process hazards. However, it should be noted that making a facility inherently safer does not automatically reduce facility risk. If making a facility inherently safer involves reducing the chemical or physical hazards of an operation, this usually translates into a lower potential severity of consequences if a loss event occurs. Since the facility risk is a function of both severity of consequences and likelihood of occurrence of a loss event (as detailed in Chapter 7), a facility change that increases the likelihood of a loss event more than it reduces its potential severity event would actually increase the overall risk. An example might be making a change to start using many 150 lb (68 kg) cylinders per day of chlorine instead of a few ton containers. This would result in connections and disconnections needing to be made 13 to 14 times more often, increasing the likelihood of a release and possibly increasing the overall risk of chlorine exposures.

Most companies integrate inherently safer (IS) design reviews into existing hazard evaluations, although some companies conduct separate reviews. Inherent safety reviews should be done at key stages of a process life cycle: during product and process development, during conceptual process design, and during routine operations.

The modern approach to process safety is to identify and evaluate the hazards posed by a process and to reduce the risk to an acceptable level consistent with the objectives of the business. Chemical process hazards come from two sources:

- Hazards that are characteristic of the material and chemistry used (e.g. flammability, toxicity, reactivity)

- Hazards that are characteristic of the process variables – the way the chemistry works in the process (e.g. pressure temperature, concentration)

Four strategies or approaches to reduce the process risk and address the hazards are:

- Inherent - Eliminate the hazard by using non-hazardous materials and process conditions (e.g., substituting water for a flammable liquid as a solvent).

- Passive - Reduce risk by process and equipment design features that reduce frequency or severity without the active functioning of any device (e.g., robust pressure vessel design; drainage and containment; blast-resistant construction).

- Active - Use controls, instrumented protective systems, and other devices such as excess flow valves, remotely actuated block valves, and safety relief valves for responding to abnormal situations (e.g., a pump that is shut off by a high level switch when the tank is 90% full) or mitigating loss event impacts. These systems are sometimes called *engineering controls*.

- Procedural - Use policies, operating procedures, administrative checks, emergency response, and other management approaches to prevent or minimize the effects of an incident (e.g., hot work permitting, emergency plans). These approaches are often called *administrative controls*.

Inherently safer design strategies are considered more reliable than passive, active or procedural approaches, although all four strategies are used to manage the risk in a process facility.

Inherently safer strategies include:

- **Minimize** (e.g., reduce the quantity of a hazardous material)

- **Substitute** (e.g., substitute the hazardous material with a less hazardous material)

- **Moderate** (e.g., use less hazardous conditions, a less hazardous form of a material, or facilities that minimize the impact of a release of a hazardous material)

- **Simplify** (e.g., design facilities that eliminate unnecessary complexity and make operating errors less likely, and that are forgiving of errors that are made)

When recommendations and action items are being developed during a hazard evaluation, these strategies and their effectiveness should be considered.

Figure 3.4 illustrates how inherent safety strategies and passive, active and procedural approaches are implemented in a process risk management system.[25] "First-order" inherent safety involves eliminating or avoiding the hazard. "Second-order" inherent safety involves reducing the frequency and/or severity of process safety incidents by applying the strategies of minimization, substitution, moderation and simplification to the hazard, such as by reducing inventory or using a hazardous material in dilute form.

Layers of protection can also be made more robust by using the principles of inherent safety when designing the safeguards. For example, a rupture disk is a simpler, more passive safeguard than a safety relief valve for providing reliable overpressure protection where a non-reclosing device can be used. An excess flow valve can more rapidly isolate a downstream line rupture than a procedural response involving detection and operator action. These approaches reduce the chance of system failure, but may not affect the presence of the hazard itself. Thus, the application of inherent safety strategies to safeguards is useful, but its application does not reduce the magnitude of the underlying hazard. Inherent safety considerations along with preventive and mitigative safeguards using passive, active, and procedural approaches are all necessary within a successful process risk management system.

Inherent Safety Reviews

Objectives. Although inherent safety thinking should be incorporated into the process development and design, inherent safety reviews are necessary to evaluate the effectiveness of the design in regard to inherent safety. Inherent safety thinking needs to be integrated into the design processes and standards. Whether inherent safety is addressed in a separate review or part of other design reviews, the objectives are the same:

- To identify and understand the hazards
- To find ways to reduce or eliminate the hazards.

Hazard evaluation procedures used. The hazard evaluation procedures most frequently used in conjunction with inherent safety reviews are Checklist Reviews, What-If Analyses, and HAZOP Studies. These hazard evaluation procedures are discussed in detail in Chapters 4 and 5 of this book. In the discussion of each of these procedures, the application of these techniques to inherently safer evaluation will also be discussed. The checklist in Appendix A4 specifically addresses inherently safer strategies and may prove useful as the basis for a Checklist Review or as a supplement to another methodology.

Preparation. As with any hazard evaluation, proper preparation is necessary. See Chapter 2 of this book for information on preparation. Information necessary for inherent safety reviews include reactivity and reaction data, flammability and toxicity data, and physical property data; see the inherent safety review process overview below for other items related to review preparation.

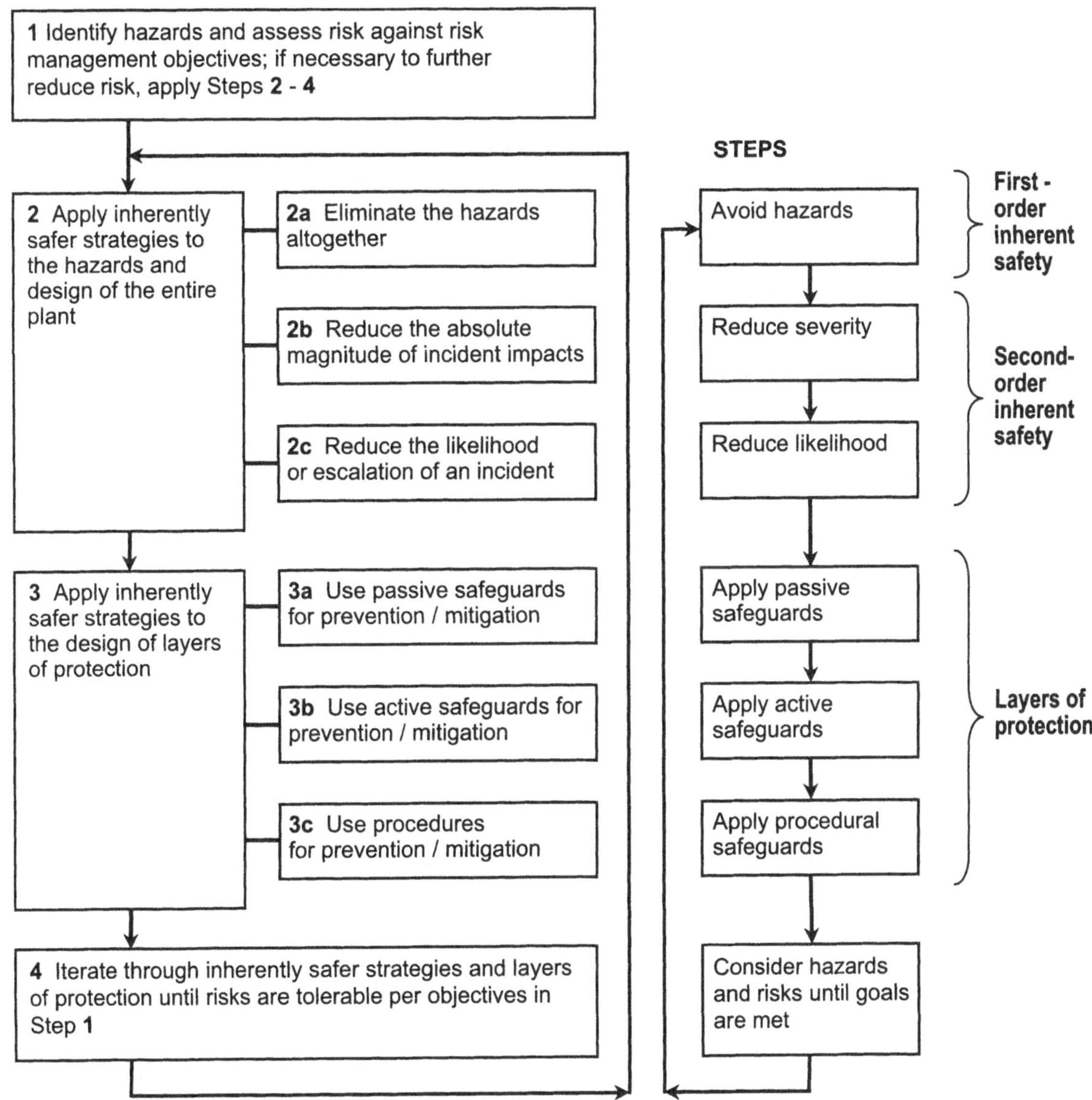

Figure 3.4 Implementation of inherently safer design within a process risk management system

Timing. Inherent safety reviews should be considered in the appropriate stages of a typical process life cycle. Specifically, reviews are most effective:

- During the research and development phase. This phase includes the chemistry synthesis stage for product/process research and development where key decisions are made on materials, chemistry, and process conditions.
- During the conceptual design phase. This phase includes scoping and development focusing on process flow, equipment selection and configuration.
- During the detailed engineering phase. This phase includes piping and instrumentation diagram development, plant layout, equipment specification, and final detailed engineering drawings.
- During the routine operation phase. During this phase the process is operating producing product. Various process changes are made to improve safety and productivity. Inherent safety must be a consideration in management of change reviews.

Team makeup. The composition of the inherent safety review team will depend on the process life cycle stage at which the review is being done. The team generally has four to seven members. The industrial hygienist/toxicologist and chemist are key members of the team to ensure understanding of the hazard associated with reactions, chemicals intermediates and products. They can help make choices among chemicals, chemistry and processes. Table 3.6 illustrates the team composition at various process life phases.

Table 3.6 Inherent safety review team composition

	Product development	Design development	Design-stage hazard evaluation	Operations
Industrial hygienist/toxicologist	√	√	√	√
Chemist	√	√	√	√
Process designer	√	√	√	√
Safety engineer	√	√	√	√
Process technologist	√	√	√	√
Environmental engineer	√	√	√	√
Control engineer		√	√	√
Operator			√	√
Operations supervisor		√	√	√
Maintenance representative			√	√

Inherent Safety Review process overview. For a typical review at the conceptual design phase, the following background information is required:

1. Define the required product.
2. Describe the optional routes to manufacture the desired product (if available) including raw materials, intermediates and waste streams.
3. Prepare simplified process flow diagram.
   — Include alternative processes
4. Develop chemical reactions
   — Desired and undesired
   — Determine potential for runaway reactions/decomposition
5. List chemicals and materials employed
   — Develop a chemical compatibility matrix
6. Define physical, chemical, and toxic properties
   — Provide NFPA hazard rating or equivalent
7. Define process conditions (pressure, temperature, etc.)
8. Estimate quantities used in each process system (tanks, reactors, etc.)
   — State plant capacity
   — Estimate quantities of waste/emissions
9. Define site-specific issues (environmental, regulatory, community, spacing, permitting, etc.).

Typical review steps are:

- Review the background information above
- Define the major hazards
- Systematically review the process flow schematic, looking at each process step and hazardous material to identify creative ways to improve the process by applying inherently safer principles to reduce or eliminate hazards. Some hazard evaluation approaches from Chapters 4 and 5, such as the use of checklists or what-if questioning, can be used for this purpose. As an example, the review team might look at a part of the process where the design calls for an intermediate storage tank, with the intermediate being a volatile and reactive material. The review might question why the intermediate storage is included in the process design, and be involved in a comparison of the hazard reduction benefit in eliminating the tank versus the process penalty expected without the surge capacity.
- During the inherent safety review at the design development phase, identify potential human factor/ergonomic issues that should be addressed by the design team.
- Document the review and follow-up actions.

Training. Training on inherently safer principles should include engineers, scientists, business leaders, operations and maintenance personnel. Training should include:

- The hierarchy of risk management strategies from procedural to active to passive to inherent levels.
- Various approaches to apply inherently safer systems including minimization, substitution, moderation, and simplification.
- Company design standards, examples, and other information that provides guidance on evaluation and implementation of inherently safer design features.
- Potential inherently safer design conflicts.

Documentation. A report of the review should be generated to document the inherent safety study, regardless of the specific type of inherent safety analysis conducted. This report should include the following information at a minimum:

- A summary of the approach used for the inherent safety review (e.g. methodology, checklist used, etc.).
- Names and qualifications of the team facilitator/leader and team makeup, including positions, names, and any relevant experience or training.
- Inherently safer alternatives considered, as well as those already implemented or included in the design.
- If an independent inherently safer systems analysis was conducted, documentation should include the method used for the analysis, what inherently safer systems were considered, and the results of each consideration. If an inherent safety checklist was used, for items that were not considered, document why those items were not considered; i.e., not applicable or considered previously.
- Documentation of rationale for rejecting potential inherently safer opportunities (e.g. cost, creation of other safety or operability problems, etc.).
- Recommendations/action plans for further evaluation or implementation of inherently safer alternatives identified during the study.

If the inherent safety review was conducted as part of a larger hazard evaluation, this information should be incorporated into the report of this activity. Electronic versions in an editable format should be maintained to facilitate future updates and revalidations.

A more complete discussion of Inherently Safer Design Reviews can be found in Chapter 8 of the *Inherently Safer Chemical Processes, Second Edition.*[25] The discussion in this section is largely a summary of Chapter 8 of that book.

Chapter 3 References

1. NFPA 704, *Standard System for the Identification of the Hazards of Materials for Emergency Response,* National Fire Protection Association, Quincy, Massachusetts, 2007.
2. R. King and J. Magid, *Industrial Hazard and Safety Handbook,* Newnes-Butterworths, London, 1979.

3. R. Pohanish, *Sittig's Handbook of Toxic and Hazardous Chemicals and Carcinogens, 4th Ed.,* Noyes Publication, New York, 2002.
4. L. Bretherick, *Handbook of Reactive Chemical Hazards, 4th Ed.,* Butterworths, London, 1990.
5. R. J. Lewis, *Sax's Dangerous Properties of Industrial Materials, 11th Ed.,* Wiley-Interscience, New York, 2005.
6. W. E. Baker et al., *Explosion Hazards and Evaluation*, ISBN 0444-42094-0, Elsevier Science Publishing Co., Inc., New York, 1983.
7. Guidelines for a Reactive Chemicals Program, 2nd Ed., Dow Chemical Company, Midland, Michigan, 1987.
8. R. L. Harris et al. (eds.), *Patty's Industrial Hygiene and Toxicology, 5th Ed.,* Wiley-Interscience, New York, 2005.
9. NFPA 491, Manual of Hazardous Chemical Reactions, in *Fire Protection Guide to Hazardous Materials, 13th Ed.*, National Fire Protection Association, Quincy, Massachusetts, 2002.
10. *Chemical Hazard Response Information System Manual (CHRIS),* (Commandant Instruction MI6465, 12A, available from U.S. Government Printing Office, Washington, DC, 20402), U.S. Coast Guard, Washington, DC, 1984, www.chrismanual.com.
11. H. K. Hatayama et al., *A Method of Determining the Compatibility of Hazardous Wastes,* EPA-600/2-80-076, Municipal Environmental Research Laboratory, U.S. Environmental Protection Agency, Cincinnati, OH, 1980. (Also available as publication PB80-221005 from the National Technical Information Service, Springfield, VA 22161.)
12. *Book of Lists for Regulated Hazardous Substances,* Government Institutes, Inc., Rockville, Maryland, 2006.
13. American Institute of Chemical Engineers, *Dow's Fire & Explosion Index Hazard Classification Guide, 7th Edition,* ISBN 0-8169-0623-8, Wiley, New York, June 1994.
14. ASTM DS51E, ASTM Computer Program for Chemical Thermodynamic and Energy Release Evaluation -- CHETAH Version 8.0, ASTM International, West Conshohocken, Pennsylvania, 2005.
15. J. C. Dove, "Process Safety—An Integral Part of Pilot Plants," *Plant/Operations Progress* 7(4), October 1988.
16. *Handbook of Chemical Hazard Analysis Procedures* (ARCHIE Manual), Federal Emergency Management Agency, Washington, DC, 1989.
17. Department of Defense, *Military Standard System Safety Program Requirements,* MIL-STD-882D, Washington, DC, 2000.
18. H. R. Greenberg and J. J. Cramer (eds.), *Risk Assessment and Risk Management for the Chemical Process Industry,* ISBN 0-442-23438-4, Van Nostrand Reinhold, New York, 1991.
19. J. Stephenson, *System Safety 2000—A Practical Guide for Planning, Managing, and Conducting System Safety Programs,* ISBN 0-0442-23840-1, Van Nostrand Reinhold, New York, 1991.
20. W. Hammer, *Handbook of System and Product Safety,* Prentice Hall, Inc., New York, 1972.
21. Center for Chemical Process Safety, *Guidelines for Consequence Analysis of Chemical Releases,* American Institute of Chemical Engineers, New York, 1999.
22. American Institute of Chemical Engineers, *Dow's Fire & Explosion Index Hazard Classification Guide, 7th Edition,* ISBN 0-8169-0623-8, Wiley, New York, June 1994.
23. American Institute of Chemical Engineers, *Dow's Chemical Exposure Index Guide*, ISBN 0-8169-0647-5, Wiley, New York, August 1988.
24. ASTM E 2012-00, "Standard Guide for the Preparation of a Binary Chemical Compatibility Chart," ASTM International, West Conshohocken, Pennsylvania, 2000.
25. Center for Chemical Process Safety, *Inherently Safer Chemical Processes: A Life Cycle Approach, 2nd Ed.,* American Institute of Chemical Engineers, New York, 2007.
26. R. W. Johnson, S. W. Rudy and S. D. Unwin, *Essential Practices for Managing Chemical Reactivity Hazards,* American Institute of Chemical Engineers, New York, 2003.
27. NOAA Chemical Reactivity Worksheet, Version 1.9, U.S. National Oceanic and Atmospheric Administration, http://response.restoration.noaa.gov/chemaids/react.html.
28. RMP*Comp Modeling Program for Risk Management Plans, U.S. Environmental Protection Agency, Washington, DC, http://yosemite.epa.gov/oswer/CeppoWeb.nsf/content/rmp-comp.htm.

4

Non-Scenario-Based Hazard Evaluation Procedures

The purpose of the next two chapters is to discuss eleven commonly used hazard evaluation techniques. Although all eleven techniques covered in the *Guidelines* are given equal treatment, not all of the methods are appropriate for every set of hazard evaluation circumstances. The techniques discussed in Chapter 4 are more appropriately used for performing general process hazard studies—usually early during the life of a process. These techniques (i.e., Preliminary Hazard Analysis, Safety Review, Relative Ranking, and Checklist Analysis) are efficient at taking a "broad brush" look at the inherent hazards of a large plant or complex process. Using these techniques before a process is commissioned can significantly improve the cost-effectiveness of subsequent safety improvement efforts.

Some of the techniques discussed in the next two chapters are experience-based and some are predictive in nature. For example, Checklist Analysis and Safety Review are based on the past experience of those who created the checklist being used or on the experience of the team members. Other techniques such as HAZOP Studies and FMEA predict incident scenarios.

For each hazard evaluation technique, the following areas are covered in early subsections for quick reference: ***Description***, ***Purpose***, ***Types of Results***, and ***Resource Requirements***. The ***Description***, ***Purpose***, and ***Types of Results*** subsections of this chapter outline what organizations can expect to achieve with particular hazard evaluation methods. This information is essential to understanding the significance of factors that can influence the selection of an appropriate hazard evaluation technique (Chapter 6).

Resource Requirements Estimates

The ***Resource Requirements*** subsections provide some basic information on the skills, materials, and effort required to perform hazard evaluations. To help users understand the magnitude of the task they are accepting when they choose a particular hazard evaluation technique, some rough estimates of the amount of effort generally required to perform a study are provided. However, estimating the time and effort needed to apply a particular hazard evaluation technique is more art than science, because the actual time to perform a study is influenced by many factors—some of which are not quantifiable.

One important factor is the complexity and size of the problem. To account for this influence and give analysts some idea of the effort that will be needed to perform hazard evaluations, estimates are based on two typical types of analysis problems—nominally, a "small system" and a "large process":

Small system - For example, a chemical unloading and storage system consisting of a rail car unloading station, transfer lines, pumps, storage tank, and pressure control/vapor return lines.

Large process - For example, a chemical reaction process consisting of a feed system, reaction section, product separation and recovery, emergency relief system, and associated connecting piping and control systems. This process may contain from 10 to 20 major vessels, including reactors, columns and accumulators.

These two examples are used to base rough estimates of the amount of time spent by each participant in a hazard evaluation. For each technique, the performance of a hazard evaluation is divided into three basic phases: *preparation, evaluation,* and *documentation.* Preparation involves all the activities discussed in Chapter 2 (e.g., collecting information, defining the analysis scope, and organizing meetings), but does not include development or updating of process safety information and operating procedures. Evaluation includes the actual analysis activity that is associated with the chosen hazard evaluation technique (e.g., for a What-If Analysis, holding the team meetings). For certain techniques that involve construction of a complex failure logic model, a *model development* phase is also included. The documentation phase includes not only recording significant results in hazard evaluation team meetings, but also developing, reviewing, and completing a formal hazard evaluation report containing a brief process description, discussion of important results, tables or logic models (if any), and a brief explanation of the significance of action items.

The technical labor estimates are given in hours, days, and weeks. For techniques involving a team, certain individuals may participate in only one or two phases, such as a HAZOP meeting (evaluation phase). Others, notably the hazard evaluation team leader, will work during all phases. Ranges are given to provide some idea of the influence that other factors can have on the time required to do the job (e.g., experience of team).

These estimates are provided only to give analysts a rough idea of the effort they should allocate for performing a hazard evaluation. ***However, because there are so many other factors that influence time and effort, analysts should use these estimates with great caution. The actual time required for a study may be much greater (or somewhat less) than these estimates indicate.*** As analysts and organizations gain experience with each hazard evaluation technique, they should become better equipped to accurately estimate the size of hazard evaluations for their facilities and become more efficient in the performance of hazard evaluations. One attempt at modeling the time required to conduct a HAZOP Study[1] considers the number of P&IDs, the complexity of the P&IDs, the effectiveness and skill of the HAZOP Study leader, and the report writing time as the major factors in the determination of the time required to perform a HAZOP.

For procedure-based operations, the number of procedures and operator-action procedural steps, as well as the approach used to develop and evaluate procedure deviations (as described in Section 9.1) are major factors in determining the time required. Both the number of operator-action procedural steps and the number and complexity of the P&IDs should be taken into account for hazard evaluations of continuous processes and their associated start-up, shutdown and emergency procedures.

---

[1] *Journal of Loss Prevention in the Process Industries, 10*(4), July 1997, pages 249-257.

Note that all Chapter 4 and 5 references pertaining to specific hazard evaluation techniques will be found at the end of each individual section; e.g., at the end of Section 4.1 for reference pertaining to Preliminary Hazard Analysis.

Analysis Procedures and Examples

Chapters 4 and 5 show how to use the eleven hazard evaluation techniques covered in these *Guidelines*. In addition to the ***Description***, ***Purpose***, ***Type of Results***, and ***Resource Requirements***, the ***Analysis Procedure*** and a simple ***Example*** is provided for each of the methods. Detailed illustrations of how these techniques can be used throughout the lifetime of a process are given in *Part II - Worked Examples for Hazard Evaluation Procedures*. Hazard analysts requiring additional details about applying hazard evaluation techniques to a particular problem should refer to the authoritative source materials mentioned in the references at the end of each methodology section and in the *Guidelines* Bibliography.

This chapter describes the following hazard evaluation techniques and summarizes how they are typically applied in the process industries:

Section 4.1 – Preliminary Hazard Analysis

Section 4.2 – Safety Review

Section 4.3 – Relative Ranking

Section 4.4 – Checklist Analysis

Analysis follow-up considerations for these methods, including the development of recommendations, prioritization of results, documentation, action item resolution and communication of findings, are all discussed in Chapter 8.

4.1 Preliminary Hazard Analysis

A Preliminary Hazard Analysis (PreHA)* is a technique that is derived from the U.S. Military Standard System Safety Program Requirements.[1]

Purpose

The Preliminary Hazard Analysis is often used to evaluate hazards early in the life of a process. A Preliminary Hazard Analysis is generally applied during the conceptual design or R&D phase of a process plant and can be very useful when making site selection decisions. It is also commonly used as a design review tool before a process P&ID is developed.

While the Preliminary Hazard Analysis technique is normally used in the preliminary phase of plant development for cases where experience provides little or no insight into potential safety problems, it may also be helpful when analyzing large existing facilities or when prioritizing hazards when circumstances prevent a more extensive technique from being used.

---

* The first and second editions of these *Guidelines* used the abbreviation "PHA" for Preliminary Hazard Analysis; however, use of this abbreviation has been changed to PreHA or no acronym is used at all, to avoid confusion with the now more common term Process Hazard Analysis which is associated with the acronym PHA.

Description

A Preliminary Hazard Analysis focuses in a general way on the hazardous materials and major process areas of a plant. It is most often conducted early in the development of a process when there is little information on design details or operating procedures, and is often a precursor to further hazard analyses. It is included in these Guidelines to illustrate a cost-effective way to identify hazards early in a plant's life. Because of its military heritage, the Preliminary Hazard Analysis technique is sometimes used to review process areas where energy can be released in an uncontrolled manner.

A Preliminary Hazard Analysis formulates a list of hazards and generic hazardous situations by considering various process characteristics. As each hazardous situation is identified, the potential causes, effects, and possible corrective and/or preventive measures are listed.

One or more hazard analysts assess the significance of process hazards and may also assign a criticality ranking to each particular situation. This criticality ranking can be used to prioritize any recommendations for improving safety that emerge from the analysis.

Types of Results

A Preliminary Hazard Analysis yields a qualitative description of the hazards related to a process design. A Preliminary Hazard Analysis also provides a qualitative ranking of hazardous situations that can be used to prioritize recommendations for reducing or eliminating hazards in subsequent phases of the life cycle of the process. A worked example showing Preliminary Hazard Analysis results can be found in Part II, Chapter 14.

Resource Requirements

Using the Preliminary Hazard Analysis technique requires that analysts have access to available plant design criteria, equipment specifications, material specifications, and other sources of information. A Preliminary Hazard Analysis can be performed by one or two people who have a process safety background. Less-experienced staff can perform a Preliminary Hazard Analysis, but the study may not be as exhaustive or as detailed, since this approach requires the analysts to use a significant amount of judgment. Table 4.1 lists general estimates of the time needed to perform a hazard evaluation using the PreHA technique.

Table 4.1 Time estimates for using the Preliminary Hazard Analysis technique

Scope	Preparation*	Evaluation	Documentation*
Small system	4 to 8 hours	1 to 3 days	1 to 2 days
Large process	1 to 3 days	4 to 7 days	4 to 7 days

* Team leader only

Analysis Procedure

Once the analysis scope has been defined, a Preliminary Hazard Analysis consists of the following steps: (1) preparing for the review, (2) performing the review, and (3) documenting the results.[2-4]

Preparing for the Review. A Preliminary Hazard Analysis requires that the team gather available information about the subject plant (or system), as well as any other relevant information (e.g., from any similar plant, or even from a plant that has a different process but uses similar equipment and materials). The hazard evaluation team should draw experience from as many sources as possible. These sources include hazard studies of similar facilities, operating experience from similar facilities, and checklists such as the one included in Appendix B.

Because a Preliminary Hazard Analysis is specifically intended to discover hazards early in the plant's life, information on the plant may be limited. However, for a Preliminary Hazard Analysis to be effective, the team must at least have a written description of the conceptual design. Thus, the basic chemicals, reactions, and process parameters involved should be known, as well as the major types of equipment (e.g., vessels, heat exchangers). In addition, the operational goals and basic performance requirements of the plant can help define the types of hazards and the operating environment at the facility. Chapter 2 provides additional advice about preparing for a hazard evaluation.

Performing the Review. Performing a Preliminary Hazard Analysis identifies major hazards and incident situations that could result in an undesired consequence. However, the Preliminary Hazard Analysis should also identify design criteria or alternatives that could eliminate or reduce those hazards. Obviously, some experience is required in making such judgments. An hazard evaluation team performing the Preliminary Hazard Analysis should consider the following factors:

- Hazardous plant equipment and materials (e.g., fuels, highly reactive chemicals, toxic substances, explosives, high pressure systems, and other energy storage systems)
- Safety-related interfaces between plant equipment items and materials (e.g., material interactions, fire/explosion initiation and propagation, and instrumented protective systems)
- Environmental factors that may influence the plant equipment and materials (e.g., earthquake, vibration, flooding, extreme temperatures, electrostatic discharge, and humidity)
- Operating, testing, maintenance, and emergency procedures (e.g., human error importance, operator functions to be accomplished, equipment layout/accessibility, and personnel safety protection)
- Facility support (e.g., storage, testing equipment, training, and utilities)
- Safety-related equipment (e.g., mitigating systems, redundancy, fire suppression, and personal protective equipment).

For each area of the subject process, hazard analysts identify hazards and evaluate the possible causes and effects of potential incidents involving these hazards. Usually, the team does not attempt to develop an exhaustive list of causes; rather, they list a sufficient number of them to judge the credibility of the incident. Then the team evaluates the "effects" of each incident; these effects should represent the reasonable-worst-case impacts associated with the potential incident. Finally, the team assigns each potential incident situation to one of the following hazard categories, based on the significance of the causes and effects of the incident:[1]

Hazard Category I	Negligible
Hazard Category II	Marginal
Hazard Category III	Critical
Hazard Category IV	Catastrophic

Organizations using the Preliminary Hazard Analysis technique should further define each of these categories for the team so that they can judge each hazard appropriately. Then the team will list any suggestions they identify for correcting or mitigating the hazards. It is not the intent of the PreHA technique to fully develop incident scenarios or evaluate scenario risks, but rather to identify the basic risk control strategies and safeguards to be developed during the later design stages.

Documenting the Results. The results of a Preliminary Hazard Analysis are conveniently recorded in a table, which displays the hazards identified, the causes, the potential consequences, hazard category, and any identified corrective or preventive measures. Table 4.2 displays the format for recording Preliminary Hazard Analysis results specified in MIL-STD-882.[1] Some organizations add other columns to record the assigned follow-up responsibilities and implementation schedule for important issues and/or to reflect the actual corrective actions that the plant staff implemented.

Table 4.2 Typical format for a Preliminary Hazard Analysis worksheet

Area: Drawing number:		Meeting date: Team members:		
Hazard	Cause	Major effects	Hazard category	Corrective / preventive measures suggested

Example

Consider a design concept that feeds hydrogen sulfide (H_2S) from a cylinder to a process unit. At this stage of the design, the analyst knows only that this material will be used in the process, nothing more. The analyst recognizes that H_2S has toxic and flammable properties, and the analyst identifies the potential release of H_2S as a hazardous situation. The analyst lists the following possible causes for such a release:

- The pressurized storage cylinder leaks or ruptures.
- The process does not consume all of the H_2S.
- The H_2S process supply lines leak/rupture.
- A leak occurs during connection of a cylinder to the process.

The analyst then determines the effects of these causes. In this case, fatalities could result from large releases. The next task is to provide guidance and design criteria by describing corrective/ preventive measures for each possible release. For example, the analyst might suggest that the designer:

- Consider a process that stores alternative, less toxic materials that can generate H_2S as needed.
- Consider developing a system to collect and destroy excess H_2S from the process.
- Provide a plant warning system for H_2S releases.
- Minimize on-site storage of H_2S, without requiring excessive delivery/handling.
- Develop a procedure using human factors engineering expertise for storage cylinder connection.
- Consider a cylinder enclosure with a water deluge system that is triggered by H_2S leak detectors.
- Locate the storage cylinder for easy delivery access, but away from other plant traffic.
- Develop a training program to be presented to all employees before start-up (and subsequently to all new employees) on H_2S effects and emergency procedures.

A sample page from the Preliminary Hazard Analysis table for the H_2S system example is shown in Table 4.3.

Table 4.3 Sample page from the H_2S system example Preliminary Hazard Analysis table

Area: **H_2S Process** Drawing number:		Meeting date: Team members:		
Hazard	Cause	Major effects	Hazard category*	Corrective / preventive measures suggested
Large inventory of high toxic hazard material	1. Failure of primary containment of H_2S in storage	Potential for fatalities from large release	IV	(a) Provide warning system (b) Minimize on-site storage (c) Develop procedure for cylinder inspection
	2. Loss of reaction control in H_2S process	Potential for fatalities from large release	III	(a) Design system to collect and destroy excess H_2S (b) Design control system to detect excess H_2S and shut down process (c) Develop procedures to ensure availability of excess destruction system prior to plant start-up

* Hazard Category: I - negligible, II - marginal, III - critical, IV - catastrophic

Section 4.1 References

1. Department of Defense, *Military Standard System Safety Program Requirements,* MIL-STD-882D, Washington, DC, 2000.
2. H. R. Greenberg and J. J. Cramer (eds.), *Risk Assessment and Risk Management for the Chemical Process Industry,* ISBN 0-442-23438-4, Van Nostrand Reinhold, New York, 1991.
3. J. Stephenson, *System Safety 2000—A Practical Guide for Planning, Managing, and Conducting System Safety Programs,* ISBN 0-0442-23840-1, Van Nostrand Reinhold, New York, 1991.
4. W. Hammer, *Handbook of System and Product Safety,* Prentice Hall, Inc., New York, 1972.

4.2 Safety Review

Undoubtedly, the Safety Review technique was the first hazard evaluation method ever used. This technique, which may also be referred to as a Process Safety Review, a Design Review, or a Loss Prevention Review can be used at any stage of the life of a process.

Purpose

Safety Reviews can be used to ensure that the plant and its operating and maintenance practices match the design intent and construction standards. The Safety Review procedure (1) keeps operating personnel alert to the process hazards; (2) reviews operating procedures for necessary revisions; (3) seeks to identify equipment or process changes that could have introduced new hazards; (4) evaluates the design basis of control systems, instrumented protective systems, and emergency relief systems; (5) reviews the application of new technology to existing hazards; and (6) reviews the adequacy of maintenance and safety inspections. The Safety Review technique is often used to perform a pre-start-up safety review of a process. The success of this technique depends strongly on the experience of the Safety Review Team.

Description

When performed on existing facilities, the Safety Review typically involves a walk-through inspection that can vary from an informal, routine visual examination to a formal team examination that takes several weeks. For processes that are still being designed, a design project team might, for example, review a set of drawings during a meeting.

Safety Reviews are intended to identify plant conditions or operating procedures that could lead to an incident and result in injuries, significant property damage, or environmental impacts. A typical Safety Review includes interviews with many people in the plant: operators, maintenance staff, engineers, management, safety staff, and others, depending upon the plant organization. Safety Reviews should be viewed as cooperative efforts to improve the overall safety and performance of the plant, rather than interfering with normal operations or as a punitive reaction to a perceived problem. Cooperation is essential; people are likely to become defensive unless considerable effort is made to present the review as a benefit to affected plant personnel and designers. Having the support and involvement of all these groups results in a thorough examination.

The Safety Review usually focuses on major risk situations. Judging general housekeeping and morale are not the normal objectives of a safety review, although they can be significant indicators of places where improvements are needed. The Safety Review should complement other process safety activities, such as routine visual inspections, as well as other hazard evaluation techniques such as Checklist Analysis and What-If Analysis.

At the end of the Safety Review, the analyst or team makes recommendations for specific actions that are needed, justifies the recommendations, assigns responsibilities, and lists goal completion dates. A follow-up evaluation or reinspection may be planned to verify that corrective actions have been completed correctly.

Types of Results

Safety Review results are qualitative descriptions of potential safety problems and suggested corrective actions. The inspection team's report includes deviations from the design intentions as well as authorized procedures and lists of newly discovered safety issues. Responsibility for implementing the corrective action remains with the plant management. Worked examples showing Safety Review results, including in a management of change context, can be found in Part II, Chapters 17 and 18.

Resource Requirements

For a comprehensive review, the team members will need access to applicable codes and standards; previous safety studies; detailed plant descriptions, such as P&IDs and flowcharts; plant procedures for start-up, shutdown, normal operation, maintenance, and emergencies; personnel injury reports; pertinent incident reports; maintenance records such as shutdown system functional checks, pressure relief valve tests, and pressure vessel inspections; and process material characteristics (i.e., toxicity and reactivity information).

The personnel assigned to Safety Review inspections must be very familiar with safety standards and procedures. Special technical skills and experience are helpful for evaluating instrumentation, electrical systems, pressure vessels, process materials and chemistry, and other special-emphasis topics. Table 4.4 lists estimates of the time needed from each team member to perform a safety review.

Analysis Procedure

A Safety Review consists of three steps: (1) preparing for the review, (2) performing the review, and (3) documenting the results.[1]

Table 4.4 Time estimates for using the Safety Review technique

Scope	Preparation*	Evaluation	Documentation*
Small system	2 to 4 hours	6 to 12 hours	4 to 12 hours
Large process	1 to 3 days	3 to 5 days	3 to 6 days

* Primarily the team leader

Preparing for the Review. The first step in preparing for a Safety Review is to define which systems, procedures, operations, and personnel will be evaluated. A sufficient number of reviewers should be assigned to cover the necessary process areas and operations in the available time. The members of the review team should have diverse backgrounds and responsibilities to promote a comprehensive, balanced study. In preparation for the review meetings, the following tasks should be completed:

- Assemble a detailed description of the plant (e.g., plot plans, P&IDs, PFDs) and procedures (e.g., operating, maintenance, emergency shutdown and response).
- Review the known hazards and process history with the review team members.
- Review all of the applicable codes, standards, and company requirements.
- Review the status of the recommendations of the previous Safety Review.
- Schedule interviews with specific individuals responsible for safe process operation.
- Request available records concerning personnel injuries, incident reports, equipment inspection, pressure relief valve testing, safety/health audits, etc.
- Plan kick-off and wrap-up visits with the plant manager or an appropriate management representative.

Chapter 2 provides additional advice about preparing for a hazard evaluation.

Performing the Review. If performed on an existing facility, the review should start with a general orientation tour of the plant and progress to specific inspections and interviews. Any part of the inspection that involves outdoor work should be scheduled early in the survey to allow time for rescheduling due to inclement weather.

The inspection team should obtain and review current copies of plant drawings as well as operating, maintenance, and emergency procedures. These serve as the basis for review of the process hazards and for discussion with operating personnel. Because many incidents are caused by failure to follow established procedures, an effort should be made to determine whether the operating staff follows the company's written operating procedures. This evaluation should also extend to the control of maintenance activities, such as routine equipment repair, welding, vessel entry, electrical lock-out, or equipment testing. Observing people as they perform their daily tasks shows how well they know procedures and how well they comply with them. Questions can further clarify why and how some actions are being performed. However, interviewers should take care to avoid inadvertently distracting workers, which could lead to unsafe situations.

Useful information can be gained by asking staff members to respond to an exercise as they would in a real emergency. The participating staff could walk through the exercise, explaining to the inspection team their sequence of actions, why they chose those actions, and how they expect the system to react. The inspection team can observe the exercise and provide a critique to the participants afterward. Alternatively, the review team can cover the relevant emergency procedures with operating personnel in a "roundtable" discussion.

Equipment inspection requires visual or diagnostic evaluation, plus a review of plant records. This effort is generally reserved for the most critical equipment. Some questions that might be addressed are:

- Is there a system for keeping important process documentation and drawings up to date?
- Is the equipment in good condition?
- Are the pressure reliefs or other safety devices properly installed, well-maintained, and properly identified?
- Do plant records show the history of inspecting and testing the equipment and safety devices?
- Are pressure vessels or critical or hazardous service equipment repaired by certified welders?
- For equipment that handles corrosive or erosive materials, have metallurgical inspections and metal-wall-thickness measurements been taken at frequent intervals?
- Does the plant have trained inspectors available whose recommendations for repair or replacement are accepted by management?
- Are safe work practices followed and permits used?

Instrumented protective systems including emergency shutdown systems deserve special attention during a Safety Review. These systems need periodic testing to verify that the controls work as intended. If possible, it is helpful to schedule a portion of the Safety Review during a planned plant shutdown so that the reviewers can witness the actual proof-testing of safety systems that could not be tested under normal operating conditions. If this is not possible, then the analyst must depend upon individual function checks of components. The practice of bypassing safety controls is a serious matter that should be reviewed and evaluated.

An inspection of fire and safety equipment is appropriate to ensure that the equipment is well maintained and that the staff is properly trained to use it. Issues involving the capability of fire protection equipment should also be covered in the review. For example, can a fire water spray monitor reach the top level of a structure? If there are sprinkler, dry chemical, or foam fire suppression systems, are they checked periodically? Can personnel demonstrate the proper donning and use of emergency breathing air equipment? Emergency response plans should be reviewed for completeness and tested, if possible.

Documenting the Results. When the inspection is completed, a report is prepared with specific recommended actions. The review team customarily provides justifications for their recommendations in the report, and they summarize their impressions about the facility or system. Finally, findings and recommendations are reviewed with appropriate plant management. Follow-up action can be planned at this time. Chapter 8 provides additional information concerning analysis follow-up considerations.

Example

In a major petrochemical complex, one of the operating units is approximately 30 years old. A business evaluation indicates that this unit could continue to operate economically for another 15 to 20 years if safety could be maintained. A Safety Review study is commissioned to evaluate the unit. The plant has been well maintained, and the plant inspection performed as part of the Safety Review indicates that the equipment is serviceable for another 15 years. However, the review team identifies several problems that need additional evaluation, including:

- The unit capacity has been increased through numerous improvements and de-bottlenecking projects, yet the sizing of the unit's safety relief valves and flare headers has never been reevaluated.
- The original pneumatic controls are still in service but the instrumented protective systems have never been reevaluated or brought up to current company standards.
- The corporate standards for equipment spacing have changed since the unit was originally built, and based on the new standard there are numerous violations of the recommended distances.

The company accepts the recommendations of the review team and takes corrective actions, including:

- A thorough design review of the pressure relief valves and flare headers. As a result of this evaluation, additional safety relief valves and a new flare header are installed.
- A design review to consider the instrumentation and control system. A modern control system is recommended and installed and instrumented protective systems are identified and evaluated. Not only is this system more reliable, but it also provides safety instrumented functions that did not originally exist.
- A review of the equipment spacing. These problems are more difficult to correct. Certain spacing requirements cannot be met with the existing unit and the adjacent facilities (the flare, the furnaces, and the control room), so a heat-activated sprinkler system is installed to address the possibility of a fire at this unit. Spacing is also a consideration in a subsequent decision to build a new consolidated control room farther away from this unit as a part of the computerization of the control systems.

Section 4.2 Reference

1. H. R. Greenberg and J. J. Cramer (eds.), *Risk Assessment and Risk Management for the Chemical Process Industry*, ISBN 0-442-23438-4, Van Nostrand Reinhold, New York, 1991.

4.3 Relative Ranking

Relative Ranking is actually an analysis strategy rather than a single, well-defined analysis method. This strategy allows hazard analysts to compare the attributes of several processes or activities to determine whether they possess hazardous characteristics that are significant enough to warrant further study. Relative Ranking can also be used to compare several process siting, generic design, or equipment layout options, and provide information concerning which alternative appears to be the "best" option. These comparisons are based on numerical values that represent the relative level of significance that the analyst gives to each hazard, potential consequence or risk depending on the approach used. Relative Ranking studies should normally be performed early in the life of a process, before the detailed design is completed, or early in the development of an existing facility's hazard analysis program. However, the Relative Ranking method can also be applied to an existing process to pinpoint the hazards of various aspects of process operation.

Purpose

The main purpose of using Relative Ranking methods is to determine the process areas or operations that are the most significant with respect to the hazard of concern, the potential severity of consequences, or the overall risk in a given study. The theory behind Relative Ranking methods has its roots in the three basic questions used in risk analysis: (1) What can go wrong? (2) How likely is it? and (3) What would the effects be? The philosophy behind Relative Ranking approaches is to address these risk analysis questions to determine the relative importance of processes and activities from a safety standpoint before performing additional and more costly hazard evaluation or risk analysis studies. (Note, however, that Relative Ranking studies that only focus on hazards or potential consequence severities may not specifically address the second question of "How likely is it?") Thus, approximate relationships of process attributes are compared to determine which areas present the greater relative hazard, consequence severity or risk. Subsequently, additional hazard evaluations may first be performed on the more significant areas of concern.

Description

Several formal Relative Ranking methods are widely used. For example, the Dow Fire and Explosion Index (F&EI) has been in existence for many years, and a booklet describing this method, published by the AIChE, is in its seventh printing. The Dow F&EI evaluates the existence and significance of fire and explosion hazards in many large areas of a process facility. The analyst divides a process or activity into separate process units and assigns indexes based on material, physical, and chemical characteristics; process conditions; plant arrangement and equipment layout considerations; and other factors. The various factors are combined into an F&EI score that can be ranked against the scores of other process units that are evaluated. The Dow F&EI can also be used by experienced analysts to gain insights on when general safety system improvements (e.g., fire protection) may be needed. Another method that is less well known and documented in the U.S.A. is the ICI Mond Index. This index is used to evaluate the chemical and toxicity hazards, as well as fire and explosion hazards, associated with a process area or operation.

Many organizations have created their own specialized indexes to rank the hazards associated with facilities, processes, and operations. For example, the Dow Chemical Company has several indexes that it uses to evaluate and manage the risk of its processes and activities. One of them is called the Dow Chemical Exposure Index (CEI). The CEI is used to rank the relative acute health hazards associated with potential chemical releases. The CEI uses a simple formula to rank the use of any toxic chemical, based on five factors: (1) a measure of toxicity, (2) quantity of volatile material available for release, (3) distance to each area of concern, (4) molecular weight of the chemical being evaluated, and (5) process variables that can affect the conditions of a release such as temperature, pressure, reactivity, and so forth.

Some specialized indexes have been developed and used by organizations to determine the application of certain recommended industry practices or regulatory requirements. For example, the U.S. Environmental Protection Agency developed a ranking method (the Threshold Planning Quantity [TPQ] Index) to help determine which materials should be considered extremely hazardous when used in emergency response planning activities associated with SARA Title III. The Occupational Safety and Health Administration and the American Petroleum Institute have suggested using a Substance Hazard Index (SHI) to help determine whether special process safety management efforts should be directed at particular processes or industrial activities.[17]

Relative Ranking methods can address fire, explosion, and/or toxicity hazards and associated safety, health, environmental, and economic effects for a process or activity. Analysts can use Relative Ranking to compare more than one process area, or several designs of the same process area. Relative Ranking techniques may be used during any phase of a plant or process lifetime to:

- Identify the individual process areas that contribute most to the anticipated overall hazard and incident attributes of a facility
- Identify the key material properties, process conditions, and/or process characteristics that contribute most to the anticipated hazard and incident attributes of a single process area or an entire facility
- Use the anticipated hazard and incident attributes to discriminate among competing design, siting, or operating options
- Compare the anticipated hazard and incident attributes of process areas or facilities to others whose attributes are better understood and/or more commonly accepted

Typically, Relative Ranking techniques do not use detailed risk analysis calculations (e.g., estimating incident frequencies using equipment failure data and consequences using complex dispersion models). Instead, most Relative Ranking methods calculate a numerical score, or index, based on a comparison of hazardous process characteristics. The calculation method for these indexes, no matter how simple, should be based on traditional concepts of risk analysis. For example, since risk is fundamentally associated with the existence of a hazard and the combination of the likelihood and effects of an incident, a Relative Ranking technique should also demonstrate a dependence on these factors. An analyst's combination of these factors using a particular Relative Ranking approach should reflect the anticipated risk characteristics of a process area or operation.

Hazard analysts electing to use the Relative Ranking approach may choose from a variety of Relative Ranking indexes.[1-7] Table 4.5 summarizes some of the well-known methods. If an organization wants to develop their own numerical Relative Ranking index, they should consider including the following factors:

- Material properties (e.g., physical state, vapor pressure, density, viscosity, toxicity level, flammability limits, flash point, autoignition temperature, reactivity)
- Process conditions (e.g., temperature, pressure, quantities of materials, relative concentrations of materials, operating environment)
- Process characteristics and support systems (e.g., purging, ventilation, utilities, cooling, heating, exothermic reactions, pressurized storage)
- System design and construction (e.g., fire proofing, explosion proofing, equipment layout and spacing, corrosion/erosion resistance, natural disaster resistance, redundancy)
- Operational activities (e.g., operator training, written procedures, error-likely situations, operating margin, operational policies)
- PSM activities (e.g., inspection and testing intervals, maintenance activities and policies, safety and hazard review programs, management of change policies)
- Exposure possibilities (e.g., operation time and frequency, number of operator activities, number of pieces of equipment, infrequent modes of operation)

Actual values of physical conditions and chemical properties may be used in the index formulas, or analysts may simply assign numerical values to the formula's variables based on their understanding of the variables' influence on the risk of the process area. For example, the degree of fire proofing present in an area could be reflected as a value between 0 and 10, while a process variable such as temperature could be used directly in numerical relationships.

Hazard analysts can then develop a formula containing the risk-sensitive process variables, producing a risk index number that models the relative risk in an area. To the extent possible, these equations should be based upon known theoretical relationships or empirical correlations among parameters.

The equations should also portray the parameters as directly and indirectly proportional variables, with appropriate exponents, to demonstrate the risk trends for each parameter. Physical laws, actual facility failure experience, previous risk assessment experience, and sound engineering judgment may be used as a basis for developing the equations. Further, these relationships should include any important aspects of the causes, consequences, and safeguards for potential incidents, since any of these may influence the risk present in a process area.

In practice, development of the risk index equations is complicated. Many of the parameters cannot be scientifically measured or assigned a value (e.g., level of fire protection). Unknown relationships could exist between some deterministic parameters (e.g., autoignition temperature of an unresearched mixture). The relative importance of each factor that contributes to the risk of the facility is subjective, relying on the technique developer's experience (e.g., "Is fire proofing equipment more important than fire fighting capability?"). Also, it is difficult to account for factors that affect frequency of incidents since little experience data are usually available and these frequency characteristics are inherently based on probabilistic factors, rather than more easily derived deterministic ones.

Table 4.5 Summary of Relative Ranking indexes

Index	Summary description
Dow Fire and Explosion Index (F&EI)[1]	Evaluates the fire and explosion hazards associated with discrete "process units" (collections of equipment containing a flammable or explosive material). Considers a variety of factors, including material properties, process conditions, process design and operating characteristics, distance from adjoining areas, the existence of safety and fire protection systems, etc. The final result is an index for each process unit analyzed. Rankings of process units can be used (1) to direct specific safety improvement efforts relating to important parameters used in the F&EI calculation or (2) to identify areas for more detailed hazard evaluation or risk analysis study.
Mond Index[2,3]	Developed by ICI's Mond Division, it is an extension of the Dow F&EI. The Mond Index specifically includes factors that address the toxicity hazards associated with materials in process units. A revised version of the method (1985) addresses some health hazards. Limited published material exists in the U.S.A. on this method and this lack of information has limited the use of the method.
Substance Hazard Index (SHI)[4]	Proposed by the Organization Resources Counselors (ORC) in their advice to OSHA on PSM as a way of ranking material hazards. Defined as "the equilibrium vapor concentration (EVC) of a material at 20 °C divided by an acute toxicity concentration." This index relates the tendency of a material if released to travel in the air and affect people through inhalation of toxic vapors. The EVC is a material's vapor pressure (mm Hg) x 106/760. Different organizations that use or recommend the SHI may use different toxicity parameters (LC_{LO}, ERPG-3, IDLH, etc.).
Material Hazard Index (MHI)[5]	Used by the State of California to determine threshold quantities of acutely hazardous materials for which risk management and prevention programs must be developed. Defined as the material vapor pressure at 25 °C divided by a level of concern. Levels of concern can be defined on the basis of toxic, fire, explosion, or other types of hazardous effects that can be propagated through the air.
Chemical Exposure Index (CEI)[6]	Developed by Dow Chemical Company. Address five types of factors that can influence the effects of release of the material: (1) acute toxicity, (2) volatile portion of material that could be released, (3) distance to areas of concern, (4) molecular weight of the substance, and (5) various process parameters such as temperature, pressure, reactivity, and so forth. The CEI is the product of values assigned for each of the factors of concern using arbitrarily defined numerical scales.
SARA Title III Threshold Planning Quantity (TPQ) Index[7]	Developed by the U.S. Environmental Protection Agency to determine which extremely hazardous substances and threshold quantities should be covered under SARA Title III. The index is a function of the material vapor pressure, the 30-minute IDLH, and a generic Gaussian dispersion calculation performed at target distance of 100 meters from the source of release.

Several options are available for overcoming the difficulties of developing numerical relationships for a Relative Ranking technique. Experimental research or historical data may be used to discover empirical relationships among parameters. This may include laboratory work to identify material and/or process condition relationships (e.g., hazardous properties of new materials). Statistical analysis of historical data to gain insight into the effects of certain procedures (e.g., emergency response actions) or design characteristics (e.g., engineered safeguards) on incident outcomes may also be valuable. This type of research will also be useful for ensuring that changes of input parameter values will produce appropriate changes in the calculated risk index number. If the numerical relationships are correctly scaled, the most important parameters will have the most significant effect on the calculated relative risk index number. An information management scheme, such as structured tabular templates or a software package that stores and manipulates the parameters, will increase the efficiency of a Relative Ranking analysis. Also, providing some guidance (e.g., ranges of values) for assigning nondeterministic parameter values whose quantification is uncertain may be helpful.

Each organization may develop its own Relative Ranking technique or select one of the accepted standard techniques.[8-16] By choosing to develop a new Relative Ranking technique, a company will be able to design the approach for their specific application. A technique using only a few important characteristics based on a simple numerical scheme could be created. This type of technique might be valuable when making design decisions among a small set of parameters for a system. The technique could also be more generic, using a larger set of parameters that apply to many different types of systems. In each case, the unique philosophy and practices of the company will be reflected. However, by developing a new ranking technique, the company will incur the cost and time of development, the need for experienced hazard analysts to develop the technique, and limited acceptance by outside organizations (i.e., regulatory agencies and other companies) until the merit of the technique is proved. A decision to develop a Relative Ranking technique should be carefully considered to ensure that a new approach is necessary and practical. A number of organizations, in addition to those mentioned in Table 4.5, have developed their own special Relative Ranking indexes.

Types of Results

All Relative Ranking methods should result in an ordered list of processes, equipment, operations, or activities. This list may have several stratified layers representing levels of significance. Other results such as indexes, scores, factor scales, graphs, etc., depend upon the particular technique used to perform the ranking. It is important to note that while these techniques all try to answer the three questions of risk analysis in some way; analysts should not consider the results of such studies as robust estimates of the risk associated with a process or activity. The Relative Ranking technique is usually not based on specific incident sequences; thus, it does not normally lend itself to developing specific safety improvement recommendations. A worked example showing Relative Ranking results can be found in Part II, Chapter 20.

Resource Requirements

The information requirements of a Relative Ranking study depend upon each ranking method's unique needs. Generally, a Relative Ranking study will require basic physical and chemical data on the substances used in the process or activity. These studies do not normally require detailed process drawings; however, information on the maximum inventories of materials, the plant's process conditions, and geographic layout of material storage areas is usually needed.

A Relative Ranking study can be carried out by a single analyst. Several analysts can work together on a large, complex process when they are experienced with the Relative Ranking technique and have access to all of the input data needed for the study. It is often better to have a trained analyst working with someone who can quickly locate and interpret the necessary material and process data needed for the analysis. Although more than one analyst may be needed, depending upon the complexity and size of the process or activity and the number and type of hazards, it is crucial that all of the analysts are "calibrated" in the same way so their judgments are consistent.

The time and cost of performing a hazard evaluation using the Relative Ranking approach will depend upon the technique chosen, the input data requirements, and the number of process areas and hazards evaluated. Table 4.6 lists estimates of the time it would take to perform a hazard evaluation using a Relative Ranking technique.

Analysis Procedure

Once a Relative Ranking approach is selected and the scope of the study defined, applying it usually involves three basic steps: (1) preparing for the review, (2) performing the review, and (3) documenting the results.

Preparing for the Review. A Relative Ranking evaluation does not require a large review team. In fact, only one analyst may be needed to perform a Relative Ranking study. The analyst will need accurate:

- Site plans
- Lists of materials, chemical properties, and quantities
- General process diagrams and equipment layout drawings
- Design and operating data
- Technical guides for the selected ranking technique

The analyst should gather this information and become familiar with it before beginning the study. Chapter 2 provides additional advice about preparing for a hazard evaluation.

Table 4.6 Time estimates for using Relative Ranking techniques

Scope	Preparation	Evaluation	Documentation
Small system	2 to 4 hours	2 to 4 hours	2 to 4 hours
Large process	1 to 3 days	4 to 8 hours	2 to 4 hours

Performing the Review. If a published Relative Ranking method (e.g., Dow F&EI) is chosen, the analyst should follow the instructions in the technical guide for that technique to perform the evaluation. Site visits and interviews to verify information and to answer process questions may be helpful. The analysis should be technically reviewed by other engineers or managers, and all assumptions should be recorded. The calculated risk index numbers (and any other factors calculated from the evaluation) should be summarized to facilitate comparisons among areas that have been reviewed.

Documenting the Results. The risk index numbers and the other factors generated by the evaluation should not be considered as accurate reflections of the absolute risks posed in process areas. Instead, these results should be considered as estimates to compare the relative risks of the areas analyzed. The results of the study may be used alone, or in conjunction with other factors, such as cost.

Additionally, the analyst may determine the most important contributors to the index numbers by reviewing the analysis documentation to help determine if corrective actions or design modifications should be undertaken to reduce the anticipated risks of the facility. By doing so, the analyst may identify the specific areas of the unit or process where the safety weaknesses exist and develop a list of action items to correct the problems.

Example

The RAJ Corporation is a diversified chemical and manufacturing company with facilities located throughout the United States. RAJ management has recently adopted a process safety management policy and is now planning the execution of the various elements of their PSM program. One of these elements involves conducting hazard evaluations on RAJ's various processes. Because RAJ has so many plants and operations that handle a variety of hazardous chemicals, they decide to perform a screening analysis of their various plant sites to determine the order in which they should perform more detailed hazard evaluations.

A hazard analyst at RAJ decides to use a Relative Ranking technique to perform the prioritization study. Because RAJ hazard analysts have little experience with hazard evaluations, RAJ decides to modify versions of a relative ranking scheme already used in industry—the substance hazard index (SHI)—rather than creating a new one. However, the RAJ analyst realizes that the SHI has some deficiencies that could affect an appropriate ranking of substances handled at RAJ facilities. First, the SHI does not take into consideration the consequences of the different amounts of material that could possibly be released to the air (i.e., a large release would cause more damage than a small release). Second, the SHI does not account for the proximity of facilities to local population centers. Finally, some of the materials that RAJ handles are flammable, and the standard SHI only addresses toxicity.

To address these concerns, the RAJ hazard analyst elects to modify the SHI formula. The analyst will multiply the SHI by (1) the square root of the mass of the contents of the largest single storage container for each material in a facility and (2) the population within a one-mile radius of the facility. RAJ decided to use the square root of the chemical inventory based on their review of simplified release rate and atmospheric dispersion correlations. In addition, to address flammability hazards, the analyst will substitute the lower flammable limit for each material for the acute toxicity concentration in the SHI formula. Thus, the following are the formulas RAJ uses for this relative ranking exercise.

$$MSHI_{toxic} = SHI \cdot M^{0.5} \cdot P = \frac{EVC \cdot M^{0.5} \cdot P}{IDLH}$$

$$MSHI_{flam} = \frac{SHI \cdot M^{0.5} \cdot P \cdot IDLH}{LFL} = \frac{EVC \cdot M^{0.5} \cdot P}{LFL}$$

where

EVC = equilibrium vapor concentration (ppm),
M = mass of hazardous substance ($\times 10^3$ kg),
P = population within a one-mile radius of facility ($\times 10^3$),
LFL = lower flammable limit (ppm),
$IDLH$ = concentration of chemical in air that is immediately dangerous to life and health (ppm).

Caution - these formulas are for illustration purposes and are not appropriate for general use by hazard analysts. For example, this illustration assumed an equivalent relationship between the threat posed by exposure to a flammable material and a toxic material.

Table 4.7 lists the various chemicals and relevant data used in this analysis. Table 4.8 lists the results from the relative ranking of these chemicals and their associated processes. From these results, RAJ elects to focus their hazard evaluation efforts on the process units at Plant C since these units have the highest composite ranking of the four facilities. In addition, RAJ will evaluate the hazards associated with the chlorine process at Plant A since this was the highest ranked process listed in Table 4.8. The remaining processes will be analyzed next year when more resources are available.

Section 4.3 References

1. American Institute of Chemical Engineers, *Dow's Fire & Explosion Index Hazard Classification Guide, 7th Edition,* ISBN 0-8169-0623-8, Wiley, New York, June 1994.
2. *The Mond Index, 2nd Ed.,* Imperial Chemical Industries, Winnington, Northwick, Cheshire, UK, 1985.
3. D. J. Lewis, *Mond Fire, Explosion and Toxicity Index: A Development of the Dow Index*, 13th Annual Loss Prevention Symposium, Houston, American Institute of Chemical Engineers, New York, 1979.
4. Organization Resources Counselors, Inc., *Process Hazards Management of Substances with Catastrophic Potential*, Process Hazard Management Task Force, Washington, DC, 1988.
5. State of California, Office of Emergency Services, *Guidance for the Preparation of a Risk Management and Prevention Program*, Sacramento, CA, 1989.
6. American Institute of Chemical Engineers, *Dow's Chemical Exposure Index Guide*, ISBN 0-8169-0647-5, Wiley, New York, August 1988.
7. 40 CFR Part 355, U.S. Environmental Protection Agency, "Extremely Hazardous Substances List," Washington, DC, 1987.
8. H. R. Greenberg and J. J. Cramer (eds.), *Risk Assessment and Risk Management for the Chemical Process Industry*, ISBN 0-442-23438-4, Van Nostrand Reinhold, New York, 1991.
9. S. A. Lapp, "The Major Risk Index System," *Plant/Operations Progress 9*(3), July 1990.
10. J. Gillett, "Rapid Ranking of Process Hazards," *Process Engineering 66*(219), 1985.
11. *Technical Guidance for Hazards Analysis: Emergency Planning for Extremely Hazardous Substances*, U.S. Environmental Protection Agency, Federal Emergency Management Agency, U.S. Department of Transportation, December 1987.

12. T. Veerman, *An Evaluation of Rapid Hazard Ranking Methods in the Process Industry*, Afd. der Scheikundige Technologie (Oct). Technischе Hogeschool, Delft, Netherlands, 1981.
13. R. H. Ross and P. Lu, *Chemical Scoring Systems Development*, Oak Ridge National Laboratory, Oak Ridge, TN, 1981.
14. K. S. Mudan, *Hazard Ranking for Chemical Processing Facilities*, American Society of Mechanical Engineers Winter Meeting, Boston, December 1987.
15. R. King and R. Hirst, *King's Safety in the Process Industries, 2nd Ed.*, Butterworth-Heinemann, London, 1998.
16. S. Mannan, ed., *Lees' Loss Prevention in the Process Industries, 3rd Ed.*, Elsevier Butterworth-Heinemann, ISBN 0-7506-7555-1, Oxford, UK, 2005.
17. Recommended Practice RP-750, "Management of Process Hazards," American Petroleum Institute, Washington, DC, January 1990, reaffirmed May 1995.

Table 4.7 Data for the Relative Ranking example

Facility	Hazardous Substance	Mass of Chemical in Largest Single Container ($\times 10^3$ kg)	SHI	Population within one mile radius of facility ($\times 10^3$)	LFL* (ppm)
Plant A	Chlorine	90	73,000	2	N/A
	Ammonia	1,000	2,400	2	N/A
Plant B	Arsine	0.01	1,000,000	0.5	N/A
	Sulfur dioxide	10	10,000	0.5	N/A
	Ammonia	90	2,400	0.3	N/A
Plant C	Hydrogen fluoride	30	50,000	3	N/A
	Chlorine	10	73,000	3	N/A
Plant D	Propylene oxide	120	3,300	7	28,000
	Sulfur dioxide	10	10,000	7	N/A

*N/A = not applicable for this example

Table 4.8 Results from the Relative Ranking example

Facility / substance	SHI ($\times 10^3$)	MSHI ($\times 10^3$)	Rank
Plant A / chlorine	73	1390	1
Plant C / hydrogen fluoride	50	822	2
Plant C / chlorine	73	693	3
Plant D / sulfur dioxide	10	221	4
Plant A / ammonia	2.4	152	5
Plant B / arsine	1,000	50	6
Plant D / propylene oxide	3.3	18	7
Plant B / sulfur dioxide	10	16	8
Plant B / ammonia	2.4	11	9

4.4 Checklist Analysis

A Checklist Analysis uses a written list of items or procedural steps to verify the status of a system. Traditional checklists vary widely in level of detail and are frequently used to indicate compliance with standards and practices. In some cases, analysts use a more general checklist in combination with another hazard evaluation method to discover common hazards that the checklist alone might miss (see Section 5.2, What-If/Checklist Analysis). The Checklist Analysis approach is easy to use and can be applied at any stage of the process's lifetime.

Purpose

Traditional checklists are used primarily to ensure that organizations are complying with standard practices. Checklists can be used to familiarize inexperienced personnel with a process by having them compare a process's attributes to various checklist requirements. Checklists also provide a common basis for management review of the analyst's assessments of a process or operation.

Description

In a traditional Checklist Analysis, the hazard analyst uses a list of specific items to identify known types of hazards, design deficiencies, and potential incident situations associated with common process equipment and operations. The Checklist Analysis technique can be used to evaluate materials, equipment, or procedures. Checklists are most often used to evaluate a specific design with which a company or industry has a significant amount of experience, but they can also be used at earlier stages of development for entirely new processes to identify and eliminate hazards that have been recognized through years of operation of similar systems.

Proper use of a checklist will generally ensure that a piece of equipment conforms with accepted standards and it may also identify areas that require further evaluation. To be most useful, checklists should be specifically tailored for an individual company, plant, or product. A Checklist Analysis of an existing process usually includes touring the subject process area and comparing the equipment to the checklist. As part of a Checklist Analysis of a process that is not yet built, experienced personnel compare the appropriate design documentation against the relevant checklists.

A detailed checklist provides the basis for a standard evaluation of process hazards. It can be as extensive as necessary to satisfy the specific situation, but it should be applied conscientiously in order to identify problems that require further attention. Generic hazard checklists are often combined with other hazard evaluation techniques to evaluate hazardous situations. Checklists are limited by their authors' experience; therefore, they should be developed by authors with varied backgrounds who have extensive experience with the systems they are analyzing. Frequently, checklists are created by simply organizing information from current relevant codes, standards, and regulations. Checklists should be viewed as living documents and should be audited and updated regularly.

Many organizations use standard checklists to control the development of a project - from initial design through plant decommissioning. The completed checklist must frequently be approved by various staff members and managers before a project can move from one stage to the next. In this way, it serves as both a means of communication and as a form of control. Checklists are used in hard copy form, although computer-based versions can also be used.

Types of Results

To create a traditional checklist, the analyst defines standard design or operating practices, then uses them to generate a list of questions based on deficiencies or differences. A completed checklist contains "yes," "no," "not applicable," or "needs more information" answers to the questions. Qualitative results vary with the specific situation, but generally they lead to a "yes" or "no" decision about compliance with standard procedures. In addition, knowledge of these deficiencies usually leads to an easily developed list of possible safety improvement alternatives for managers to consider. A worked example showing Checklist Analysis results can be found in Part II, Chapter 17.

Resource Requirements

To properly perform this technique, required resources include an appropriate checklist, an engineering design procedures and operating practices manual, and someone to complete the checklist who has basic knowledge of the process being reviewed. If a relevant checklist is available from previous work, analysts should be able to use it as long as they have the necessary guidance. If no relevant checklist exists, one person (sometimes several people) must prepare a checklist and perform the evaluation. An experienced manager or staff engineer should then review the Checklist Analysis results and direct the next action. It is important that checklists are reviewed periodically to include the latest codes, standard, regulations or practices.

The Checklist Analysis method is versatile. The type of evaluation performed with a checklist can vary; it can be used quickly for simple evaluations or for more expensive in-depth evaluations. It is a highly cost-effective way to identify customarily recognized hazards. Table 4.9 is an estimate of the time it takes to perform a hazard evaluation using the Checklist Analysis technique. The checklist technique is experience based and its success depends on the experience of the checklist developer as well as those who use the checklist.

Analysis Procedure

Once the scope of the analysis has been defined, a Checklist Analysis consists of three main steps: (1) selecting or developing an appropriate checklist, (2) performing the review, and (3) documenting the results.[1-8]

Table 4.9 Time estimates for using the Checklist Analysis technique

Scope	Preparation*	Evaluation	Documentation*
Small system	2 to 4 hours	4 to 8 hours	4 to 8 hours
Large process	1 to 3 days	3 to 5 days	2 to 4 days

* Primarily the team leader

Selecting a checklist. A Checklist Analysis is an experience-based approach. The hazard analyst should select an appropriate checklist from available resources (e.g., internal standards, consensus codes, industry guidelines). If no specific, relevant checklist is available, then the analyst must use his or her own experience and the information available from authoritative references to generate an appropriate checklist. Appendix A gives checklists that can be used in managing change, addressing chemical reactivity hazards, and identifying inherently safer approaches. Appendix B provides a number of checklist items for specific types of equipment and operating situations.[9-15]

A checklist should be prepared by an experienced engineer who is familiar with the general plant operation and the company's standard policies and procedures. A checklist is developed so that aspects of a system's design or operation that do not comply with company or common industrial standard practices will be discovered through responses to the questions in the checklist. Once a checklist has been prepared, it can be applied by less experienced engineers as an independent evaluation or as part of another hazard review study. Detailed checklists for a particular process should be augmented by generic checklists to help ensure thoroughness. Chapter 2 provides additional guidelines for preparing for a hazard evaluation.

Performing the review. The analysis of an existing system should include tours and visual inspections of the subject process areas by the hazard evaluation team members. During these tours, the analysts compare the process equipment and operations to the checklist items. The reviewer responds to the checklist issues based on observations from site visits, system documentation, interviews with operating personnel, and personal perceptions. When the observed system attributes or operating characteristics do not match the specific desired features on the checklist, the analyst notes the deficiency. A Checklist Analysis of a new process, prior to construction, is usually performed by the team members in a meeting and focuses on review of the process drawings, completion of the checklist, and discussion of the deficiencies.

Checklist Analyses and inherent safety reviews. Checklist Analyses can be used either independently or in combination with the What-If Analysis technique for inherent safety reviews. The approach is the same as other checklist analyses. The use of a checklist to address inherent safety issues was discussed in Section 3.7.

A checklist of inherent safety considerations is given in Appendix A4, and a more detailed checklist on inherent safety issues can be found in Reference 16. Example checklist questions include:

Substitution

- ❑ Can a flammable solvent be replaced with water?
- ❑ Is it possible to completely eliminate hazardous raw materials, process intermediates, or by-products by using an alternative process or chemistry?

Minimization

- ❑ Can hazardous finished product inventory be reduced?
- ❑ Can alternative equipment with reduced hazardous material inventory be used, such as
  - Centrifugal extractors in place of extraction columns?
  - Flash dryers in place of tray dryers?
  - Continuous in-line mixers (static mixers) in place of mixing vessels or reactors?

Moderation

- ❑ Is it possible to limit the supply pressure of raw materials to less than the maximum allowable working pressure of the vessels to which they are delivered?
- ❑ Is it possible to make the reaction conditions (temperature, pressure) less severe by using a catalyst or a better catalyst (e.g., structured or monolithic vs. packed-bed)?

Simplification

- ❑ Can equipment be designed such that it is difficult or impossible to create a potentially hazardous situation due to an operating or maintenance error, such as by
  - Easy access and operability of valves to prevent inadvertent errors?
  - Elimination of all unnecessary cross-connections?
  - Designing temperature-limited heat transfer equipment to prevent exceeding maximum process or equipment design temperatures?

Documenting the results. The hazard evaluation team performing the Checklist Analysis should summarize the deficiencies noted during the tours and/or meetings. The report should contain a copy of the checklist that was used to perform the analysis. Any specific recommendations for safety improvement should be provided along with appropriate explanations.

Example

A proposed continuous process is shown in Figure 4.1. In this process, a phosphoric acid solution and an ammonia solution are provided through flow control valves to an agitated reactor. The ammonia and phosphoric acid react to form diammonium phosphate (DAP), a nonhazardous product. The DAP flows from the reactor to an open-top storage tank. Relief valves are provided on the storage tanks and the reactor with discharges to outside of the enclosed work area.

If too much phosphoric acid is fed to the reactor (compared to the ammonia feed rate), an off-specification product is created, but the reaction is safe. If the ammonia and phosphoric acid flow rates both increase, the rate of energy release may accelerate, and the reactor, as designed, may be unable to handle the resulting increase in temperature and pressure. If too much ammonia is fed to the reactor (as compared to the normal phosphoric acid feed rate), unreacted ammonia may carry over to the DAP storage tank. Any residual ammonia in the DAP tank will be released into the enclosed work area, causing personnel exposure. Ammonia detectors and alarms are provided in the work area.

A Checklist Analysis is scheduled for the system using a standard company checklist. A sample of the analysis documentation is included in Table 4.10. The appropriate decision makers review the documentation and implement corrective actions to eliminate deficiencies indicated by the analysis.

Section 4.4 References

1. S. Mannan, ed., *Lees' Loss Prevention in the Process Industries, 3rd Ed.,* Elsevier Butterworth-Heinemann, ISBN 0-7506-7555-1, Oxford, UK, 2005.
2. H. R. Greenberg and J. J. Cramer (eds.), *Risk Assessment and Risk Management for the Chemical Process Industry*, ISBN 0-442-23438-4, Van Nostrand Reinhold, New York, 1991.
3. A. W. M. Balemans et al., "Check-list: Guidelines for Safe Design of Process Plants," *Loss Prevention 1*, American Institute of Chemical Engineers, New York, 1974.
4. S. G. Hettig, "A Project Checklist of Safety Hazards," *Chemical Engineering 73*(26), 1966.

5. R. King and J. Magid, *Industrial Hazard and Safety Handbook,* Newnes-Butterworths, London, 1979.
6. M. J. Marinak "Pilot Plant Prestart Safety Checklist," *Chemical Engineering Progress 63*(11), 1967.
7. J. U. Parker, "Anatomy of a Plant Safety Inspection," *Hydrocarbon Processing 46*(1), 1967.
8. V. J. Whitehorn and H. W. Brown, "How to Handle a Safety Inspection," *Hydrocarbon Processing 46*(4 and 5), 1967.
9. Center for Chemical Process Safety, *Guidelines for Hazard Evaluation Procedures, Second Edition with Worked Examples,* American Institute of Chemical Engineers, New York, 1992.
10. C. H. Vervalin, ed., *Fire Protection Manual for Hydrocarbon Processing Plants, 2nd Ed.*, ISBN 87201-286-7, Gulf Publishing Co., Houston, 1973.
11. API 14C, Recommended Practice for Analysis, Design, Installation, and Testing of Basic Surface Safety Systems for Offshore Production Platforms, 4th ed., American Petroleum Institute, Dallas, September 1986.
12. OSHA Instruction CPL 2-2.45A, *Process Safety Management of Highly Hazardous Chemicals – Compliance Guidelines and Enforcement Procedures*, Occupational Safety and Health Administration, Washington, DC, September 1992.
13. American Institute of Chemical Engineers, *Dow's Fire & Explosion Index Hazard Classification Guide, 7th Edition,* ISBN 0-8169-0623-8, Wiley, New York, June 1994.
14. D. C. Hendershot, "Design of Inherently Safer Chemical Processing Facilities," presented at the Texas Chemical Council Safety Seminar, Galveston, TX, June 11, 1991.
15. H. W. Fawcett and W.S. Wood, *Safety and Accident Prevention in Chemical Operations, 2nd Ed.,* John Wiley & Sons, New York, 1982.
16. Center for Chemical Process Safety, *Inherently Safer Chemical Processes: A Life Cycle Approach, 2nd Ed.,* American Institute of Chemical Engineers, New York, 2007.

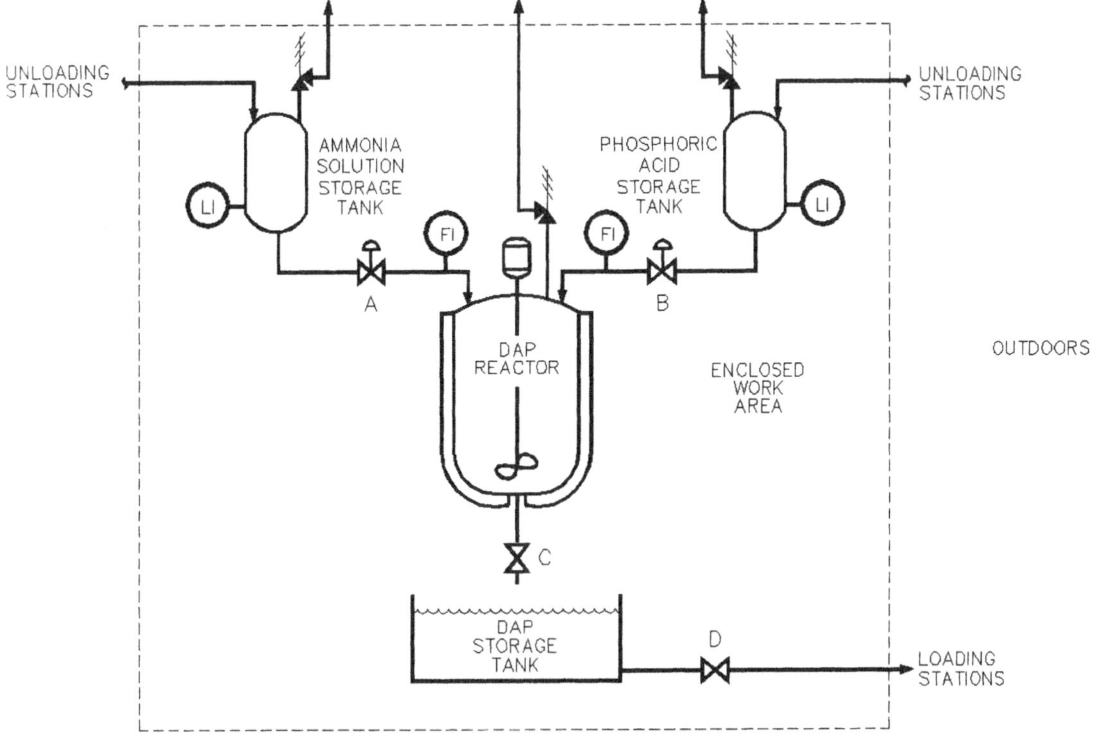

Figure 4.1 DAP process schematic for the Checklist Analysis example

Table 4.10 Sample Items from the checklist for the DAP process example

Materials	
Do all raw materials continue to conform to original specifications?	No. The concentration of ammonia in the ammonia solution has been increased to require less frequent purchase of ammonia. The relative flow rates to the reactor have been adjusted for the higher ammonia concentration.
Is each receipt of material checked?	Yes. The supplier has proven to be very reliable in the past. The labeling of the truck and the driver invoice is verified before unloading is permitted, but no sampling for the type of material or actual concentration is performed.
Does the operating staff have access to MSDSs?	Yes. MSDSs are available 24 hours a day at the processing location and in the administration building in the safety office.
Is fire fighting and safety equipment properly located and maintained?	No. The fire fighting and safety equipment location has not changed, but a new interior wall was constructed in the processing area. Because of the new wall, some places in the processing area cannot be adequately protected with existing fire fighting equipment. The existing equipment is in good condition and is inspected and tested monthly.
Equipment	
Has all equipment been inspected as scheduled?	Yes. The maintenance crew has inspected the equipment in the processing area according to the company inspection standards. However, failure data and maintenance department concerns suggest that inspections of the acid handling equipment may be too infrequent.
Have pressure relief valves been inspected as scheduled?	Yes. The inspection schedule has been followed.
Have safety systems and shutdowns been tested at an appropriate frequency?	Yes. There has been no deviation from the inspection schedule. However, inspection and maintenance of safety systems and shutdowns have been performed during processing operations, which is against company policy.
Are the proper maintenance materials (i.e., spare parts) available?	Yes. The company maintains a low inventory of replacement parts as an economic policy, although preventive maintenance and short-life items are readily available. Other items, except major pieces of equipment, are available by agreement with a local distributor within four hours.
Procedures	
Are the operating procedures current?	Yes. The written operating procedures were updated six months ago after some minor changes to operating steps were made.
Are the operators following the operating procedures?	No. The recent changes to operating steps have been slowly implemented. Operators feel that one of the changes may have not considered the personal safety of the operator.
Are new operating staff trained properly?	Yes. An extensive training program with periodic reviews and testing has been implemented, and training performance has been documented for all employees.
How are communications handled at shift change?	Operator shifts overlap 30 minutes to allow the next shift to learn the current operating status of the process from the previous shift.
Is housekeeping acceptable?	Yes. Housekeeping appears to be satisfactory.
Are safe work permits being used?	Yes. But they do not necessarily require shutdown of the process for some activities (e.g., testing or maintaining safety system components).

5

Scenario-Based Hazard Evaluation Procedures

This chapter continues the same format as Chapter 4 for describing commonly employed hazard evaluation techniques. But whereas the basic methods in Chapter 4 focused on experience-based and broad-brush approaches to evaluating hazards, the methods in this chapter focus on predictive, analytical methods that have as their common denominator the *incident scenario*, as described in Chapter 1. Hence, each method uses a different approach, but all can be used to systematically determine what can go wrong and what safeguards are in place to provide hazard containment and control.

Chapter 5 describes the following hazard evaluation techniques and summarizes how they are typically applied in the process industries.

Section 5.1 – What-If Analysis

Section 5.2 – What-If/Checklist Analysis

Section 5.3 – Hazard and Operability (HAZOP) Studies

Section 5.4 – Failure Modes and Effects Analysis (FMEA)

Section 5.5 – Fault Tree Analysis (FTA)

Section 5.6 – Event Tree Analysis (ETA)

Section 5.7 – Cause-Consequence Analysis (CCA) and Bow-Tie Analysis

Section 5.8 – Other Techniques

The hazard evaluation techniques covered first in Chapter 5 (What-If Analysis, What-If/Checklist Analysis, HAZOP Study, and FMEA) are excellent choices for performing detailed analyses of a wide range of hazards during the design phase of the process and during routine operation. Although each of these methods can be used as purely qualitative studies, they need to be more carefully structured to document incident scenarios if used as the basis for scenario risk analysis or Layer of Protection Analysis (LOPA). This more careful structuring is discussed as part of each methodology description.

Some of the hazard evaluation techniques covered in these *Guidelines* should be reserved for use in special situations requiring detailed analysis of one or a few specific hazardous situations of concern. These techniques (i.e., Fault Tree Analysis, Event Tree Analysis, Cause-Consequence Analysis, and Human Reliability Analysis) require specially trained and skilled practitioners. Analysts are cautioned to use these methods on tightly focused problems, since they require significantly more time and effort to perform than do the more broad-brush approaches.

Some of the hazard evaluation techniques have naturally evolved and been improved over the years. For example, more efficient means for implementing particular methods have been developed such as better approaches for solving fault trees and shortcuts for identifying deviations in a HAZOP Study. The documentation format for many of the techniques has also improved. For example, it is now customary to see a "Safeguards" or similarly named column explicitly listed in a HAZOP table, whereas the safeguards of the process may only have been implicitly considered in the analysis as originally documented. This does not necessarily mean that studies performed using the original approaches are no longer valid; rather, it means that the techniques have developed to meet the changing needs of the industry.

Another cautionary note – the techniques described herein are not ***all*** of the methods that are available for performing hazard evaluations. Section 5.8 lists some other hazard evaluation techniques that are less commonly used. These other methods are often hybrids or extensions of standard techniques that have been modified (and/or renamed) to suit an organization's special needs. These other approaches may be quite effective and, in some cases, even better than the standard methods included in these *Guidelines*. However, the standard methods presented in Chapters 4 and 5 are used in the vast majority of hazard evaluations performed in the process industries.

Readers should not view these Guidelines as a barrier to stop them from developing and using novel or specialized hazard evaluation techniques. For example, the *Third Edition* contains information on the Layer of Protection Analysis technique. This technique was not considered in the second edition, but it has since evolved and is now frequently employed. Ultimately, using qualified and experienced personnel to perform hazard evaluations will have a much greater bearing on the success of a study than will the specific hazard evaluation technique that is employed.

5.1 What-If Analysis

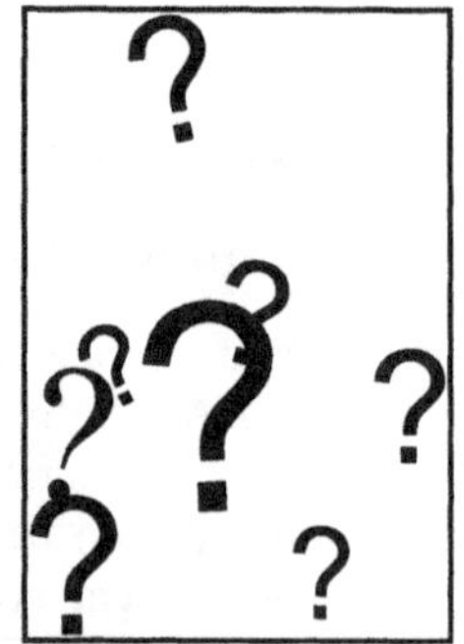

The What-If Analysis technique is a brainstorming approach in which a group of experienced people familiar with the subject process ask questions or voice concerns about possible undesired events.

Purpose

The purpose of a What-If Analysis is to identify hazards, hazardous situations, or specific event sequences that could produce undesirable consequences. An experienced group of people identifies possible abnormal situations, their consequences, and existing safeguards, then suggests alternatives for risk reduction where obvious improvement opportunities are identified or where safeguards are judged to be inadequate. The method can involve examination of possible deviations from the design, construction, modification, or operating intent. It requires a basic understanding of the process intention, along with the ability to mentally combine possible deviations from the design intent that could result in an incident. This is a powerful technique if the staff is experienced; otherwise, the results are likely to be incomplete.

Description

What-If Analysis is not as inherently structured as some other techniques such as HAZOP Studies and FMEA. Instead, it requires the analyst to adapt the basic concept to the specific application. Very little information has been published on the What-If Analysis method or its application. However, it is frequently used by industry at nearly every stage of the life cycle of a process and has a good reputation among those skilled in its use.

The What-If Analysis concept encourages the hazard evaluation team to think of questions that begin with "What-If." However, any process safety concern can be voiced, even if it is not phrased as a question. For example:

- I'm concerned about having the wrong material delivered.
- What if Pump A stops running during start-up?
- What if the operator opens valve B instead of A?

Usually, the scribe records all of the questions on a chart pad, marking board, or computer. Then the questions may be divided into specific areas of investigation (usually related to consequences of concern), such as electrical safety, fire protection, or personnel safety. Each area is subsequently addressed by a team of knowledgeable people. The questions are formulated based on experience and applied to existing drawings and process descriptions. For an operating plant, the investigation may include interviews with plant staff not represented on the hazard evaluation team. There may be no specific pattern or order to these questions, unless the leader provides a logical pattern such as dividing the process into functional systems. The questions can address any off-normal condition related to the plant, not just component failures or process variations.

Types of Results

In its simplest form, the What-If Analysis technique generates a list of questions and answers about the process. It may also result in a tabular listing of hazardous situations (generally with no ranking of or quantitative implication for the identified potential incident scenarios), as well as their consequences, safeguards, and possible options for risk reduction. A worked example showing What-If Analysis results can be found in Part II, Chapter 13.

Resource Requirements

Since What-If Analysis is so flexible, it can be performed at any stage of the life cycle of a process, using whatever process information and knowledge is available. For each area of the process, two or three people may be assigned to perform the analysis; however, a larger team may be preferred. It is sometimes better to use a larger group for a large process, dividing the process into smaller segments, than to use a small group for a long time on the whole process.

The time and cost of a What-If Analysis are proportional to the plant complexity and number of areas to be analyzed. Once an organization has gained experience with it, the What-If Analysis method can become a cost-efficient means for evaluating hazards during any project phase. Table 5.1 lists basic estimates of the time needed to perform a hazard evaluation using the What-If Analysis technique.

Analysis Procedure

After the scope of the study is defined, the What-If Analysis consists of the following steps: first, preparing for the review; second, performing the review; and third, documenting the results.[1-3]

Preparing for the review. The information needed for a What-If Analysis includes chemical data, process descriptions, drawings, and operating procedures. Other information may also be needed, depending on the life cycle phase of the facility, what regulatory requirements need to be met, and whatever specific information is needed to provide definitive answers to the team's what-if questions. It is important that all the information is up to date and available to the hazard evaluation team, preferably in advance of the team meetings.

If an existing plant is being reviewed, the review team may want to interview personnel responsible for operations, maintenance, utilities, or other services. In addition, if the team is having the What-If meeting off-site, they should visit the plant to get a better idea of its layout, construction, and operation. Thus, before the review begins, visits and interviews should be scheduled.

The last part of this step is the preparation of some preliminary What-If questions to "seed" the analysis meetings. If this analysis is an update of a previous review or an examination of a plant modification, any questions listed in the report for the previous study can be used. For a new plant or a first-time application, preliminary questions should be developed by the team leader and/or team members before the meetings, although additional questions formulated during the meetings are essential. The cause and effect thought process used in other types of studies described in the chapter (such as HAZOP Studies) can help formulate questions. Chapter 2 provides additional advice about preparing for a hazard evaluation.

Performing the review. The review meetings should begin with a basic explanation of the process, given by plant staff having overall plant and process knowledge plus expertise relevant to the team's investigation area. The presentation should also describe the plant's safety precautions, safety equipment, and health control procedures.

Table 5.1 Time estimates for using the What-If Analysis technique

Scope	Preparation*	Evaluation	Documentation*
Small system	4 to 8 hours	4 to 8 hours	1 to 2 days
Large process	1 to 3 days	3 to 5 days	1 to 3 weeks

* Primarily, team leader and scribe; may be lower using computer software

The subject process is reviewed by members of the team who vocalize potential safety concerns. However, the team should not limit themselves to the prepared What-If questions or restrict themselves to making their statements into questions. Rather, they should use their combined expertise and team interaction to articulate whatever issues they believe necessary to ensure that the investigation is thorough. The team should not be rushed and should not work too many hours consecutively. Ideally, a team should meet no more than four to six hours per day. What-If team meetings that run for more than a week of consecutive days are not desirable.

There are two ways that the meetings can be conducted. One way preferred by some is to first list on a chart pad or marking board all of the safety issues and questions, then begin considering them. Another way is to consider each question and issue one at a time, with the team determining the significance of each situation as it is brought up before soliciting other questions or issues from the team. Both ways can work, but it is usually preferable to list all the questions before answering them to prevent interruption of the team's "creative momentum." If the process is complex or large, it should be divided into smaller segments so that the team does not spend several consecutive days just listing questions. Often, the team will think of additional questions as a result of considering earlier issues.

Initially, the team leader should outline the proposed scope of the study and the team should agree with it. The team generally proceeds from the beginning of the subject process to its end, although the hazard evaluation team leader can order the analysis in any logical way he or she sees fit. Then the team answers each question and addresses each issue (or indicates a need for more information) and identifies the hazard, potential consequences, engineered safeguards, and possible solutions. In the process, they add any new What-If questions that become apparent during the analysis. (Sometimes the proposed answers are developed by individuals outside the initial meeting, and then presented to the team for endorsement or modification.)

Inherent safety reviews. The What-If Analysis technique can be used to evaluate hazards in an inherent safety review. For this technique to be effective, team members must be well-versed in the principles of inherent safety discussed in section 3.7. The team would brainstorm process deviation scenarios and identify inherent safety improvements for reducing or eliminating the potential for the scenario to develop. The strategies of substitution, minimization, moderation and simplification can be used to determine the safety improvements.[4]

For example, a team might ask "What if the chlorine ton container at the cooling tower developed a leak in the tubing?" The analysis might determine that the vaporized chlorine could affect a nearby school. After discussing existing safeguards, a recommendation might be developed to consider replacing the chlorine cylinder with a solid biocide (substitution). The potential for chemical reactivity hazards associated with the use of the biocide would of course need to be evaluated as part of the determination of residual risk.

Another analysis might evaluate the question of "What if the sulfuric acid was inadvertently loaded into the 50% caustic tank?" The team may conclude, after evaluation of the chemical reactivity hazard, the severity of consequences, and existing safeguards that might include procedural verifications, unique connections, and emergency relief protection that no action is warranted. Discussion might nevertheless proceed to whether opportunities exist to substitute, minimize, moderate, or simplify.

Documenting the results. As with any study, documentation is the key to transforming the team's findings into measures for hazard elimination or reduction. Table 5.2 is an example What-If Analysis worksheet. Such a table makes the documentation easier and more organized. In addition to completed tables, the hazard evaluation team usually develops a list of suggestions for improving the safety of the analyzed process based on the tabular What-If Analysis results. Some companies document their What-If Analyses with a narrative-style format rather than a table.

Example

A team is assigned to investigate personnel hazards from the reactor section of the DAP example system (Section 4.4, Figure 4.1) using the What-If Analysis technique. Table 5.3 is a partial list of What-If questions that were developed during the team meeting.

Table 5.2 Typical format for a What-If Analysis worksheet

Area: Drawing Number:			Meeting Date: Team Members::	
What If	Hazard	Consequence	Safeguards	Recommendation

Table 5.3 What-If questions for the DAP process example

What If...

- ❑ the wrong feed material is delivered instead of phosphoric acid?
- ❑ the phosphoric acid concentration is too low?
- ❑ the phosphoric acid is contaminated?
- ❑ valve B is closed or plugged?
- ❑ too high a proportion of ammonia is supplied to the reactor?
- ❑ vessel agitation stops?
- ❑ valve C is closed or plugged?

The team addresses the first question and would probably consider what other materials could be mixed with ammonia and produce a hazard. If one or more such materials are known, then their availability at the plant is noted, along with the possibility that the vendor could deliver something marked phosphoric acid that is actually another material. The hazard of incorrect combinations of materials is identified if it endangers plant employees or the community. The team also identifies pertinent hardware or procedural safeguards that could prevent these scenarios. Remedies are suggested to guard against the wrong materials being used in place of phosphoric acid. The team continues through the questions in this manner until they reach the output of the process. In this case, the team noted that the process is located in a cinderblock building at the plant and that extreme weather conditions may exceed the capabilities of the heating or air conditioning system. Thus, they add two (perhaps more) questions.

What if the outside temperature is less than -20°F?

What if the outside temperature is greater than 100°F?

The responses to all of the What-If questions are recorded in a table as the team proceeds with the evaluation. A sample page from this documentation is shown in Table 5.4.

Table 5.4 Sample page from the What-If Analysis table for the DAP process example

Process: DAP Reactor Topic Investigated: Toxic Releases			Analysts: Date: ##/##/##	
What If	Hazard	Consequence	Safeguards	Recommendation
Wrong feed material is delivered instead of phosphoric acid?	Contaminant incompatibility	Potentially hazardous phosphoric acid or ammonia reactions with contaminants, or production of off-specification product	[] Reliable vendor [] Plant material handling procedures	Ensure adequate material handling and receiving procedures and labeling exist
Phosphoric acid concentration is too low?	Ammonia inhalation toxicity	Unreacted ammonia carryover to the DAP storage tank and release to the work area	[] Reliable vendor [] Ammonia detector and alarm	Verify phosphoric acid concentration before filling storage tank
Phosphoric acid is contaminated?	Contaminant incompatibility	Potentially hazardous phosphoric acid or ammonia reactions with contaminants, or production of off-specification product	[] Reliable vendor [] Plant material handling procedures	Ensure adequate material handling and receiving procedures and labeling exist
Valve B is closed or plugged?	Ammonia inhalation toxicity	Unreacted ammonia carryover to the DAP storage tank and release to the work area	[] Periodic maintenance [] Ammonia detector and alarm [] Flow indicator in phosphoric acid line	Alarm/shutoff of ammonia (valve A) on low flow through valve B
Too high a proportion of ammonia is supplied to the reactor?	Ammonia inhalation toxicity	Unreacted ammonia carryover to the DAP storage tank and release to the work area	[] Flow indicator in ammonia solution line [] Ammonia detector and alarm	Alarm/shutoff of ammonia (valve A) on high flow through valve A

What-If Analyses Used for Further Scenario Evaluations

When the scenarios developed in a What-If Analysis are to be used as the starting point for scenario risk evaluation or for a Layer of Protection Analysis (Chapter 7), the What-If Analysis and its documentation may need to be more explicitly structured and documented.

First, the "What-If" condition should represent an initiating cause (or have a separate "Cause" column); for example, if the question is "What if the unloading connection comes apart when the transfer begins?" then this represents the initiating cause of Unloading connection comes apart when the transfer begins. If a "What-If" question involves failure of a standby safety system such as "What If the relief valve inlet is blocked," then this is not an initiating cause, since the process can continue to operate normally even with the relief inlet blocked. The "Consequence" in this case could be documented as "Less protection against vessel overpressurization" but the effectiveness of the relief system would be discussed in association with each actual overpressurization scenario.

Next, the "Consequence" should be a description of the events that would ensue if the initiating cause occurred and no preventive safeguards intervened, all the way through to a loss event. Depending on how the impacts of the loss event are documented, the Consequence description may also need to include a description of the range of impacts of the loss event. (In some What-If Analysis worksheets, the Hazard and Consequence columns are combined. This may be appropriate for a more basic What-If Analysis, such as the worked example in Chapter 13, but having two separate columns allows for the more explicit distinction between the Hazard, which is the underlying physical or chemical condition that has the potential for causing harm, and the Consequence, which are the effects following the What-If condition traced all the way to the loss event. Considering the Hazard separately also allows for better consideration of inherently safer approaches, which would reduce or eliminate the underlying Hazard.)

Finally, having a separate "Safeguards" column is necessary for evaluating scenario risk. The "Safeguards" should be only those preventive safeguards that would come into effect after the initiating cause and before the loss event, for the specific initiating cause / loss event combination, plus source-mitigative safeguards (as discussed in Section 1.4). Containment and control measures and receptor-mitigative safeguards are better noted in a Comment field or separate from the analysis worksheet.

Section 5.1 References

1. S. Mannan, ed., *Lees' Loss Prevention in the Process Industries, 3rd Ed.,* Elsevier Butterworth-Heinemann, ISBN 0-7506-7555-1, Oxford, UK, 2005.
2. H. R. Greenberg and J. J. Cramer (eds.), *Risk Assessment and Risk Management for the Chemical Process Industry*, ISBN 0-442-23438-4, Van Nostrand Reinhold, New York, 1991.
3. *Chemical Process Hazard Review*, ACS Symp. Series 274, American Chemical Society, Washington, DC, 1985.
4. Center for Chemical Process Safety, *Inherently Safer Chemical Processes: A Life Cycle Approach, 2nd Ed.,* American Institute of Chemical Engineers, New York, 2007.

5.2 What-If/Checklist Analysis

The What-If/Checklist Analysis technique combines the creative, brainstorming features of the What-If Analysis method (Section 5.1) with the systematic features of the Checklist Analysis method (Section 4.4).

Purpose

The purpose of a What-If/Checklist Analysis is to identify hazards, consider the general types of incidents that can occur in a process or activity, evaluate in a qualitative fashion the effects of these incidents, and determine whether the safeguards against these potential incident situations appear adequate. Frequently, the hazard evaluation team members will suggest ways for reducing the risk of operating the process.

Description

This hybrid method capitalizes on the strengths and compensates for the individual shortcomings of the separate approaches. For example, the Checklist Analysis method is an experience-based technique, and the quality of a hazard evaluation performed using this approach is highly dependent on the experience of the checklist's authors. If the checklist is not complete, then the analysis may not effectively address a hazardous situation. The What-If Analysis portion of the technique encourages the hazard evaluation team to consider potential abnormal situations and consequences that are beyond the experience of the authors of a good checklist, and thus are not covered on the checklist. Conversely, the checklist portion of this technique lends a more systematic nature to the What-If Analysis. The What-If/Checklist Analysis technique may be used at any stage of a process's life. This technique is being discussed as a stand alone technique because of its popularity and effectiveness.

Like most other hazard evaluation techniques, the method works best when performed by a team experienced in the subject process. This technique is generally used to analyze the most common hazards that exist in a process. Although it is able to evaluate the significance of incidents at almost any level of detail, the What-If/Checklist Analysis method usually focuses on a less-detailed level of resolution than, for example, the FMEA technique. Often, a What-If/Checklist Analysis is the first hazard evaluation performed on a process, and as such, it is a precursor for subsequent, more detailed studies.

Types of Results

A hazard evaluation team using the What-If/Checklist Analysis technique usually generates a table of What-If questions (initiating causes), effects, safeguards and action items. The results from such a study may also include a completed checklist. However, some organizations use a narrative style to document the results of such studies. A worked example showing What-If/Checklist Analysis results can be found in Part II, Chapter 22.

Resource Requirements

Most What-If/Checklist Analyses are performed by a team of personnel experienced in the design, operation, and maintenance of the subject process. The number of people needed for such a study depends upon the complexity of the process, and to some extent, the stage of life at which the process is being evaluated. Table 5.5 lists estimates of the time needed to perform a hazard evaluation using the What-If/Checklist Analysis technique. What-If/Checklist is an experienced based technique that relies on the experience of the checklist author and the experience of the review team.

Analysis Procedure

A What-If/Checklist Analysis consists of the following steps: (1) preparing for the review, (2) developing a list of What-If questions and issues, (3) using a checklist to cover any gaps, (4) evaluating each of the questions and issues, and (5) documenting the results.[1-3] A variation of this procedure is for the team to reverse the order of steps 2 and 3 or to develop What-If questions concurrently as they progress through a detailed checklist.

Preparing for the review. Chapter 2 of these *Guidelines* provides general guidance for preparing for team-based hazard evaluations. For a What-If/Checklist Analysis, the hazard evaluation team leader assembles a qualified team, determines the physical and analytical scope for the proposed study, and, if the process/activity is rather large, divides it into functions, physical areas, or tasks to provide some order to the review of the process. (See section 4.4 for the checklist portion of this analysis) The hazard evaluation team leader should obtain or develop an appropriate checklist for the team to use in conjunction with the What-If Analysis. The checklist should focus on general hazardous characteristics of the process or operation. Appendix B contains an example of a detailed checklist that a hazard evaluation team leader could use as the basis for constructing checklists appropriate for almost any analysis.

Developing a list of What-If questions and issues. Section 5.1 describes the approach a hazard evaluation team uses when meeting to develop questions and issues involving potential incident situations.

Using a checklist to cover any gaps. Once the team has identified all of the questions and issues it can in a particular area or step of the process or activity, the hazard evaluation team leader will use the checklist he or she previously obtained (or prepared). The team considers each checklist item to see whether any other potential incident situations or concerns arise. If so, these are evaluated in the same way as the original What-If questions (the checklist is reviewed for each area or step in the process or activity). In some cases it may be more desirable to have the hazard evaluation team brainstorm the hazards and potential incident situations of a process before using the checklist.[1] In other situations, effective results can be obtained by beginning with a checklist and using items in it to create What-If questions and issues that might not otherwise have been considered. However, if the checklist is used first, leaders should take precautions to avoid letting the checklist restrict the creativity and imagination of the team.

Table 5.5 Time estimates for using the What-If/Checklist Analysis technique

Scope	Preparation*	Evaluation	Documentation*
Small system	6 to 12 hours	6 to 12 hours	4 to 8 hours
Large process	1 to 3 days	4 to 7 days	1 to 3 weeks

* Primarily, team leader and scribe; may be lower using computer software

Evaluating each of the questions and issues. After developing questions and issues involving potential incident situations, the team considers each incident situation or safety concern; qualitatively determines the potential effects of the incident implied by the situation or concern; and lists existing safeguards to prevent, mitigate, or contain the effects of the potential incident. The team then evaluates the significance of each situation and determines whether a particular safety improvement option should be recommended. This process is repeated for each area or step of the process or activity. Sometimes this evaluation is performed by specific team members outside the team meeting and is subsequently reviewed by the team.

Documenting the results. The results of a What-If/Checklist Analysis are documented in the same way as the results for a What-If Analysis (see Section 5.1). Usually the scribe will use a marking board, chart pad, or computer linked to a projection device to record questions, issues, effects, safeguards, action items, etc., during the meeting. Following the meeting the hazard evaluation team leader and scribe usually summarize these results in a tabular form similar to that shown in Table 5.2. For a What-If/Checklist Analysis, the hazard evaluation team may also document the completion of the checklist to help illustrate the completeness of the study.

Example

To increase production, the K. R. Mody Chemical Company has installed a new transfer line between its existing 90-ton chlorine storage tank and its reactor feed tank. Before each batch, the operator must transfer one ton of chlorine into the feed tank; the new line will allow this to be done in about one hour (with the old line it took about three hours). Nitrogen pressure will be used to force the liquid chlorine through the mile-long, uninsulated, welded pipeline in the elevated rack between the barge terminal and the process unit. Both the storage tank and the reactor feed tank operate at ambient temperature. Figure 5.1 is a schematic of the new feed line configuration.

To transfer chlorine, the operator sets PCV-1 to the desired pressure, opens HCV-1, and verifies that the level in the feed tank is rising. When the high level alarm in the feed tank signals that one ton of chlorine has been transferred, the operator closes HCV-1 and PCV-1. HCV-2 is normally left open between batches so liquid chlorine will not be trapped in the long pipeline.

A hazard evaluation team is chartered to perform a What-If/Checklist Analysis of this process modification and meets to consider potential incidents of concern and whether there is adequate protection against them. The What-If portion of the meeting generates the issues listed in Table 5.6. Subsequently, the team uses the two checklists shown in Figure 5.2 and Table 5.7 to supplement their What-If questions.[1] Table 5.8 lists the additional safety concerns the team identifies using the checklists; these concerns may have gone unnoticed had only the What-If Analysis method been used.

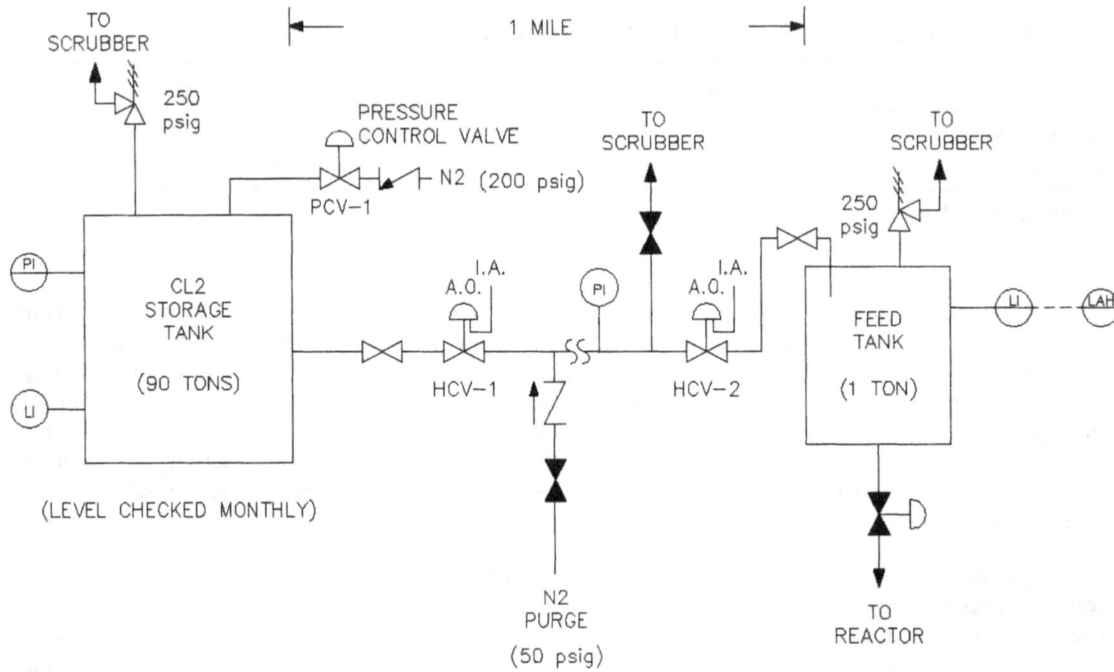

Figure 5.1 Schematic for the chlorine feed line example

Table 5.6 What-If questions for the chlorine feed line example

- ❑ What if there is moisture left in the line?
- ❑ What if the operator transferred a double batch of chlorine?
- ❑ What if HCV-1 is left closed?
- ❑ Will the pipeline overpressure and rupture if left full of liquid chlorine during the summer?
- ❑ Is the piping material rated for temperatures below - 20 °F? (- 29 °C)
- ❑ What if there is reverse flow from the feed tank to the storage tank?
- ❑ Can chlorine leak into and contaminate the nitrogen system?
- ❑ What can be done to isolate the header if there is a major rupture such as caused by vehicle impact?
- ❑ What is the design basis of the scrubber? Can it handle all of the chlorine if the pipeline needed to be depressured quickly in an emergency?
- ❑ What if HCV-2 is inadvertently closed?
- ❑ What if the level indicator/alarm fails in the feed tank?
- ❑ What if air enters the system? Can it cause an incident in the reactor?

Figure 5.2 Example of a simplified checklist for hazard evaluation (adapted from Reference 1)

Storage of raw materials, products, and intermediates		
Storage tanks	Design, separation, inerting, materials of construction	____
Dikes	Capacity, drainage	____
Emergency valves	Remote control—hazardous materials	____
Inspections	Flash arresters, relief devices	____
Procedures	Contamination prevention, analysis	____
Specifications	Chemical, physical, quality, stability	____
Limitations	Temperature, time, quantity	____
Materials handling		
Pumps	Relief, reverse rotation, identification, materials of construction	____
Ducts	Explosion relief, fire protection, support	____
Conveyors, mills	Stop devices, coasting, guards	____
Procedures	Spills, leaks, decontamination	____
Piping	Ratings, codes, cross-connections, materials of construction	____
Process equipment, facilities, and procedures		
Procedures	Start-up, normal, shutdown, emergency	____
Conformance	Job audits, shortcuts, suggestions	____
Loss of utilities	Electrical, heating, coolant, air, inerts, agitation	____
Vessels	Design, materials, codes, access, materials of construction	____
Identification	Vessels, piping, switches, valves	____
Relief devices	Reactors, exchangers, glassware	____
Review of incidents	Plant, company, industry	____
Inspections, tests	Vessels, relief devices, corrosion	____
Hazards	Runaways, releases, explosions	____
Electrical	Area classification, conformance, purging	____
Process	Description, test authorizations	____
Operating ranges	Temperature, pressure, flows, ratios, concentrations, densities, levels, time, sequence	____
Ignition sources	Peroxides, acetylides, friction, fouling, compressors, static electricity, valves, heaters	____
Compatibility	Heating media, lubricants, flushes, packing	____
Safety margins	Cooling, contamination	____
Personal protection		
Protection	Barricades, personal, shower, escape aids	____
Ventilation	General, local, air intakes, rate	____

Figure 5.2 Example of a simplified checklist for hazard evaluation (adapted from Reference 1)

Exposures	Other processes, public, environment	____
Utilities	Isolation: air, water, inerts, steam	____
Hazards manual	Toxicity, flammability, reactivity, corrosion, symptoms, first aid	____
Environment	Sampling, vapors, dusts, noise, radiation	____
Sampling facilities		
Sampling points	Accessibility, ventilation, valving	____
Procedures	Plugging, purging	____
Samples	Containers, storage, disposal	____
Analysis	Procedures, records, feedback	____
Maintenance		
Decontamination	Solutions, equipment, procedures	____
Vessel openings	Size, obstructions, access	____
Procedures	Vessel entry, welding, lockout, work permits	____
Fire protection		
Fixed protection	Fire areas, water demands, distribution system, sprinklers, deluge, monitors, inspection, testing, procedures, adequacy	____
Extinguishers	Type, location, training	____
Fire walls	Adequacy, condition, doors, ducts	____
Drainage	Slope, drain rate	____
Emergency response	Fire brigades, staffing, training, equipment	____
Controls and emergency devices		
Controls	Ranges, redundancy, fail-safe	____
Calibration, inspection	Frequency, adequacy	____
Alarms	Adequacy, limits, fire, fume	____
Shutdown systems	Tests, bypass procedures	____
Relief devices	Adequacy, vent size, discharge, drain, support	____
Emergencies	Dump, drown, inhibit, dilute	____
Process isolation	Block valves, fire-safe valves, purging	____
Instruments	Air quality, time lag, reset windup, materials of construction	____
Waste disposal		
Ditches	Flame traps, reactions, exposures, solids	____
Vents	Discharge, dispersion, mists	____
Characteristics	Sludges, residues, fouling materials	____

Table 5.7 Example of a hazard checklist

Acceleration (uncontrolled — too much, too little)
- Inadvertent motion
- Sloshing of liquids
- Translation of loose objects

Deceleration (uncontrolled — too much, too little)
- Impacts (sudden stops)
- Failures of brakes, wheels, tires, etc.
- Falling objects
- Fragments or missiles

Chemical Reaction (non-fire; can be subtle over time)
- Disassociation, product reverts to separate components
- Combination, new product formed from mixture
- Water-reactive material
- Corrosion, rust, etc.

Electrical
- Shock
- Burns, arc flash
- Overheating
- Static electricity
- Inadvertent activation
- Explosion, electrical

Explosions
- Commercial explosive or shock-sensitive material
- Containment (confined vapor/dust/mist explosion or bursting vessel explosion potential)
- Confinement/congestion (vapor cloud explosion potential)
- Liquefied gas or liquid above boiling point (BLEVE potential)

Flammability and Fires
- Presence of fuel — solid, liquid, gas, dust
- Presence of strong oxidizer — oxygen, peroxide, etc.
- Presence of strong ignition force — welding torch, heaters
- Pyrophoric or spontaneously combustible material

Heat and Temperature
- Source of heat, non-electrical
- Hot surface burns
- Very cold surface burns
- Increased gas pressure caused by heat
- Increased flammability caused by heat
- Increased volatility caused by heat
- Increased reactivity caused by heat

Mechanical
- Sharp edges or points
- Rotating equipment
- Reciprocating equipment
- Pinch points
- Weights to be lifted
- Stability/toppling tendency
- Ejected parts or fragments

Pressure
- Compressed gas
- Compressed air tool
- Pressure system exhaust
- Unintentional pressure release
- Objects propelled by pressure
- Water hammer
- Flex hose whipping
- Hydrostatic pressure

Leak of Material
- Flammable
- Toxic
- Corrosive
- Slippery

Radiation
- Ionizing radiation
- Ultraviolet light
- High intensity visible light
- Infrared radiation
- Electromagnetic radiation
- Laser radiation

Toxicity
- Gas or liquid
  - — Asphyxiant
  - — Irritant
  - — Systemic poison
  - — Carcinogen
  - — Mutagen
- Combination product
- Combustion product

Vibration
- Vibrating tools
- High noise source level
- Mental fatigue
- Flow or jet vibration
- Supersonics

Miscellaneous
- Contamination
- Lubricity

Table 5.8 Additional safety issues generated by using hazard checklists in the chlorine feed line example

- ❑ What if the line is contaminated with oil during maintenance?
- ❑ What if the nitrogen header pressure regulator fails?
- ❑ Is the chlorine tank rated for full vacuum?
- ❑ What if there is a leak in the line during a night transfer operation?
- ❑ Have previous chlorine release incidents in industry been reviewed?
- ❑ Does this equipment meet the recommendations of the Chlorine Institute?
- ❑ Are there any sampling or drain points at the low spots in the pipeline?
- ❑ Has the correct metallurgy been specified for this equipment?
- ❑ If inert material-lined piping is being used, how will its integrity be periodically tested?
- ❑ What emergency notification systems exist for alerting the community?

What-If/Checklist Analyses Used for Further Scenario Evaluations

When the scenarios developed in a What-If Analysis are to be used as the starting point for scenario risk evaluation or for a Layer of Protection Analysis (Chapter 7), the What-If Analysis and its documentation may need to be more explicitly structured and documented. The same considerations that apply to What-If Analyses (Section 5.1) also apply to What-If/Checklist analyses.

Section 5.2 References

1. A.F. Burk, "What-If/Checklist—A Powerful Process Hazards Review Technique," presented at the AIChE 1991 Summer National Meeting, Pittsburgh, August, 1991.
2. H. R. Greenberg and J. J. Cramer (eds.), *Risk Assessment and Risk Management for the Chemical Process Industry*, ISBN 0-442-23438-4, Van Nostrand Reinhold, New York, 1991.
3. *Chemical Process Hazard Review*, ACS Symp. Series 274, American Chemical Society, Washington, DC, 1985.

5.3 Hazard and Operability Studies

The Hazard and Operability (HAZOP) Study or HAZOP Analysis technique was developed to identify and evaluate safety hazards in a process plant, and to identify operability problems which, although not hazardous, could compromise the plant's ability to achieve design productivity.

Purpose

The purpose of a HAZOP Study is to carefully review a process or operation in a systematic fashion to determine whether deviations from the design or operational intent can lead to undesirable consequences. This technique can be used for continuous or batch processes and can be adapted to evaluate written procedures. The HAZOP team lists potential causes and consequences of the deviation as well as existing safeguards protecting against the deviation. When the team determines that inadequate safeguards exist for a credible deviation, it usually recommends that action be taken to reduce the risk.

Description

Although originally developed to anticipate hazards and operability problems for technology with which organizations have little experience, it has been found to be very effective for use with existing operations. Use of the HAZOP Study technique requires a detailed source of information concerning the design and operation of a process. Thus, it is most often used to analyze processes during or after the detailed design stage. Several variations of the HAZOP Study technique are in practice in the chemical industry.

In a HAZOP Study, an interdisciplinary team uses a creative, systematic approach to identify hazard and operability problems resulting from deviations from the process's design intent that could lead to undesirable consequences. An experienced team leader systematically guides the team through the plant design using a fixed set of words (called "guide words"). These guide words are applied at specific points or "study nodes" in the plant design and are combined with specific process parameters to identify potential deviations from the plant's intended operation.

For example, the guide word "No" combined with the process parameter "Flow" results in the deviation "No Flow." Sometimes, a leader will use checklists or process experience to help the team develop the necessary list of deviations that the team will consider in the HAZOP meetings. The team then agrees on possible causes of the deviations (e.g., operator error blocks in pump), the consequences of deviations (e.g., pump overheats), and the safeguards applicable to the deviations (e.g., pressure relief valve on the pump discharge line). If the causes and consequences are significant and the safeguards are inadequate, the team may recommend a follow-up action for management consideration. In some cases, the team may identify a deviation with a realistic cause but unknown consequences (e.g., an unknown reaction product) and recommend follow-up studies to determine the possible consequences.

The essence of the HAZOP Study approach is to review process drawings and/or procedures in a series of meetings, during which a multidisciplinary team uses a prescribed protocol to methodically evaluate the significance of deviations from the normal design intention. Imperial Chemical Industries (ICI) originally defined the HAZOP Study technique to require that HAZOP Studies be performed by an interdisciplinary team.[1-3] Thus, while it is possible for one person to use the HAZOP Study thought process, such a study is not properly called a HAZOP Study.

The primary advantage of the brainstorming associated with HAZOP Study is that it stimulates creativity and generates new ideas. This creativity results from the interaction of a team with diverse backgrounds. Consequently, the success of the study requires that all participants freely express their views, but participants should refrain from criticizing each other to avoid stifling the creative process. This creative approach combined with the use of a systematic protocol for examining hazardous situations helps improve the thoroughness of the study.

The HAZOP Study focuses on specific points of the process or operation called "study nodes," which are process sections or operating steps. One at a time, the HAZOP team examines each section or step for potentially hazardous process deviations that are derived from a set of established guide words. One purpose of the guide words is to ensure that all relevant deviations of process parameters are evaluated. Sometimes, teams consider a fairly large number of deviations (i.e., up to 10 to 20) for each section or step and identify their potential causes and consequences. Normally, all of the deviations for a given section or step are analyzed by the team before it proceeds further.

For an effective HAZOP Study to be conducted, it is essential to consider the process deviations as abnormal situations that represent deviations from a well-specified design or operational intention. The documented statement of intention should include all relevant safety limits and operating procedure steps, since these limits define the boundaries of normal operation. The HAZOP Study facilitator should ensure by some means that all meaningful deviations from all relevant safety limits and procedure steps are evaluated by the study team. It is well worth the up-front time for the design intention to be carefully and thoroughly developed before proceeding with the study of a given node or segment.

HAZOP Studies can be performed on new projects as well as on existing facilities.[4-6] For new projects, it is best to conduct a HAZOP Study when the process design is fairly firm. Normally, the system P&IDs are available so the team can formulate meaningful answers to the questions raised in the HAZOP Study process, but it is still possible to change the design without incurring major costs. However, HAZOP Studies can also be performed at earlier process life cycle stages as long as the team members have adequate process documentation and knowledge upon which to base their analysis. But a HAZOP Study performed at this early stage should not be a substitute for a thorough design review.

Although the basic HAZOP Study approach is well-established, the way that it is employed may vary from organization to organization. Table 5.9 lists terms and definitions that are commonly used in HAZOP Studies. The guide words shown in Table 5.10 are the original ones developed by ICI for use in a HAZOP Study and are applied to process parameters such as those shown in Table 5.11. Some organizations have modified this list to be specific to their operations and to guide teams more quickly to the areas where significant process safety problems may exist. Other organizations have created specialized lists of guide words or specific deviations for analyzing batch operations and procedural steps.[4] See References 10 and 11 for further information on the basic HAZOP Study approach and variations.

Table 5.9 Common HAZOP Study terminology

Term	Definition
Study nodes (process sections; segments)	Sections of equipment with definite boundaries (e.g., a line between two vessels) within which process parameters are investigated for deviations. The locations on P&IDs at which the process parameters are investigated for deviations
Operating steps	Discrete actions in a batch process or a procedure analyzed by a HAZOP Study team. May be manual, automatic, or software-implemented actions. The deviations applied to each step are somewhat different than the ones used for a continuous process
Intention (design intent)	Definition of how the plant is expected to operate in the absence of deviations. Takes a number of forms and can be either descriptive or diagrammatic (e.g., process description, flowsheets, line diagrams, P&IDs)
Guide words	Simple words that are used to qualify or quantify the design intention and to guide and stimulate the brainstorming process for identifying process hazards
Process parameter	Physical or chemical property associated with the process. Includes general items such as reaction, mixing, concentration, pH, and specific items such as temperature, pressure, phase, and flow
Deviations	Departures from the design intention that are discovered by systematically applying the guide words to process parameters (flow, pressure, etc.) resulting in a list for the team to review (no flow, high pressure, etc.) for each process section. Teams often supplement their list of deviations with ad hoc items
Causes	Initiating causes; reasons why deviations might occur. Once a deviation has been shown to have a credible cause, it can be treated as a meaningful deviation. These causes can be hardware failures, human errors, unanticipated process states (e.g., change of composition), external disruptions (e.g., loss of power), etc.
Consequences	Results of deviations (e.g., release of toxic materials). Normally, the team assumes active preventive safeguards can fail and examine the consequences through the loss event; impacts may also be described. Minor consequences, unrelated to the study objective, are not considered
Safeguards	Engineered systems or administrative controls designed to prevent the cause, protect against the deviation progressing to a loss event, or mitigate the immediate loss event consequences (e.g., process alarms, shutdowns, automatic isolation)
Actions (recommendations, action items)	Suggestions for design changes, procedural changes, or areas for further study (e.g., adding a redundant pressure alarm or reversing the sequence of two operating steps)

Table 5.10 Original HAZOP Study guide words and meanings

Guide words	Meaning
NO *or* NOT	Negation of the design intent
LESS *or* LESS OF	Quantitative decrease
MORE *or* MORE OF	Quantitative increase
PART OF	Qualitative decrease
AS WELL AS *or* MORE THAN	Qualitative increase
REVERSE	Logical opposite of the intent
OTHER THAN	Complete substitution

Table 5.11 Common HAZOP Study process parameters

Flow	Time	Frequency	Mixing
Pressure	Composition	Viscosity	Addition
Temperature	pH	Voltage	Separation
Level	Speed	Information	Reaction

In the original ICI approach, each guide word is combined with relevant process parameters and applied at each point (study node, process section, or operating step) in the process that is being examined. The following is an example of creating deviations using guide words and process parameters:

Guide Words		**Parameter**		**Deviation**
NO	+	FLOW	=	NO FLOW
MORE	+	PRESSURE	=	HIGH PRESSURE
AS WELL AS	+	ONE PHASE	=	TWO PHASE
OTHER THAN	+	OPERATION	=	MAINTENANCE

Guide words are applied to both the more general parameters (e.g., react, mix) and the more specific parameters (e.g., pressure, temperature). With the general parameters, it is not unusual to have more than one deviation from the application of one guide word. For example, "more reaction" could mean either that a reaction takes place at a faster rate, or that a greater quantity of product results. On the other hand, some combinations of guide words and parameter will yield no sensible deviation (e.g., "as well as" with "pressure").

With the specific parameters, some modification of the guide words may be necessary. In addition, analysts often find that some potential deviations are irrelevant because of a physical limitation. For example, if temperature parameters are being considered, the guide words "more" or "less" may be the only possibilities.

The following are other useful alternative interpretations of the original guide words:

- Sooner or Later for OTHER THAN when considering time
- Where Else for OTHER THAN when considering position, sources, or destination
- Higher and Lower for MORE and LESS when considering levels, temperature, or pressure

When dealing with a design intention involving a complex set of interrelated plant parameters (e.g., temperature, reaction rate, composition, and pressure), it may be better to apply the whole sequence of guide words to each parameter individually than to apply each guide word across all of the parameters as a group. Also, when applying the guide words to an operating instruction (e.g., procedural step), it may be more useful to apply the sequence of guide words to each word or phrase separately, starting with the key part that describes the activity. These parts of the sentence usually are related to some impact on the process parameters. For example, in the procedural step "The operator starts flow A when pressure B is reached," the guide words would be applied to:

- Starts flow A (no, more, less, etc.)
- When pressure B is reached (sooner, later, etc.)

The guide-word-based HAZOP Study method is the originally defined HAZOP technique. However, several variations of this basic method have been developed. These variations will be discussed later in this section *(HAZOP Study Variations)*. In many situations, these variations may be more effective than the original guide-word approach.

Types of Results

The results of a HAZOP Study are the team's findings, which include identification of hazards and operating problems; recommendation's for changes in design, procedures, etc., to improve the system; and recommendations to conduct studies of areas where no conclusion was possible due to a lack of information. The results of team discussions concerning the causes, effects, and safeguards for deviations for each node or section of the process are recorded in a column-format table. Worked examples showing HAZOP Study results, including in the contexts of cyclic reviews and batch process evaluations, can be found in Part II, Chapters 15, 19 and 20.

Resource Requirements

The HAZOP Study requires accurate, up-to-date P&IDs or equivalent drawings, and other detailed process information, such as operating procedures. A HAZOP Study also requires considerable knowledge of the process, instrumentation, and operation; this information is usually provided by team members who are experts in these areas. Trained and experienced leaders are an essential part of an efficient, high quality HAZOP.

The HAZOP team for a large, complex process may consist of five to seven people with a variety of experience: design, engineering, operations maintenance, and so forth. One team member leads the analysis and another (the scribe) typically records the results of the team's deliberations. Experienced facilitators sometimes also serve as the scribe. For a simple process or in a limited scope review, a team can have as few as three or four people if the people have the necessary technical skills and experience. Table 5.12 lists estimates of the time needed to perform a hazard evaluation using the HAZOP Study technique.

Analysis Procedure

The concepts presented above are put into practice in the following steps: (1) preparing for the review, (2) performing the review, and (3) documenting the results. Figure 5.3 illustrates the concept of the HAZOP Study technique. It should be noted that the activities listed as "Follow-up" in Figure 5.3 are not actually part of the HAZOP Study methodology, and are not necessarily the responsibility of the HAZOP Study team.

It is important to recognize that some of these steps can take place concurrently. For example, in some cases the team may review the design, record the findings, and perform follow-up over the same period of several weeks or months. Nonetheless, the steps are discussed separately as though they are executed one at a time.

Preparing for the review. Chapter 2 describes the various tasks that hazard evaluation team leaders must perform to prepare for hazard evaluations. This section amplifies some of these items because of their importance to the success of a HAZOP Study. The amount of preparation depends upon the size and complexity of the process being analyzed.

Table 5.12 Time estimates for using the HAZOP Study technique

Scope	Preparation*	Evaluation	Documentation**
Small system	8 to 12 hours	2 to 6 days	2 to 6 days
Large process	2 to 4 days	2 to 6 weeks	2 to 6 weeks

* Primarily team leader and scribe, although others may work some during this phase

** Team leader and scribe only; may be lower for experienced scribes using computer software in the team meetings

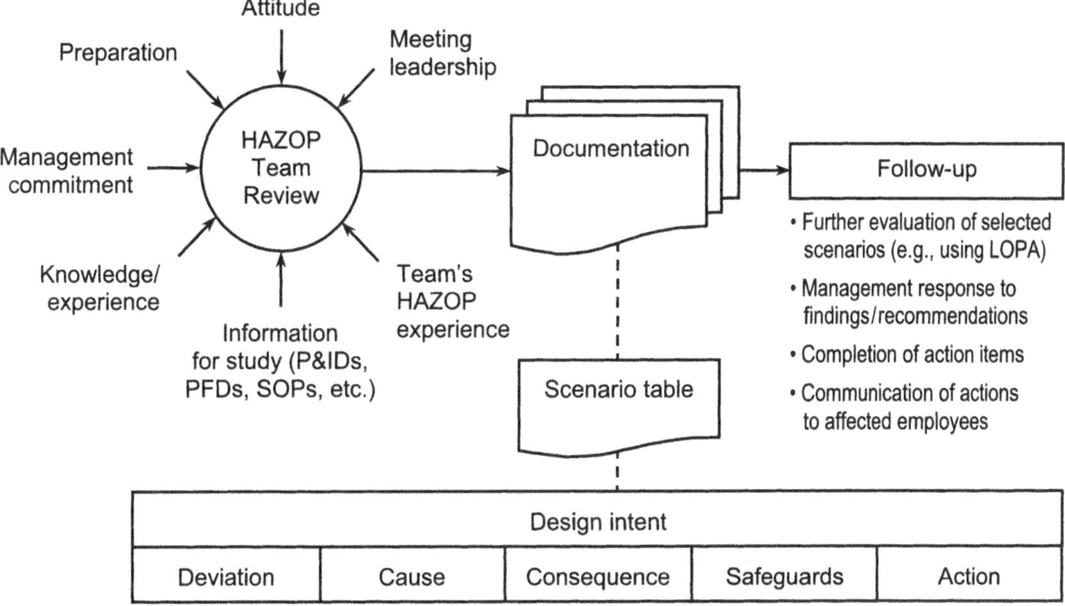

Figure 5.3 Overview of the HAZOP Study technique *(see text regarding Follow-up items)*

Define the purpose, objectives, and scope of the study. The purpose, objectives, and scope of the study should be made as explicit as possible. The objectives are normally set by the person who is responsible for the plant or project; this person is assisted by the HAZOP Study leader. It is important that people work together to provide the proper direction and focus for the study. It is also important to define what specific consequences are to be considered. For example, a HAZOP Study might be conducted to determine where to build a plant to have the minimal impact on public safety. In this case, the HAZOP Study should focus on deviations that result in off-site effects.

Select the team. The hazard evaluation team leader should ensure the availability of an adequately sized and skilled HAZOP Study team. A HAZOP Study team, at a minimum, should consist of a person or persons serving as the leader and scribe, a person with experience operating the actual facility being studied or one as similar as possible to it, and at least one other individual who has an understanding of the intended, engineered design and operation of the process. Ideally, the team consists of five to seven members, although a smaller team could be sufficient for a simpler, less hazardous plant. If the team is too large, the group approach can be difficult. On the other hand, if the group is too small, it may lack the breadth of knowledge needed to assure thoroughness. Section 2.6 provides more detail on team composition.

Obtain the necessary data. Typically, the data consist of various drawings in the form of P&IDs, flowsheets, and plant layout schematics. Additionally, there may be operating instructions, instrument sequence control charts, logic diagrams, and computer programs. Occasionally, there may be plant manuals and equipment manufacturers' manuals. Important drawings and data should be provided to HAZOP team members well before the meetings.

Convert the data into a suitable form and plan the study sequence. The amount of work required in this stage depends on the type of process. With continuous processes, preparation can be minimal. Study nodes or process sections may be identified before the meetings using up-to-date flowsheets and P&IDs. Sufficient copies of each drawing should be available for team members to see during the meeting(s).

Sometimes, team leaders may also develop a preliminary list of deviations to be considered in the meeting and prepare a worksheet on which to record the team's responses. However, the leader should avoid using a previously assembled list as the only deviations to be considered. This could stifle the creative synergism of the team when identifying process hazards and could result in missing some hazardous deviations due to complacency. It is to be expected that, due to the learning process that accompanies the study, some changes will be made as the study progresses.

With batch processes, preparation is usually more extensive, primarily due to the more complicated operations and procedures (e.g., heat the reactor in step 3, cool the reactor in step 8). Analyzing procedures is a large part of the HAZOP Study for batch processes (see Section 9.1). In some circumstances (e.g., when two or more batches of material are being processed at the same time), it may be necessary to prepare a display indicating the status of each vessel at each step of the process. If operators are physically involved in the process (e.g., in charging vessels rather than simply controlling the process), their activities should be represented by flow charts.

To ensure the team approaches the plant and its operation methodically, the leader will usually prepare a plan before the study begins. This means the team leader must spend some time before the meetings to determine the "best" study sequence, based on how the specific plant is operated.

Arrange the necessary meetings. Once the data and drawings have been assembled, the team leader is in a position to plan the review meetings. The first requirement is to estimate the meeting time needed for the study. As a general rule, each process section or study node will take an average of 20-30 minutes. For example, a vessel with two inlets, two exits, and a vent should take about three hours. Thus, a leader can estimate the HAZOP meeting time required by considering the number of process sections or nodes. Another way to make a rough estimate is to allow about two to three hours for each major piece of equipment. Fifteen minutes should also be allowed for each simple verbal statement in operating procedures, such as "switch on pump," "motor starts," or "pump starts." After estimating the meeting time required, the team leader can arrange the review meetings. Ideally, each session should last no more than four to six hours (preferably in the morning). Longer sessions are undesirable because the team's effectiveness usually begins to diminish. In extreme cases, sessions may be held on consecutive days with longer hours, but such a program should be attempted only in exceptional circumstances (e.g., when team members are from out of town and cannot travel to the meeting site every day).

With large projects, one team may not be able to analyze all of the subject processes within the allotted time; it may be necessary to use several teams and team leaders (one of the team leaders should act as a coordinator). The coordinator will divide the processes into logical sets, allocate portions of the process to different teams, and prepare schedules for the study as a whole.

Performing the Review. The HAZOP Study technique requires that a process drawing or procedure be divided into study nodes, process sections, or operating steps and that the hazards of the process be addressed using the guide words. Figure 5.4 illustrates the typical flow of activities in a HAZOP meeting. As the team applies all of the relevant guide words to each process section or step, they record either (1) the deviation with its causes, consequences, safeguards, and actions, or (2) the need for more complete information to evaluate the deviation. As hazardous situations are detected, the team leader should make sure that everyone understands them. As mentioned in Chapter 2, it is important for the HAZOP team leader to control the degree of problem solving that occurs during the team meetings. To minimize inappropriate problem solving, the leader can:

- Complete the study of one process deviation and associated suggested actions before proceeding to the next deviation
- Evaluate all hazards associated with a process section before considering suggested actions for improving safety.

In practice, HAZOP leaders should strike a compromise, allowing the team enough time to consider solutions that are easy to resolve, yet not allowing the team to spend too much time "designing solutions." It may not be appropriate, or even possible, for a team to find a solution during a meeting. On the other hand, if the solution is straightforward, a specific recommendation should be recorded immediately. To ensure effective meetings, the team leader must keep several factors in mind: (1) do not compete with the members; (2) take care to listen to all of the members; (3) during meetings, do not permit anyone to be put on the defensive; and (4) keep the energy level high by taking breaks as needed.

Although the team leader will have prepared for the study, the HAZOP technique may expose gaps in the available plant operating information or in the knowledge of the team members. Sometimes calling a specialist for information on some aspect of plant operation or deciding to postpone certain parts of the study to obtain more information may be necessary.

Documenting the Results. The recording process is an important part of the HAZOP Study. The person assigned to scribe the meetings must be able to distill the pertinent results from the myriad of conversations that occur during the meetings. It is impossible to manually record all that is said during the meetings, yet it is very important that all important ideas are preserved. Some analysts may decide to minimize their documentation effort by not pursuing (and not documenting) the causes of deviations for which there are no significant safety consequences. It may be helpful to have the team members review the final report and reconvene for a report review meeting. Reviewing key issues will often fine-tune the findings and uncover other problems. Normally, the results of HAZOP meetings are recorded in a tabular format (Table 5.13); however, action items may be recorded separately.

Table 5.13 Typical format for a HAZOP Study worksheet

Team: HAZOP Team #3 Meeting Date: ##/##/##				Drawing Number: 70-0BP-57100 (Figure 5.5) Revision Number: 3	
Item	Deviation	Causes	Consequences	Safeguards	Actions
Study node, process section, or operating step description. Definition of design intention.					
1.1					

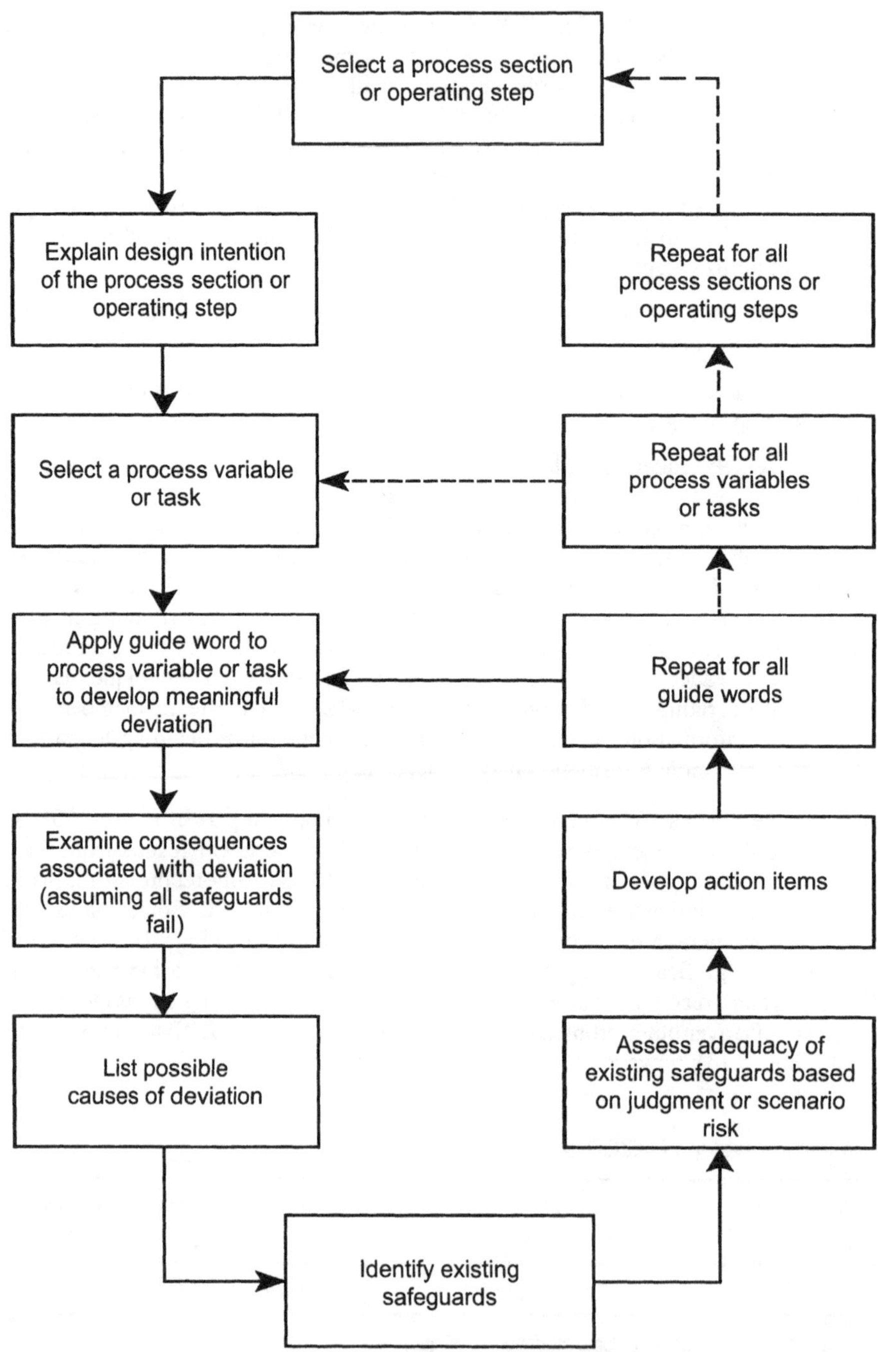

Figure 5.4 HAZOP Study method flow diagram

HAZOP Study Variations

As mentioned at the beginning of this section, the original guide-word-based HAZOP approach was developed by ICI. Over the years many organizations have modified the approach to suit their special needs.[8] Also, a number of novel improvements have been suggested, and some are in routine practice. These various industry-, company- or facility-specific approaches may be quite appropriate in the applications for which they are intended. The purpose of this section is to outline relatively common variations of the HAZOP Study technique that preserve the essential elements of the original approach.

There are two basic variations of the guide-word HAZOP Study approach:

- Different ways to identify deviations for consideration by the HAZOP team
- Different ways to document the HAZOP team meeting results

Other variations involving ways of ranking the HAZOP results also exist. However, these are not discussed here; see Section 8.2 concerning alternatives for prioritizing the results of hazard evaluations. The following sections outline several alternatives that organizations are using in each of the two categories listed above.

Identifying relevant process deviations. One of the strengths of the guide-word approach is its ability to exhaustively identify process deviations for each study node, process section, or operating step. Systematically applying the various guide words to a complete list of process parameters in a "brute force" fashion can lead to the development of an overwhelming list of deviations for the team to evaluate. The advantage of this approach is that an inexperienced HAZOP leader and team can be fairly confident that they have considered all of the ways that a process can malfunction. The disadvantage is that an experienced team may be burdened by this rigorous, yet ponderous, approach if many of the guide-word-based deviations lead to "dead ends" from a hazard significance standpoint.

Two alternatives have emerged through practice as efficient ways for relatively experienced leaders and teams to generate the list of deviations they consider for each study node, process section, or operating step: the ***library-based approach*** and the ***knowledge-based approach***. Both variations seek to increase the efficiency of the team meetings by minimizing the time spent in identifying the causes, effects, and safeguards of deviations that would obviously not result in an effect of interest.

The ***library-based approach*** is the most closely related alternative to the original guide-word approach. Before the HAZOP team meeting, the HAZOP leader or scribe should survey a standard library of potential deviations to determine which ones are relevant for each node, section, or step. Each node, section, or step usually focuses on a major piece of equipment. Depending upon the type of equipment involved (e.g., reactor, column, pump, tank, piping, heat exchanger), some potential deviations will not be relevant. For example, high level in a normally liquid-filled vessel would not be expected to create any significant hazard.

In addition, depending upon how skilled the HAZOP leader or scribe is at dividing the process into nodes or sections, some hazardous deviations may be omitted from consideration if the analyst believes that evaluating upstream or downstream sections will identify these effects. For example, consider three process sections involving a feed line, a set of redundant centrifugal pumps, and a discharged line leading to a reactor. There is nothing gained by a team considering "High Flow," or other such deviations in the pump section if this same deviation and its effects have been considered elsewhere. The location of the effect of "High Flow" would normally be at the reactor. Since there are no credibly different causes and effects of "High Flow" in the pump section itself, time can be saved if the team does not have to reconsider the causes and effects of deviations in sections that are otherwise identified in upstream or downstream sections.

Table 5.14 is an example of a library that HAZOP leaders and scribes could use to select relevant deviations of interest before the HAZOP meetings.[9] Using this approach can save as much as 20% to 50% in meeting time, depending upon the specific process, hazards of interest, and types of equipment involved. However, analysts choosing this variation should recognize the potential for inadvertently missing some deviations because the team did not help create the library.

The ***knowledge-based approach*** is a specialization of the guide-word HAZOP Study in which the guide words are supplemented or partially replaced by both the team's knowledge and specific checklists. This knowledge base is used to compare the design to well-established basic design practices that have been developed and documented from previous plant experience. The implicit premise of this version of the HAZOP Study technique is that the organization has extensive design standards that the team members are familiar with. An important advantage of this method is that the lessons learned over many years of experience are incorporated into the company's practices and are available for use at all stages in the plant's design and construction. Thus, the knowledge-based HAZOP Study can help ensure that the company's practices, and therefore its experience, have indeed been incorporated in the design.

Comparing a process design to codes and practices will generate additional questions that are different from the guide-word HAZOP Study deviations. For example, questions might be:

- "Shouldn't the design be like...?"
- "Will the proposed design change control the hazard so that the risk of operating the process does not increase?"

As a more specific example, consider the discharge from a centrifugal pump. The guide-word HAZOP approach would apply the guide word "Reverse" to identify the need for a check valve. The knowledge-based HAZOP approach might also identify the need for a check valve because an actual problem was experienced with reverse flow, and the use of check valves on a centrifugal pump discharge has been adopted as a standard practice.

Sometimes a checklist can be used to help a team develop deviations for a particular study. Appendix B contains a comprehensive checklist that readers can use to supplement their knowledge when leading a HAZOP team.

It is also important to note that the experience-based checklists used in this variation of HAZOP may be of little value compared to the guide-word approach when portions of the process involve major changes in equipment technology or involve new chemistry.

Inherent Safety Reviews. Inherent safety reviews can be conducted as separate studies using a form of HAZOP Study, or they can be incorporated into HAZOP Studies conducted for other purposes. The inherent safety strategies of substitute, minimize, moderate, and simplify can be used as HAZOP guide words (Table 5.15), or similar guide words can be employed. Besides the use of these additional guide words, the HAZOP Study would proceed as usual. The guide words can either be applied node by node or to larger sections of the process. Another approach utilizes the typical HAZOP guide words to identify scenarios and evaluate the adequacy of existing safeguards, but team discussions for scenarios where further risk control is warranted might be structured to intentionally focus on whether opportunities exist to substitute, minimize, moderate, or simplify. Section 3.7 describes the concept of an inherent safety review in more detail; see also References 12 and 13.

Table 5.14 Example library of relevant deviations for process section types

	Process section type*				
Deviation	Column	Tank/vessel	Line	Heat exchanger**	Pump**
High flow			X		
Low/no flow			X		
High level	X	X			
Low level	X	X			
High interface		X			
Low interface		X			
High pressure	X	X	X		
Low pressure	X	X	X		
High temperature	X	X	X		
Low temperature	X	X	X		
High concentration	X	X	X		
Low concentration	X	X	X		
Reverse/misdirected flow			X		
Tube leak				X	
Tube rupture				X	
Leak	X	X	X	X	X
Rupture	X	X	X	X	X

* This library was developed for a specific strategy for defining process sections. Readers are cautioned not to blindly use this typical library or any other library without carefully reviewing it for relevance and completeness.

** This library assumes that other deviations (e.g., low flow, high temperature) are considered when examining the line downstream of these equipment items.

Table 5.15 Inherent safety strategies as HAZOP Study guide words

Guide word	Description
Substitute	Replace a substance with a less hazardous material or processing route with one that does not involve hazardous material. Replace a procedure with one that eliminates or presents less of a hazard.
Minimize	Use smaller quantities of hazardous materials when the use of such materials cannot be avoided. Perform a hazardous procedure as few times as possible when the procedure is unavoidable.
Moderate	Use hazardous materials in their least hazardous forms or identify processing options that involve less severe processing conditions.
Simplify	Design processes, processing equipment, and procedures to eliminate opportunities for errors; e.g., by removing the possibility that two chemicals that may be incompatible under certain conditions can react unfavorably, such that the chemicals will not be together in the process. Designing equipment that cannot be overpressured by process conditions.

Documentation Options. Many organizations have developed specialized ways of documenting the results of a HAZOP Study team meeting. The following are a few of the more common variations known to the authors.

Cause-by-cause HAZOP table. In the cause-by-cause (CBC) approach, the table explicitly correlates the consequences, safeguards and actions to each particular cause of a deviation. A team may identify as many causes of a deviation as are appropriate, and every cause will have an independent set of consequences and safeguards related to it. For example, consider the deviation "pump leak." If the first initiating cause of that deviation is seal failure, the table would list all consequences, safeguards and actions related to seal failures. The team would then proceed to the next cause, which might be pump casing flange failure. Thus, by its very nature, CBC is more precise in the treatment of data than the deviation-by-deviation (DBD) approach described below. CBC may reduce ambiguity in some instances. For example, a malfunctioning level controller may cause high level, but that same level controller may be a safeguard against high feed flow causing high level. Displaying the same item as both a cause and a safeguard, as in DBD, can be confusing. If data for a particular table are potentially confusing, or if personal or company needs require that explicit safeguards be clearly defined for each cause, a hazard evaluation team should consider using CBC for its documentation approach.

Deviation-by-deviation HAZOP table. In the deviation-by-deviation (DBD) approach, all causes, consequences, safeguards, and actions are related to a particular deviation. However, no correlation between individual causes, consequences, and safeguards for that deviation is expressed. Thus, all causes listed for a deviation do not necessarily result in all of the listed consequences, and specific cause/consequence/safeguard/action relationships are not explicitly identified. For example, high steam flow and external fire may both cause high temperature, but a sprinkler system is only an effective safeguard against the fire. Documenting a table using the DBD approach assumes that the correlation(s) among causes, consequences, safeguards, and actions can be inferred by anyone reading the HAZOP table. The DBD documentation approach is widely used because the table construction requires less time and its length is greatly reduced compared with the CBC approach.

Exception-only HAZOP table. In this approach, the table includes only those deviations for which the team believes there are credible causes and significant effects. The advantage of this approach is that the resulting HAZOP meeting time and table length are greatly reduced. A major disadvantage is that it is almost impossible to audit such an analysis for completeness. This is especially important if the report could be subject to the scrutiny of a regulatory agency or corporate audit team. Hence, documentation by exception would be appropriate primarily for a study such as a preliminary design review. The exception-only approach can be used with either the DBD or CBC format.

Action item-only HAZOP table. In this variation, only the suggestions that a team makes for safety improvements are recorded. These action items can then be passed on for risk management decision-making purposes. The advantage of this variation is that it can save meeting time and documentation time outside the meeting, since no detailed table is prepared as a result of the team meetings. The obvious disadvantage is that there is no documentation to audit the analysis for quality assurance, revalidation, management of change, or other purposes. Documenting just the HAZOP Study action items would only be appropriate for purposes such as preliminary reviews.

Example

Using the DAP reaction system presented in the Checklist Analysis example, a guide-word HAZOP Study is performed to address personnel hazards. The DAP process schematic is repeated here as Figure 5.5; readers should refer to Section 4.4 for the description of the DAP process. The team leader applies the guide words to the process parameters using the deviation-by-deviation approach described above. A sample guide word application is presented for the phosphoric acid solution line to the DAP reactor.

Process Section: Phosphoric acid feed line to the DAP reactor

Design Intention: Feed phosphoric acid at a controlled rate to the DAP reactor

Guide Word: No **Process Parameter**: Flow

Deviation: No Flow

Consequences:

(1) Excess ammonia in the reactor, leading to...
(2) Unreacted ammonia in the DAP storage tank, with subsequent...
(3) Release of unreacted ammonia from the DAP storage tank to the enclosed work area, with possible personnel injury
(4) Loss of DAP production

Causes:
(1) No feed material in the phosphoric acid storage tank
(2) Flow indicator/controller fails high
(3) Operator sets the flow controller too low
(4) Phosphoric acid control valve B fails closed
(5) Plugging of the line
(6) Leak or rupture of the line

Safeguards:
(1) Periodic maintenance of valve B

Actions:
(1) Consider adding an alarm/shutdown of the system for low phosphoric acid flow to the reactor
(2) Ensure that periodic maintenance and inspection for valve B are adequate
(3) Consider using a closed tank for DAP storage

This process is repeated with other combinations of guide words and process parameters for each section of the design. Every process section is evaluated, and the relevant information is recorded in a HAZOP Study table. The resulting HAZOP Study table, showing only a few selected sections and deviations, is presented in Table 5.16.

Section 5.3 References

1. *Hazard and Operability Studies, Process Safety Report 2*, Imperial Chemical Industries Limited, London, 1974.
2. *A Guide to Hazard and Operability Studies*, Chemical Industries Association, Alembic House, London, 1977.
3. R. E. Knowlton, *Hazard and Operability Studies, The Guide Word Approach,* Chemetics International Company, Vancouver, British Columbia, 1981.
4. H. R. Greenberg and J. J. Cramer (eds.), *Risk Assessment and Risk Management for the Chemical Process Industry*, ISBN 0-442-23438-4, Van Nostrand Reinhold, New York, 1991.
5. "HAZOP Studies and other PHA Techniques for Process Safety and Risk Management," AIChE/ASME Professional and Technical Training Courses, American Society of Mechanical Engineers, New York, 2007.
6. R. Dupont et al., *Accident and Emergency Management*, ISBN 978-0-471-18804-9, John Wiley & Sons, Inc., New York, 1991.
7. W. E. Bridges et al., "Integrating Human Reliability Analysis with Process Hazard Evaluations," *International Conference on Hazard Identification and Risk Analysis, Human Factors and Human Reliability in Process Safety,* Center for Chemical Process Safety, Orlando, FL, January, 1992.
8. K. A. Ford and W. H. Brown, "Innovative Applications of the HAZOP Technique," presented at the AIChE Spring National Meeting, Orlando, March 1990.
9. *Technical Specifications for Performing a HAZOP Analysis*, JBF Associates, Inc., Knoxville, TN, 1990.
10. S. Mannan, ed., *Lees' Loss Prevention in the Process Industries, 3rd Ed.,* Elsevier Butterworth-Heinemann, ISBN 0-7506-7555-1, Oxford, UK, 2005.
11. F. Crawley, M. Preston and B. Tyler, *HAZOP: Guide to Best Practice,* ISBN 0-85295-427-1, Institution of Chemical Engineers, Rugby, UK, 2000.
12. P. Amyotte et al., "Incorporation of Inherent Safety Principles in Process Safety Management," *Proceedings of 21st Annual CCPS International Conference – Process Safety Challenges in a Global Economy,* American Institute of Chemical Engineers, April 2006.
13. Center for Chemical Process Safety, *Inherently Safer Chemical Processes: A Life Cycle Approach, 2nd Ed.,* American Institute of Chemical Engineers, New York, 2007.

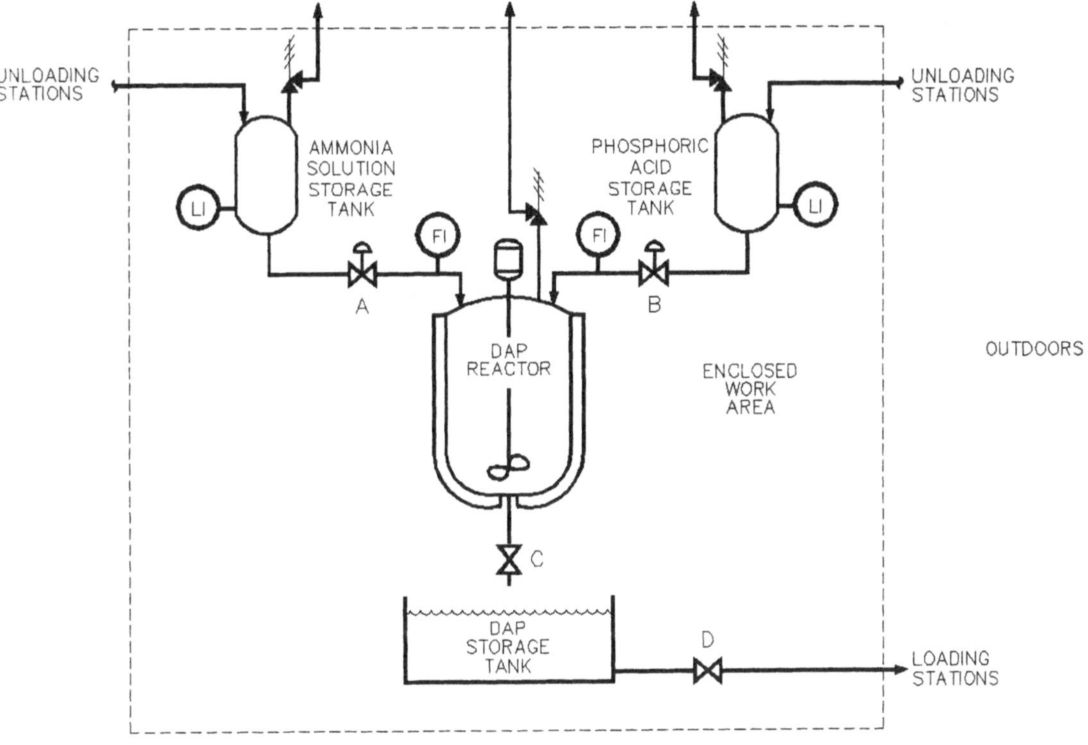

Figure 5.5 DAP process schematic for HAZOP Study example

Table 5.16 Sample deviations from the HAZOP Study table for the DAP process example

Team: HAZOP Team #3 Meeting date: ##/##/##			Drawing number: 70-0BP-57100 (Figure 5.5) Rev. 3 Study approach: Deviation by deviation		
Item	Deviation	Causes	Consequences	Safeguards	Actions
1.0 Vessel – Ammonia solution storage tank. Intent – Contain, at ambient temperature and atmospheric pressure, an inventory of 20% ammonium hydroxide solution (ammonia solution) corresponding to a tank level between 10% and 80% full.					
1.1	High level (>80%)	Unloading ammonia solution from the unloading station without adequate space in the storage tank Ammonia storage tank level indicator fails low	Potential release of ammonia vapors to the atmosphere	Level indicator on the storage tank Ammonia storage tank relief valve to the atmosphere	Review ammonia unloading procedures to ensure adequate space in tank before unloading Consider sending the relief valve discharge to a scrubber Consider adding an independent high level alarm for the ammonia storage tank

Table 5.16 Sample deviations from the HAZOP Study table for the DAP process example (continued)

Item	Deviation	Causes	Consequences	Safeguards	Actions
Team: HAZOP Team #3 Meeting date: ##/##/##				Drawing number: 70-0BP-57100 (Figure 5.5) Rev. 3 Study approach: Deviation by deviation	
2.0 Line – Ammonia feed line to the DAP reactor. Intent – Deliver 20% ammonia solution to the reactor at y gpm and z psig.					
2.1	High flow (>y gpm)	Ammonia feed line control valve A fails open Flow indicator fails low Operator sets ammonia flow rate too high	Unreacted ammonia solution carryover to the DAP storage tank and release to the work area	Periodic maintenance of valve A Ammonia detector and alarm	Consider adding an alarm/shutdown of the system for high ammonia solution flow to the reactor Ensure periodic maintenance and inspection for valve A is adequate Ensure adequate ventilation exists for enclosed work area and/or consider using an enclosed DAP storage tank
2.9	Leak	Corrosion Erosion External impacts Gasket and packing failures Maintenance errors	Small, continuous leak of ammonia to the enclosed work area	Periodic maintenance of line Periodic inspection tours by operator in the DAP process area	Ensure adequate ventilation exists for enclosed work area
3.0 Vessel – Phosphoric acid solution storage tank. Intent – Contain, at ambient temperature and atmospheric pressure, an inventory of 85% phosphoric acid feed solution corresponding to a tank level between 10% and 85% full.					
3.7	Low concen-tration of phosphoric acid (<85%)	Low phosphoric acid concentration supplied by the vendor Error in charging phosphoric acid to the supply tank	Unreacted ammonia in the reactor carried over to the DAP storage tank and released to the enclosed work area	Acid unloading and transfer procedure Ammonia detector and alarm	Ensure existence of adequate material handling and receiving procedures and labeling Consider verifying the phosphoric acid concentration in the storage tank before operation Ensure adequate ventilation exists for enclosed work area and/or consider using an enclosed DAP storage tank
4.0 Line – Phosphoric acid feed line to the DAP reactor. Intent – Deliver 85% phosphoric acid feed solution to the reactor at a rate of x gpm and y psig.					

Table 5.16 Sample deviations from the HAZOP Study table for the DAP process example (continued)

Team: HAZOP Team #3 Meeting date: ##/##/##			Drawing number: 70-0BP-57100 (Figure 5.5) Rev. 3 Study approach: Deviation by deviation		
Item	Deviation	Causes	Consequences	Safeguards	Actions
4.2	Low/no flow* (< x gpm)	No feed material in the phosphoric acid storage tank Flow indicator fails high Operator sets acid flow rate too low Phosphoric acid feed line control valve B fails closed Plugging of line Leak or rupture of line	Unreacted ammonia in the reactor carried over to the DAP storage tank and released to the enclosed work area	Periodic maintenance of valve B Ammonia detector and alarm	Consider adding an alarm/shutdown of the system for low phosphoric acid flow to the reactor Ensure periodic maintenance and inspection for valve B is adequate Ensure adequate ventilation exists for enclosed work area and/or consider using an enclosed DAP storage tank
5.0 Vessel - DAP reactor. Intent – Contain the complete reaction of ammonia solution and phosphoric acid at x °C and y psig by providing agitation and residence time.					
5.10	Loss of agitation	Agitator motor fails Agitator mechanical linkage fails Operator fails to activate the agitator	Unreacted ammonia in the reactor carried over to the DAP storage tank and released to the enclosed work area	Ammonia detector and alarm	Consider adding an alarm/ shutdown of the system for loss of agitation in the reactor Ensure adequate ventilation exists for enclosed work area and/or consider using an enclosed DAP storage tank
6.0 Line – DAP reactor outlet line to the DAP storage tank. Intent – Deliver product flow to the storage tank at x gpm and y psig.					
6.3	Reverse/mis-directed flow	No credible causes identified			
7.0 Vessel - DAP storage tank. Intent – Store, at ambient temperature and pressure, an inventory of DAP product corresponding to a 0% to 85% tank level.					
7.1	High level (>85%)	Excess flow from the reactor No flow to the loading stations	Operability problems caused by overfilling the DAP storage tank to the enclosed area (DAP does not pose an acute health hazard to personnel)	Operator observation of the DAP storage tank level	Consider adding a high level alarm for the DAP storage tank Consider providing a dike around the storage tank

* These deviations were discussed separately but were combined in this table

5.4 Failure Modes and Effects Analysis

A Failure Modes and Effects Analysis (FMEA) tabulates failure modes of equipment and their effects on a system or plant.

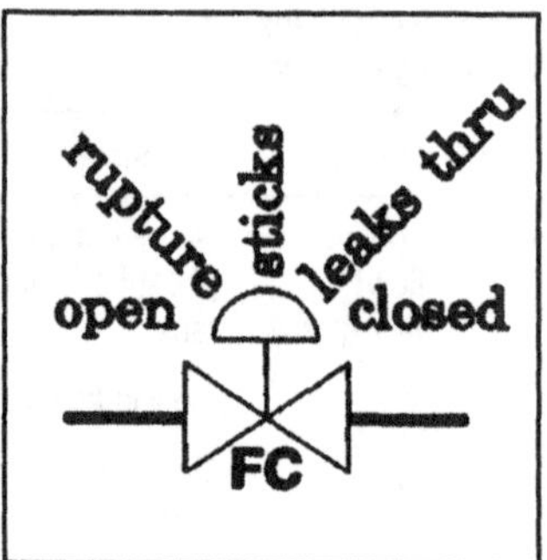

Purpose

The purpose of an FMEA is to identify single equipment and system failure modes and each failure mode's potential effect(s) on the system or plant. This analysis typically generates recommendations for increasing equipment reliability, thus improving process safety.

Description

The failure mode describes how equipment fails (open, closed, on, off, leaks, etc.). The effect of the failure mode is determined by the system's response to the equipment failure. An FMEA identifies single failure modes that either directly result in or contribute significantly to an incident. Human operator errors are usually not examined directly in an FMEA; however, the effects of inadequate design, improper installation, lack of maintenance, or improper operation are usually manifested as an equipment failure mode.

Failure Modes and Effects Analysis evaluates how equipment can fail (or be improperly operated) and the effects these failures can have on a process.[1-7] These failure descriptions provide analysts with a basis for determining where changes can be made to improve a system design. During a FMEA, hazard analysts describe potential consequences and relate them only to equipment failures; they rarely investigate damage or injury that could arise if the system operated successfully. An FMEA is not as efficient as other methods such as HAZOP Studies in identifying an exhaustive list of combinations of equipment failures that lead to incidents, since it examines all failure modes that result in safe outcomes as well as those that can lead to or contribute to loss events.

Each individual failure is considered as an independent occurrence, with no relation to other failures in the system, except for the subsequent effects that it might produce. However, in special circumstances, common cause failures of more than one system component may be considered.

The results of an FMEA are usually listed in tabular format, equipment item by equipment item. Generally, hazard analysts use FMEA as a qualitative technique, although it can be extended to give a priority ranking based on failure severity. See Section 8.2 for a description of the Failure Modes, Effects, and Criticality Analysis (FMECA) method.[8,9]

Types of Results

An FMEA generates a qualitative, systematic reference list of equipment, failure modes and effects. A worst-case estimate of consequences resulting from single failures is included. The FMEA may be easily updated for design changes or system/plant modifications. FMEA results are usually documented in a column-format table. Hazard analysts usually include suggestions for improving safety in appropriate items in the table. A worked example showing FMEA results is in Part II, Chapter 21.

Resource Requirements

Using the FMEA approach requires the following data and information sources: a system or plant equipment list or P&ID, knowledge of equipment function and failure modes, and knowledge of system or plant function and responses to equipment failures.

An FMEA can be performed by a single analyst, but such an analysis should be reviewed by others to help ensure completeness. Staff requirements will vary with the size and complexity of equipment items being analyzed. All analysts involved in the FMEA should be familiar with the equipment functions and failure modes and how the failures might affect other portions of the system or plant.

The time and cost of an FMEA is proportional to the size of the process and number of components analyzed. On the average, an hour is sufficient for analyzing two to four equipment items. As with any hazard evaluation of systems with similar equipment performing similar functions, the time requirements are reduced significantly due to the repetitive nature of the evaluations. Table 5.17 lists estimates of the time needed to perform a hazard evaluation using the FMEA technique.

Analysis Procedure

The FMEA procedure contains three steps: (1) defining the study problem, (2) performing the review, and (3) documenting the results.[10-13]

Defining the study problem. This step identifies the specific items to be included in the FMEA and the conditions under which they are analyzed. Defining the problem involves (1) establishing an appropriate level of resolution for the study and (2) defining the boundary conditions for the analysis. A detailed problem definition is a necessary ingredient to performing a thorough and efficient FMEA.

The level of resolution determines the extent of detail included in the FMEA. If a plant-level hazard is being addressed, the FMEA should focus on the failure modes of individual systems and their subsequent effects, with respect to the plant-level hazard. For example, the FMEA might focus on the plant's feed system, batch mixing system, oxidizing system, product separation system, and various support systems. When a system-level hazard is being addressed, the FMEA should focus on the failure modes and effects of the individual pieces of equipment that make up the system, while keeping in mind the effect on the overall system. For a system-level hazard, such as loss of temperature control in an oxidizing system, the FMEA might focus on the feed pump, cooling water pump, cooling water flow control valve, and temperature sensors and alarms that exist in the oxidizing system. Of course, effects identified at the system or equipment level may subsequently be related to potential plant-level hazards.

Table 5.17 Time estimates for using the FMEA technique

Scope	Preparation	Evaluation	Documentation*
Small system	2 to 6 hours	1 to 3 days	1 to 3 days
Large process	1 to 3 days	1 to 3 weeks	2 to 4 weeks

* May be lower using computer software

Defining the analysis boundary conditions includes:

- Identifying the plant and/or systems that are the subject of the analysis.

- Establishing the physical system boundaries for the FMEA. This includes the interfaces with other processes and utility/support systems. One way to indicate the physical system boundaries is to mark them on a system drawing that encompasses all equipment within the scope of the FMEA. These boundary conditions should also state the operating conditions at the interfaces.

- Establishing the system analytical boundaries, including: (1) the failure modes, operating consequences, causes, or existing safeguards that will not be considered and (2) the initial operating condition or position of equipment. As an example of effects beyond the scope of the study, an analyst may choose not to consider airplane crashes, earthquakes, or tornadoes as causes of failure modes. An example of an initial condition is specifying whether a valve is normally open or closed.

- Collecting up-to-date reference information that identifies the equipment and its functional relationship to the plant/system. This information is needed for all equipment included within the system boundary and appropriate interfaces with the rest of the plant.

Chapter 2 provides some additional advice about preparing for a hazard evaluation.

Performing the review. The FMEA should be performed in a deliberate, systematic manner to reduce the possibility of omissions and to enhance the completeness of the FMEA. One way to help ensure a thorough and efficient review is to develop a consistent format for recording the FMEA results. Having a standard FMEA table format helps make the information contained in the FMEA tables consistent and helps maintain the defined level of resolution. Table 5.18 shows an example format for an FMEA table. Additional information, such as failure mode causes, may be added to make it easier for the team to prioritize the results (see Section 8.2) for a particular application. An FMEA table can be produced by beginning at a system boundary on a reference drawing and systematically evaluating the items in the order that they appear in the process flow path. Each equipment item can then be checked off on the reference drawing or equipment list when its failure modes have been evaluated. All failure modes should be evaluated for each component or system addressed in the FMEA before proceeding to the next component. The following items should be standard entries in an FMEA table:

Equipment identification. A unique equipment identifier that relates the equipment to a system drawing, process, or location. This identifier distinguishes between similar pieces of equipment (e.g., two motor-operated valves) that perform different functions within the same system. Equipment numbers or identifiers from system drawings, such as P&IDs, are usually available and provide a reference to existing system information. Any systematic coding scheme is acceptable if the identifiers are (1) meaningful to the analysts who must work with the FMEA results and (2) traceable back to drawings or equipment lists for those who must later use the FMEA results.

Table 5.18 Typical format for an FMEA worksheet

DATE: PLANT: SYSTEM:					PAGE: of REFERENCE: ANALYST(S):	
Item	Identification	Description	Failure Mode	Effects	Safeguards	Actions

Equipment description. The equipment description should include the equipment type, operating configuration, and other service characteristics (such as high temperature, high pressure, or corrosive service) that may influence the failure modes and their effects. For example, a valve might be described as a "motor-operated valve, normally open, in a three-inch sulfuric acid line." These descriptions need not be unique for each piece of equipment.

Failure modes. The analyst should list ***all*** of the failure modes for each component that are consistent with the equipment description. Considering the equipment's normal operating condition, the analyst should consider all conceivable malfunctions that alter the equipment's normal operating state. For example, the failure modes of a normally closed valve may include:

- The valve sticks closed (or fails to open when required).
- The valve inadvertently moves to an open position.
- The valve leaks to the environment.
- The valve leaks internally.
- The valve body ruptures.

Table 5.19 contains additional examples of equipment failure modes. The analyst should include all postulated failure modes so that their effects can be addressed. The failure modes that would a directly result in a process deviation (abnormal situation) are initiating causes, and these failure modes can be used as the starting point for developing incident scenarios such as can be used in a Layer of Protection Analyses. If so, the Effects for these failure modes need to be developed all the way to a loss event as defined in Section 1.3. Other failure modes, such as a standby safety shutoff valve sticking open, represent safeguard failures. These are not initiating causes, since the process can continue to operate normally with the safety shutoff valve in the failed-open position. The "Effects" in this case could be documented as less protection against the loss event against which it is protecting. The effectiveness of the shutoff system should then be discussed in association with each initiating-cause failure mode to which it pertains.

Table 5.19 Examples of equipment failure modes used in an FMEA

Equipment description	Example failure modes
Pump, normally operating	▪ Fails on (fails to stop when required) ▪ Transfers off (stops when required to run) ▪ Seal leak/rupture ▪ Pump casing leak/rupture
Heat exchanger, high pressure on tube side	▪ Leak/rupture, tube side to shell side ▪ Leak/rupture, shell side to external environment ▪ Tube side, plugged ▪ Shell side, plugged ▪ Fouling

Effects. For each identified failure mode, the analyst should describe both the immediate effects of a failure at the failure location and the anticipated effects of the failure on other equipment, as well as on the overall system or process. For example, the immediate effect of a pump seal leak is a spill in the area of the pump. If the fluid is flammable, the spill could ignite because the pump is a potential ignition source. Subsequently, the fire may damage nearby equipment as well as threaten the safety of people in the area. The key to performing a consistent FMEA is ensuring that the effects of all equipment failures are analyzed using a common basis. Typically, analysts evaluate effects on a reasonable worst case basis, assuming that existing safeguards do not work. However, more optimistic assumptions may be satisfactory as long as all equipment failure modes are analyzed on the same basis.

Safeguards. For each identified failure mode, the analyst should describe any safety features or procedures associated with the system that can reduce the likelihood of a specific failure occurring or that can mitigate the consequences of the failure. For example, a high pressure shutdown for a reactor may reduce the likelihood of high pressure events damaging the reactor, while a properly sized relief valve may mitigate the consequences of any excessively high pressure events in the reactor. In this regard, the FMEA identifies safeguards and evaluates their effectiveness in the same manner as for a What-If Analysis, What-If/Checklist Analysis, or HAZOP Study.

Actions. For each identified failure mode, the analyst should list any suggested corrective actions for reducing the likelihood of effects associated with the failure mode. For instance, installation of a redundant high pressure alarm may be suggested for a reactor. The corrective actions for a particular piece of equipment may focus on the causes or effects of specific failure modes or may apply to all of the failure modes collectively.

Documenting the results. The documentation of the FMEA review is a systematic and consistent tabulation of the effects of equipment failures within a process or system. The equipment identification in the FMEA provides a direct reference between the equipment and system P&IDs or process flow diagrams.

Example

Using the DAP reaction system presented in the Checklist Analysis example (Section 4.4), an FMEA study is performed to address safety hazards to plant personnel. The DAP process schematic is repeated here as Figure 5.6; refer to Section 4.4 for the description of the DAP process. Each component of the reaction system is evaluated with the relevant information recorded in an FMEA table. The section of the FMEA table for Control Valve B in the phosphoric acid solution line is presented in Table 5.20.

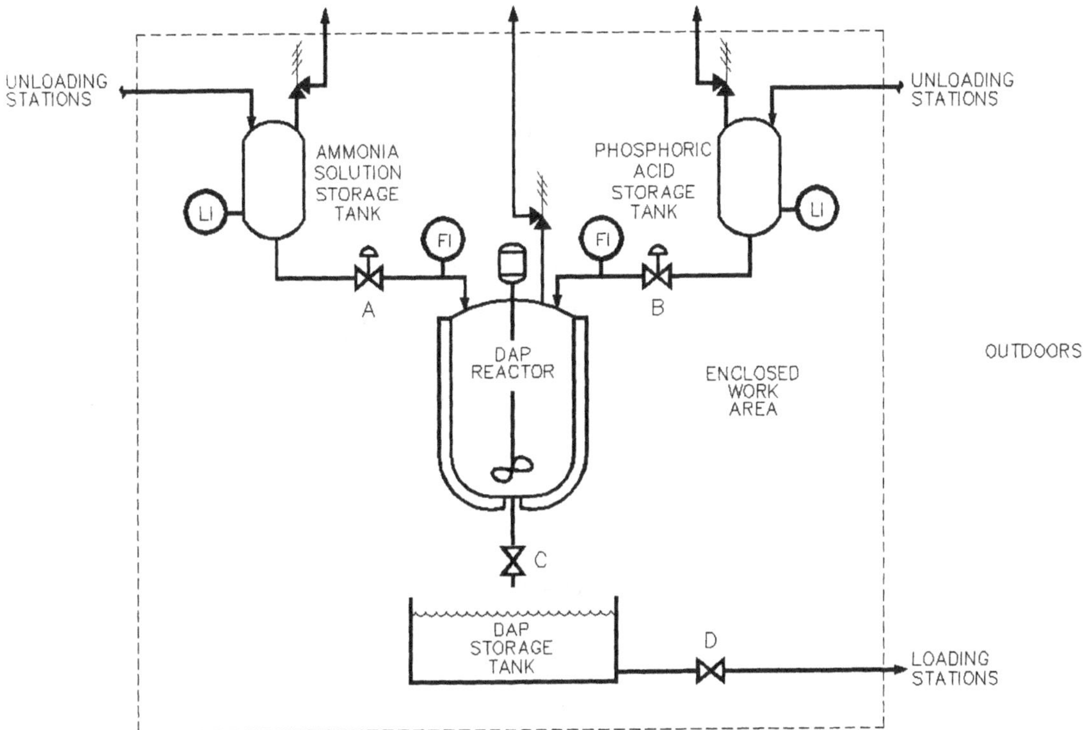

Figure 5.6 DAP process schematic for the FMEA example

Table 5.20 Sample page from the FMEA table for the DAP process example

DATE: ##/##/## PAGE: 5 of 20
PLANT: DAP Plant REFERENCE: Figure 5.6
SYSTEM: Reaction System ANALYST(S):

Item	Identification	Description	Failure Mode	Effects	Safeguards	Actions
4.1	Valve B on the phosphoric acid solution line	Motor-operated, normally open, phosphoric acid service	Fails open	Excess flow of phosphoric acid to the reactor High pressure and high temperature in the reactor if the ammonia feed rate is also high May cause a high level in the reactor or the DAP storage tank Off-specification production (i.e., high acid concentration)	Flow indicator in the phosphoric acid line Reactor relief valve vented to the atmosphere Operator observation of the DAP storage tank	Consider alarm/shutdown of the system for high phosphoric acid flow Consider alarm/shutdown of the system for high pressure and high temperature in the reactor Consider alarm/shutdown of the system for high level in the DAP storage tank
4.2	Valve B on the phosphoric acid solution line	Motor-operated, normally open, phosphoric acid service	Fails closed	No flow of phosphoric acid to the reactor Ammonia carry-over to the DAP storage tank and release to the enclosed work area	Flow indicator in the phosphoric acid line Ammonia detector and alarm	Consider alarm/shutdown of the system for low phosphoric acid flow Consider using a closed tank for DAP storage and/or ensure adequate ventilation of the enclosed work area
4.3	Valve B on the phosphoric acid solution line	Motor-operated, normally open, phosphoric acid service	Leak (external)	Small release of phosphoric acid to the enclosed work area	Periodic maintenance Valve designed for acid service	Verify periodic maintenance and inspection is adequate for this valve
4.4	Valve B on the phosphoric acid solution line	Motor-operated, normally open, phosphoric acid service	Rupture	Large release of phosphoric acid to the enclosed work area	Periodic maintenance Valve designed for acid service	Verify periodic maintenance and inspection is adequate for this valve

Section 5.4 References

1. H. R. Greenberg and J. J. Cramer (eds.), *Risk Assessment and Risk Management for the Chemical Process Industry*, ISBN 0-442-23438-4, Van Nostrand Reinhold, New York, 1991.
2. J. Stephenson, *System Safety 2000—A Practical Guide for Planning, Managing, and Conducting System Safety Programs*, ISBN 0-0442-23840-1, Van Nostrand Reinhold, New York, 1991.
3. E. J. Henley and H. Kumamoto, *Reliability Engineering and Risk Assessment*, Englewood Cliffs, New Jersey, Prentice-Hall, 1981.
4. *PRA Procedures Guide: A Guide to the Performance of Probabilistic Risk Assessments for Nuclear Power Plants,* NUREG/OR 2300, Section 3.6.1, January 1983.
5. K. B. Klaassen and J. E. L. van Peppen, *System Reliability, Concepts and Applications*, Edward Arnold, Routledge, Chapman and Hall, Inc., New York, 1989.
6. N. J. McCormick, *Reliability and Risk Analysis,* Academic Press, Inc., New York, 1981.
7. S. Mannan, ed., *Lees' Loss Prevention in the Process Industries, 3rd Ed.,* Elsevier Butterworth-Heinemann, ISBN 0-7506-7555-1, Oxford, UK, 2005.
8. Procedures for Performing a Failure Mode, Effects and Criticality Analysis, MIL-STD-1629A, Department of Defense, Washington, DC, November 1980.
9. W. E. Jordan, "Failure Modes, Effects and Criticality Analysis," proceedings of the Annual Reliability and Maintainability Symposium, San Francisco, Institute of Electrical and Electronics Engineers, New York, 1982.
10. Procedures for Performing a Failure Mode and Effect Analysis, MIL-STD-1629A, U.S. Navy, 1977.
11. W. Hammer, *Handbook of System and Product Safety,* Prentice Hall, Inc., New York, 1972.
12. R. I. Wagoner, *Hazards Analysis of Petroleum Systems (HAPS): An Adaptation of the Failure Modes and Effects Analysis (FMEA) Technique,* AREYE Corporation, Firendswood, TX, 1988.
13. H. E. Lambert, *Failure Modes and Effects Analysis,* NATO Advanced Study Institute, 1978.

5.5 Fault Tree Analysis

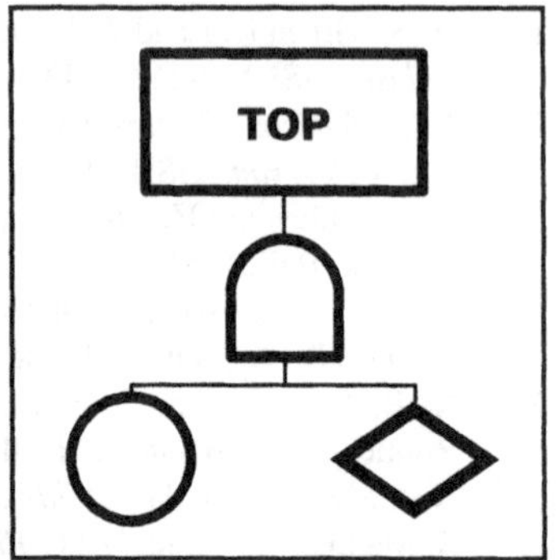

Fault Tree Analysis (FTA) is a deductive technique that focuses on one particular incident or main system failure, and provides a method for determining causes of that event.

Purpose

The purpose of an FTA is to identify combinations of equipment failures and human errors that can result in an incident. FTA is well suited for analyses of highly redundant systems. For systems particularly vulnerable to single failures that can lead to incidents, it is better to use a single-failure-oriented technique such as FMEA or HAZOP Study. FTA is often employed in situations where another hazard evaluation technique (e.g., HAZOP Study) has pinpointed an important incident of interest that requires more detailed analysis.

Description

The fault tree is a graphical model that displays the various combinations of equipment failures and human errors that can result in the main system failure of interest (called the Top event). The strength of FTA as a qualitative tool is its ability to identify the combinations of basic equipment failures and human errors that can lead to an incident. This allows the hazard analyst to focus preventive or mitigative measures on significant basic causes to reduce the likelihood of an incident.

A fault tree is a graphical model that illustrates combinations of failures that will cause one specific failure of interest, called a *Top event.*[1-6] Fault Tree Analysis is a deductive technique that uses Boolean logic symbols (i.e., AND gates, OR gates) to break down the causes of a Top event into basic equipment failures and human errors (called *basic events*). The analyst begins with an incident or undesirable event that is to be avoided and identifies the immediate causes of that event. Each of the immediate causes (called *fault events*) is further examined in the same manner until the analyst has identified the basic causes of each fault event or reaches the boundary established for the analysis. The resulting fault tree model displays the logical relationships between basic events and the selected Top event.

Top events are specific hazardous situations that are typically identified through the use of a more broad-brush hazard evaluation technique (e.g., What-If Analysis, HAZOP Study). A fault tree model can be used to generate a list of the failure combinations (failure modes) that can cause the Top event of interest. These failure modes are known as cut sets. A minimal cut set (MCS) is a smallest combination of component failures which, if they all occur or exist simultaneously, will cause the Top event to occur.[7,8] Such combinations are the "smallest" combinations in that all of the failures in a MCS must occur if the Top event is to occur as a result of that particular MCS. For example, a car will not operate if the cut set "no fuel" and "broken windshield" occurs. However, the MCS is "no fuel" because it alone can cause the Top event; the broken windshield has no bearing on the car's ability to operate. Sometimes analysts may include special conditions or circumstantial events in a fault tree model (e.g., the existence of a certain plant operating condition). Thus, a list of minimal cut sets represents the known ways the undesired consequence can occur, stated in terms of equipment failures, human errors, and associated circumstances.

Table 5.21 Logic and event symbols used in fault trees

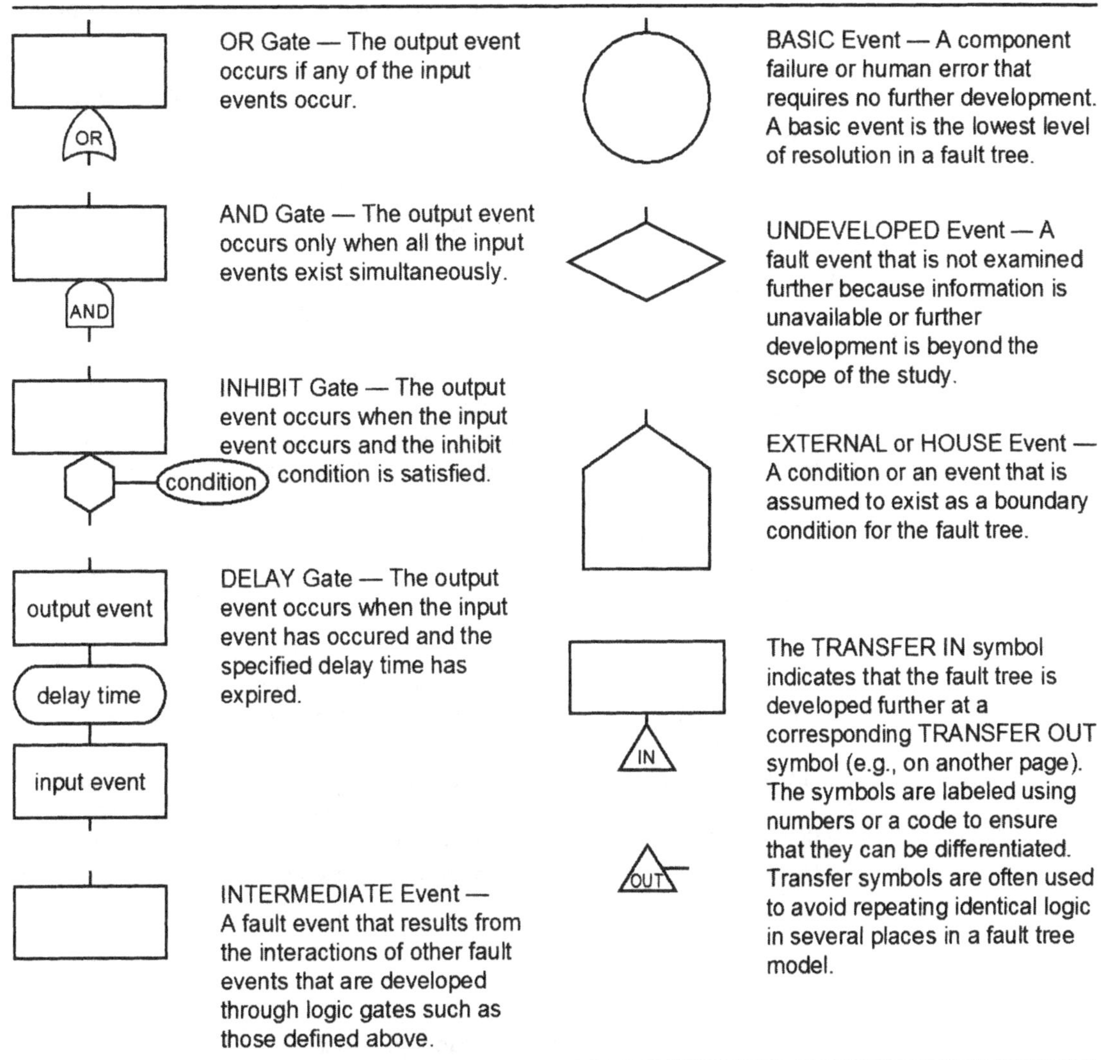

OR	OR Gate — The output event occurs if any of the input events occur.
AND	AND Gate — The output event occurs only when all the input events exist simultaneously.
condition	INHIBIT Gate — The output event occurs when the input event occurs and the inhibit condition is satisfied.
output event / delay time / input event	DELAY Gate — The output event occurs when the input event has occured and the specified delay time has expired.
	INTERMEDIATE Event — A fault event that results from the interactions of other fault events that are developed through logic gates such as those defined above.
	BASIC Event — A component failure or human error that requires no further development. A basic event is the lowest level of resolution in a fault tree.
	UNDEVELOPED Event — A fault event that is not examined further because information is unavailable or further development is beyond the scope of the study.
	EXTERNAL or HOUSE Event — A condition or an event that is assumed to exist as a boundary condition for the fault tree.
IN / OUT	The TRANSFER IN symbol indicates that the fault tree is developed further at a corresponding TRANSFER OUT symbol (e.g., on another page). The symbols are labeled using numbers or a code to ensure that they can be differentiated. Transfer symbols are often used to avoid repeating identical logic in several places in a fault tree model.

Logic and event symbols for Fault Tree Analysis. The fault tree is a graphical representation of the relationships between failures and a specific consequence. Table 5.21 lists standard symbols used in fault tree construction to display these relationships.[9-11]

Definitions. Fault events and basic events representing failures of equipment or humans (hereafter, both equipment and humans are referred to as components) can be divided into failures and faults. A component *failure* is a malfunction that requires the component to be repaired before it can successfully function again. For example, when a pump shaft breaks, it is classified as a component failure. A component *fault* is a malfunction that will "heal" itself once the conditions causing the malfunction are corrected. An example of a component fault is a switch whose contacts fail to operate because they are wet — when the contacts are dried they will operate properly.

Whether a component malfunction is classified as a fault or a failure, a basic assumption of Fault Tree Analysis is that all components are in either a failed state or a working state. Analysis of several degraded operating states is generally not practical. Analysts must define the conditions of failure and success for each event used in a fault tree model.

Faults and failures described in a fault tree can be grouped into three classes: (1) primary faults and failures, (2) secondary faults and failures, and (3) command faults and failures.[1,7,8] Primary faults and failures are malfunctions that occur when the component is operating in the environment for which its operation was intended. For example, when a pressure vessel ruptures at a pressure within its design limits due to a defective weld, this is a primary failure. Primary faults and failures (basic events) are usually attributable to a defect in the failed component and not to an external force or condition. Thus, components experiencing a primary fault or failure are assignable to their own malfunction.

Secondary faults and failures are equipment malfunctions that occur in an environment for which operation of the equipment was not intended. For example, when a pressure vessel ruptures because some other system fault or failure causes the internal pressure to exceed the vessel's design limits, this is a secondary failure. Secondary faults and failures are not the responsibility of the equipment that failed, but can be attributed to some external force or condition.

Command faults and failures are equipment malfunctions in which the component functions as it was designed; its operation is called a malfunction because the component function was not desired. For example, a temperature alarm failing to announce high temperature in a process because the temperature sensor has failed is a command fault for the alarm. The alarm experiences this fault because the sensor fails to command the alarm to sound when high temperature occurs; thus, no repair action is necessary for the alarm. Command faults and failures are not the responsibility of the equipment that fails, but the fault of the equipment that was supposed to command it.

All three classes of faults and failures will normally appear in a fault tree. One of the objectives of Fault Tree Analysis is to identify the basic events that result in an incident. These basic events are primary faults and failures that identify the component that is responsible for the failure. Secondary and command faults and failures are the intermediate fault events that are further defined by the primary faults and failures.

Types of Results

An FTA produces system failure logic models that use Boolean logic gates (i.e., AND, OR) to describe how equipment failures and human errors can combine to cause a main system failure. Many fault tree models may result from the analysis of a large process; the actual number of models depends on how selective the fault tree analyst was in choosing the Top event or events of concern. The analyst usually solves each logic model to generate a list of failures called *minimal cut sets* that can result in the Top event. These lists of minimal cut sets can be qualitatively ranked by the number and type (e.g., hardware or procedural) of failures in each cut set. Cut sets containing more failures are generally less likely than those containing fewer failures. Inspection of these lists of minimal cut sets reveals system design/operation weaknesses for which the analysts may suggest possible safety improvement alternatives. A worked example showing Fault Tree Analysis results can be found in Part II, Chapter 16.

Resource Requirements

Using FTA requires a detailed understanding of how the plant or system functions, detailed process drawings and procedures, and knowledge of component failure modes and their effects. Organizations wanting to perform an FTA should use well-trained and experienced analysts to ensure an efficient and high quality analysis that avoids the pitfalls commonly encountered when conducting a Fault Tree Analysis.[13]

Qualified analysts can develop fault trees by themselves, but they must have a detailed understanding of the process and, even then, the models should be reviewed with the engineers, operators, and other personnel who have operating experience with the systems and equipment that are included in the analysis. A single analyst/single fault tree approach promotes continuity within the fault tree, but the analyst must have access to all of the information needed to define the failures that contribute to the Top event. A team approach may be used if the subject process is extremely complex or more than one fault tree is needed, with each qualified team member concentrating on one individual fault tree. Interaction among team members and other experienced personnel is necessary to ensure consistency in the development of related or linked models.

Time and cost requirements for an FTA depend on the complexity of the systems involved in the analysis and the level of resolution of the analysis. Modeling a single Top event involving a simple process with an experienced team could require a day or less. Complex systems and large problems with many top events could require many weeks or months, even with an experienced analysis team. Table 5.22 lists estimates for the time needed to perform a hazard evaluation using the FTA technique. Note that these time estimates are only for a qualitative Fault Tree Analysis as described in this section; considerable additional time could be required if the resulting scenarios were quantified and the quantitative assumptions documented.

Analysis Procedure

There are four steps an analyst must take to perform a Fault Tree Analysis: (1) defining the problem, (2) constructing the fault tree, (3) analyzing the fault tree model qualitatively, and (4) documenting the results.[1,9]

Defining the problem. To define the problem, both a Top event and boundary conditions for the analysis must be selected.[7,8] These boundary conditions include:

Table 5.22 Time estimates for using the Fault Tree Analysis technique

Scope	Preparation	Model Construction	Qualitative Evaluation	Documentation
Small system	1 to 3 days	3 to 6 days	2 to 4 days	3 to 5 days
Large process	4 to 6 days	2 to 3 weeks	1 to 4 weeks	3 to 5 weeks

- System physical bounds
- Level of resolution
- Initial conditions
- Not allowed events
- Existing conditions
- Other assumptions

Defining the ***Top event*** is one of the most important aspects of the first step. The Top event is the consequence (or undesired event) that is the subject of the Fault Tree Analysis (this event is normally identified through previous hazard evaluations). Top events should be precisely defined for the system or plant being evaluated because analyzing broadly scoped or poorly defined Top events can often lead to an inefficient analysis. For example, a Top event of "fire at the plant" is too general for a Fault Tree Analysis. Instead, an appropriate Top event should be "runaway reaction in process oxidation reactor during normal operation." This event description is well-defined and properly scoped because it tells "what," "where," and "when." The what (runaway reaction) tells us the type of incident, the where (process oxidation reactor) tells us the system or process equipment involved in the incident, and the when (during normal operation) tells us the overall system configuration. Even better is to define the Top event as a specific loss event (irreversible, physical event), such as a vessel rupture due to an unrelieved runaway reaction.[13]

The ***physical system boundaries*** encompass the equipment, the equipment's interfaces with other processes, and the utility/support systems that will be included in the Fault Tree Analysis. Along with the physical system boundaries, the analyst should specify the level of resolution for the fault tree events (the ***level of resolution*** simply states the amount of detail to be included in the fault tree). For example, a motor-operated valve can be included as a single piece of equipment, or it can be described as several hardware items (e.g., the valve body, valve internals, and the motor operator). This breakdown could also include the necessary switchgear, power supply, and the human operator needed for the valve to perform. One factor that should be considered in deciding on the level of resolution is the amount of detailed failure information that is available to the analyst, perhaps from an FMEA or previous safety study. The resolution of the fault tree should be limited to the detail needed to satisfy the analysis objective and should parallel the resolution of the available information.

Another boundary condition is the ***initial equipment configuration*** or ***initial operating conditions***. This information states the configuration of the system and equipment that is included in the Fault Tree Analysis. The analyst specifies which valves are open, which valves are closed, which pumps are on, which pumps are off, etc., for all equipment within the physical system boundaries. These boundary conditions describe the system in its normal, unfailed state.

Not allowed events are, for the purposes of the Fault Tree Analysis, events that are considered to be incredible or that, for some other reason, are not to be considered in the analysis. For example, wiring failures might be excluded from the analysis of an instrument system. ***Existing conditions*** are (also for the purposes of the Fault Tree Analysis) events or conditions considered certain to occur. Often the unallowed and existing events do not appear in the finished fault tree, but their effects must be considered in developing other fault events as the fault tree is constructed.

The analyst may specify ***other assumptions*** as necessary to define the system for the Fault Tree Analysis. For example, the analysis may assume that the system is operating at 50 percent of normal capacity. After the problem definition is complete and all boundary conditions are established, these additional assumptions should clarify any uncertainties that remain about the state of the system.

Constructing the Fault Tree. Fault tree construction begins at the Top event and proceeds, level by level, until all fault events have been traced to their basic contributing causes (basic events). The analyst begins with the Top event and, for the next level, uses deductive cause and effect reasoning to determine the immediate, necessary, and sufficient causes that result in the Top event. Normally, these are not basic causes, but are intermediate faults that require additional development. If the analyst can immediately determine the basic causes of the Top event, the problem may be too simple for Fault Tree Analysis and could be evaluated by other methods (such as FMEA).

Figure 5.7 shows an example fault tree that was created using the symbols defined previously. The immediate causes of the Top event are shown in the fault tree in relation to the Top event. If any one of the immediate causes results directly in the Top event, it is connected to the Top event with an OR logic gate. If all of the immediate causes are required for the Top event to occur (as they are in Figure 5.7), they are connected to the Top event with an AND logic gate. Each of the intermediate events is treated in the same manner as the Top event. For each intermediate event, the causes are determined and shown on the fault tree with the appropriate logic gate. The analyst follows this procedure until all intermediate basic events have been developed to their fault causes.

Table 5.23 lists several basic rules that have evolved to promote consistency and completeness in the fault tree construction process. These guidelines are intended to emphasize the importance of systematic and methodical fault tree construction. ***Using the shortcuts prohibited by these rules often leads to an incomplete fault tree that overlooks potentially important combinations of failures.*** These shortcuts also limit the use of the fault tree as a communication tool because only the analyst who developed the fault tree will be able to decipher the logic model.

Analyzing the Fault Tree model. The completed fault tree provides useful information by displaying how failures interact to result in an incident. However, even an experienced analyst cannot identify directly from the fault tree all of the combinations of failures that can lead to the incident of interest (unless he or she is looking at a very simple tree). Therefore, this section discusses one method of obtaining these combinations (minimal cut sets) for the fault tree (this process is also known as "solving" the fault tree).[12] The minimal cut sets are all of the combinations of failures that can result in the fault tree Top event. They are logically equivalent to the information displayed in the fault tree. The minimal cut sets are useful for ranking the ways in which the incident may occur, and they allow quantification of the fault tree if appropriate data are available. There are several ways, manual and computerized, to solve a fault tree model for its minimal cut sets.[1,8,9] Large fault trees require computer programs to determine their minimal cut sets; however, the method described here will allow the analyst to solve many simple fault trees encountered in practice.

The fault tree solution method has four steps: (1) uniquely identify all gates and basic events, (2) resolve all gates into sets of basic events, (3) remove duplicate events within sets, and (4) delete all supersets (sets that contain other sets). The result of the procedure is a list of minimal cut sets for the fault tree. This procedure is demonstrated with an example using the fault tree shown in Figure 5.8.

> ***Step 1.*** The first step is to uniquely identify all gates and basic events in the fault tree. In Figure 5.8, the gates are identified with letters and the basic events with numbers. Each identification is unique, and if a basic event appears more than once in the fault tree, it must have the same identifier each time. For example, basic event 2 appears twice in Figure 5.8, each time with the same identifier.

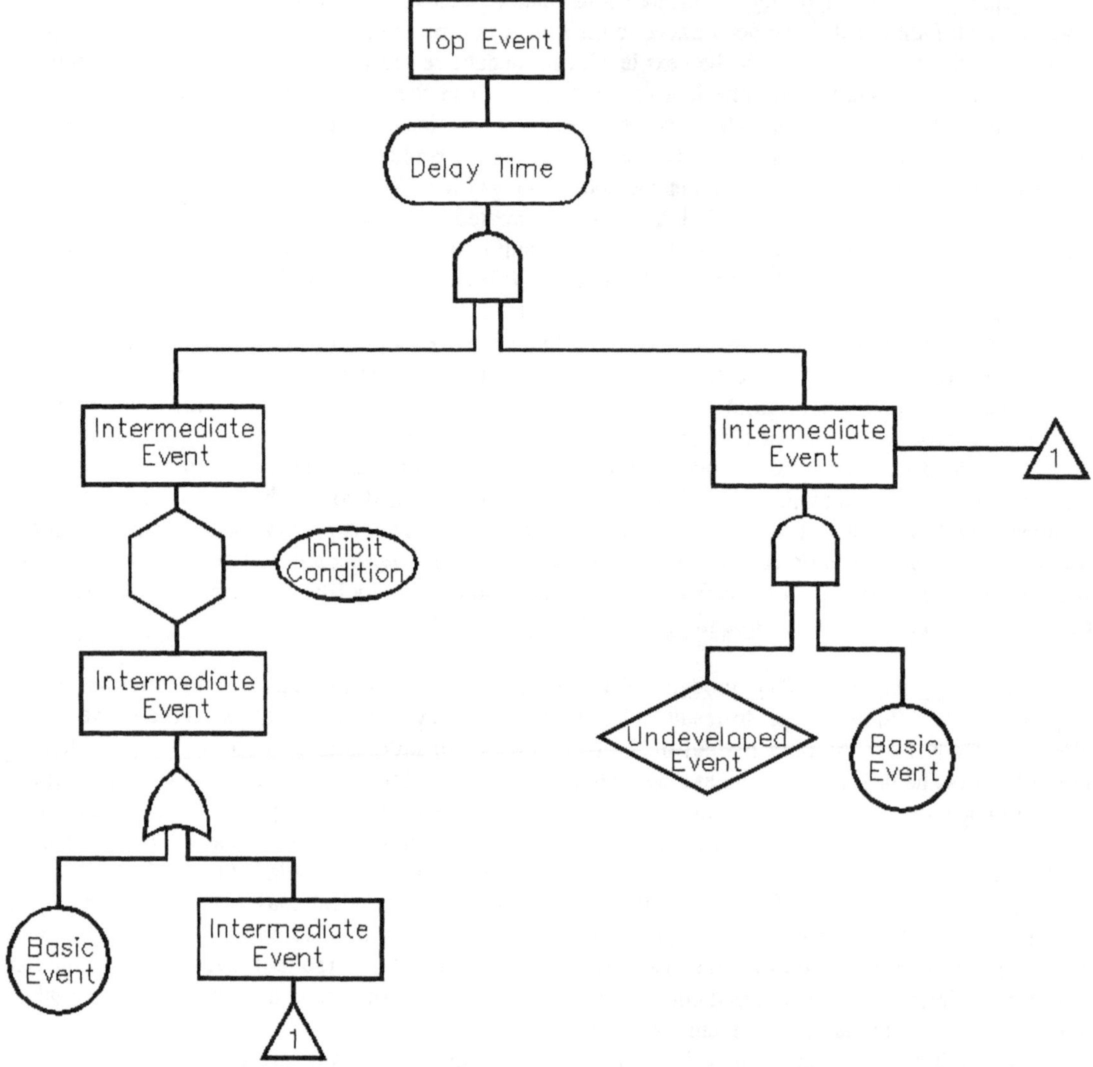

Figure 5.7 Example fault tree structure

Step 2. The second step is to resolve all the gates into basic events. This is done in a matrix format, beginning with the Top event and proceeding through the matrix until all gates are resolved. Gates are resolved by replacing them in the matrix with their inputs. The Top event is always the first entry in the matrix and is entered in the first column of the first row (see Figure 5.9, Part [a]). There are two rules for entering the remaining information in the matrix: the OR-gate rule and the AND-gate rule.

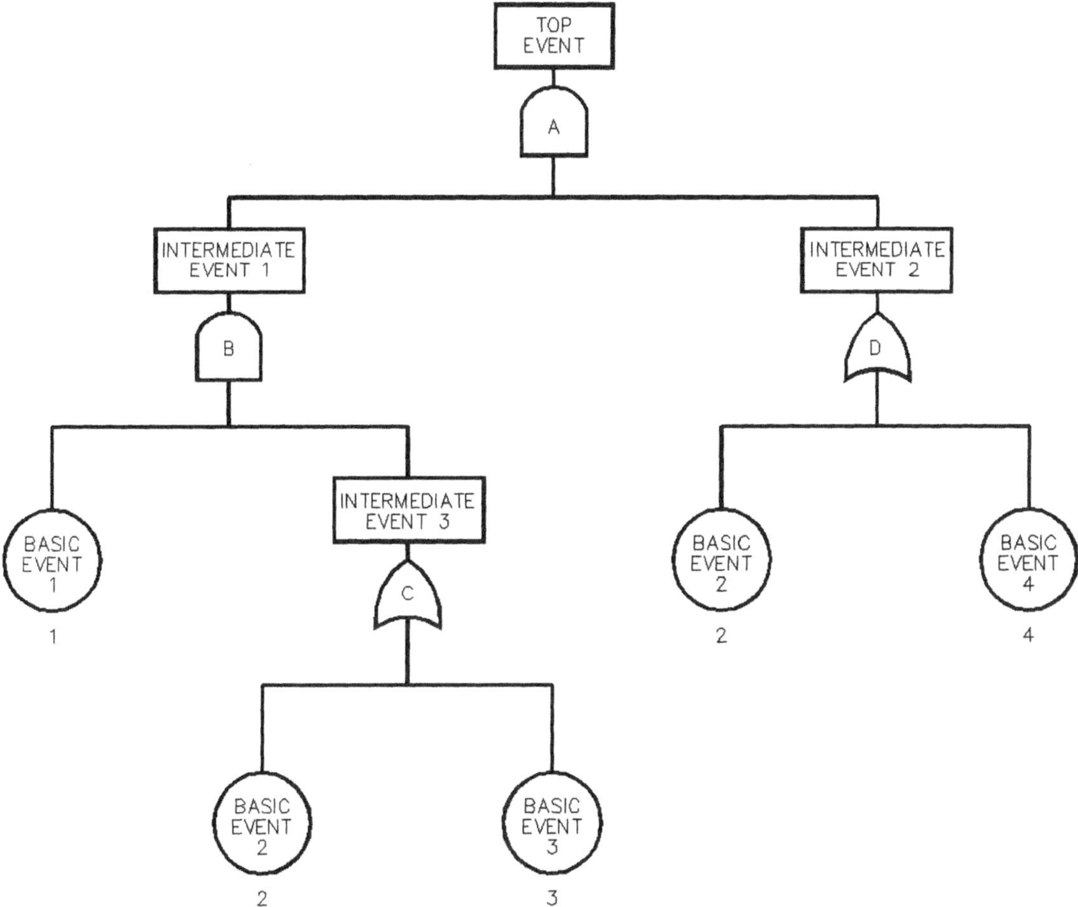

Figure 5.8 Sample fault tree with gates and basic events identified

The OR-gate rule - The first input to an OR gate replaces the matrix's gate identifier, and all other inputs are inserted in the next empty rows of the matrix, one input per row. In addition, if there are other entries in the row where the OR gate appeared, these entries must be repeated in all the rows that contain the other gate inputs.

The AND-gate rule - When resolving an AND gate in the matrix, the first input to the AND gate replaces the gate identifier in the matrix, and the other inputs to the AND gate are inserted in the next available column, one input per column, on the same row that the AND gate appeared on. Each subsequent gate is resolved and all other entries to an AND gate are included on each new row created. INHIBIT and DELAY gates are resolved as if they were AND gates.

These two rules are repeated as necessary until only basic event identifiers remain in the matrix.

Table 5.23 Rules for constructing fault trees

Fault Event Statements	Write the statements that are entered in the event boxes and circles as malfunctions. State precisely a description of the component and the failure mode of the component. Making these statements as precise as possible is necessary for complete description of the fault event. The "where" and "what" portions specify the equipment and its relevant failed state. The "why" condition describes the state of the system with respect to the equipment, thus telling why the equipment state is considered a fault. These statements must be as complete as possible; the analyst should resist the temptation to abbreviate them during the fault tree construction process
Fault Event Evaluation	When evaluating a fault event, ask the question "Can this fault consist of an equipment failure?" If the answer is "yes," classify the fault event as a "state-of-equipment fault." If the answer is "no," classify the fault event as a "state-of-system fault." This classification aids in the continued development of the fault event. If the event is a state-of-equipment fault, add an OR gate to the fault event and look for primary, secondary, and command failures that can result in the event. If the fault event is a state-of-system fault, look for the causes of the fault event
No Miracles	If the normal functioning of equipment propagates a fault sequence, assume that the equipment functions normally. Never assume that the miraculous and totally unexpected failure of some equipment interrupts or prevents an incident from occurring
Complete Each Gate	All inputs to a particular gate should be completely defined before further analysis of any other gate. For simple models, the fault tree should be completed in levels, and each level should be completed before beginning the next level. However, experienced analysts may find this rule to be unwieldy when developing large fault trees
No Gate-to-Gate	Gate inputs should be properly defined fault events; that is, gates should not be directly connected to other gates. Short-cutting the fault tree development leads to confusion because the outputs of the gates are not specified

The fault tree in Figure 5.8 is solved using these two rules. Part (a) of Figure 5.9 shows the first entry in the matrix, gate A, which is the Top event in the sample fault tree. Gate A is an AND gate, so the AND-gate rule is applied to resolve gate A into its inputs, gates B and D, as shown in Part (b). Now the next gate is chosen to be resolved; for example, gate B. Gate B is also an AND gate, so its inputs are entered on the same row as gate B. This replacement is shown in Part (c).

Next, gate D is resolved. Gate D is an OR gate, so its first input replaces D, and its second input is entered in the next available row, as shown in Part (d). Notice that Gate C is now the only gate left in the matrix, appearing on both row 1 and row 2. Each occurrence of gate C is resolved separately. First, on row 1, \the OR-gate rule is applied to gate C as shown in Part (e), resulting in a new set of entries in row 3. Similarly, we resolve the second occurrence of gate C is resolved as shown in Part (f). This completes the resolution of the gates in the matrix. The results of this step are four sets of basic events:

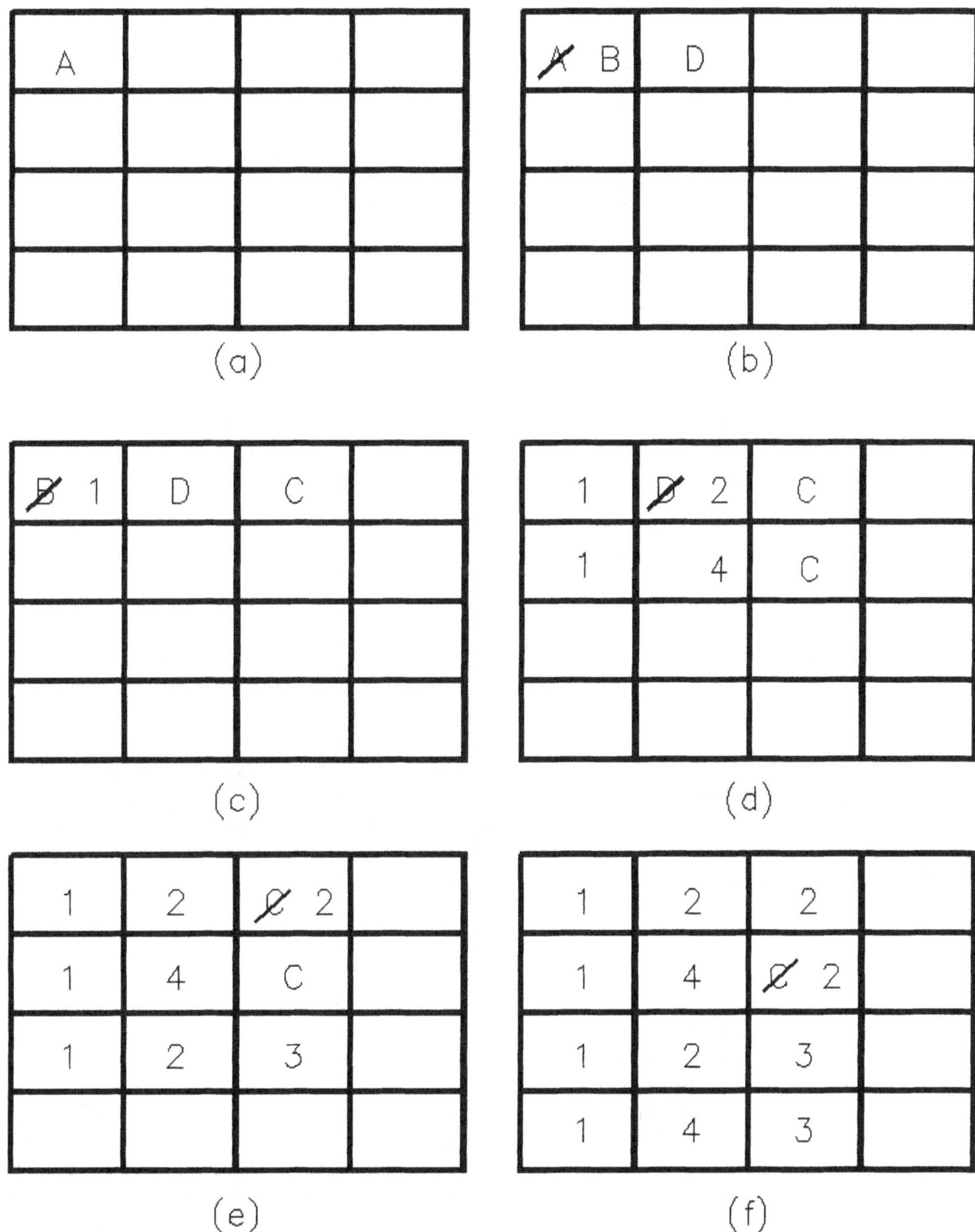

Figure 5.9 Matrix for resolving gates of the sample fault tree

Cut Set I:	1, 2, 2
Cut Set II:	1, 2, 4
Cut Set III:	1, 2, 3
Cut Set IV:	1, 3, 4

Step 3. The third step of the fault tree solution procedure is to remove duplicate events within each set of basic events identified. Only Cut Set 1 has a repeated basic event in the results: basic event 2 appears twice. When this repeated event is removed, the sets of basic events are:

Cut Set I:	1, 2
Cut Set II:	1, 2, 4
Cut Set III:	1, 2, 3
Cut Set IV:	1, 3, 4

Step 4. The fourth step of the fault tree solution procedure is to delete all supersets that appear in the sets of basic events. In these sets, there are two supersets. Both II and III are supersets of I; that is, II and III each contain I. Once these supersets are deleted; the remaining sets are the minimal cut sets for the example fault tree:

Minimal Cut Set I:	1, 2
Minimal Cut Set IV:	1, 3, 4

Once the list of minimal cut sets for a particular Top event is found, an analyst can evaluate the constituent failures to determine the "weak links" in the system. Section 8.3 describes a procedure for prioritizing the results of Fault Tree Analysis and similar cause-specific hazard evaluation techniques. Using the results of this qualitative analysis, fault tree analysts can propose suggestions for improving the safety of the system being studied.

Documenting the results. The final step in performing a Fault Tree Analysis is to document the results of the study. The hazard analyst should provide a description of the system analyzed, a discussion of the problem definition, a list of assumptions, the fault tree model(s) that were developed, lists of minimal cut sets, and an evaluation of the significance of the MCSs. In addition, any recommendations that arise from the FTA should be presented.

Example

Figure 5.10 shows an example reaction system. The system consists of a process reactor for a highly unstable process that is sensitive to small increases in temperature. The system is equipped with a deluge water system for emergency cooling to protect against an uncontrolled reaction. To prevent a runaway reaction during an increase in temperature, the inlet flow of process material to the reactor must be stopped or the deluge system must be activated. The reactor temperature is monitored by a sensor (T1) that automatically activates the deluge system by opening the deluge water supply valve when a temperature rise is detected. At the same time, sensor T1 sounds an alarm in the control room to alert the operator of the temperature rise. When the alarm sounds, the operator pushes the inlet valve close button to close the inlet valve and stop inlet flow to the reactor. The operator also pushes the deluge system open button in the control room if the deluge system is not activated by sensor T1. If the inlet valve closes or the deluge system is activated, system damage due to an uncontrolled reaction is averted.

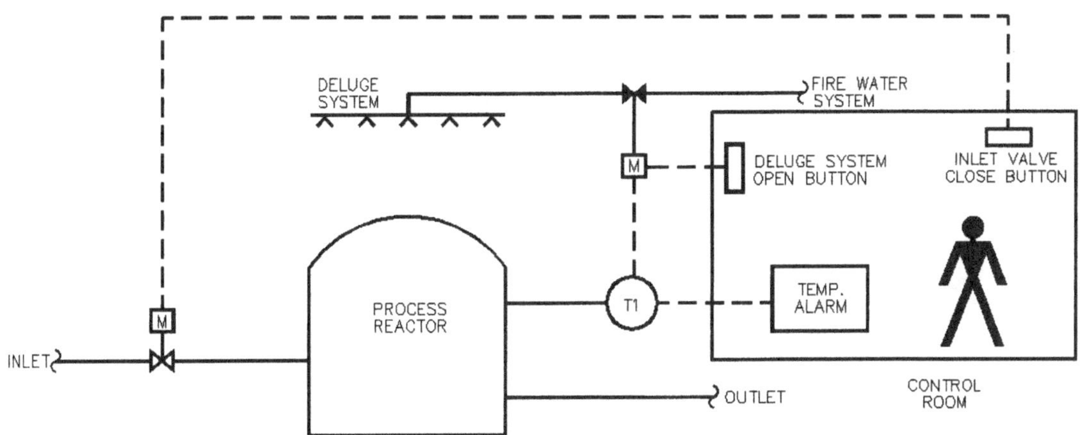

Figure 5.10 Emergency cooling system schematic for the Fault Tree Analysis example

The first step in the Fault Tree Analysis is to define the problem. For this example, the problem definition is as follows.

Top Event:	Damage to reactor due to high process temperature
Existing Event:	High process temperature
Not Allowed Events:	Electric power failures, push button failures, and wiring failures
Physical Bounds:	As shown in Figure 5.10. Process components upstream or downstream of the reactor are not considered
System Configuration:	Inlet valve open, firewater supply valve closed
Level of Resolution:	Equipment as shown in Figure 5.10

This problem definition completely describes the system and conditions to be developed in the fault tree.

The fault tree construction begins with the Top event and proceeds level by level until all faults have been traced to their basic causes. To begin the fault tree, first determine the immediate, necessary, and sufficient causes of the Top event and identify the logic gate that defines the relationship of those causes to the Top event. From the example system description, two necessary conditions are identified for the occurrence of the Top event, given that high temperature exists in the reactor:

- No flow from deluge system
- Reactor inlet valve remains open

Since both of these events must simultaneously exist to result in the Top event, the development requires an AND logic gate as shown in Figure 5.11. The existing condition of high temperature in the reactor is shown in Figure 5.11 as a house event.

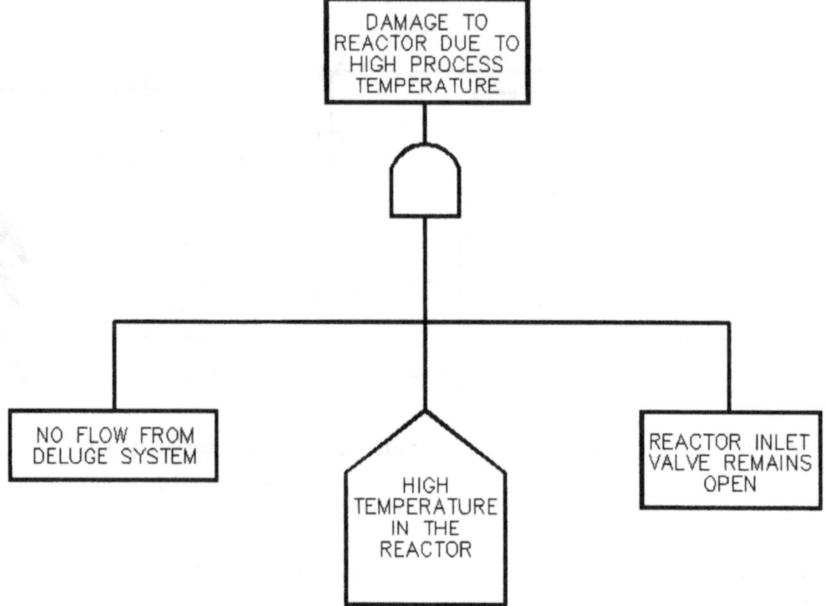

Figure 5.11 Development of the Top event for the emergency cooling system example

Development now continues to the next level; that is, each of the fault events is developed by determining its immediate, necessary, and sufficient causes. For the event "no flow from deluge system," there are two possible causes:

- Loss of firewater supply
- Deluge system valve does not open

Since either of these causes results in no flow, they are added to the fault tree with an OR logic gate. Figure 5.12 shows this development and the continued development of the event "reactor inlet valve remains open.' Figure 5.12 also shows the "loss of firewater supply" event represented by the symbol for an undeveloped event (diamond). There may be several causes for the event, but they are outside the problem definition and are not developed in the fault tree.

Fault tree development continues until all of the events are solved to their basic events or until a system boundary is reached. The completed fault tree is shown in Figure 5.13. As an exercise, the reader is encouraged to review the development of Figure 5.13 and to understand the thought process that leads to the completed fault tree.

The next step in Fault Tree Analysis is to determine the minimal cut sets for the fault tree by following the procedure described earlier. To allow the reader to check the solution, Table 5.24 contains the list of minimal cut sets. The minimal cut sets are ranked based on the analyst's judgment of the relative importance of the events that make up each minimal cut set. (See Section 8.2 for a complete description of prioritizing results.)

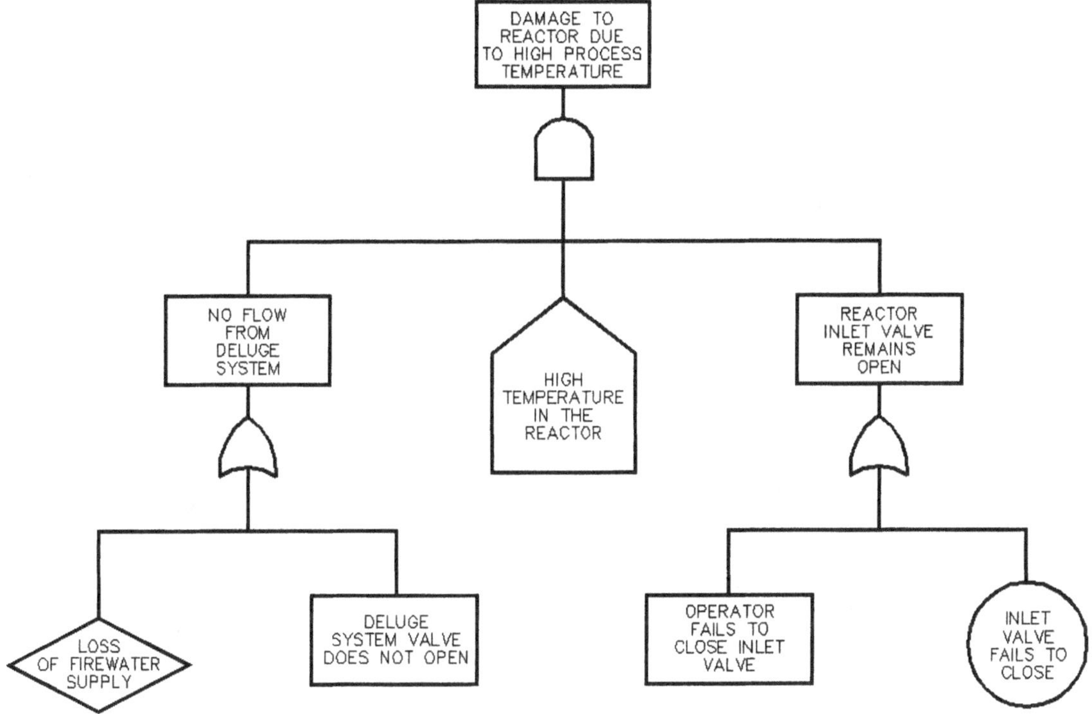

Figure 5.12 Development of the first two intermediate events for the emergency cooling example

If the fault tree Top event is a loss event, and the fault tree is constructed properly, the minimal cut sets can be used in the same way as scenarios from other hazard evaluation procedures to determine a scenario frequency (Section 7.3 through 7.5). Each minimal cut set should have one initiating cause with an associated frequency. The other events in the minimal cut set are failures of preventive safeguards or other events or conditions to which dimensionless probabilities can be assigned. The overall scenario frequency is then the initiating cause frequency multiplied by all of the other event probabilities. For the example given, the initiating cause is high temperature in the reactor (which oftentimes would be further developed), and each of the other events are basic events or undeveloped events associated with protective system failures. The first minimal cut set, Event 8 and Event 3, is the combination of "High temperature in the reactor" and "Temperature sensor fails to detect temperature rise." These two events together are sufficient to result in the Top event, which is a loss event that would have associated impacts. The scenario frequency for this minimal cut set would be the frequency of "High temperature in the reactor" events times the probability of failure on demand (PFD) of the temperature sensor failing to detect the temperature rise. The severity of consequences of reactor damage would need to be assessed separately from the Fault Tree Analysis and combined with this minimal cut set frequency to obtain an overall scenario risk estimate.

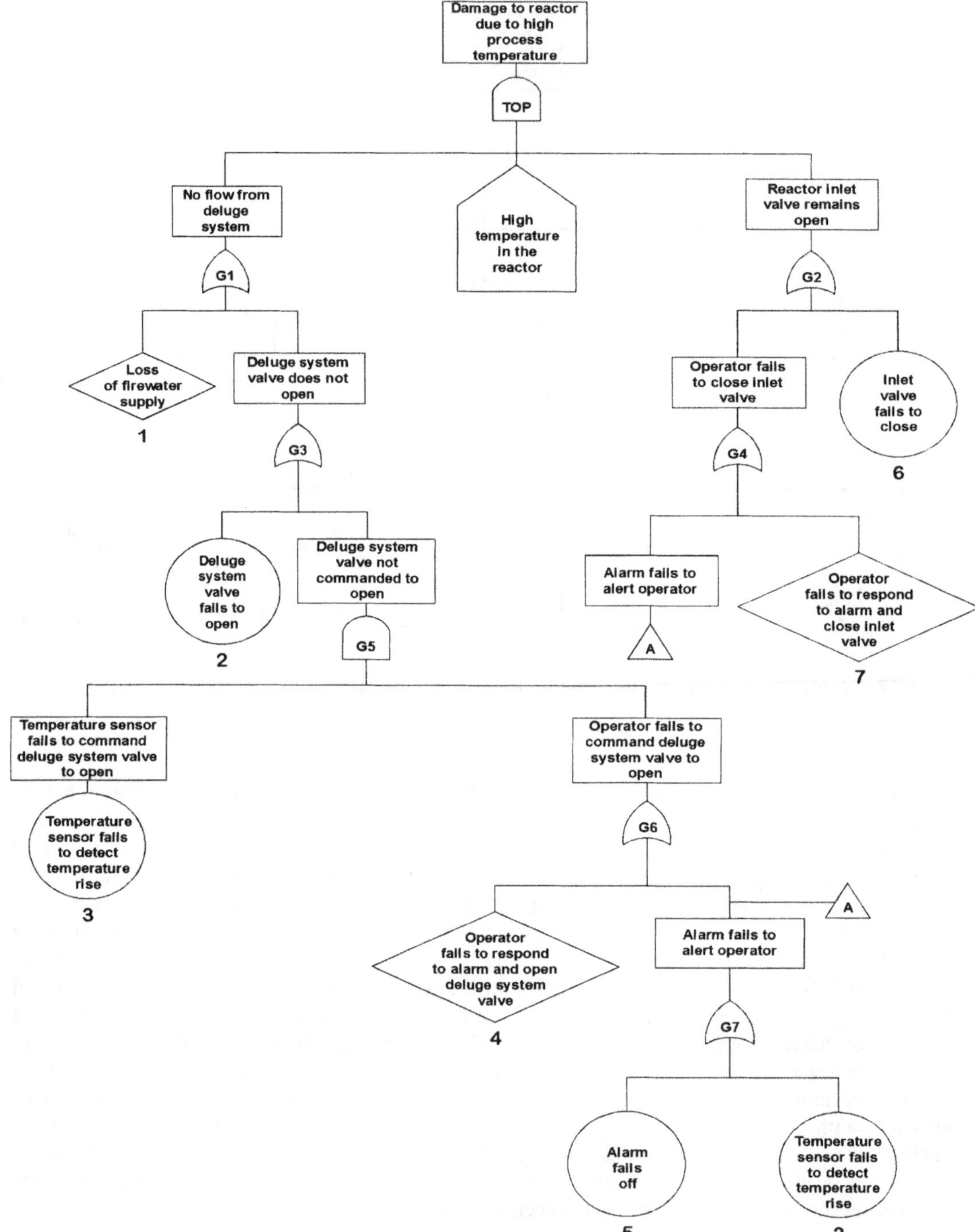

Figure 5.13 Completed fault tree for the emergency cooling system example

Table 5.24 Minimal cut sets for the emergency cooling system example fault tree

MCS no.	Events	Type of events
1	8, 3	Initiating cause, active equipment failure
2	8, 1, 7	Initiating cause, active equipment failure, human error
3	8, 2, 7	Initiating cause, active equipment failure, human error
4	8, 1, 5	Initiating cause, active equipment failure, active equipment failure
5	8, 1, 6	Initiating cause, active equipment failure, active equipment failure
6	8, 2, 5	Initiating cause, active equipment failure, active equipment failure
7	8, 2, 6	Initiating cause, active equipment failure, active equipment failure

Section 5.5 References

1. Center for Chemical Process Safety, *Guidelines for Chemical Process Quantitative Risk Analysis, 2nd Ed.*, American Institute of Chemical Engineers, New York, 1999.
2. H. R. Greenberg and J. J. Cramer (eds.), *Risk Assessment and Risk Management for the Chemical Process Industry*, ISBN 0-442-23438-4, Van Nostrand Reinhold, New York, 1991.
3. J. Stephenson, *System Safety 2000—A Practical Guide for Planning, Managing, and Conducting System Safety Programs*, ISBN 0-0442-23840-1, Van Nostrand Reinhold, New York, 1991.
4. R. Dupont et al., *Accident and Emergency Management*, ISBN 978-0-471-18804-9, John Wiley & Sons, Inc., New York, 1991.
5. S. Mannan, ed., *Lees' Loss Prevention in the Process Industries, 3rd Ed.,* Elsevier Butterworth-Heinemann, ISBN 0-7506-7555-1, Oxford, UK, 2005.
6. A. E. Green, ed., *High Risk Safety Technology*, ISBN 471-10153-2, John Wiley & Sons Ltd., New York, 1982.
7. J. B. Fussell, "Fault Tree Analysis: Concepts and Techniques," *Generic Techniques in System Reliability Assessment*, NATO, Advanced Study Institute, 1976.
8. W. E. Vesely et al., *Fault Tree Handbook*, NUREG-0492, U.S. Nuclear Regulatory Commission, Washington, DC, 1981.
9. *PRA Procedures Guide—A Guide to the Performance of Probabilistic Risk Assessments for Nuclear Power Plants*, NUREG/CR-2300, Vols. 1 and 2, USNRC, Washington, DC, January 1983.
10. I. S. Sutton, *Process Reliability Using Risk Management Techniques*, ISBN 0-442-00174-6, Van Nostrand Reinhold, New York, 1991.
11. N. J. McCormick, *Reliability and Risk Analysis*, Academic Press, New York, 1981.
12. J. B. Fussell et al., *Mocus: A Computer Program to Obtain Minimal Cut Sets from Fault Trees,* ANCR-1156, Idaho National Engineering Laboratory, Idaho Falls, 1974.
13. L.A. Minton and R.W. Johnson, "Fault Tree Faults," *International Conference on Hazard Identification and Risk Analysis, Human Factors and Human Reliability in Process Safety,* American Institute of Chemical Engineers, New York, 1992.

5.6 Event Tree Analysis

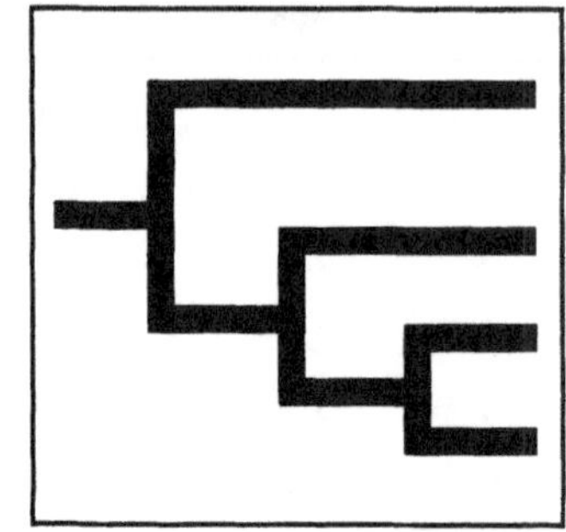

An event tree graphically shows all of the possible outcomes following the success or failure of protective systems, given the occurrence of a specific initiating cause (equipment failure or human error).[1-6] Event trees are also used to study other events, such as starting at a loss event and evaluating mitigation systems. The use of event trees for performing Human Reliability Analyses is discussed as a special application in Section 9.5.

Purpose

Event trees are used to identify the various incidents that can occur in a complex process. After these individual event sequences are identified, the specific combinations of failures that can lead to the incidents can then be determined using Fault Tree analysis.

Description

The results of the Event Tree Analysis are event sequences; that is, sets of failures or errors that lead to an incident. An Event Tree Analysis is well suited for analyzing complex processes that have several layers of safety systems or emergency procedures in place to respond to specific initiating events.

Types of Results

The results of an Event Tree Analysis are the event tree models and the safety system successes or failures that lead to each defined outcome. Event sequences depicted in an event tree represent logical AND combinations of events; thus, these sequences can be put into the form of a fault tree model for further qualitative analysis. Analysts use these results to identify design and procedural weaknesses, and normally provide recommendations for reducing the likelihood and/or consequences of the analyzed potential incidents. A worked example showing Event Tree Analysis results can be found in Part II, Chapter 16.

Resource Requirements

Using ETA requires knowledge of potential initiating causes (that is, equipment failures or system upsets that can potentially cause an incident), and knowledge of safety system functions or emergency procedures that potentially mitigate the effects of each initiating cause.

An Event Tree Analysis can be performed by a single analyst as long as the analyst has a detailed knowledge of the system, but a team of two to four people is often preferred. The team approach promotes brainstorming, which results in a more complete event tree. The team should include at least one member with knowledge of Event Tree Analysis, and the remaining members should have knowledge of the processes and experience working with the systems included in the analysis.

Time and cost requirements for an Event Tree Analysis depend on the number and complexity of initiating causes and safeguards included in the analysis. Several days should be sufficient for the team to evaluate several initiating causes for a simple process; complex processes could require many weeks.

Table 5.25 lists estimates of the time needed to perform a hazard evaluation using the ETA technique. Note that these estimates are only for a qualitative Event Tree Analysis as described in this section; significant additional time would be required if the resulting scenarios were quantified and the quantitative assumptions documented.

Analysis Procedure

The general procedure for Event Tree Analysis contains six steps: (1) identifying the initiating causes or loss events of interest that can result in the type of incident or impact of concern, (2) identifying the safeguards designed to respond to the initiating cause or loss event, (3) constructing the event tree, (4) describing the resulting event sequence outcomes, (5) determining the event sequence minimal cut sets, and (6) documenting the results.[7-9] Each of these steps is discussed below. Chapter 2 provides additional advice about preparing for a hazard evaluation.

Identifying a starting event of interest. Selecting an appropriate initiating cause (generally termed the *initiating event* when performing Event Tree Analyses) or loss event (if studying mitigation safeguards) is an important part of Event Tree Analysis. The event of interest, an initiating cause (initiating event) if a traditional event tree or a loss event if a mitigation event tree, will be referred to as the *starting event.* The starting event could also be an intermediate event, such as a process upset condition. If the starting event is an initiating cause, it should be a system or equipment failure or human error that could result in the effects of interest, depending on how well the system or operators respond to the event. If the selected event results directly in a specific incident, a Fault Tree Analysis is better suited to determine its causes. In most applications of Event Tree Analysis, the initiating cause is "anticipated"; that is, the plant design includes systems, barriers, or procedures that are intended to respond to and mitigate the effects of the initiating cause.

Identifying the safeguards designed to respond to the starting event. The safeguards that respond to the initiating cause or loss event can be thought of as the plant's defenses against the potential consequences of the starting event. These safeguards may include, but are not limited to:

- Alarms that alert the operator when the initiating cause occurs
- Operator actions to be performed in response to alarms or as required by procedures
- Protective systems that automatically respond to the initiating cause
- Last-resort safety systems such as quench systems, pressure relief systems and scrubber systems
- Automatic isolation or other mitigation safeguards intended to limit loss event impacts.

Table 5.25 Time estimates for using the Event Tree Analysis technique

Scope	Preparation	Model Construction	Qualitative Evaluation	Documentation
Small system	1 to 2 days	1 to 3 days	1 to 2 days	3 to 5 days
Large process	4 to 6 days	1 to 2 weeks	1 to 2 weeks	3 to 5 weeks

In particular, these safeguards influence the ultimate effects of any incident resulting from the starting event. The analyst should identify, in the chronological order in which they are expected to respond, all safeguards that can protect against or mitigate the effect of the starting event. The descriptions of these safeguards should state their intended purpose. The successes and failures of the safeguards are accounted for in the event tree.

Constructing the Event Tree. The event tree displays the development of event sequences, beginning with the starting event and proceeding to the control and safety system responses. The results are clearly defined incidents that can result from the initiating cause. An analyst tries to lay out actions of the safeguards chronologically, although many times the events may occur almost simultaneously. The analyst should carefully factor in the normal process control response to upset conditions when evaluating the safety system response to upsets.

The first step in constructing the event tree is to enter the starting event and safeguards that apply to the analysis. The initiating cause or loss event is listed on the left-hand side of the page, and the safeguards are listed across the top of the page. Figure 5.14 shows the first completed step for a generic incident starting with an initiating cause. The line underneath the initiating cause description represents the progression of the incident path from the occurrence of the initiating cause to the first safeguard.

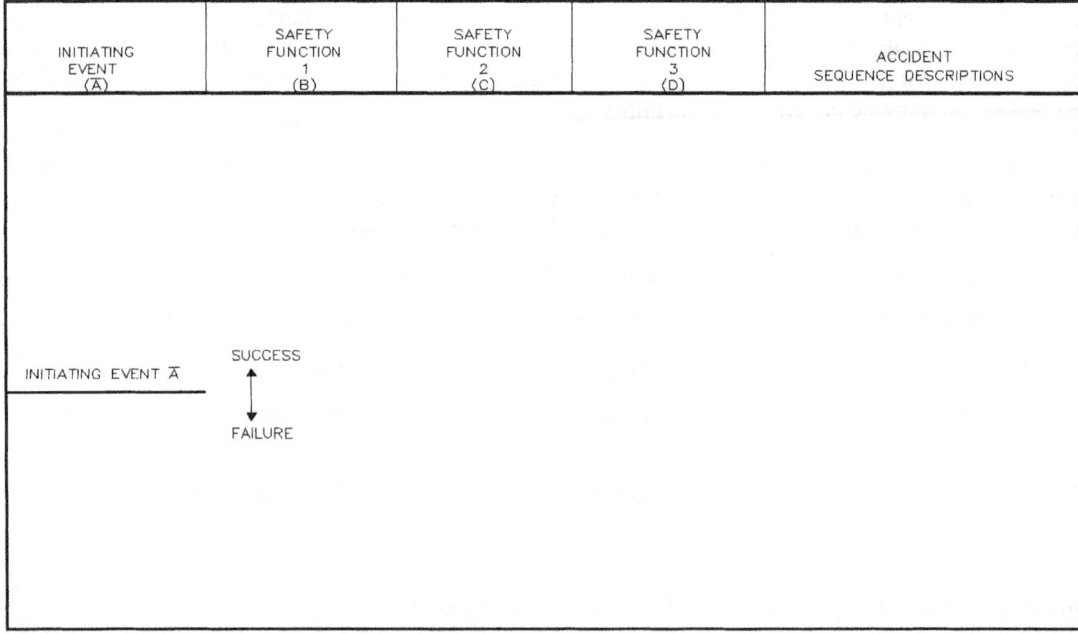

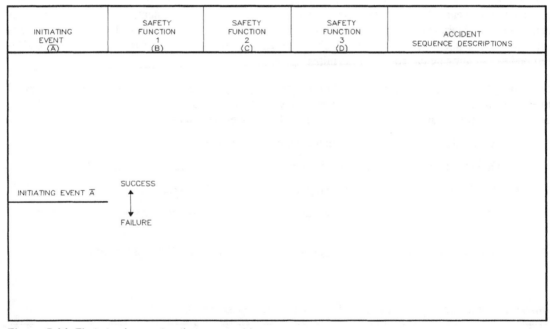

Figure 5.14 First step in constructing an event tree

The next step is to evaluate the safeguard. Normally only two possibilities are considered: success or failure of the safeguard. The analyst should assume that the initiating cause has occurred, define the success/failure criteria for the safeguard, and decide whether the success or failure of the safeguard affects the course of the incident. If the incident is affected, the event tree divides (i.e., at a branch point) into two paths to distinguish between the success and failure of the safeguard. Normally, success of the function is denoted by an upward path, and failure of the function, by a downward path. If the safeguard does not affect the course of the incident, the incident path proceeds, with no branch point, to the next safeguard. Letters (for example, A, B, C, or D) are used to indicate success of the safeguard, and "bars" over the letters indicate failure of the function (for example, $\bar{A}$, $\bar{B}$, $\bar{C}$, or $\bar{D}$). For this example, the first safeguard does affect the course of the incident, as shown by the branch point depicted in Figure 5.15.

Every branch point developed in the event tree creates additional incident paths that must be evaluated individually for each of the subsequent safety systems. When evaluating a safeguard on an incident path, the analyst must assume the previous successes and failures have occurred as dictated by the path. This can be seen in the example when the second safeguard is evaluated (Figure 5.16.). The upper path requires a branch point because the first safeguard was successful, but the second safeguard can still affect the course of the incident. The lower path allows the second safeguard no opportunity to affect the course of the incident if the first safeguard fails. The lower incident path proceeds directly to the third safeguard.

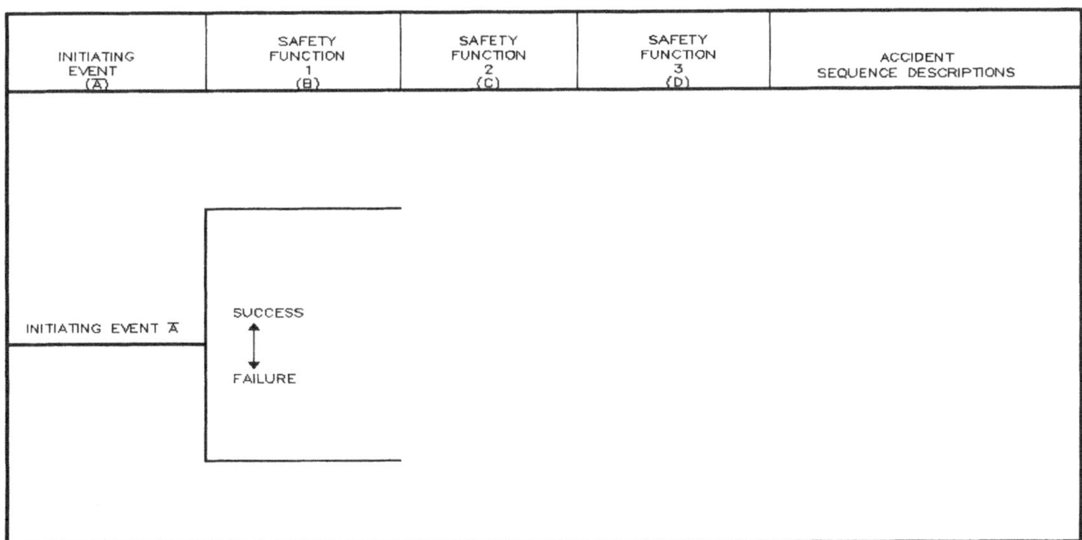

Figure 5.15 Developing the first safeguard in the sample event tree

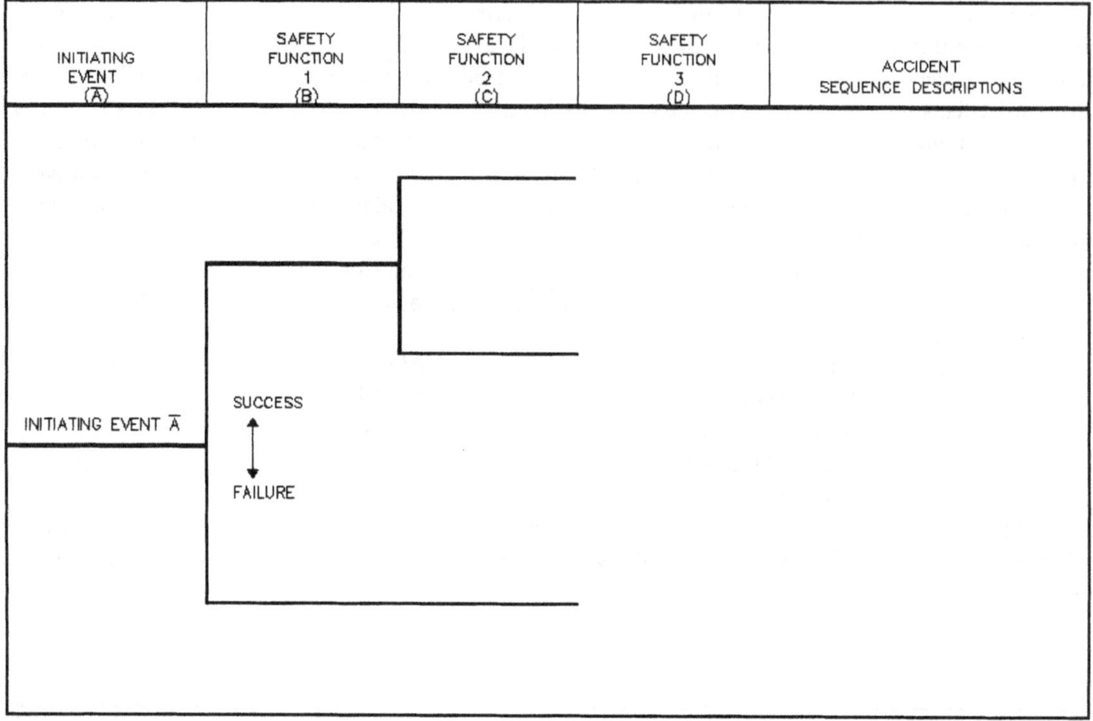

Figure 5.16 Developing the second safeguard in the sample event tree

Figure 5.17 shows the completed example event tree. The uppermost incident path has no branch point for the third safeguard because, in the design of this system, an upset does not challenge the third function if the first and second safeguards were successful. The other incident paths contain branch points for the third safeguard because it can still affect the outcome of the incident paths.

Describing the resulting incident sequence outcomes. The next step of the Event Tree Analysis procedure is to describe the various outcomes of the incident sequences. The sequences will represent the variety of outcomes that can follow the initiating cause. One or more of the sequences may represent a safe recovery and a return to normal operations, or an orderly shutdown. The sequences of importance, from a safety viewpoint, are those that result in consequences of concern.

Determining incident sequence minimal cut sets. Incident sequences in an event tree can be analyzed in the same way that fault trees are analyzed to determine their minimal cut sets. Each incident sequence represents a logical "ANDing" of the initiating cause and subsequent safety system failures. Thus, each sequence can be thought of as a separate fault tree with the incident sequence description as the Top event, followed by an AND gate containing the initiating cause and all of the contributing safety system failures. The safety system failures and their associated logic models must assume that the defined successes of subsequent safeguards have occurred.

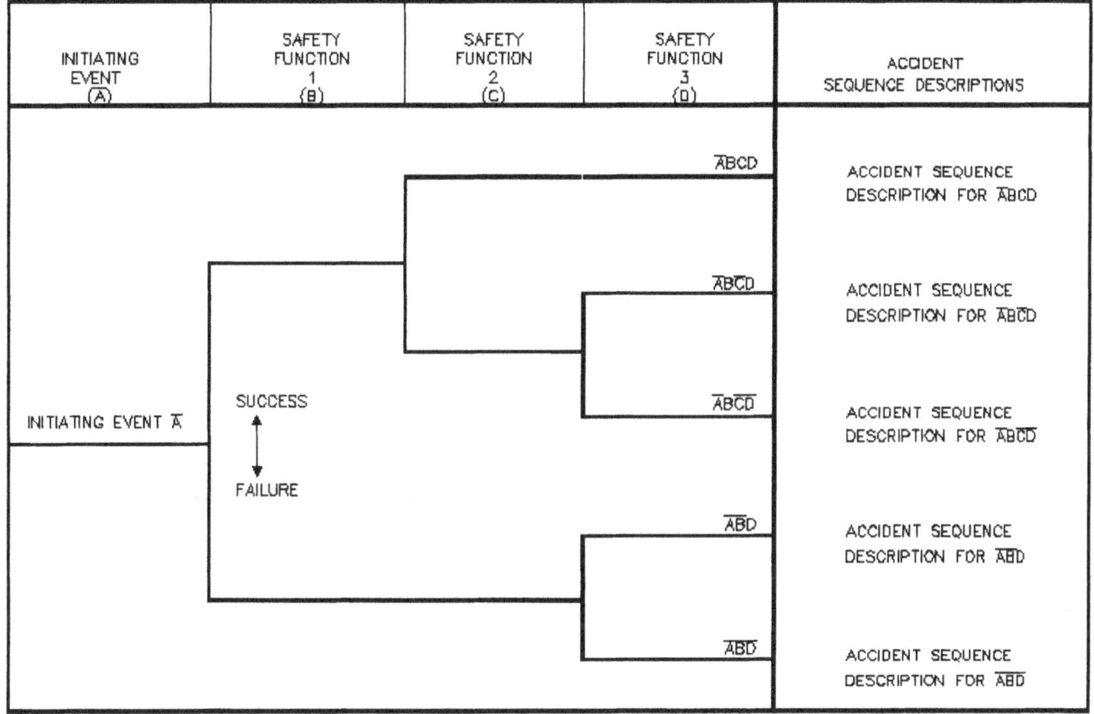

Figure 5.17 Developing the third safeguard in the sample event tree

For example, Figure 5.18 is the fault tree representation of the $\overline{A}\,\overline{B}\,\overline{D}$ incident sequence from the event tree in Figure 5.17. Often, when the failure logic of the initiating cause or safety system is complex, analysts build separate fault trees for these individual system failures and use event trees to define the various combinations that can lead to incident consequences of concern. Then analysts combine the individual initiating cause and safety system fault trees into various (larger) incident sequence fault trees. Using standard qualitative analysis methods described in Section 5.5, analysts solve these incident sequence fault trees to determine the minimal cut sets for each incident sequence of interest. These results can be prioritized using the approaches discussed in Section 8.2.

Documenting the results. The final step in performing an Event Tree Analysis is to document the results of the study. The hazard analyst should provide a description of the system analyzed, a discussion of the problem definition, including the incident initiating causes analyzed, a list of assumptions, the event tree model(s) that were developed, lists of incident sequence minimal cut sets, a discussion of the consequences of the various incident sequences, and an evaluation of the significance of the incident sequence MCS. In addition, any recommendations that arise from the Event Tree Analysis should be presented.

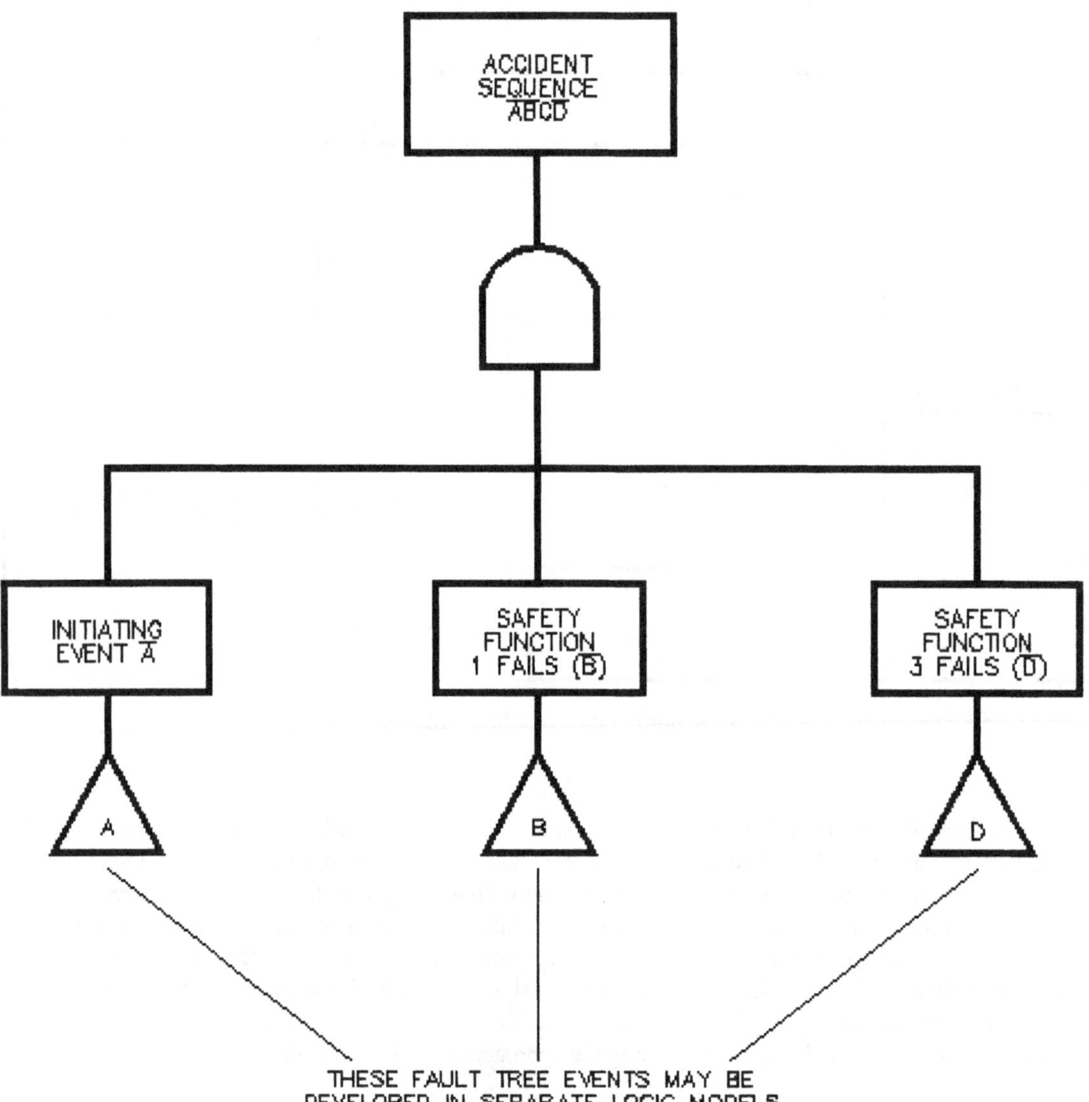

Figure 5.18 Example of an incident sequence fault tree

Example

The starting event is defined as "loss of cooling water to an oxidation reactor." (Note that this is probably not an actual initiating cause if there are specific operational errors or equipment failures such as valve or pump failures that could cause loss of cooling water to the reactor. Nevertheless, it can be used as an ETA starting point.) The following safeguards are designed to respond to this starting event:

- Oxidation reactor high temperature alarm alerts operator at temperature T1
- Operator reestablishes cooling water flow to the oxidation reactor
- Automatic shutdown system stops reaction at temperature T2

These safeguards are listed in the order they are intended to occur. The alarm and the shutdown system each have their own temperature sensors, and the temperature alarm is the only alarm that alerts the operator to the problem. Figure 5.19 presents the event tree for this starting event and these safeguards.

The first safeguard (high temperature alarm) affects the course of the incident by alerting the operator, if the alarm performs correctly. Therefore, a branch point (A) is necessary at the first safeguard. Because the operator may or may not react to the high temperature alarm, a branch point (B) is established for the second safeguard (operator reaction) on the success path from the first safeguard (high temperature alarm). Since the operator cannot react to the initiating cause if the high temperature alarm does not operate, no branch point development at the second safeguard (operator reaction) is included on the failure path from the first safeguard (high temperature alarm). The uppermost event sequence path has no branch point for the third safeguard (automatic shutdown) because the alarm and operator were successful. The third safeguard (C) will be needed if either of the first two safeguards fails. Branch points are developed for the lower paths because the shutdown system will affect the outcome of the event sequence paths.

The analyst should examine the successes and failures in each resulting sequence and provide an accurate description of its expected outcome. This description should be as detailed as necessary to describe the incident. Figure 5.19 has sequence descriptions entered for each event sequence path in the event tree of the example system. This figure also shows a shorthand notation often employed in Event Tree Analysis. Each failure and success event in the diagram is assigned a symbol (in this case, $\overline{A}$, $\overline{B}$, $\overline{C}$ and $\overline{D}$), and the failure sequences are represented by the symbols (these symbols show the combinations of successes and failures that cause a particular incident). For example, in Figure 5.19 the uppermost sequence is simply labeled "$\overline{A}$BC." This sequence is interpreted as "the initiating cause occurs and safeguards B and C operate successfully."

Once the sequences are described, the analyst can rank the incidents based on the number and types of failures and/or severity of their outcomes. The structure of the event tree, clearly showing the progression of the incident, helps the analyst specify where additional procedures or safety systems will be most effective in protecting against these incidents. Section 8 discusses ways of prioritizing the results of hazard evaluation techniques.

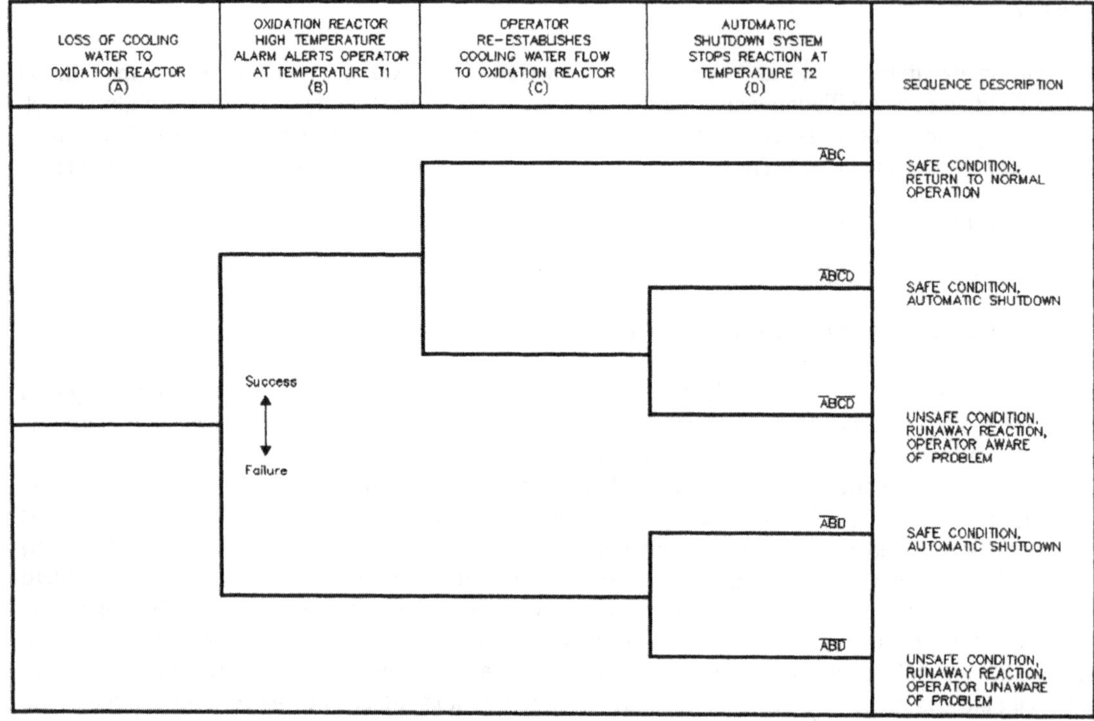

Figure 5.19 Event tree for the example initiating cause "Loss of cooling water to the oxidation reactor"

Section 5.6 References

1. N. J. McCormick, *Reliability and Risk Analysis*, Academic Press, Inc., New York, 1981.
2. H. R. Greenburg and J. J. Cramer (eds.), *Risk Assessment and Risk Management for the Chemical Process Industry*, ISBN 0-442-23438-4, Van Nostrand Reinhold, New York, 1991.
3. S. Mannan, ed., *Lees' Loss Prevention in the Process Industries, 3rd Ed.,* Elsevier Butterworth-Heinemann, ISBN 0-7506-7555-1, Oxford, UK, 2005.
4. I. S. Sutton, *Process Reliability Using Risk Management Techniques*, ISBN 0-442-00174-6, Van Nostrand Reinhold, New York, 1991.
5. R. Dupont et al., *Accident and Emergency Management*, ISBN 978-0-471-18804-9, John Wiley & Sons, Inc., New York, 1991.
6. M. Cornell, "Fault Trees vs. Event Trees in Reliability Analysis," *Risk Analysis 4*(3), 1981.
7. Center for Chemical Process Safety, *Guidelines for Chemical Process Quantitative Risk Analysis, 2nd Ed.*, American Institute of Chemical Engineers, New York, 1999.
8. *PRA Procedures Guide—A Guide to the Performance of Probabilistic Risk Assessments for Nuclear Power Plants*, NUREG/CR-2300, Vols. 1 and 2, USNRC, Washington, DC, January 1983.
9. *Reactor Safety Study*, WASH-1400, Atomic Energy Commission, Washington, DC, 1975.
10. N. Limnious and J. P. Jeannette, "Event Trees and Their Treatment on PC Computers," Vol. 18, 1987.

5.7 Cause-Consequence Analysis and Bow-Tie Analysis

A Cause-Consequence Analysis (CCA) is a blend of the Fault Tree Analysis and Event Tree Analysis techniques that were discussed in the preceding sections.

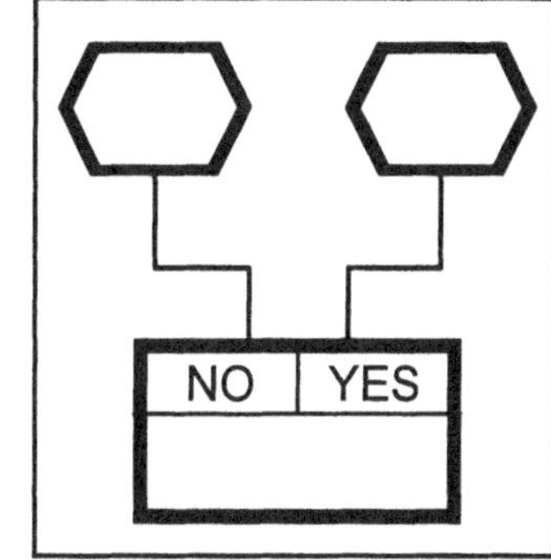

Purpose

As the name suggests, the purpose of a Cause-Consequence Analysis is to identify the basic causes and consequences of potential incidents.

Description

Cause-Consequence Analysis (CCA) combines the inductive reasoning features of Event Tree Analysis with the deductive reasoning features of Fault Tree Analysis.[1-4]

A major strength of a Cause-Consequence Analysis is its use as a communication tool. The cause-consequence diagram displays the relationships between the incident outcomes (consequences) and their basic causes. This technique is most commonly used when the failure logic of the analyzed incidents is rather simple, since the graphical form, which combines both fault trees and event trees on the same diagram, can become quite detailed.

Types of Results

A Cause-Consequence Analysis generates diagrams portraying incident sequences and qualitative descriptions of potential incident outcomes.

Resource Requirements

Using CCA requires knowledge of the following data and information sources: knowledge of component failures or process upsets that could cause incidents, knowledge of safety systems or emergency procedures that can influence the outcome of an incident, and knowledge of the potential impacts of all of these failures.

A Cause-Consequence Analysis is best performed by a small team (two to four people) with a combined range of experience. One team member should be experienced in CCA (or Fault Tree Analysis and Event Tree Analysis), while the remaining members should have experience with the design and operation of the systems included in the analysis.

Time and cost requirements for a CCA are highly dependent on the number, complexity and level of resolution of the events included in the analysis. Scoping-type analyses for several initiating causes can usually be accomplished in a week or less. Detailed CCA studies may require many weeks, depending on the complexity of any supporting fault trees. Table 5.26 lists estimates of the time needed to perform a hazard evaluation using the CCA technique.

Table 5.26 Time estimates for using the Cause-Consequence Analysis technique

Scope	Preparation	Model Construction	Qualitative Evaluation	Documentation
Small system	1 to 2 days	1 to 3 days	1 to 3 days	3 to 5 days
Large process	4 to 6 days	1 to 2 weeks	1 to 2 weeks	3 to 5 weeks

Analysis Procedure

A general procedure for CCA contains six steps: (1) selecting an event or type of incident situation to be evaluated, (2) identifying the safeguards (systems, operator actions, etc.) that influence the course of the incident resulting from the event, (3) developing the event sequence paths resulting from the event (Event Tree Analysis), (4) developing the combinations of intermediate events and safeguard failures to determine their basic causes (Fault Tree Analysis), (5) evaluating the event sequence minimal cut sets, and (6) documenting the results.[1,2,6,7] Chapter 2 provides general advice on how to prepare for hazard evaluations.

Selecting an event to be evaluated. The events analyzed in Cause-Consequence Analysis can be defined in two ways:

- Top event (as in a Fault Tree analysis)
- Intermediate event (e.g., a *deviation* as defined for HAZOP Studies)

The event of interest is a fault event or equipment failure that would otherwise be suited for a Fault Tree Analysis or Event Tree Analysis. Additional information for selecting these events can be found in the appropriate sections of the Fault Tree Analysis and Event Tree Analysis procedures (Sections 5.4 and 5.5, respectively).

Identifying safeguards and developing event sequence paths. These steps are the same as those performed in Event Tree Analysis. The various event sequence paths are constructed based on the chronological successes and failures of the appropriate safeguards. The primary difference between Event Tree and Cause-Consequence Analysis is the symbols used in the diagram. Figure 5.20 shows the symbol most often used for the event tree branch point in the cause-consequence diagram. This symbol contains the safeguard description normally written over the event tree branch point. Figure 5.21 shows the symbol used in a cause-consequence diagram to represent the resulting consequence. No corresponding symbol is normally used in the event tree diagram.

Developing the intermediate and safeguard failure events to determine basic causes. In this step, the analyst actually applies Fault Tree Analysis techniques to the starting event and safeguard failure events represented in the event tree portion of the cause-consequence diagram. Each fault description is treated as if it were a fault tree Top event or intermediate event. The analysts should describe the outcomes of each incident sequence in the Cause-Consequence model. Section 5.5 provides procedures for fault tree construction.

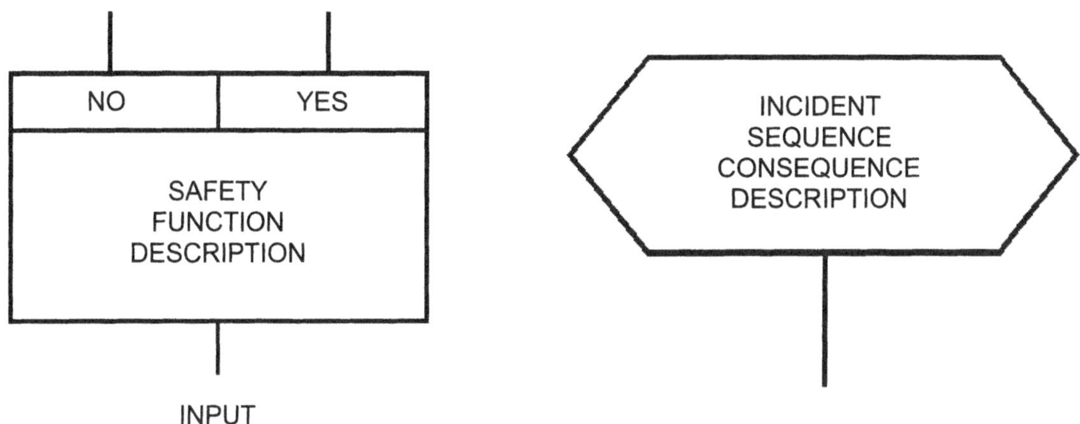

Figure 5.20 Branch point symbol used In Cause-Consequence Analysis

Figure 5.21 Consequence symbol used in Cause-Consequence Analysis

Evaluating the incident sequence minimal cut sets. The incident sequence minimal cut sets are determined in a manner similar to fault tree minimal cut sets. The incident sequence is composed of a sequence of events, each of which is a Top event for a fault tree that is part of the cause-consequence diagram. For an incident sequence to occur, all of the events in the sequence must occur. An incident sequence fault tree is constructed by connecting all the safeguard failures to an AND logic gate, with the incident sequence occurrence as the new Top event. The standard fault tree solution technique (discussed in Section 5.5) can then be used to determine the incident sequence minimal cut sets. This process can be repeated for all incident sequences identified in the CCA. Evaluating the results of the CCA is a two-step process. First, the incident sequences are ranked based on their severity and importance to plant safety. Then, for each important incident sequence, the incident sequence minimal cut sets can be ranked to determine the most important basic causes. Section 8.2 discusses methods for ranking minimal cut sets.

Documenting the results. The final step in performing a Cause-Consequence Analysis is to document the results of the study. The hazard analyst should provide a description of the system analyzed, a discussion of the problem definition including the incident initiating causes analyzed, a list of assumptions, the cause-consequence diagrams that were developed, lists of incident sequence minimal cut sets, a discussion of the consequences of the various incident sequences, and an evaluation of the significance of the incident sequence MCS. In addition, any recommendations that arise from the CCA should be presented.

Example

A CCA can be performed on the "loss of cooling water to the oxidation reactor" intermediate event described in the Event Tree Analysis example (Section 5.6). As before, the safety features related to this event are:

- Oxidation reactor high temperature alarm alerts operator at temperature T1
- Operator reestablishes cooling water flow to the oxidation reactor
- Automatic shutdown system stops reactions at temperature T2

These safeguards are listed in the order in which they are intended to occur. The alarm and shutdown system each have their own temperature sensors, and the temperature alarm is the only thing that alerts the operator to the problem. The cause-consequence diagram is represented in Figure 5.22. For simplicity, the fault tree logic associated with each branching operation is not shown.

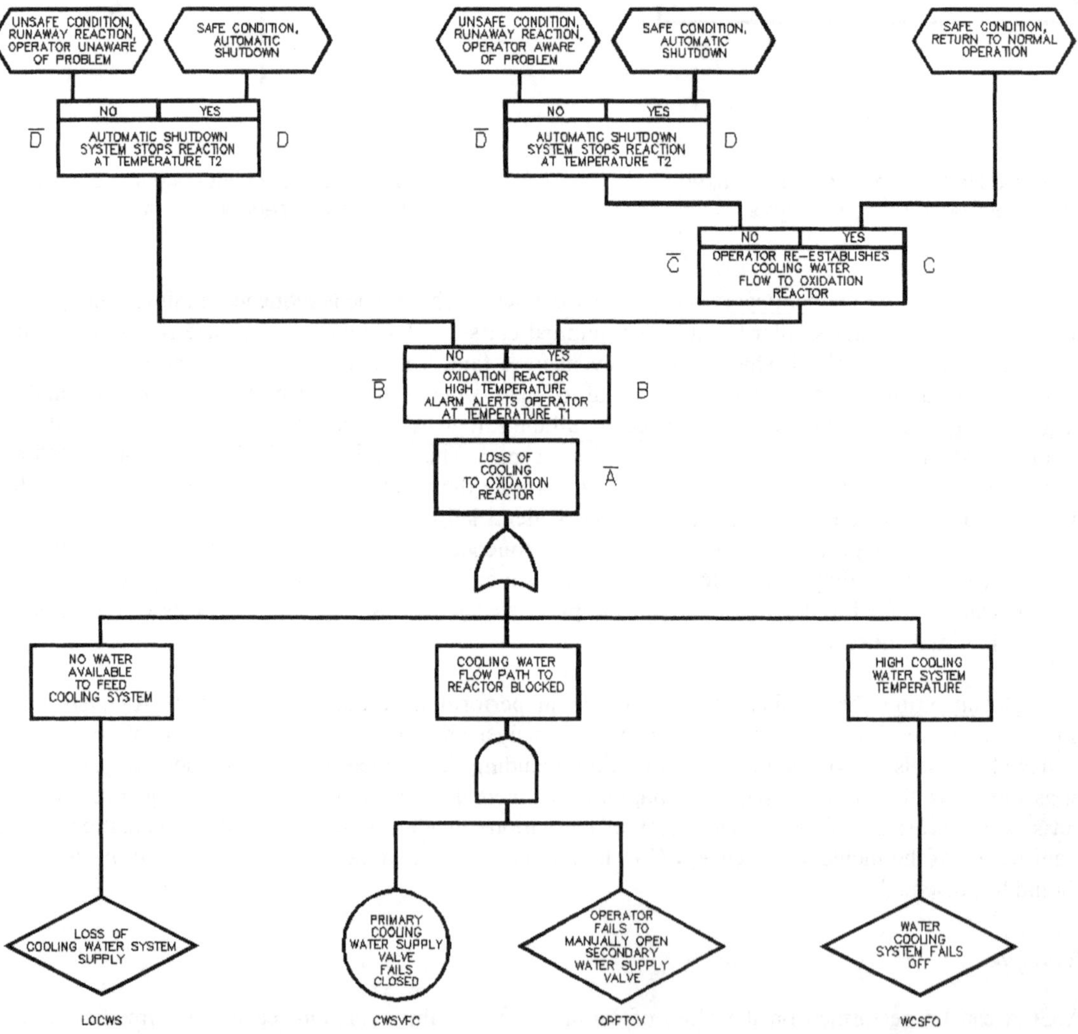

Figure 5.22 Cause-consequence diagram for the example intermediate event "loss of cooling water to the oxidation reactor"

The CCA reveals two incident sequences that may have undesirable effects:

- Unsafe condition, runaway reaction, operator unaware of problem
- Unsafe condition, runaway reaction, operator aware of problem

For each of these two incident sequences, the minimal cut sets are determined to provide insight into each incident's causes. The cut sets for the two undesirable incident sequences are presented in Table 5.27, ranked by the number and the types of events in each cut set.

Bow-Tie Analysis

A less formal variation of Cause-Consequence Analysis is the "Bow-Tie" technique. It similarly combines two methodologies presented in earlier sections, Fault Tree Analysis and Event Tree Analysis, and uses the format of an incident investigation and root cause analysis technique known as Causal Factors Charting.[4,9] The Bow-Tie analysis offers a cost-effective approach for a screening hazard evaluation of processes that are well understood. This approach is a qualitative hazard evaluation technique ideally suited for the initial analysis of an existing process, or application during the middle stages of a process design.[8] It has also been used as a structured approach for hazard evaluation within European safety case studies when quantification is not possible or not desirable.[10] Effective application of this technique depends on the experience and skills of the hazard evaluation team.

The analysis representation resembles a bow tie, as in Figure 5.23, because hazards and initiating causes appear on the on one side and impacts on the other, with the initiating causes and impacts being tied to a specific loss event as the focal point in the middle of the diagram. Typical loss event scenarios are identified and shown on the pre-event (left) side of the diagram, while credible consequences and scenario outcomes are presented on the post-event (right) side.

Table 5.27 Incident sequence minimal cut sets for "loss of cooling water to the oxidation reactor"

Unsafe condition, runaway reaction, operator unaware of problem		
1	LOCWS, $\bar{B}$, $\bar{D}$	Active equipment failure, active equipment failure, active equipment failure
2	WCSFO, $\bar{B}$, $\bar{D}$	Active equipment failure, active equipment failure, active equipment failure
3	CWSVFC, OPFTOV, $\bar{B}$, $\bar{D}$	Active equipment failure, human error, active equipment failure, active equipment failure
Unsafe condition, runaway reaction, operator aware of problem		
1	LOCWS, $\bar{C}$, $\bar{D}$	Active equipment failure, human error, active equipment failure
2	WCSFO, $\bar{C}$, $\bar{D}$	Active equipment failure, human error, active equipment failure
3	CWSVFC, OPFTOV, $\bar{C}$, $\bar{D}$	Active equipment failure, human error, human error, active equipment failure

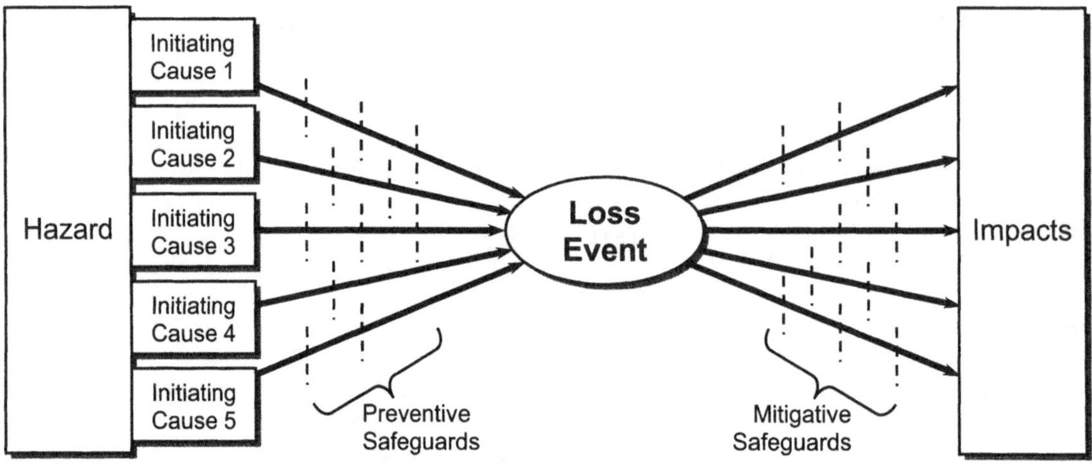

Figure 5.23 Generic "bow-tie" diagram

Associated preventive and mitigative safeguards are also shown on the diagram. The Bow-Tie technique in its visual form makes the analysis easy to understand, and can show what safeguards protect against particular initiating causes and loss event consequences. In the Bow-Tie method, the loss event is the equivalent of both (1) a Fault Tree Analysis Top event for deductively determining initiating causes and identifying preventive safeguards, and (2) an Event Tree Analysis starting point for studying loss event consequences and identifying mitigative safeguards. A more detailed bow-tie diagram for simple systems may actually show a fault tree to the left of the loss event and an event tree to the right, and may also include linkages to safety management system activities and procedures.[10]

Section 5.7 References

1. D. S. Nielsen, *The Cause/Consequence Diagram Method as a Basis for Quantitative Accident Analysis*, RISO-M-1374, Atomic Energy Commission Research Establishment, Riso, Denmark, 1971.
2. Center for Chemical Process Safety, *Guidelines for Chemical Process Quantitative Risk Analysis, 2nd Ed.*, American Institute of Chemical Engineers, New York, 1999.
3. H. R. Greenberg and J. J. Cramer (eds.), *Risk Assessment and Risk Management for the Chemical Process Industry*, ISBN 0-442-23438-4, Van Nostrand Reinhold, New York, 1991.
4. S. Mannan, ed., *Lees' Loss Prevention in the Process Industries, 3rd Ed.*, Elsevier Butterworth-Heinemann, ISBN 0-7506-7555-1, Oxford, UK, 2005.
5. D. S. Nielsen et al., "A Cause-Consequence Chart of a Redundant Protection System," *IEEE Transactions on Reliability 24*, 1975.
6. J. R. Taylor, *Cause-Consequence Diagrams,* NATO Advanced Study Institute, 1978.
7. D. S. Nielsen, "Use of Cause-Consequence Charts in Practical Systems Analysis," *Reliability and Fault Tree Analysis*, SIAM, Philadelphia, 1975.
8. J. Philley, "Collar Hazards with a Bow-Tie," www.chemicalprocessing.com, January 2006.
9. Center for Chemical Process Safety, *Guidelines for Investigating Chemical Process Incidents, 2nd Ed.*, American Institute of Chemical Engineers, New York, 2003.
10. Det Norske Veritas, "Marine Risk Assessment," Offshore Technology Report OTO 2001/063, prepared for the UK Health and Safety Executive, HSE Books, Sudbury, Suffolk, UK, 2002, www.hse.gov.uk.

5.8 Other Techniques

These *Guidelines* present the hazard evaluation techniques most commonly used for analyzing process systems. Even so, some organizations have chosen to use other techniques that may be just as effective in satisfying their process safety objectives.[1-5] Many of these other methods are actually hybrids, extensions, or derivatives of the major approaches discussed in Chapters 4 and 5. Others are much different from any of the standard techniques.

A variety of approaches has been developed in industries other than the chemical process industry. The following are the names of a few of these methods. Readers seeking more information should refer to the specific references for these techniques.

- Digraph Analysis[6,7]
- Management Oversight Risk Tree (MORT) Analysis[1,8-10]
- Hazard Warning Structure[11,12]
- Multiple Failure/Error Analysis[1]
- Energy Trace Analysis[13]
- Loss-of-Containment Analysis[14]

The exclusion of these and other methods from detailed consideration in these *Guidelines* should not be viewed as a rejection of their applicability, usefulness or efficiency.

Section 5.8 References

1. H. R. Greenberg and J. J. Cramer (eds.), *Risk Assessment and Risk Management for the Chemical Process Industry*, ISBN 0-442-23438-4, Van Nostrand Reinhold, New York, 1991.
2. J. Stephenson, *System Safety 2000—A Practical Guide for Planning, Managing, and Conducting System Safety Programs*, ISBN 0-0442-23840-1, Van Nostrand Reinhold, New York, 1991.
3. Z. Nivolianitou et al., "Reliability Analysis of Chemical Processes by the DYLAM Approach," *Reliability Engineering 14*, 1986.
4. H. Kumamoto and E. J. Henley, "Safety and Reliability Synthesis of Systems with Control Loops," *AIChE Journal 25*, January 1979.
5. S. Mannan, ed., *Lees' Loss Prevention in the Process Industries, 3rd Ed.*, Elsevier Butterworth-Heinemann, ISBN 0-7506-7555-1, Oxford, UK, 2005.
6. G. J. Powers and S. A. Lapp, "Computer-Aided Fault Tree Synthesis," *Chemical Engineering Progress*, April 1976.
7. J. D. Andrews and J. M. Morgan, "Application of the Digraph Method of Fault Tree Construction to Process Plant," *Reliability Engineering 14*, 1986.
8. N. W. Knox and R. W. Eicher, *MORT User's Manual*, SSDC-4 (Revision 2), U.S. Department of Energy, Idaho Falls, ID, 1983.
9. W. G. Johnson, *MORT Safety Assurance Systems*, Marcel Dekker, New York, 1980.
10. W. G. Johnson, *MORT, the MANAGEMENT OVERSIGHT AND RISK TREE*, U.S. Atomic Energy Commission, Washington, DC, 1973.
11. F. P. Lees, "The Hazard Warning Structure of Major Hazards," *Transactions of the Institution of Chemical Engineers 60*(211), London, 1982.

12. F. P. Lees, *Hazard Warning Structure: Some Illustrative Examples Based on Actual Cases,* Paper 4B/1 presented at the 4th National Reliability Conference at the National Centre for Systems Reliability, Culcheth, UK, 1983.
13. C. J. Hocevar and C. M. Orr, "Hazard Analysis by the Energy Trace Method," *Proceedings of the Ninth International System Safety Conference,* pp. H-59 to H-70, July 1989.
14. R. W. Johnson, “Analyze Hazards, Not Just Risks,” *Chemical Engineering Progress 96*(7), July 2000, 31-40.

6

Selection of Hazard Evaluation Techniques

The ability to ensure process safety at a facility is influenced by many things: for example, employing appropriate technology in design and construction, anticipating the effects of external circumstances, understanding and dealing with human behavior, and having effective management systems. One of the cornerstones of an effective process safety management system is a successful hazard evaluation program.

A successful hazard evaluation program requires tangible management support; sufficient, technically competent people (some of whom must be trained to use hazard evaluation techniques); an adequate, up-to-date information database; and the right tools to perform hazard evaluations. Fortunately, a variety of flexible hazard evaluation techniques exist. Each technique presented in these *Guidelines* has been applied in the chemical process industry and is appropriate for use in a wide variety of situations.

In an effective hazard evaluation program, excellent performance is based on successfully executing individual hazard evaluations. A successful hazard evaluation can be defined as one in which (1) the need for risk information has been met, (2) the results are of high quality and are easy for decision makers to use, and (3) the study has been performed with the minimum resources needed to get the job done. Obviously, the technique selected has a great bearing on each hazard evaluation's success.

Many factors can affect which hazard evaluation technique is chosen. Before dealing with the technical aspects of this decision, it is worthwhile to address an often overlooked question that can also significantly influence the success of a hazard evaluation: who should decide which hazard evaluation technique is used? It is appropriate and necessary that management define the basic charter for a hazard evaluation: the main objective of the study, the type of decision-making information (results) needed, and the initial resources and deadlines for performing the work. But the technical project manager and/or hazard evaluation team leader should select the most appropriate hazard evaluation method or methods to fulfill the study's charter.

Many organizations have developed policies that specify that analysts use certain types of hazard evaluation techniques.[1-4] Usually, providing this guidance does not present a problem as long as the hazard analyst can use an alternate hazard evaluation method if it can better satisfy the study's charter. For example, suppose a corporate safety group has decided that facilities under their jurisdiction must use the HAZOP Study technique to perform the majority of hazard evaluations. The hazard evaluation team leader for a major process modification project is requested to perform an analysis of the distributed control system (DCS) that is being installed in conjunction with the project. In this case the team leader believes, based on his experience, that the HAZOP Study approach is not the most efficient method to investigate the ways in which the DCS can cause hazardous situations. Instead, the leader wants to use the FMEA technique, which, for this type of analysis problem, he has seen work more efficiently than HAZOP Study. Management listens to the leader's recommendation and allows him to use the FMEA technique.

Hazard evaluation specialists should be allowed some freedom to select one or more proper methods for the job. Selecting the most appropriate hazard evaluation method is a critical step in ensuring the success of a hazard evaluation. The hazard evaluation of a particular process may involve switching between more than one methodology to study different parts of the process, in order to achieve both a thorough and an efficient review.

Since selecting an appropriate hazard evaluation technique is more an art than a science, there may be no "best" method for a particular application. And it may turn out that a plant's need for better risk information cannot be met by hazard evaluation techniques alone (see Section 1.2). Nonetheless, assuming that it is appropriate to use hazard evaluation techniques, this chapter discusses a strategy for selecting a method that is likely to contribute to the success of a study or particular part of a study.

6.1 Factors Influencing the Selection of Hazard Evaluation Techniques

Each hazard evaluation technique has its unique strengths and weaknesses. Chapters 4 and 5 describe many of the attributes of the hazard evaluation techniques covered in these *Guidelines*. Understanding these attributes is prerequisite to selecting an appropriate hazard evaluation technique. The process of selecting an appropriate hazard evaluation technique may be a difficult one for the inexperienced practitioner because the "best" technique may not be apparent. As hazard analysts gain experience with the various hazard evaluation methods, the task of choosing an appropriate technique becomes easier and somewhat instinctive.

The thought process behind selecting hazard evaluation techniques is complex, and a variety of factors can influence the decision-making process.[5-8] Table 6.1 lists six categories of factors that analysts should consider when selecting a hazard evaluation technique for a specific application. The importance that each of these categories has on the selection process may vary from facility to facility, company to company, and industry to industry. However, the following general observations about the relative significance of these factors should be true for nearly every situation.

1. The ***motivation for the study*** and ***type of results needed*** should be the most important factors that analysts consider. The selected technique should be the most effective way to provide the information required to satisfy the reasons for the study. Other factors should not overshadow these concerns; otherwise, the product of the study may not meet an organization's risk management needs. The motivation for the study and the need for particular results usually determine how important other selection factors are.

Table 6.1 Categories of factors that could influence the selection of hazard evaluation techniques

- Motivation for the study
- Type of results needed
- Type of information available to perform the study
- Characteristics of the analysis problem
- Perceived risk associated with the subject process or activity
- Resource availability and analyst/management preference

2. The ***type of information available, characteristics of the analysis problem,*** and ***perceived risk*** associated with the process or activity are factors dealing with the inherent boundary conditions of the subject analysis; these factors represent conditions over which the analyst typically has no control. If these factors dominate the analyst's choice, he or she may not be able to choose any techniques except those allowed by these factors.

   For example, if all other factors lead an analyst to believe that the Fault Tree Analysis technique should be used for a particular situation, but there are no detailed process drawings available from which to define the system and its failure characteristics, then the analyst must either correct the information deficiency (i.e., get detailed drawings) or choose another technique. If the drawings are not available because they have not been updated, then it may be possible (although time-consuming and expensive) to develop updated drawings. On the other hand, if the subject process is in the conceptual design phase, then it will be impossible to obtain detailed drawings. Another technique should be selected unless the hazard evaluation team leader believes that the Fault Tree Analysis technique can be applied in a less detailed fashion and still satisfy the objectives of the hazard evaluation.

3. The last category involves ***resource availability*** and ***analyst/management preference***. Although important considerations, they should not dominate the selection of hazard evaluation techniques. Too often, though, these factors are the main ones the analyst considers. Choosing a hazard evaluation technique based solely on its low cost or because the particular technique is frequently used can lead to inefficient, low-quality, or unsuitable results.

The following sections discuss each category and provide examples of factors that analysts should consider when selecting an appropriate hazard evaluation technique.

Motivation for the Hazard Evaluation

This category of factors should be the most important to every hazard analyst. Performing a hazard evaluation without understanding its motivation and without having a well-defined purpose is likely to waste safety improvement resources. A number of issues can shape the purpose of a given study. For example, what is the impetus for doing the study in the first place? Is the study being chartered as part of a policy for performing hazard evaluations of new processes? Are insights needed to make risk management decisions concerning the improvement of a mature, existing process? Or is the study being done to satisfy a regulatory or legal requirement?

Hazard analysts responsible for selecting the most appropriate technique and assembling the necessary human, technical, and physical resources must be provided a well-defined, written purpose so that they can efficiently execute the study's charter.

Type of Results Needed

Depending on the motivation for a hazard evaluation, a variety of results could be needed to satisfy the study's charter. Defining the specific type of information needed to satisfy the objective of the hazard evaluation is an important part of selecting the most appropriate hazard evaluation technique. The following are five categories of information that can be produced from hazard evaluations:

- List of hazards
- List of potential incident situations
- List of alternatives for reducing risk or areas needing further study
- Prioritization of results
- Input for a quantitative risk analysis

As described in Chapter 3, some hazard evaluation techniques can be used solely to identify the hazards associated with a process or activity. If that is the only purpose of the study, then a technique can be selected that will provide a list or a "screening" of areas of the process or operation that possess a particular hazardous characteristic.

Nearly all hazard evaluation techniques can provide lists of potential incident situations and possible risk reduction alternatives (i.e., action items); a few of the hazard evaluation techniques can also be used to prioritize the action items based on the team's perception of the level of risk associated with the situation that the action item addresses. Chapter 8 discusses the use of hazard evaluation techniques for prioritizing the results of a hazard evaluation. If an organization can anticipate that their need for risk management information is not likely to be satisfied by a qualitative analysis, then a hazard analyst may elect to use a hazard evaluation technique that provides more definitive input as a basis for performing a quantitative risk analysis, in the event that such an analysis is needed.

Type of Information Available to Perform the Study

There are two conditions that define what information is available to the hazard evaluation team: (1) the stage of life the process or activity is in when the study needs to be performed and (2) the quality and currentness of the available documentation. The first condition is fixed for any hazard evaluation, and the analyst cannot do anything to change it. Table 6.2 shows what information becomes available through the plant's evolution.

The stage of life of the process establishes the practical limit of detailed information available to the hazard evaluation team. For example, if a hazard evaluation is to be performed on the conceptual design of a process, it is highly unlikely that an organization will have already produced a P&ID for the proposed process. Thus, if the analyst must choose between HAZOP Study and What-If Analysis, then this "phase-of-life" factor would dictate that the What-If Analysis method should be used, since there is not enough information to perform an adequate HAZOP Study. Figure 6.1 illustrates which techniques are commonly used for hazard evaluations at various phases of a process life cycle.

An analyst cannot accelerate the development of a process just so he or she can use a certain hazard evaluation technique. However, analysts may use a detailed technique in a less detailed way. ***Ultimately, if the analysts believe that, because of the lack of information, the objectives of the study cannot be met using an appropriate hazard evaluation technique, they should recommend to management that their objectives be reexamined or the study be delayed until sufficient information becomes available.***

Table 6.2 Typical information available to hazard analysts

Type of information	Increasing level of detail	Time when information becomes available from project inception
▪ Specific operating experience	↑	↑
▪ Operating procedures		
▪ Existing equipment		
▪ Piping and instrumentation diagrams		
▪ Process flow diagrams (PFDs)		
▪ Experience with similar processes		
▪ Material inventories		
▪ Basic process chemistry		
▪ Material, physical, and chemical data		

The second condition deals with the quality and currentness of the documentation that does exist. For a hazard evaluation of an existing process, hazard analysts may find that the P&IDs are not up-to-date or do not exist in a suitable form. Using any hazard evaluation technique on out-of-date process information is not only futile; it is a waste of time and resources. Thus, if all other factors point to using a technique (e.g., the HAZOP Study technique) for the proposed hazard evaluation that requires such information, then the analysts should ask management to have the necessary, up-to-date process drawings created. ***As discussed in Chapter 2, an important part of an overall hazard evaluation program is establishing a foundation to support hazard evaluations. Good planning in the creation of this information can help avoid delays in the performance of hazard evaluations.***

Characteristics of the Analysis Problem

To choose a hazard evaluation technique, an analyst should look at certain characteristics of the plant or process being studied. These characteristics can be divided into five areas: (1) the complexity and size of the problem, (2) the type of process, (3) the type of operation(s) included in the process, (4) the nature of the inherent hazards, and (5) the incidents or situations of concern.

The ***complexity and size*** of the problem are important because some hazard evaluation techniques can get bogged down when used to analyze extremely complicated problems. The complexity and size of a problem are functions of the number of processes or systems being analyzed, the number of pieces of equipment in each process or system, the number of operating steps, and the number and types of hazards and effects being analyzed (e.g., toxic, fire, explosion, economic, or environmental).

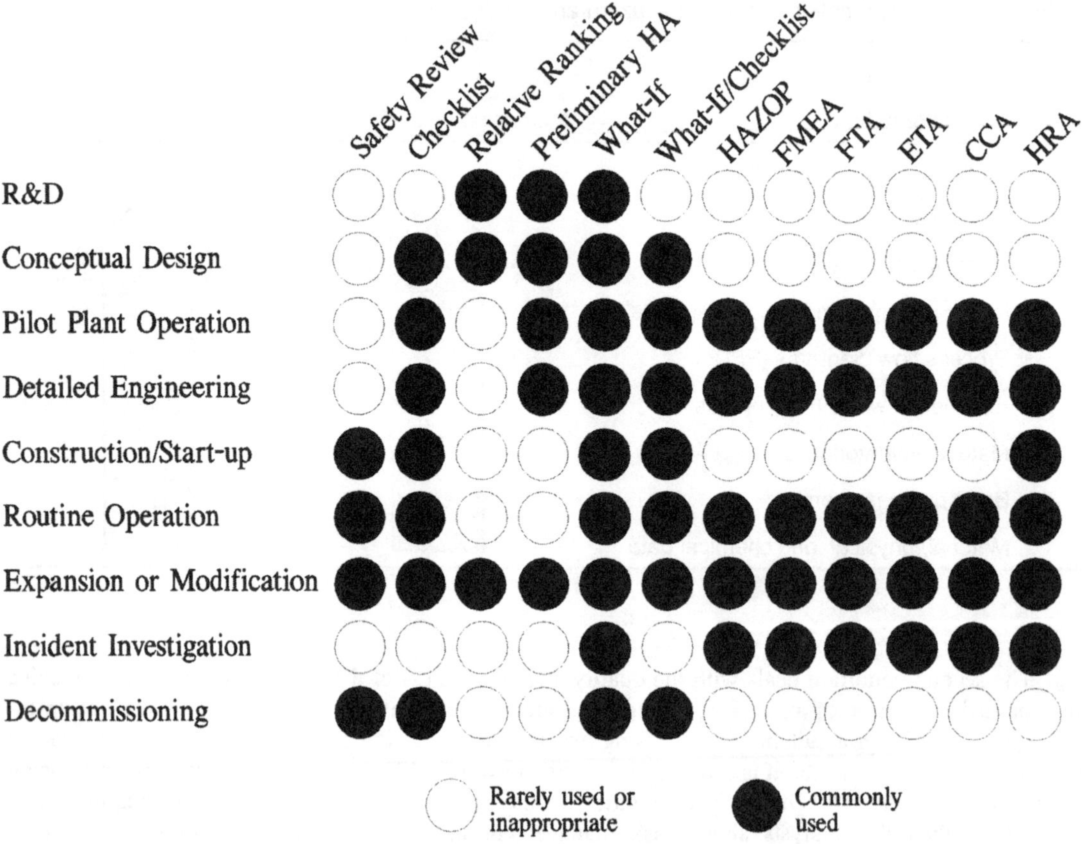

Figure 6.1 Typical uses for hazard evaluation techniques

It is particularly important that hazard analysts select a level of resolution that is compatible with the purpose of the study. For example, if a large facility is to be analyzed, a prudent hazard evaluation team leader should divide the facility into as many smaller pieces as necessary for analysis. Different techniques may be used to analyze each part of the process, depending upon the characteristics of each analysis problem. However, if the purpose of the hazard evaluation is primarily to screen hazards (e.g., develop emergency response plans), analysts should choose a level of resolution that looks at systems rather than individual components. For emergency planning purposes, an analyst might use a What-If Analysis to identify general types of incident sequences that can have an impact on the plant population.

For many hazard evaluation techniques, considering a larger number of equipment items or operating steps will increase the time and effort needed to perform a study. For example, using the FMEA technique will generally take five times more effort for a process containing 100 equipment items than for a process containing 20 items. The HAZOP meeting time for analyzing a batch reactor system consisting of 50 operating steps will take about twice as long as for a batch process with 25 steps. Thus, the types and number of hazards and effects being evaluated is proportional to the effort required to perform a hazard evaluation, although in some cases it may not be a linear relationship.

Analysts should carefully consider the extra time and effort it will take to analyze a variety of hazards in a complex system. For example, analyzing all types of hazards in a complex process at the same time may make it difficult for the analysts to focus consistently on the significance of incident situations involving one class of the analyzed hazards. On the other hand, complex systems that have many similar or redundant pieces of equipment or features may not take as long to analyze.

The ***type of process*** (Table 6.3) also affects the selection of a hazard evaluation technique. Individual processes can be composed of one or more of these process types. Most of the hazard evaluation techniques covered in these *Guidelines* can be used for almost any type or combination of process types. However, certain hazard evaluation techniques are better suited for particular processes than others. For example, the FMEA approach has a well-deserved reputation for efficiently analyzing the hazards associated with electronic and computer systems, whereas the HAZOP Study approach may not work as well for these types of systems.

The ***type of operations*** included in the subject process also influences the selection of hazard evaluation techniques. Whether an operation is (1) a fixed facility or a transportation system; (2) permanent or transient; and (3) continuous, semi-batch, or batch can affect the selection of techniques. All of the techniques in these *Guidelines* can be used for analyzing fixed facilities or for transportation operations. Because potential incidents involving transportation systems typically involve single, discrete events (e.g., vehicle failures due to impact), single-failure analysis methods such as FMEA, What-If Analysis, or What-If/Checklist Analysis are used more often than FTA. However, sometimes ETA is used to consider the combination of circumstances surrounding a spill from a transport vehicle.

Table 6.3 Types of processes

Process type	Example
Chemical	Catalytic reaction in a chemical reactor vessel
Physical	Separation of a chemical mixture in a distillation column
Mechanical	Dry material handling in a screw conveyor
Biological	Fermentation in an incubation chamber
Electrical	480 VAC power supply system
Electronic	Integrated circuits in a PLC
Computer	Microprocessor-based digital control system
Human	Manual mixing of two chemicals in an open vat

The permanency of the process can affect the selection decision in the following way: if all other factors are equal, analysts may use a more detailed, exhaustive approach if they know that the subject process will operate continuously over a long period of time. The more detailed, and perhaps better documented, analysis of the permanent operation could be used to support many process safety management activities. For example, a HAZOP Study table listing the detailed evaluation of types of process upsets, causes, consequences, safeguards, etc., could be used in an operator training program.

On the other hand, analysts may choose a less extensive technique if the subject activity is a one-time operation. An analyst may be better served using the Safety Review method to evaluate a one-time maintenance activity. However, analysts are cautioned to recognize that a temporary operation can present significant hazards and could justify the use of a more detailed hazard evaluation technique such as Fault Tree Analysis.

Finally, some methods such as What-If Analysis, What-If/Checklist Analysis, HAZOP Studies, Event Tree Analysis and Human Reliability Analysis are better able to analyze batch processes than others (e.g., Fault Tree Analysis, FMEA, Cause-Consequence Analysis) because the latter methods cannot easily deal with the need to evaluate the time-dependent nature of batch operations.

The ***nature of the hazards*** associated with the process has a minor influence on the selection of a hazard evaluation technique. Toxicity, fire, explosion, and reactivity hazards can all be analyzed with any of the hazard evaluation techniques covered in these *Guidelines*, although some of the published Relative Ranking indexes will cover only certain hazards (e.g., the Dow F&EI only covers fire and explosion hazards).

The charter of a hazard evaluation may address a variety of ***types of failures, events, or situations of concern***. Whether a study focuses on (1) single failures versus multiple failures; (2) simple loss of containment events; (3) loss of function events; (4) process upsets; or (5) hardware, procedure, software, or human failures can affect the technique selection decision. The biggest influence in this category of factors is whether the analysis is directed at evaluating complex, multiple failure situations. Fault Tree Analysis, Event Tree Analysis, Cause-Consequence Analysis, and Human Reliability Analysis techniques are primarily used for these situations. Single-failure-oriented methods such as HAZOP Studies and FMEA are not normally used for this purpose, although they can be extended to evaluate a few simple incident situations involving more than one event. The remaining factors in this category have a relatively minor impact on the selection process.

Perceived Risk of the Subject Process or Activity

If all hazard evaluations were perfect, then it would not matter which hazard evaluation technique is used or who performs the analysis. But, unfortunately, the techniques, analysts, and studies can never be perfect. An important contributor to this imperfection is one of the important limitations of hazard evaluation techniques and studies discussed in Chapter 1—the issue of completeness. Neither a hazard evaluation technique nor an analyst can guarantee that all possible incident situations involving a process have been identified.

Organizations deal with the limitation of completeness in two main ways. First, they use interdisciplinary teams to perform the analysis, capitalizing on the team members' combined experience. This "many heads are better than one" strategy is the key to performing high-quality hazard evaluations when using certain techniques (e.g., HAZOP Studies, What-If/Checklist Analysis). Second, organizations tend to use more systematic techniques for those processes that they believe pose higher risk (or, at least, for situations in which incidents are expected to have severe consequences). Thus, the greater the perceived risk of the process, the more important it is to use hazard evaluation techniques that minimize the chance of missing an important incident situation.

Recall from Chapter 1 that an organization has several types of information at its disposal to help the analysts understand the inherent risk of a process or activity:

- The amount of experience with the process
- The nature of the experience with the process
- The continued relevance of process experience

The most important experience factor is the length of time over which the experience is gained. Has the process been operating for over 30 years, and are there many such processes operating within the organization and the industry? Or is the process relatively new? For a new process involving first-of-its-kind technology that is still in the design phase, an organization may have absolutely no experience with the subject process. Sometimes, there may be some similar company or industry experience that members of an organization can draw upon to derive their understanding of risk.

The next experience factor deals with the actual operating record of the process. Have there been frequent, high-consequence incidents? Or have there only been a few minor incidents and near misses? Sometimes a process will have operated for many years and never have experienced a major incident, even though the potential has always existed. The most immediate "experience" that has the greatest impact on an organization's risk perception is a recent incident that motivates managers to perform a hazard evaluation as a part of the follow-up investigation.

The last experience factor deals with the current relevance of the experience base to the subject process. There may have been many changes to the process that invalidated the operating experience as a current indicator of process risk. Or there may have been only a few minor changes over the years that have been adequately dealt with by the organization's management of change policy. In the latter case the organization may justifiably have confidence that the existing experience base may be a good predictor of the subject process's future safety performance.

All of these factors contribute to the level of confidence or concern that an organization has about process risk. Typically, when (1) the subject process has operated relatively free of incidents over a long time and the potential for a high-consequence incident is perceived to be low, and (2) there have been few changes to the process that would invalidate this experience base, then organizations will tend to select less exhaustive, less systematic, more experience-based hazard evaluation techniques, such as Safety Review and Checklist Analysis. When the opposite is perceived, more rigorous, predictive techniques are generally preferred, such as HAZOP Studies, What-If/Checklist Analysis, Fault Tree Analysis, and so forth.

Resource Availability and Preference

A variety of other factors can influence the selection of hazard evaluation techniques. Some factors that customarily affect technique selection are: (1) availability of skilled and knowledgeable personnel, (2) target dates by which to perform the study, (3) financial resources, (4) preference of the hazard analysts, and (5) preference of the manager(s) that charters the hazard evaluation.

Generally, two types of personnel must be available for the hazard evaluation: (1) skilled leaders and practitioners of the particular hazard evaluation technique chosen, and (2) people knowledgeable in the process or activity being analyzed. Chapter 2 describes the types of technical knowledge that may be needed for a hazard evaluation. If the necessary design engineers, operators, maintenance personnel, etc., are not available, then the quality of the hazard evaluation is in jeopardy. People skilled in the use of particular hazard evaluation techniques are also important for the effective performance of hazard evaluations. Some techniques, because of their inherent systematic nature, may require less practiced leaders than would other techniques. However, experience has shown that having a hazard evaluation team leader who has the proper training and has facilitated and completed many such studies increases the chance of having a successful study.

Many hazard evaluation techniques require the creative interaction of participants on a team. Team meetings can typically last for days, weeks, or months, depending upon the complexity of the subject process. Other techniques (e.g., Fault Tree Analysis) may be performed primarily by individuals working alone. However, these detailed, single-analyst approaches require a "gestation period" to enable the analyst to create realistic models of the causes of potential incidents. Team situations may not be as helpful when using these techniques; however, these models may be constructed based on information derived from a team meeting or may efficiently be reviewed in a team meeting environment. Altogether, schedule constraints should take a back seat to other technical concerns. However, if the decision concerning selection of a technique comes down to two otherwise equal approaches, then the hazard analyst should select the one that will produce the desired results using the fewest resources in the least time.

Realistic estimates for funding hazard evaluations are necessary for an organization to adequately plan for the performance of all process safety management activities. Table 6.4 summarizes the technical effort estimates presented in Chapters 4 and 5 for performing hazard evaluations of small systems and large processes. These estimates are provided only to give analysts a rough idea of the relative scale of effort for which they should plan when performing a hazard evaluation. In addition, because there are so many other factors that influence time and effort, analysts should use these estimates with great caution. The actual time required for a study may be much longer (or somewhat shorter) than these estimates indicate. Documentation times for some methods can be significantly lower with the effective use of computer software programs and by completing as much as possible of the analysis documentation as the study progresses. As analysts and organizations gain experience with each hazard evaluation technique, they should become better equipped to accurately estimate the time and resource requirements for hazard evaluations conducted on their facilities and become more efficient in performing hazard evaluations.

Hazard evaluations done with an inadequate budget, marginally staffed, and under tight or unrealistic schedule constraints are usually not destined for success. ***The quality of the results from a hazard evaluation is inevitably a strong function of the quality of the team and its efforts.*** If adequate in-house personnel are unavailable to lead hazard evaluations, then an organization should acquire training for its prospective hazard analysts. Under tighter schedule constraints, outside consultants can be used to lead and document hazard evaluations. In the end, organizations should try to do as many of these studies as possible using in-house personnel to better capitalize on the overall learning experience.

Ideally, hazard evaluations should be performed using techniques that are most familiar to the hazard evaluation team leader and other study participants. In addition, management may sometimes have a preference for one technique over another. However, management preference should not overshadow other technical reasons for selecting a particular method. To avoid unproductive disputes concerning selection of hazard evaluation techniques, hazard analysts should help educate management with tangible examples of the benefits, strengths, limitations, and relative costs of each technique (see Chapters 4 and 5).

Table 6.4 Summary of typical staff effort estimates for hazard evaluation techniques

	Phase of hazard evaluation							
	Preparation		Modeling		Evaluation		Documentation	
Technique	Small system	Large process	Small system	Large process	Small system	Large process	Small system	Large process
Safety Review	2 to 4h*	1 to 3d*	n/a*	n/a	4 to 8h	3 to 5d	4 to 8h	3 to 6d
Checklist Analysis	2 to 4h	1 to 3d	n/a	n/a	4 to 8h	3 to 5d	4 to 8h	2 to 4d
Relative Ranking	2 to 4h	1 to 3d	n/a	n/a	4 to 8h	3 to 5d	4 to 8h	3 to 5d
Preliminary Hazard Analysis	4 to 8h	1 to 3d	n/a	n/a	1 to 3d	4 to 7d	1 to 2d	4 to 7d
What-If Analysis	4 to 8h	1 to 3d	n/a	n/a	4 to 8h	3 to 5d	1 to 2d	1 to 3w
What-If / Checklist Analysis	6 to 12h	1 to 3d	n/a	n/a	6 to 12h	4 to 7d	4 to 8h	1 to 3w
HAZOP Study	8 to 12h	2 to 4d	n/a	n/a	2 to 6d	2 to 6w	2 to 6d	2 to 6w
FMEA	2 to 6h	1 to 3d	n/a	n/a	1 to 3d	1 to 3w	1 to 3d	2 to 4w
Fault Tree Analysis	1 to 3d	4 to 6d	3 to 6d	2 to 3w	2 to 4d	1 to 4w	3 to 5d	3 to 5w
Event Tree Analysis	1 to 2d	4 to 6d	1 to 3d	1 to 2w	1 to 2d	1 to 2w	3 to 5d	3 to 5w
Cause-Consequence Analysis	1 to 2d	4 to 6d	1 to 3d	1 to 2w	1 to 3d	1 to 2w	3 to 5d	3 to 5w
Human Reliability Analysis	4 to 8h	1 to 3d	1 to 3d	1 to 2w	1 to 2d	1 to 2w	3 to 5d	1 to 3w

* h = hours, d = days, w = weeks, m = months, n/a = not applicable

6.2 Decision-Making Process for Selecting Hazard Evaluation Techniques

Each hazard evaluation technique has unique strengths and weaknesses. Moreover, each industry, organization, facility, and process/activity will have unique objectives and needs when it comes to performing hazard evaluations. The six categories of factors discussed in Section 6.1 may have varying degrees of importance, depending upon the circumstances for each particular application of hazard evaluation techniques. Thus, it is difficult to construct a universal decision-making flowchart that would be correct for every organization and facility. However, it is possible to suggest a logical order for considering the factors discussed in Section 6.1.

Figure 6.2 illustrates a general order for considering the various factors that could influence which hazard evaluation technique is used for a given study. Certainly, the factors involving motivation and type of results should be most important to every organization; these factors provide the basic definition for satisfying the need for greater risk understanding, which likely precipitated the charter for a hazard evaluation. The information available, characteristics of the problem, and perceived risk may have varying degrees of importance placed upon them, depending upon the culture of the sponsoring organization and facility. The amount of resources needed to support a hazard evaluation team should be the last factor considered in technique selection, although it should be wisely used to select between otherwise equal analysis technique alternatives.

6.3 Example Using the Proposed Selection Criteria

Even though it is difficult to develop a universally applicable decision logic for organizations to use in selecting hazard evaluation techniques, it is appropriate to illustrate the process one would use to develop such a framework. Figure 6.3 is a detailed flowchart that organizations may use (1) to directly select hazard evaluation techniques or (2) to help develop their own internal guidelines and philosophies regarding the use of hazard evaluation techniques. Borderline cases may indicate a situation where the use of more than one methodology is appropriate. The following abbreviations for the hazard evaluation techniques are used in the flowchart.

PreHA	Preliminary Hazard Analysis	**HAZOP**	Hazard and Operability Study
SR	Safety Review	**FMEA**	Failure Modes and Effects Analysis
RR	Relative Ranking	**FTA**	Fault Tree Analysis
CL	Checklist Analysis	**ETA**	Event Tree Analysis
WI	What-If Analysis	**CCA**	Cause-Consequence Analysis
WI/CL	What-If/Checklist Analysis	**HRA**	Human Reliability Analysis

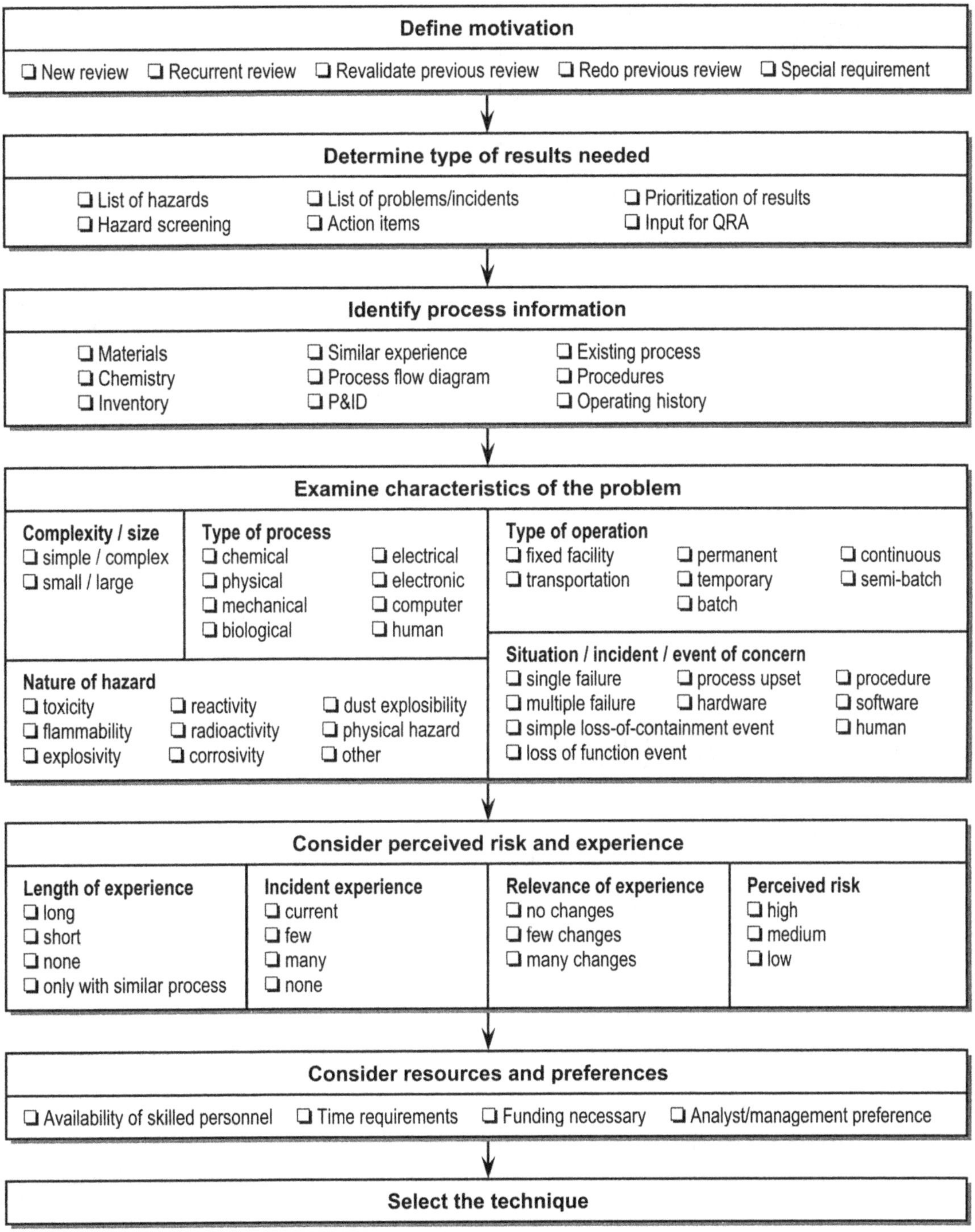

Define motivation

❑ New review ❑ Recurrent review ❑ Revalidate previous review ❑ Redo previous review ❑ Special requirement

↓

Determine type of results needed

❑ List of hazards
❑ Hazard screening
❑ List of problems/incidents
❑ Action items
❑ Prioritization of results
❑ Input for QRA

↓

Identify process information

❑ Materials
❑ Chemistry
❑ Inventory
❑ Similar experience
❑ Process flow diagram
❑ P&ID
❑ Existing process
❑ Procedures
❑ Operating history

↓

Examine characteristics of the problem

Complexity / size
❑ simple / complex
❑ small / large

Type of process
❑ chemical ❑ electrical
❑ physical ❑ electronic
❑ mechanical ❑ computer
❑ biological ❑ human

Type of operation
❑ fixed facility ❑ permanent ❑ continuous
❑ transportation ❑ temporary ❑ semi-batch
❑ batch

Nature of hazard
❑ toxicity ❑ reactivity ❑ dust explosibility
❑ flammability ❑ radioactivity ❑ physical hazard
❑ explosivity ❑ corrosivity ❑ other

Situation / incident / event of concern
❑ single failure ❑ process upset ❑ procedure
❑ multiple failure ❑ hardware ❑ software
❑ simple loss-of-containment event ❑ human
❑ loss of function event

↓

Consider perceived risk and experience

Length of experience
❑ long
❑ short
❑ none
❑ only with similar process

Incident experience
❑ current
❑ few
❑ many
❑ none

Relevance of experience
❑ no changes
❑ few changes
❑ many changes

Perceived risk
❑ high
❑ medium
❑ low

↓

Consider resources and preferences

❑ Availability of skilled personnel ❑ Time requirements ❑ Funding necessary ❑ Analyst/management preference

↓

Select the technique

Figure 6.2 Criteria for selecting hazard evaluating techniques

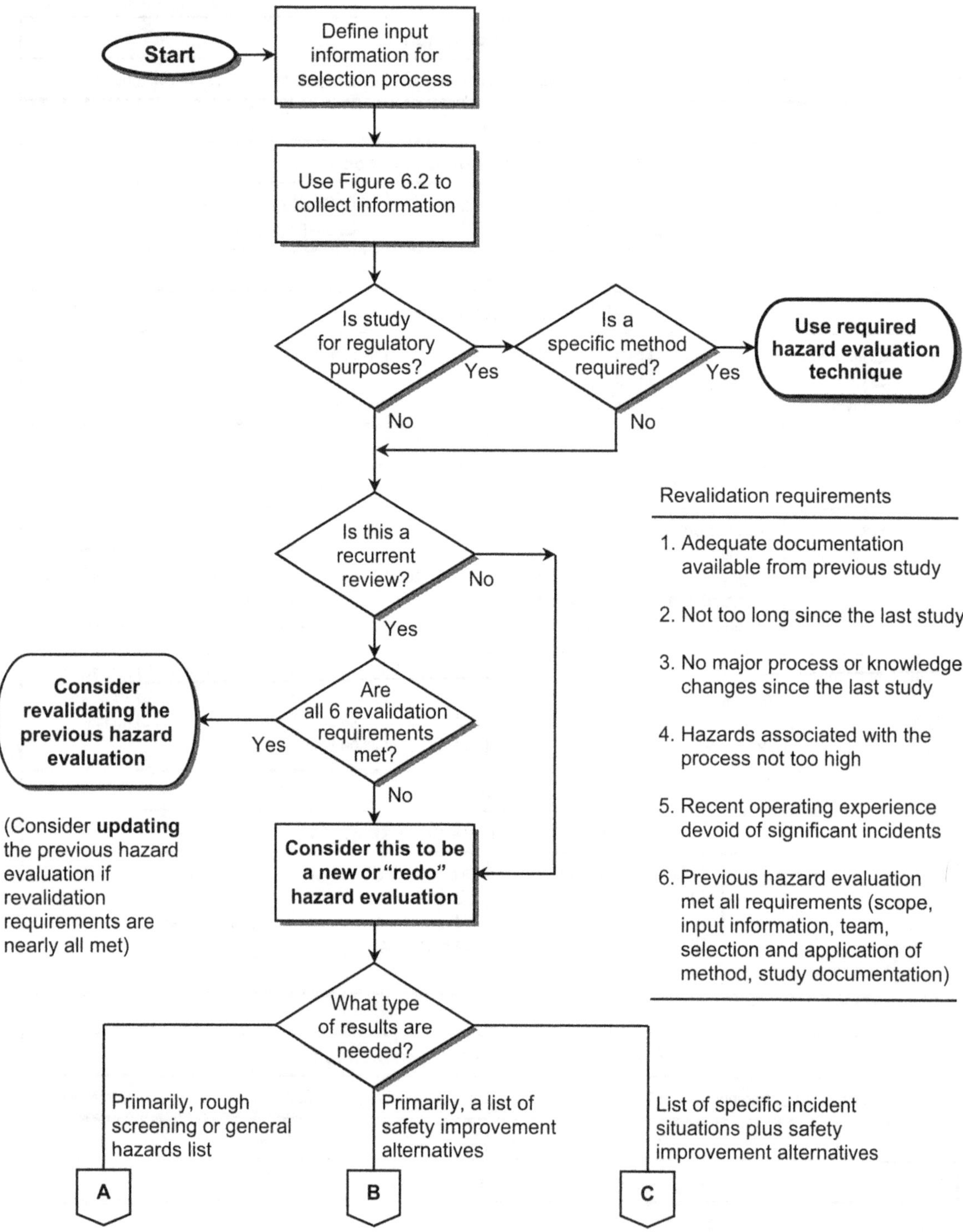

Figure 6.3 Example flowchart for selecting a hazard evaluation technique (page 1 of 7)

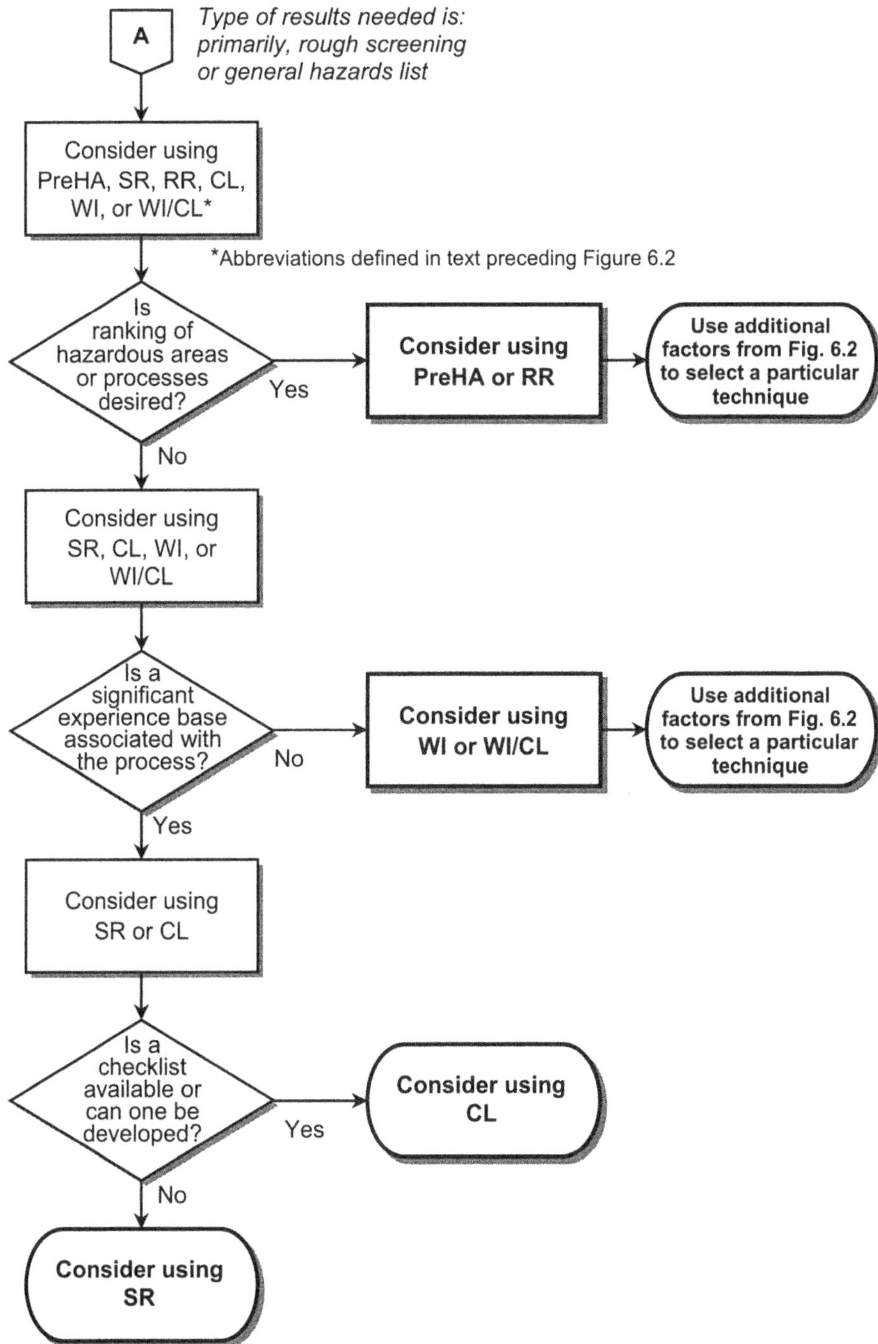

Figure 6.3 Example flowchart for selecting a hazard evaluation technique (page 2 of 7)

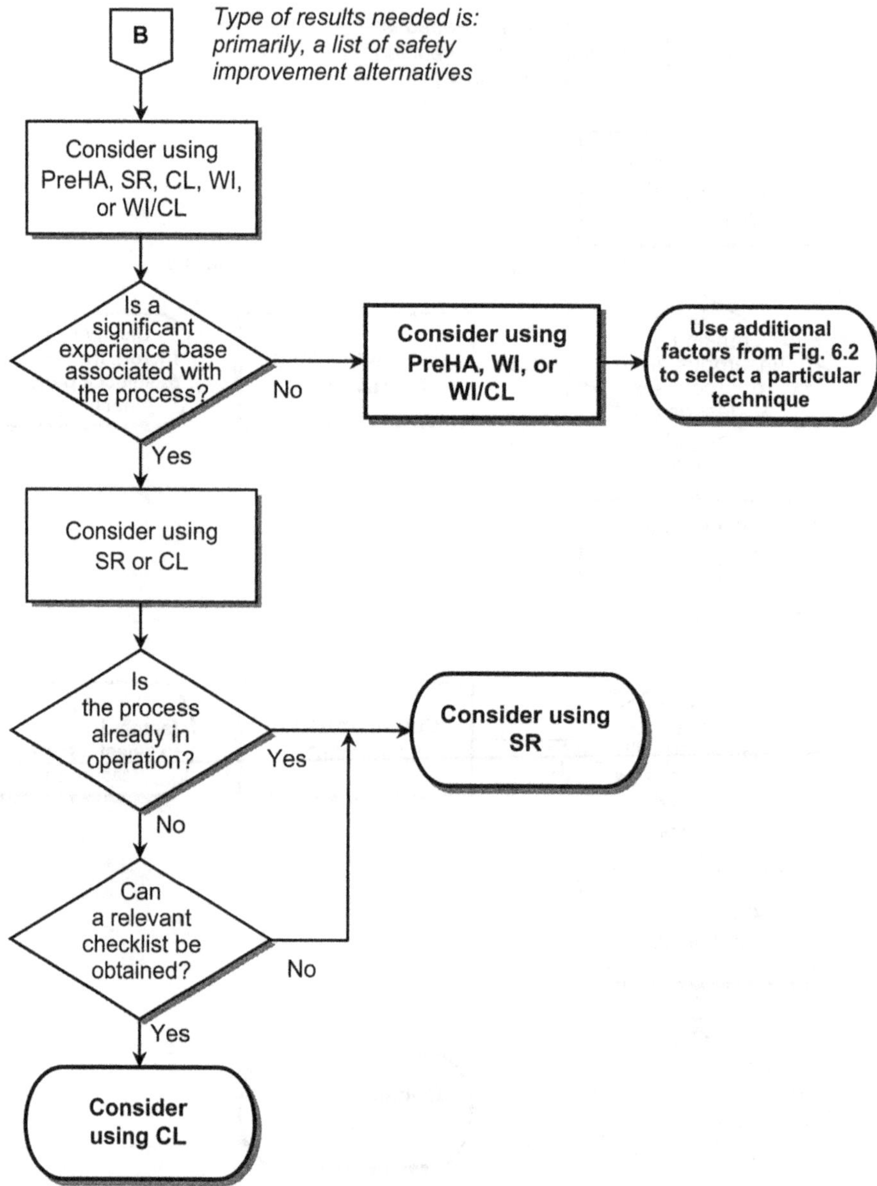

Figure 6.3 Example flowchart for selecting a hazard evaluation technique (page 3 of 7)

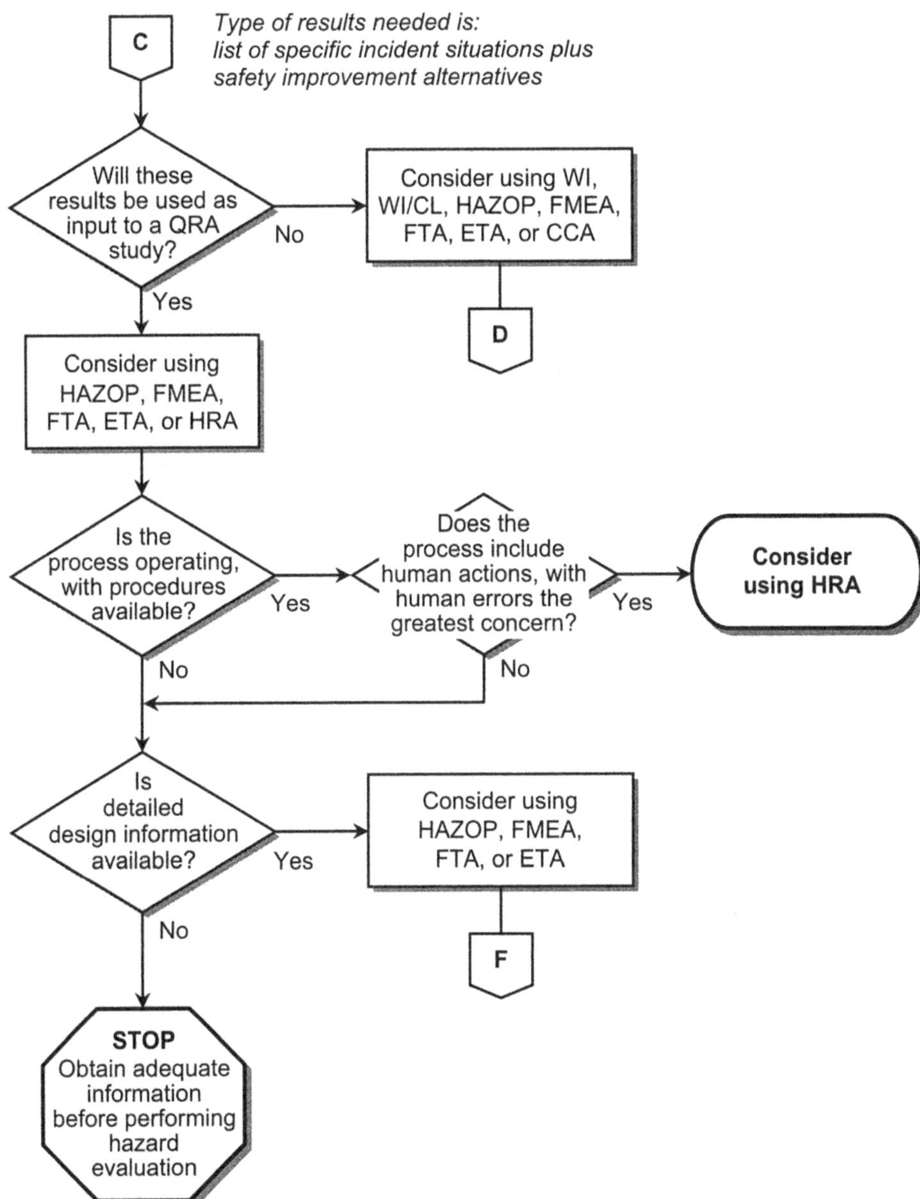

Figure 6.3 Example flowchart for selecting a hazard evaluation technique (page 4 of 7)

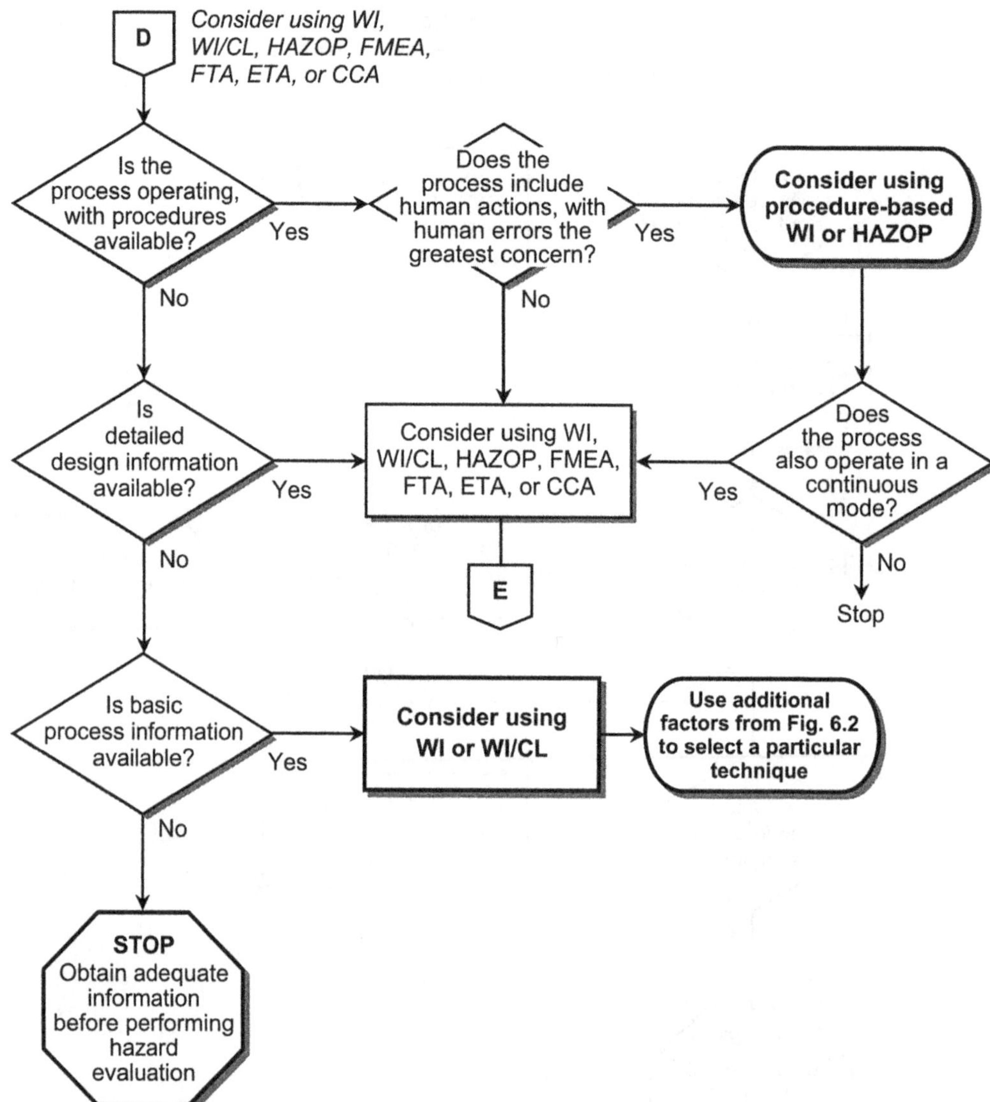

Figure 6.3 Example flowchart for selecting a hazard evaluation technique (page 5 of 7)

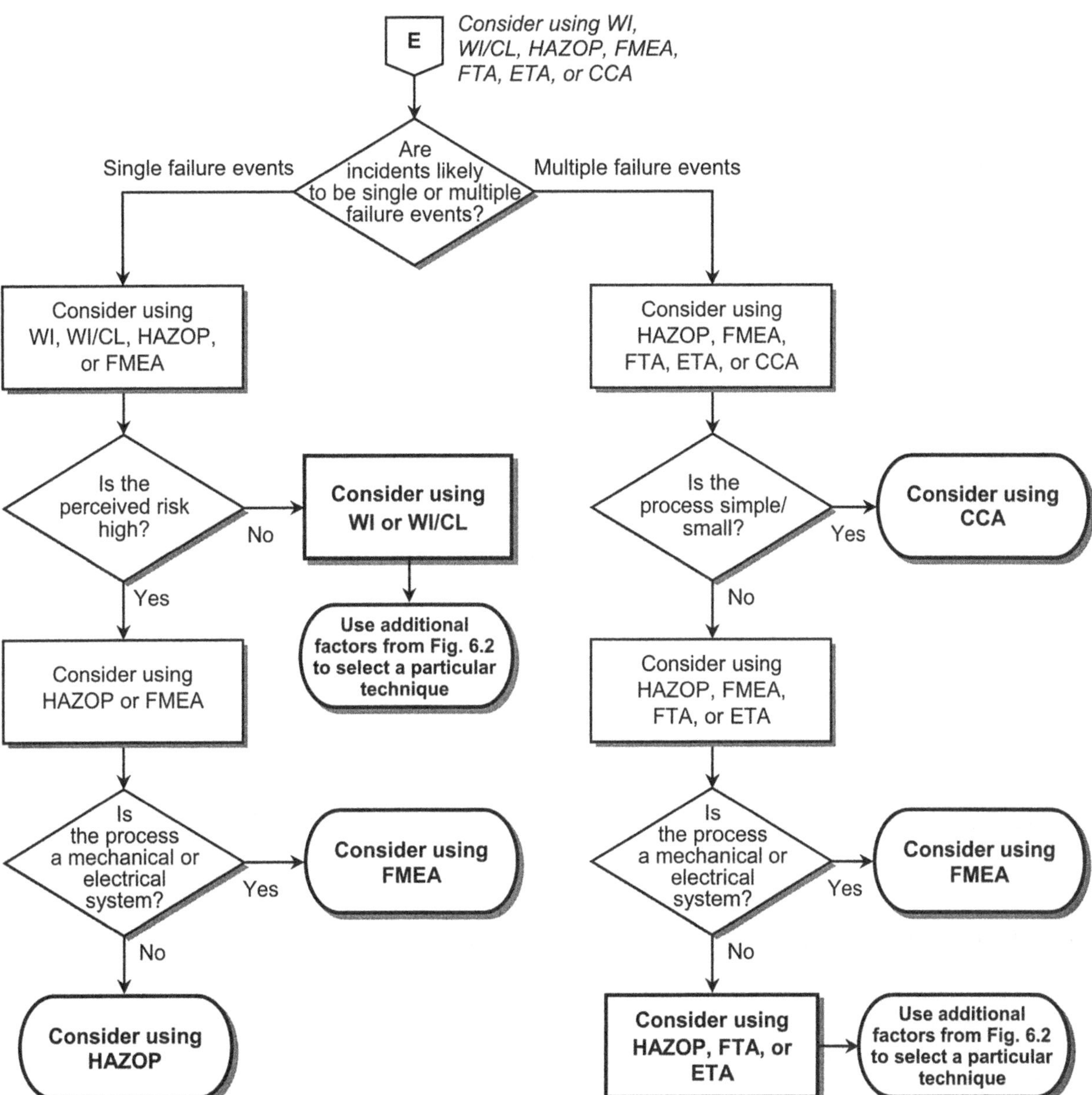

Figure 6.3 Example flowchart for selecting a hazard evaluation technique (page 6 of 7)

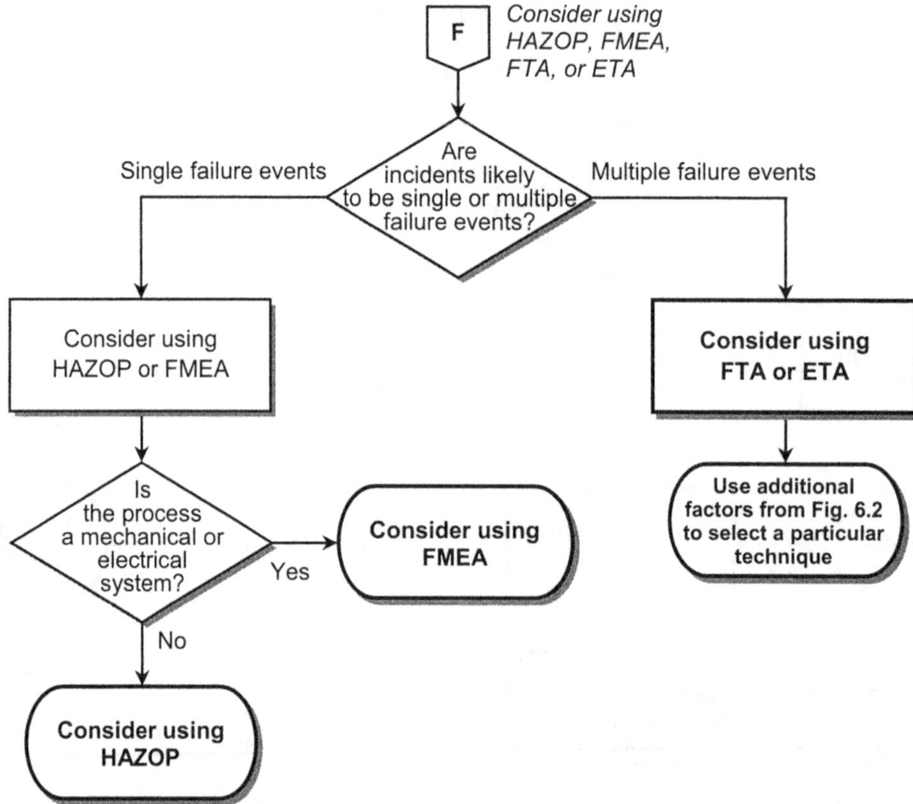

Figure 6.3 Example flowchart for selecting a hazard evaluation technique (page 7 of 7)

6.4 Hazard Reviews for Management of Changes

As a practical manifestation of continuing operations, change is an inevitable and necessary feature for all organizations. When changes occur in an operation that is hazardous or contains hazards of any sort, it is imperative that the change process be managed to understand and control those hazards. Most organizations have some management of change (MOC) policies and procedures in place to address the wide range of issues related to changes. The objective of this section is to provide guidance in how to effectively include hazard evaluation in the MOC process.

The management of change process addresses the entire range of process safety issues. The scope of this section is limited to guidance in evaluating the impact of the change on (acute) health and environmental impacts. Several CCPS *Guidelines* books include chapters related to the general practical implementation and application of MOC,[9-11] and one comprehensive text is devoted solely to the topic.[12]

Change in an operating processing plant is expressed in many ways. Of critical concern is the way safety and health evaluations are recognized and addressed. Every change to the process must be examined for impact on the basic process control system, operating and maintenance procedures, the integrity of primary and secondary containment systems, and all passive, active, and administrative safeguards. One must start with the assumption that any change is going to affect process safety, then devise a procedure to specifically identify adverse effects on process risks.

It is common to use a checklist to ensure that all safety and regulatory items are addressed and documented during a process change. Examples of lists are available in the referenced texts and are typically modified to meet the needs of the enterprise. A checklist for management of change review can contain 20 to 50 items. Table 6.5 has a list of items that are related to environment, health and safety that should be on every MOC review document.

The left column of Table 6.5 contains items that have direct impact on how the new operation runs as a result of the change. The right column contains items that address the documentation of the review process.

A process change can range from simple to complex. Large changes in processes are recognized as major or complex and generally lead to adherence to established MOC policies. However, small changes to processes, such as to facility maintenance, can sometimes be overlooked as an item that requires a MOC review. As such, small but significant compromises to the process safety systems may be inadvertently introduced. Thus, maintenance, or other similar tasks, should include a method to remind facility personnel of potential management of change issues. If the work constitutes a true "replacement in kind," no further action would be necessary. However, if the work causes a replacement or change that is not "in-kind," the management of change process for the facility must be completed to ensure that safeguards remain adequate. Where safety and environmental implications of the change are minor and well understood, a checklist reviewed and approved by an authorized person may suffice. Chapter 18 in Part II gives a worked example of a Safety Review conducted as part of a management of change process.

Table 6.5 MOC review documents related to environment, health and safety

▪ Hazard communication review (MSDS update)	▪ Safety systems and their functions update
▪ Safe upper and lower limits for process parameters	▪ New/revised operating procedures
▪ Change in maximum intended inventory	▪ Relief or ventilation system design update
▪ Re-evaluation of consequences of deviation	▪ Electrical classification drawing update
▪ Process hazard analysis (risk analysis)	▪ Process flow diagram update
▪ Pre-start-up safety review	▪ Piping & instrumentation diagrams update

It is possible that during an emergency, it may become necessary to implement a process change outside the enterprise MOC procedures. To prevent creation of unacceptable risk of harm or loss, procedures need to be in place to address such emergency changes. The following is a brief outline of an example procedure. If an emergency change is to be implemented, three people (one of whom is a supervisor or manager) knowledgeable in the process discuss the proposed change. The discussion must consider the safety aspects of making the change. If the team concurs that the change can be done safely, the change may be implemented immediately. The supervisor responsible for this decision documents the parameters that were considered in connection with the change (i.e., impact on safety and health, risk of waiting). As soon as feasible after the emergency is under control, the routine enterprise MOC procedures are followed.

Temporary changes have been the cause of numerous incidents. An example of a temporary change is exceeding an established operating limit during controlled testing and development of a process. During the testing, a number of changes could occur over the short duration of the test. It is important that a time limit for these temporary changes be established and monitored since, without control, these changes have a tendency to become permanent. In the case of process testing, the management of change procedure needs to include a provision that the equipment and procedures are returned to their original condition at the end of the temporary change. It is imperative that a review of these changes include safety, health, and environmental considerations and incorporate those into the temporary operating procedures and processes. ***If the process changes are to become a permanent part of the process, then the impact of the process change on the previous process hazard review needs to be evaluated to determine the adequacy of the existing hazard review.*** A revalidation or redo may be needed.

Crucial to any change is a safety systems and functions review, both for engineered and administrative safeguards. Before a new or changed process is started, a readiness review is needed. One important item to include in the review checklist is verification and validation of safety systems.

In-Depth Change Reviews

If a more thorough review needs to be done to evaluate a change, one or more of the techniques in Chapter 4 and 5 can be used for an in-depth safety analysis of the change, which may be in addition to a previous hazard evaluation completed for the operating facility. As for all hazard evaluations, proper consideration needs to be given to the appropriate skills and number of persons needed for an in-depth change review and sufficient unconstrained time allowed for creative, thoughtful review.

Appendix A1 contains an example of a What-If checklist that has been used in location-led hazard evaluations of facility or operational changes. Note that this particular checklist focuses on mechanical failures and may need to be expanded to fully cover operational issues. Another example of how to document a hazard review for management of change is given in Appendix A2. The more complete listing of safety questions in Appendix B can also be used.

6.5 Combined Hazard Reviews

With the large commitment of time and resources required to conduct thorough hazard evaluations, it is natural to seek more efficient means of completing the hazard evaluations (while still gaining the same benefits). Discussed in this section are valid "short-cuts" that may apply to some types of systems. The use of such combined hazard reviews must always be done with caution, as each approach discussed here has its drawbacks and pitfalls.

Types of Combined Hazard Reviews

Templating similar unit operations. Some types of process facilities have many similar unit operations. For example, refineries have fractionation and purification trains with many distillation columns, reboilers, and overhead condensers. The incident scenarios for similar unit operations can likewise be expected to have many similarities, and this is generally quite true. Loss of steam heat to the reboiler, loss of bottoms level control, loss of reflux, loss of condensing, and various loss of containment scenarios may all have common elements in such a system. The templating approach takes the hazard evaluation completed for one operation and uses it as the starting point for studying a similar unit operation. Adjustments are then made as necessary to account for subtle (but often very important) differences between the systems.

Hazard evaluation of generic technologies. Likewise, manufacturers of certain kinds of packaged units or standardized systems may conduct a hazard evaluation on one basic system, then apply the findings and recommendations from that study to the generic design and consider the hazard evaluation to cover any number of systems supplied to different customers. Examples of such technologies may include ammonia refrigeration systems, gas separation plants, cryogenic nitrogen storage and vaporization systems, and fired equipment such as boilers or off-gas incinerators.

Batch processes and product families. A third category of process facilities that may employ combined hazard reviews includes batch plants that use the same basic equipment, controls, and procedures to manufacture a whole range of similar products by changing raw materials within a family of compounds (e.g., using the same organic compound but with a different carbon chain length) or changing catalysts, process conditions, batch times, or additives. A single hazard evaluation might be used to cover the entire family of similar production runs, by using what is considered the most hazardous recipe or process conditions and applying the results to the full range of products.

Benefits of Combined Hazard Reviews

The most obvious benefit of combined hazard reviews is the savings in time and effort gained by reducing the level of effort needed to complete all required hazard evaluations. Side benefits might include promoting consistency between hazard evaluations conducted on similar units, and giving incentive to focus a more detailed and in-depth review on the standard process.

Potential Pitfalls of Combined Hazard Reviews

Many pitfalls must be avoided when employing combined hazard reviews. Templating similar operations may lead to complacency when quickly covering similar processes from a starting template, with the tendency to not employ the same degree of investigation and careful review to the similar system. Subtle changes between systems can easily be overlooked, as can important differences in such factors as how the systems are operated by different crews, how they are maintained, where they are located, what people and property might be affected by an incident, and how likely ignition of flammable vapors might be. Generic technologies used in different locations might have drastic differences in how they are operated and maintained over time, especially if in different countries and cultures, and the effects of slight modifications over time can cause the generic technologies to diverge from one another. Subtleties in process chemistry can have very important implications for product families being produced in the same batch facilities; for example, a very rapid reaction might be considered the "worst case" and be selected as the basis for the entire review, whereas a slower reaction might actually be more hazardous by allowing unreacted material to accumulate before suddenly initiating a rapid reaction.

Strategies for Combined Hazard Reviews

One strategy that can be used to avoid some of the potential pitfalls of using templates or generic hazard evaluations is to use only the scenarios from the initial study and not the assessment of cause likelihood, safeguards effectiveness, or consequence severity. These latter factors generally have more variation between facilities than the basic underlying scenarios, due to differences in how equipment is tested and maintained, what the operating environment might be including extreme conditions, and what people and adjacent processes might be affected by a fire, explosion or hazardous material release involving the facility. Batch processes could use both upper and lower bounds of e.g. key ingredient reactivity to more confidently cover the family of batch reactions, rather than choosing only one extreme or only the "average" process. It may also be advisable to have a different review team apply a template or generic hazard evaluation to the similar facility, and gain the benefit of a different experience set rather than the same team reviewing many similar processes.

6.6 Hazard Evaluation at Different Plant Lifetime Stages

Process hazards exist from the onset of a project through its end or assimilation into another project. Previous sections address hazard evaluation technique selection(s) and the application of the methods to projects as change occurs. This section will discuss the necessary adjustments in how a hazard evaluation is conducted as a project moves from one phase to another. The message to take from this section is that safety and safe operation must always be kept in mind during all phases of a project.

The following plant lifetime stages will be used in this section:

1) Process development

2) Detailed design

3) Construction and start-up

4) Operating lifetime

5) Extended shutdowns

6) Decommissioning

Each stage is made of a number of sub-stages. For example, process development can include conceptual design, lab studies, and a pilot plant stage. Detailed design can include process specification and design, safety system development, and instrument selection for control. Each stage and sub-stage will vary in the information available for a stage-appropriate hazard evaluation. The Worked Examples in Part II illustrate how hazard evaluations can be employed at different plant lifetime stages.

Process Development

Research and development. A project starts somewhere with a need or opportunity identified. It is not important how or where the project starts, but it is important that environmental, health and safety effects be considered early in the benefit/risk justification. It is at this stage that an initial hazard evaluation is performed. Relatively few hazard evaluation techniques are applicable primarily due to the preliminary nature of the project. However, as reviews of the hazards and refinements are made to the concept through lab and literature work, options become available to consider an inherently safer design, operation, and procedures for the project. This can include exploring alternative chemistry or processes. It is at this early stage that evaluation and implementation of justifiable safety designs will be the least costly and most effective in creating a product or operating a process at the minimum health and environmental risk.

A particularly important element to consider in the process development stage is how heat generation, both the absolute amount and rate, will be addressed by heat control/removal methodologies. An especially valuable bit of data to generate is the adiabatic temperature rise via lab data or thermodynamic calculations. Flammable material management can be examined in this early development stage to minimize, eliminate, or replace flammable materials with less hazardous components.

In addition to the development lab investigations that go into bringing a new product/process along the commercialization path, basic data need to be acquired with respect to runaway reaction potential. A variety of controlled condition tests, such as ARC, ARSST, DSC, etc., are available that help establish runaway potential. More sophisticated reaction calorimetry tests may be needed to help define the safe operating limits for temperature, pressure, and composition, as well as reaction time or residence time.

At some point in the process development stage, a process flow diagram is created to establish an event sequence, material balance, heat balance, and the like. A hazard evaluation update is essential, as critical reaction or process information was likely not available at an earlier hazard evaluation or has since been refined. There is also an opportunity to revisit inherently safer (IS) design options.[14] As the process flow becomes more defined, the safety system development needs to be formalized with initial set points, allowable and safe deviations, and an initial statement of overall health, safety, and environmental risk tolerance.

Pilot plant operations act as segue to the implementation phase. Here the earlier hazard assessments are confirmed and refined. Previous heat management scenarios are played out under plant-like operating conditions. The added benefit is the exposure of the process to a range of start-up, shutdown, and upset conditions. The expected range for the materials of construction can be tested in a controlled manner for effects on the process and associated impact on process safety.

Project Implementation. In a likely parallel effort with the pilot plant work, the front-end engineering would have begun with process, equipment, and instrument specifications as the P&IDs are developed from the process flow diagrams and control strategies. A Preliminary Hazard Analysis is appropriate at this juncture. The results will form the basis for the implementation of a safety system consisting of passive and active safeguards.

At this point in the front-end engineering process, a "Layer of Protection Analysis" (LOPA, Section 7.6) may be applied to quantify the health and safety risk and assess the need for further safeguards. Reference would be made to the initial statement of overall health, safety, and environmental risk tolerance created earlier in the process development stage. If the LOPA indicates that process risk is unacceptable, then additional safeguard(s) would need to be incorporated. The recognition for the need of additional safeguards would lead to functional specification of items such as a safety instrumented system (SIS). CCPS has published guidance on the selection, design and implementation of such instrumented protective systems.[13]

The operating procedures created and used in the pilot plant work would form the framework of the initial operating procedures. Batch operations may require few modifications for full-scale operation. However, the transition of operating procedures for continuous operations from pilot plant to full scale will likely require a complete rewrite. Administrative safeguards, such as preventive verifications and protective responses to upset conditions, can begin to be built into the operating procedures at this stage.

Detailed Design

When the commitment is made to launch the detailed design, major elements of the process are "locked" subject to minor changes. This would imply that the hazards are also "locked". It is appropriate and desirable that a full hazard evaluation be completed as soon as the process P&IDs and major equipment specifications are well-established. It is critical that the hazard evaluation be done before heavy construction is started and major equipment is ordered. This gives the opportunity to adjust, at a minimal cost, to any critical health and safety flaws discovered in the review of the process design and/or procedures, including facility siting considerations (Section 9.6).

Construction and Start-up

Any project manager will recognize the need to control field change orders during construction. The primary driver is construction cost control. Change orders can also have an impact on the safety systems design and integrity. All changes that occur following the full hazard evaluation must be documented and noted on safety documents. Near the end of construction and prior to start-up, the hazard evaluation ~~is~~ may need to be revalidated to incorporate any changes to the process and/or procedures.

Following construction, there are a number of validation tasks to perform. One of the tasks is a readiness review as referenced in Section 6.4. Table 6.6 contains a list of some items to consider when completing the readiness review. Further guidance on the subject has been published by CCPS.[15]

Table 6.6 Some items to consider in a readiness review

- ❑ Construction and equipment meets design specifications
- ❑ Written safety, operating, maintenance, and emergency procedures in place
- ❑ Hazard evaluation performed and all action items completed
- ❑ Training completed and documented for all affected personnel
- ❑ Equipment and area access appropriate for routine operations as well as an emergency
- ❑ Possible body positions acceptable (for valve operation, sampling, maintenance, etc.)
- ❑ Spill protection requirements met according to company guidelines
- ❑ Noise protection devices in place and appropriate signs posted
- ❑ MSDSs in place
- ❑ PFDs / P&IDs / electrical and other appropriate drawings updated / red-lined
- ❑ Safety systems tested and validated
- ❑ Insulation / protection installed for hot or cold operations
- ❑ Tripping and/or head knocking hazards resolved or addressed
- ❑ Relief devices properly installed and directed to a safe location
- ❑ Hazard / warning labels in place

Operating Lifetime

Production operations. Following the initial start-up, production units are restarted following:

- a normal shutdown
- an emergency (abnormal) shutdown
- a "turnaround" or maintenance phase.

Each has safety issues associated that could be unique to the shutdown event. Start-up from a normal or scheduled shutdown (without maintenance action) should already have been addressed as a procedure-based hazard evaluation in the hazard evaluation of normal operations. Less likely to have had an established safety review is the start-up from an emergency or abnormal shutdown, since not all possible emergency scenarios would have been predicted. These are frequently considered to be operated like a normal start-up once the system has reached a safe state. A careful restart review is advised, since by the (hopefully) unique nature of the emergency, experience with the specific type of restart is limited. The start-up following a turnaround has historically been problematic. Planned turnarounds and shutdowns for maintenance may intentionally change the equipment, such as by cleaning or modification. However, safety reviews are as important before a restart following a turnaround as they are before a new commission, and their value must not be underestimated.

The health and safety concerns during planned shutdown procedures are part of a complete hazard evaluation. What may be lacking is the evaluation of safety issues during a response to an emergency. Focus of analysis during emergencies is often solely on causal events inside the fenceline. Frequently, causal events initiated outside the fenceline are discussed almost as an afterthought. Floods, power loss, high winds, earth movement, and other forces of nature, while unpredictable in timing and force, can have a more or less predictable impact on the process. Predictable effects such as these can be addressed and controlled to protect people, property, and the environment.

Another type of shutdown scenario is where batch operations are suspended due to the operating hours of the plant. For example, when a batch plant operates less than 24/7, then processes may be suspended at various stages of the batch sequence. A batch may be heeled over the weekend, or a batch charged but the heat-up stage postponed until the next manned shift, which could be hours or days later. The procedure-based hazard evaluation review is capable of addressing these issues with respect to suspended operations.

Cyclic Reviews. Even as process changes never end during the life of a facility, there will always be the necessity to continue hazard evaluations. Periodic updating or revalidation of the hazard study to incorporate facility changes (Section 6.4) is the method used to maintain adequate safeguards.[21] The timing of these cyclic reviews depends on factors such as regulations, the rate of process changes, and the nature of those changes. Chapter 19 contains a worked example of a HAZOP Study performed for a cyclic review.

A significant change outside the fenceline can also trigger the need for a hazard review. Examples of such changes are:

- Population changes such as new residential housing nearby
- Land / water / air traffic pattern changes
- Necessary community emergency and security adjustments
- New buildings such as schools or commercial establishments
- Demolished buildings.

Other examples could be cited, but the point is made that the plant operation and the associated risk may be affected by events and changes beyond the plant boundaries.

Extended Shutdowns

Mothballing a plant or a unit within a plant site goes beyond the steps taken to shutdown a process to an established safe state. Tanks, lines, and valves must all be drained and any residual reactive materials neutralized. Most of these operations will likely be performed only once during the lifetime of the operation, and hence will have no history to help guide the safe implementation of the mothballing effort. A potentially riskier activity is a restart from an extended shutdown. In that event, the condition and intended operation of all equipment and instruments must be checked. In particular, the safety systems must be re-verified and validated. The restart of a mothballed unit within and perhaps attached to an active production site should include a review of the hazard evaluation for the connected active units.

Decommissioning

An active plant or unit that is slated for decommissioning would go through the stages of a normal shutdown, then cleanout in preparation for mothballing, followed by disassembly. In the refining industry as well as in other process industries, a hazard review is conducted before decommissioning. Health, safety, and environmental issues would be related to uncontrolled pressure releases, workers potentially exposed to noxious or toxic vapors, and spills during line separation.

A plant or unit that has been previously mothballed would have all the issues associated with the decommissioning of an active plant or unit, with the added potential hazard associated with corrosion products. Over a period of time, residues can change such that these new compounds may be unknown and have unknown health and environmental effects or thermal decomposition sensitivities.

Hazard Evaluation of Mega-Projects

The hazard analysis of extremely large projects is complex by definition. These reviews will take weeks to months to complete. Just as construction, commissioning, and start-up would be performed by multiple teams, so should the hazard evaluation. In this case, the review team would be made of a group of team leaders, each with the skill and knowledge of the hazard evaluation technique used for the area of responsibility. The primary project manager would likely task one individual with the responsibility to coordinate the entire hazard evaluation effort, to ensure consistency among all units. A "divide and conquer" approach is taken so the reviews can be completed in a timely manner.

Hazard Evaluation during Mergers and Acquisitions

Mergers and acquisitions (M&As) among the process industries occur frequently both within countries and internationally. Each M&A has its own unique challenges, but one issue common to most if not all chemical process M&As is the need to unify the hazard evaluation cultures of all parties. When M&A sites have fairly well-established programs involving hazard evaluations, the main hurdles are having the acquired sites conform to common risk tolerance criteria, and the merging of existing hazard evaluation data into the central company resolution tracking database.

If sufficient due diligence was performed, the quality (or lack thereof) of hazard evaluation in M&A activities need not come as a surprise to the company or department responsible for the hazard evaluation. The due diligence checklist needs to contain items that investigate the quality of hazard evaluations in the affected organization.

A strategy for maintaining operations while hazard evaluation issues are resolved is for the lead organization to complete a preliminary hazard evaluation as soon as possible, followed by a full hazard evaluation within three to six months. This is to identify and address the hazards with the greatest risk, while operating procedures, P&IDs, and safety systems are updated to the higher corporate standard and/or are validated.

The opposite of mergers and acquisitions are divestitures. Sites should be sold with complete and well-documented hazard evaluation reports, including documentation of the status of all action item resolutions.

6.7 Integrating Occupational Safety, Environment, Reliability, Maintainability, Quality, and Security into Hazard Evaluations

Process plants must continuously meet many objectives to remain viable. These include operating facilities and maintaining equipment in such a way as to reliably meet production and quality requirements while complying with regulations and avoiding negative impacts on personnel, the public, or the environment. This involves having many different programs in place that are often thought of as separate entities, but must nevertheless be all overseen and resourced by plant management.

Both the potentials and the limitations of using hazard evaluations as one of the means to integrate the diverse but complementary considerations of personnel safety, environmental protection, plant reliability/ maintainability, product quality, and site security are addressed in this section. Further help in integrating the management of process safety, environmental, health, and quality programs has been published by CCPS.[13] The variety of considerations discussed here that can be tied into hazard evaluations highlights the importance of carefully defining the scope of a hazard evaluation and ensuring there are no gaps in the coverage of these considerations for a project, whether incorporated into the hazard evaluation or considered separately.

Occupational Safety Considerations

Occupational hazards that are found in general industrial workplaces are usually considered outside the scope of most process safety hazard evaluations. These include such hazards as potential exposure to stored or connected electrical energy, potential contact with rotating or reciprocating equipment, performing work above ground level, on-site vehicle traffic in pedestrian areas, and compressed air systems. Occupational health considerations that are also outside the scope of most hazard evaluations include those such as the presence of elevated levels of noise, radiation, heat, asbestos, silicon, atmospheric chemical contaminants, or carcinogens in the workplace. The health effects that usually result from exposure to these physical or chemical agents are from continuous or longer-term exposures, whereas hazard evaluations as described in this book pertain to acute or episodic incidents.

Occupational hazards are evaluated and controlled by safety and health programs that are managed differently than for process hazards. Their evaluation may involve a similar type of up-front hazard assessment, such as those for determining the proper personal protective equipment or the location of confined spaces. However, how such hazards are controlled is often explicitly prescribed by regulation and/or company standards. For small facilities with relatively minor or very common process hazards, it may be possible to combine evaluation of both process hazards and occupational hazards in the same periodic team-based Checklist Reviews where comprehensive checklists are available or can be readily developed.

Even for facilities with major process hazards, it will often be important to have one or more persons on the hazard review teams that are familiar with the occupational safety and health practices at the facility. This is due to their being some overlap between process safety and occupational safety considerations, especially in areas such as the following:

- Materials handling: when evaluating the likelihood of an operational error or equipment failure in the handling of material such as with vehicles, cranes or hoists resulting in impact on process equipment and loss of containment of process material or energy.
- Hot work: when evaluating the likelihood of an ignition source being present upon release of a flammable or heated combustible material or presence of an explosible dust or mist.

- Personal protective equipment: when evaluating the impact of a fire, explosion, or chemical release on site personnel or their ability to either escape or safely remain and shut down an out-of-control process.
- Fire protection equipment: when evaluating the likelihood of a small fire escalating to engulf process equipment, or when evaluating the impact of a fire on personnel or property including possible domino effects.

Other areas of overlap between occupational safety and health and process safety include chemical labeling, ergonomics, personnel training, work permitting, energy isolation / initial opening of lines and equipment, recommissioning / start-up checks, ventilation systems, contractor safety, safe work practices, confined space entry, emergency planning and response, and incident investigation.

Evaluation of occupational safety hazards does not usually involve rigorous techniques such as Fault Tree Analysis. Checklist and safety reviews are more commonly employed. One common approach is to conduct a Job Safety Analysis. Performing a JSA is a good approach for examining a work procedure one step at a time to identify and document recognized hazards and determine appropriate hazard control measures.[16] Other techniques are used by industrial hygienists to identify and evaluate exposure potentials,[18,19] and system safety principles can likewise be applied to industrial hygiene applications.[20]

Environmental Protection Considerations

In the context of hazard evaluations, considering environmental impacts is similar to considering personnel safety impacts with respect to their being a distinction made between low-level, long-term effects of chronic exposures and higher-level, short-term effects that can result from acute, episodic events such as fires and large toxic releases. The acute effects are taken into account in assessing the severity of consequences in hazard evaluation scenarios, and these may include environmental impacts such as a toxic vapor cloud causing damage to sensitive vegetation or a liquid spill or firefighting runoff resulting in a fish kill or contamination of an aquifer.

Some regulatory authorities require environmental impacts to be considered when conducting hazard evaluations of high-hazard facilities. The environmental impacts can be documented by indicating the magnitude of effects using a predetermined set of severity categories, or by describing the likely effects in word form such as in the Consequences column of a What-If Analysis or HAZOP Study worksheet. The anticipated environmental impact can then be taken into account as the hazard evaluation team assesses whether the scenario under study has adequate safeguards. When further risk reduction is warranted, and the elevated risk is attributable to potential environmental impacts, measures such as the following might be implemented:

- Additional measures to reduce the likelihood of the initiating cause, which will most often be a loss-of-containment cause
- Additional or enhanced means to rapidly detect and isolate a release
- Additional or enhanced secondary containment measures
- Additional or enhanced emergency response, dispersion, cleanup, neutralization, or other mitigative measures to reduce the impacts on environmental receptors.

It can be seen that these are many of the same types of risk reduction measures that would be employed for safeguarding on-site and/or off-site populations.

Reliability, Maintainability, and Quality Considerations

Reliability, maintainability, and quality considerations are usually considered outside the scope of most hazard evaluations, except as they may also have personnel safety, environmental damage, or property loss consequences. However, the reason for their exclusion is generally in an attempt to reduce the scope and time requirements for the hazard evaluations. Most scenario-based hazard evaluation methodologies *per se* are good tools for studying reliability, maintainability, and quality issues in process facilities. This has always been the case, for example, with the hazard and operability study method, by its very name. The original development and usage of HAZOP Studies was for the purpose of studying both hazard and operability considerations, and the cost-effectiveness of using the method to find and correct (or avoid) operability problems was demonstrated many times over.

A variation on HAZOP Studies known as HAZROP (for hazard, reliability and operability) has been used to combine HAZOP Studies with reliability-centered maintenance.[17] Likewise, modifications to the FMEA methodology known as FMECA (as described in Section 8.2) and FMEDA (Failure Modes, Effects and Diagnostic Analysis) are used for product quality and reliability purposes and as six-sigma tools. FMEA can be used with a quality, reliability and/or safety focus, making it possible to meet multiple objectives with one FMEA.

The most effective means of considering reliability, operability/maintainability, and quality impacts in the context of hazard evaluations is to quantify the consequences of abnormal situations in the same terms or categories that property damage impacts from fires and explosions would be considered. For example, if consequence severity categories of order-of-magnitude total costs were used (such as $10,000, $100,000, $1 million, etc.), then the same scale should be used for assessing the loss potential of events such as compressor bearing failures or product batches being off-quality. The impact should also take into consideration mitigating factors such as the availability of a spare compressor or the ability to rework bad product.

Recommendations for reliability, quality, or operability/maintainability improvements are often recorded and tracked separately from safety and environmental hazard evaluation findings. In this way, their benefit can be assessed by the business unit on each recommendation's own cost-benefit merits, taking feasibility and resource constraints into account. Many recommendations will be made not on a risk-reduction basis but as straightforward changes or improvements to a final design (e.g., installing a low-point drain where one was omitted) or an operating procedure (e.g., adding a means of improving batch-to-batch consistency).

Site Security Considerations

Site security considerations have not generally been integrated into process safety hazard evaluations, but have rather been assessed in separate security vulnerability assessments that are generally qualitative and very subjective. However, significant parallels exist between chemical plant security and process safety considerations. These similarities, as well as significant differences, are compared in Table 6.7. Some inferences that can be drawn by examining Table 6.7 are as follows.

- Security-related and safety-related incident scenarios are structured essentially the same, with hazards, initiating causes, consequences, impacts, and safeguards. Hence, it is reasonable to expect that potential security incidents could be evaluated by the same basic scenario-based approach as employed by various hazard evaluation methodologies.

Table 6.7 Comparison between site security and process safety scenario elements *(differences italicized)*

Consideration	Site security	Process safety
Hazards requiring containment and control	Hazardous process materials and energies and potential chemical interactions	Hazardous process materials and energies and potential chemical interactions
Containment and control systems	Various means of making abnormal situation initiating causes less likely, including e.g. *deterrence, vigilance, site access controls, perimeter guards and barriers*	Various means of making abnormal situation initiating causes less likely, including e.g. *operating discipline, mechanical integrity program, equipment guards and barriers*
Abnormal situation initiating cause	*Facility intrusion by unauthorized person or weapon with malevolent intent*	*Unintentional, unplanned human error, mechanical failure, or external event*
Initial detection systems	*Intrusion detection*	*Process deviation detection*
Preventive safeguards	*Means to delay intruder until sufficiently potent response force can arrive to stop intruder before consequence occurs*	*Means to bring process back under control or safely shut down process before consequence occurs*
Loss events	Fire, explosion, toxic release, unplanned shutdown, *chemical theft, vandalism*	Fire, explosion, toxic release, unplanned shutdown
Mitigative safeguards	Fire fighting, blast shielding, secondary containment, vapor release countermeasures, site and community emergency response	Fire fighting, blast shielding, secondary containment, vapor release countermeasures, site and community emergency response
Impacts	Injuries/fatalities, environmental damage, property damage, business interruption, *fear/panic*	Injuries/fatalities, environmental damage, property damage, business interruption

- Since the underlying hazards are essentially the same, means of reducing the hazards and thus making a facility inherently safer will benefit both site security and process safety.

- Significantly different means must be employed to prevent, detect, and interrupt malevolent human actions than to prevent and detect unintentional human errors, mechanical failures, and external events and interrupt process deviations.

- Fire, explosion, and hazardous material release consequences and mitigative measures are the same or similar for security and safety incidents. Hence, it is reasonable to expect that the potential impacts of chemical plant security incidents could be evaluated by the same basic consequence modeling methods as employed by chemical hazard evaluations.

It is also reasonable to expect that the same approach can be used for determining the adequacy of safeguards in both cases, by using the combined likelihood and severity of each scenario to judge where risks are higher than warranted. The primary difficulty encountered when using scenario-based approaches to evaluate security risks is assessing the likelihood of malevolent human actions, including by insiders such as disgruntled employees. Approaches have been used that address this issue by assigning relative likelihoods based on factors of facility access, past history and/or known intention for targeting the same general type of facility, past history and/or known intention for targeting the particular facility, and capability to successfully carry out the malevolent action. In addition, evaluating the effectiveness of safeguards may need to take into account safeguards being intentionally disabled by the same attacker. For example, an arsonist intending to inflict damage by a fire scenario may also attempt to disable the facility's firewater system.

Chapter 6 References

1. J. Stephenson, *System Safety 2000—A Practical Guide for Planning, Managing, and Conducting System Safety Programs*, ISBN 0-0442-23840-1, Van Nostrand Reinhold, New York, 1991.
2. *Risk Analysis in the Process Industries*, European Federation of Chemical Engineering Publication No. 45, Institution of Chemical Engineers, Rugby, England, 1985.
3. *Responsible Care Management System® Guidance and Interpretations (RCMS 102)*, American Chemistry Council, Arlington, VA, 2004.
4. F. Crawley, M. Preston and B. Tyler, *HAZOP: Guide to Best Practice,* ISBN 0-85295-427-1, Institution of Chemical Engineers, Rugby, UK, 2000.
5. D. F. Montague, "Process Risk Evaluation—What Method to Use?," *Reliability Engineering and System Safety*, Vol. 29, Elsevier Science Publishers Ltd., England, 1990.
6. S. L. Nicolosi and F. L. Leverenz, "Selection of Hazard Evaluation Methods," AIChE Summer National Meeting, Denver, August 1988.
7. P. F. McGrath, "Using Qualitative Methods to Manage Risk," *Reliability Engineering and System Safety*, Vol. 29, 1990.
8. M. G. Gressel and J. H. Gideon, "An Overview of Process Hazard Evaluation Techniques," *American Industrial Hygiene Association Journal 52*(4), April 1991.
9. Center for Chemical Process Safety, *Plant Guidelines for Technical Management of Chemical Process Safety, Revised Edition*, American Institute of Chemical Engineers, New York, 1995, Chapter 7.
10. Center for Chemical Process Safety, *Guidelines for Technical Management of Chemical Process Safety*, American Institute of Chemical Engineers, New York, 1989.
11. Center for Chemical Process Safety, *Guidelines for Process Safety Documentation,* ISBN 0-8169-0625-4, American Institute of Chemical Engineers, New York, 1995, Chapter 10.
12. Center for Chemical Process Safety, *Guidelines for Management of Change for Process Safety*, American Institute of Chemical Engineers, New York, 2007.
13. Center for Chemical Process Safety, *Guidelines for Integrating Process Safety Management, Environment, Safety, Health and Quality,* American Institute of Chemical Engineers, New York, 1997.
14. Center for Chemical Process Safety, *Inherently Safer Chemical Processes: A Life Cycle Approach, 2nd Ed.,* American Institute of Chemical Engineers, New York, 2007.
15. Center for Chemical Process Safety, *Guidelines for Performing Effective Pre-Start-up Safety Reviews,* American Institute of Chemical Engineers, New York, 2007.
16. OSHA Publication 3071, Job Hazard Analysis, U.S. Department of Labor, Occupational Safety and Health Administration, Washington, DC, 2002, http://www.osha.gov/Publications/osha3071.pdf.

17. R. L. Post, "HAZROP: An Approach to Combining HAZOP and RCM." *Hydrocarbon Processing 80*(5), May 2001, pp. 69-76.
18. M. Jayjock, J. Lynch, and D. Imel (eds.), *Risk Assessment Principles for the Industrial Hygienist,* ISBN 13-978-0-932627-97-1, American Industrial Hygiene Association, Fairfax, Virginia, 2000.
19. J. R. Mulhausen and J. Damiano (eds.), *A Strategy for Assessing and Managing Occupational Exposures, 2nd Ed.,* ISBN 13-0-932627-86-2, American Industrial Hygiene Association, Fairfax, Virginia, 1998.
20. P. L. Clemens, "System Safety Principles Applied in Industrial Hygiene Practice," *Hazard Prevention,* November/December 1987.
21. W. L. Frank and D. K. Whittle, *Revalidating Process Hazard Analyses*, ISBN 0-8169-0830-3, American Institute of Chemical Engineers, New York, 2001

7

Risk-Based Determination of the Adequacy of Safeguards

This chapter examines the next step widely taken in the use of the scenario-based hazard evaluation methods of Chapter 5. So far, the various methods have different approaches but all get around to answering the questions ***"What can go wrong?"*** and ***"What safeguards are in place?"*** and, in the process, come up with a set of possible incident scenarios for the facility being studied. The next question to be asked is ***"Are the safeguards adequate?"*** with the implication that something will need to be done if they are found to be inadequate. This chapter discusses approaches taken for determining the adequacy of safeguards in the context of hazard evaluations. Chapter 8 will then present information regarding how to prioritize and implement recommended actions for reducing hazards and risks.

Two basic approaches are used to determine whether further actions are needed to control risks. One approach relies on accumulated experience; the second approach uses scenario risk estimates. Both are likely to be employed in most hazard evaluations, particularly those in which scenario-based methods (as presented in Chapter 5) are used.

Experience-Based Action Decisions

Action decisions based on accumulated experience are relatively straightforward. For some systems, the existing or proposed layout, materials, equipment, instrumentation and controls, procedures, operating parameters, and maintenance strategies can be compared against established practices and standards and, where they are found to fall short, recommendations are made that will bring the facility into compliance with expectations. There may be more than one means by which the compliance can be achieved, but whether or not action is warranted is a straightforward decision.

An example of this approach might be a situation in which a company has many fired heaters, with hundreds of facility-years of experience in their operation. Lessons have been learned over the years, mostly from process incidents including near misses, and this accumulated knowledge has been embodied in a standardized checklist for the design and operation of fired heaters within the company. Alternatively, the company may have adopted industry codes and standards for fired heaters, taking advantage of an even greater breadth of experience. A hazard evaluation of one of the company's fired heaters, either by itself or as part of a larger study, might consist of little more than a review of process safety information, a walkaround inspection, discussion of experience with the system including inspection results, review of the operating and maintenance procedures, and comparison of the system's design and operation to the company checklist (Checklist Analysis). Where shortcomings are found, actions are not debated, risk-ranked, or further analyzed, but automatically decided to be necessary to address the inadequacies.

Risk-Based Action Decisions

The remainder of this chapter is devoted to strategies for determining the adequacy of existing safeguards on the basis of scenario risk, rather than solely on the basis of accumulated experience. *Risk,* as defined for the purposes of process safety, is the combination of the expected frequency (events/year) and impact (effects/event) of a single incident or a group of incidents. Determining the adequacy of safeguards as part of a hazard evaluation is generally done by somehow analyzing the risk associated with one scenario (a single potential incident) at a time. The scenario risk is most often determined by the product of the scenario likelihood (estimated loss event frequency) and its total severity of consequences (scenario impact):

Scenario Frequency x ***Scenario Impact*** = ***Scenario Risk***

(loss events/year) x (impact/loss event) = (impact/year)

Various measures of scenario risk are possible depending on the effects considered. Typical measures are injuries per year, fatalities per year, and total monetary losses per year.

Sections 7.1 through 7.5 of this chapter discuss the elements that are common to various approaches employed for determining scenario risk, including where the different kinds of safeguards come into the picture. Once a scenario risk estimate is made, a judgment can then be made as to whether that scenario risk is too high. (The meaning of "too high" is discussed in Section 7.5.) If the scenario risk is too high, by implication this means that the existing safeguards are not adequate and something needs to be done to reduce the risk, unless in some cases it can be demonstrated that the risk is as low as reasonably practicable (ALARP).

For a complete hazard evaluation using scenario-based approaches, this same risk-estimation process must be done for every identified scenario having a consequence of concern. Section 15.7 includes a worked example of a small part of a HAZOP Study that illustrates scenario risk analysis.

It is also possible to combine individual scenario risks to get a measure of risk for a whole process or even an entire facility. Some companies require their hazard evaluations to determine whether the total process risk is too high, as well as determining where individual scenario risks are too high.

7.1 Scenarios from Scenario-Based Hazard Evaluations

The starting point for determining scenario risk is the identification of incident scenarios using one or more of the scenario-based hazard evaluation procedures described in Chapter 5. Figure 7.1 summarizes the sequence in which several of these common hazard evaluation procedures answer this first basic question of ***"What can go wrong?"***

Fault Tree Analysis (FTA) is a deductive approach that starts with the consequence of concern (the "Top event" of the fault tree) and traces its contributing factors, including enabling conditions and necessary safeguard failures, back in time until causes (basic initiating causes) are reached. The analyst then resolves the basic events of the fault tree along with their AND/OR logic into "minimal cut sets," which are the scenarios of interest. What-If Analysis (including What-If/Checklist Analysis), Failure Modes and Effects Analysis (FMEA), and Event Tree Analysis (ETA) are all inductive approaches that identify initiating causes by various means, and then follow possible sequences of events forward in time through to loss events and impacts. The HAZOP Study method is unique in that it starts in the middle at process deviations, then identifies the deviation initiating causes, and then determines what loss events and impacts could ensue.

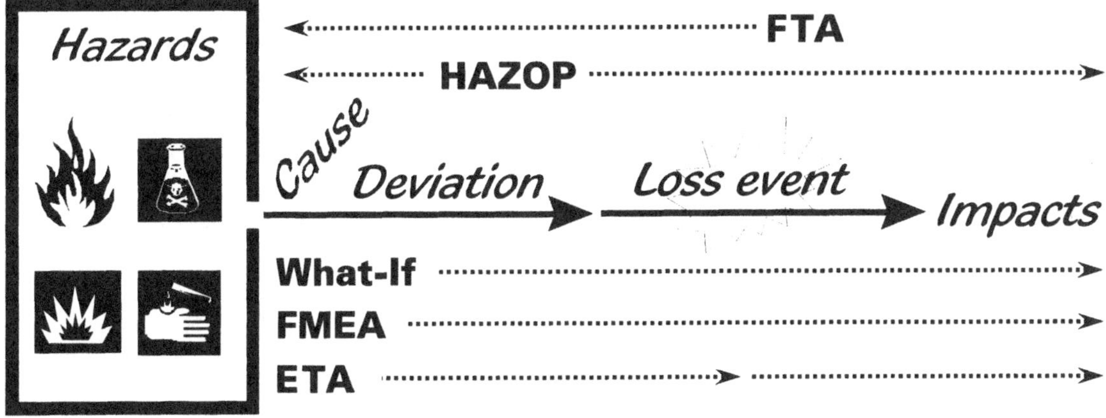

Figure 7.1 Summary of commonly used approaches to identifying incident scenarios

The common element to these approaches is the development of incident scenarios that are more or less detailed time-sequence descriptions of unique initiating cause / loss event combinations, sometimes termed "cause-consequence pairs." Each initiating cause might lead to more than one loss event (e.g., loss of reactor cooling might result in a bad batch, or under different conditions it could result in a runaway reaction and vessel rupture explosion); likewise, a given loss event (e.g., methanol pool fire) might have various possible initiating causes. However, it is unique initiating cause/loss event combinations that form the scenarios of interest (Table 7.1). The ultimate objective of the scenario development part of a hazard evaluation using one or more of the scenario-based approaches is to come up with a complete and comprehensive set of scenarios that relate to process safety consequences of concern. This is the goal to strive toward when conducting scenario-based hazard evaluations.

7.2 Severity of Consequences

Both the scenario impact (severity of the scenario consequences) and the scenario likelihood must be estimated in order to determine scenario risk. It is often found desirable to first estimate the consequence severity before the scenario likelihood. If the impact is found to be below a threshold of concern with respect to the scope of the hazard evaluation (e.g., a minor quality problem with no safety-related impact), the scenario as a whole can be passed over.

It is helpful to keep in mind a distinction between *loss event* and *impact* (see Figure 7.1) when estimating scenario risks. A *loss event* can be thought of as an undesirable physical event such as an explosion, a fire, or a sudden release of a toxic or corrosive material. A loss event can usually be associated with a specific moment in time, and generally involves a significant, irreversible step-change increase in the entropy (degree of disorder) of the system. It can be thought of as the "point of no return" for an incident scenario. Up to this point in time, it may be possible to safely shut down or regain control of the process without significant loss or harm. After this point in time, some degree of loss and/or harm will result, although the impact may not be significant to the scope of the hazard evaluation (e.g., a product quality impact where the off-specification product can be reworked or blended, or a production interruption that can be managed by available product inventory).

Table 7.1 Scenarios are unique initiating cause / loss event combinations *(safeguards, impacts not listed)*

Scenario	Initiating cause	Intermediate events	Loss event
1	Formaldehyde solution storage tank level transmitter LT111 gives false low reading	Tank overfilled	Formaldehyde solution released from storage tank into diked area
2	Formaldehyde solution storage tank level transmitter LT111 gives false low reading	Tank overfilled into dike with open drain valve	Formaldehyde solution released from storage tank into nearby stream
3	Caustic unloaded into wrong storage tank	Caustic mixed with formaldehyde solution; polymerization reaction initiated; vapors generated; tank vapor space pressurized	Toxic formaldehyde vapors released from tank vent
4	Caustic unloaded into wrong storage tank	Caustic mixed with formaldehyde solution; polymerization reaction initiated; vapors generated; tank vapor space pressurized; ignition source present	Flash fire
5	Caustic unloaded into wrong storage tank	Caustic mixed with formaldehyde solution; polymerization reaction initiated; vapors generated; tank overpressurized	Tank rupture explosion with no fire; contents energetically released to surrounding area
6	Caustic unloaded into wrong storage tank	Caustic mixed with formaldehyde solution; polymerization reaction initiated; vapors generated; overpressurized; ignition source present or created by failing tank	Tank rupture explosion and fire; contents energetically released to surrounding area
7	etc.		

Companies often predefine graduated scenario impact categories that are to be used in conjunction with their hazard evaluations. One example of impact categories is given in the next chapter (Table 8.7). Table 7.2 shows another example of a type of impact scale used to define the severity of consequences in terms of various environmental, health and safety (EHS) impact categories. This scale uses a rough order-of-magnitude difference between each level of severity.[1-2] Layer of Protection Analysis (LOPA, Section 7.6) also uses order-of-magnitude scales.

It can be noted that this type of impact scale is not limited to a fixed number of severity magnitudes. For some facilities with major hazards, consequences with severity magnitudes of 6 or higher are possible. (For example, off-site impacts from the December 1984 methyl isocyanate release in Bhopal, India would be somewhere around a severity magnitude of 8 on this scale.) This type of impact scale avoids the pitfall of having a very limited number of impact levels, where severe consequences and truly catastrophic consequences would both be indicated as having the same maximum-severity ("high") level of impact.

Table 7.2 Example of EHS impact categories and magnitudes used in hazard evaluations

	Impact magnitude				
Impact category	**1**	**2**	**3**	**4**	**5**
On-site (worker) health effects	Recordable injury	Lost-time injury	Multiple or severe injuries	Permanent health effects	Fatalities
Off-site (public) effects	Odor; exposure below limits	Exposure above limits	Injury	Hospitalization or multiple injuries	Severe injuries or permanent effects
Environmental impacts	Reportable release	Localized and short-term effects	Intermediate effects	Widespread or long-term effects	Widespread and long-term effects
Accountability; attention/ concern/response	Plant	Division; regulators	Corporate; neighborhood	Local/state	State/national

Estimating the degree of loss and/or harm for each scenario consequence of concern is done in many different ways, ranging from a qualitative team judgment to a fully quantitative consequence analysis using methods such as those described in Reference 3. An intermediate approach is often taken, with the hazard evaluation team making judgments where possible based on the team members' collective knowledge and experience, possibly supplemented by determinations that may already have been performed as initial assessments of worst-case consequences (Section 3.6) or for other purposes such as worst-case calculations for complying with regulatory requirements. When the team is unable to make an estimate, the team may recommend that a consequence analysis be performed on the identified scenario or group of scenarios.

Mitigative Safeguards and Scenario Impacts

Mitigative safeguards act to reduce the loss event impact if they are successfully employed. Hence, their effectiveness affects the scenario risk and is thus a valid part of a hazard evaluation team's discussions.

Some companies take the approach of using worst-case impacts, without taking credit for any mitigative safeguards, in evaluating scenario risk. This has the advantage of simplicity, and may be appropriate particularly for design-stage hazard evaluations. However, for an operating facility, it does not give credit for mitigative safeguards that the facility already has in place and is effectively managing, and will thus overstate the risk where such safeguards are actually employed.

Some companies go to the other extreme and assume mitigative safeguards will always work, with loss event impacts being evaluated on that basis. The integrity of mitigative safeguards is managed to ensure their proper functioning should a loss event occur. This again has the advantage of simplicity for the hazard evaluation team, but the disadvantage is that it will give a non-conservative estimate of scenario risk for situations where weaknesses might exist in the mitigative safeguard systems or where the mitigation will not be effective against a particular scenario.

A fully quantitative risk analysis would look at all possible incident outcomes for a given loss event, such as by the use of event trees to evaluate the probability of success or failure of each applicable mitigative safeguard and the overall risk of each resulting scenario. An experienced facilitator may make use of such tools when a particular loss event needs to be more closely evaluated during a qualitative study, but it is generally not feasible to discuss and document every possible incident outcome when conducting hazard evaluations such as What-If Analyses. Hence, some sort of simplified treatment of mitigative safeguards is nearly always employed in most hazard evaluations. Nevertheless, the hazard evaluation team should not only discuss the pertinent mitigative safeguards, but also document assumptions that are made with respect to mitigative safeguards when determining scenario impacts.

One type of mitigative safeguards can effectively be treated in a more explicit manner in scenario-based hazard evaluations. This type was defined in Section 1.4 as ***source-mitigative safeguards***. These mitigative measures are designed to act after the loss event occurs, but affect the "source term" that would be used in a consequence analysis (e.g., rate and/or duration of material or energy release to the surroundings). Examples include excess flow valves, dry-break connections on unloading hoses, automatic release detection and isolation systems, and engineered vapor release mitigation systems such as deluges and water curtains. These source-mitigative safeguards may be designed as independent protection layers (IPLs) against more-severe consequences. A good way to treat source-mitigative safeguards is to consider as two different hazard evaluation scenarios (1) the case where the source-mitigative measure works and reduces the scenario impact and (2) the case where the source-mitigative measure does not work and the worst-case consequence ensues. In this way, the measure can be documented as a safeguard and its probability of failure can be explicitly assessed. For example, if an excess flow valve is suitable for a given service, regularly tested or replaced, and identified and managed as an independent protection layer, it might be assessed to be 90% effective in isolating a line break in a bottom outlet line from a large storage tank of hazardous material. One scenario would be the more likely case where the excess flow valve worked and the amount of material released was limited to the contents of the transfer line. The second scenario, which would be an order of magnitude less likely but might have much more severe impacts, would be where the excess flow valve failed to work and the entire tank contents would be released through the failed bottom outlet line, assuming no other means of isolating the release are possible.

The other type of mitigative safeguards, ***receptor-mitigative safeguards***, include those measures related to features such as facility siting, fire- and blast-resistant design, personal protective equipment, and emergency response that serve to reduce loss event impacts if they are effective at intervening between the source of the released material or energy and the receptors (people, property, environment) that could be affected. These mitigative safeguards are often treated in a more general way in hazard evaluations. For example, if a scenario involves an unexpected, energetic release of a volatile hazardous material during a line opening operation, then the worker health impact might be assessed taking into account the personal protective equipment expected to be worn at the time, if appropriate PPE is required to always be used for line breaks in this part of the process and implementation of this requirement is effectively managed. It would still be appropriate for the hazard evaluation team to discuss whether the required PPE would protect against the particular scenario being reviewed, and to bring to light any known weaknesses in the implementation of the PPE requirements.

7.3 Frequency of Initiating Causes

For those scenarios having a consequence of concern, the next step is to estimate how likely it is for the loss event to occur. In the risk equation at the beginning of the chapter, this was denoted as the *scenario frequency,* with events per year as the unit of measure. This scenario frequency represents the likelihood per year of operation that the loss event will occur. For example, if the loss event is a vapor cloud explosion following a loss-of-containment cause in a particular process segment, the scenario frequency would be vapor cloud explosions per year. For most process hazard consequences of concern, the scenario frequency is a fractional number, and hopefully a very low number that is well outside the realm of the facility's experience. This makes it difficult for a review team to do what needs to be done for this aspect of a hazard evaluation; i.e., for each scenario with a consequence of concern, to make a judgment as to the overall likelihood of occurrence of a major loss event.

However, the scenario frequency can be broken down into more readily estimated parts that are more likely to be within the realm of the facility's and review team's experience. This separating of the whole (scenario frequency) into its component parts is the very definition of "analysis", and is a key to understanding process risk estimates, even in qualitative hazard evaluations. The two primary component parts of the scenario frequency are the initiating cause frequency and the total failure probability of the preventive safeguards:

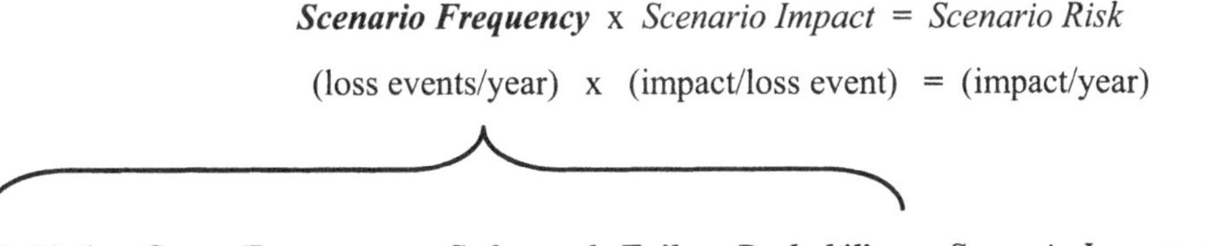

Initiating Cause Frequency x ***Safeguards Failure Probability*** x *Scenario Impact* = *Scenario Risk*

(initiating events/year) x (dimensionless probability) x (impact/loss event) = (impact/year)

To reiterate the safeguards terminology introduced in Section 1.4, *preventive* safeguards are those that are capable of keeping the loss event from occurring, given that an initiating cause takes place (Figure 7.2). The failure probability of the preventive safeguards is discussed in Section 7.4.

The ***initiating cause frequency*** is the likelihood that the scenario initiating cause (initiating event) will occur. For most scenarios, it is much easier to estimate an initiating cause frequency than the overall scenario frequency, since the likelihood of typical initiating causes such as mechanical failures of individual components, utility losses, operational errors, and external events are more easily estimated and more often within the realm of experience than the likelihood of the scenario loss events such as fires and explosions.

Companies often predefine graduated cause frequency categories to be used for their hazard evaluations. One example of initiating cause frequency categories is given in Table 7.3, which employs an order-of-magnitude difference between each likelihood category. This type of likelihood scale is not limited to a fixed number of magnitudes, but the most commonly used categories are listed. Most hazard evaluation teams with operational experience can readily distinguish cause frequencies at an order-of-magnitude resolution, such as deciding whether cooling tower water supply will be lost once a month, once a year, or once every ten years. Some companies allow frequency estimates to be more refined than on an order-of-magnitude basis, since a factor of ten difference in frequency covers such a broad range. For example, if a hazard evaluation team determined by historical experience that the frequency of a given operational error was about once every three years, an initiating cause frequency magnitude of **-0.5** on the scale in Table 7.3 (1/3 years = $10^{-0.5}$/year) might be used.

Cause frequencies that are more difficult to estimate are those associated with rare initiating causes such as loss-of-containment causes (e.g., flange/connection failures) that may lead directly to a scenario consequence. Generic or company-wide failure rate data can be employed that give failure frequencies on the basis of per connection, per length of piping, or per vessel or other fixed or rotating equipment component.

7.4 Effectiveness of Safeguards

As illustrated in Figure 7.2, *preventive safeguards* are all the risk-reduction provisions that can keep a particular loss event from occurring, given a particular initiating cause is realized; i.e., the layers of protection that pertain to each unique initiating cause / loss event combination ("cause-consequence pair"). Preventive safeguards only come into play after a scenario initiating cause occurs and before a scenario loss event (irreversible event, as described in Section 7.2) is realized. By contrast, containment and control measures affect the cause frequency and mitigative safeguards affect the loss event impacts.

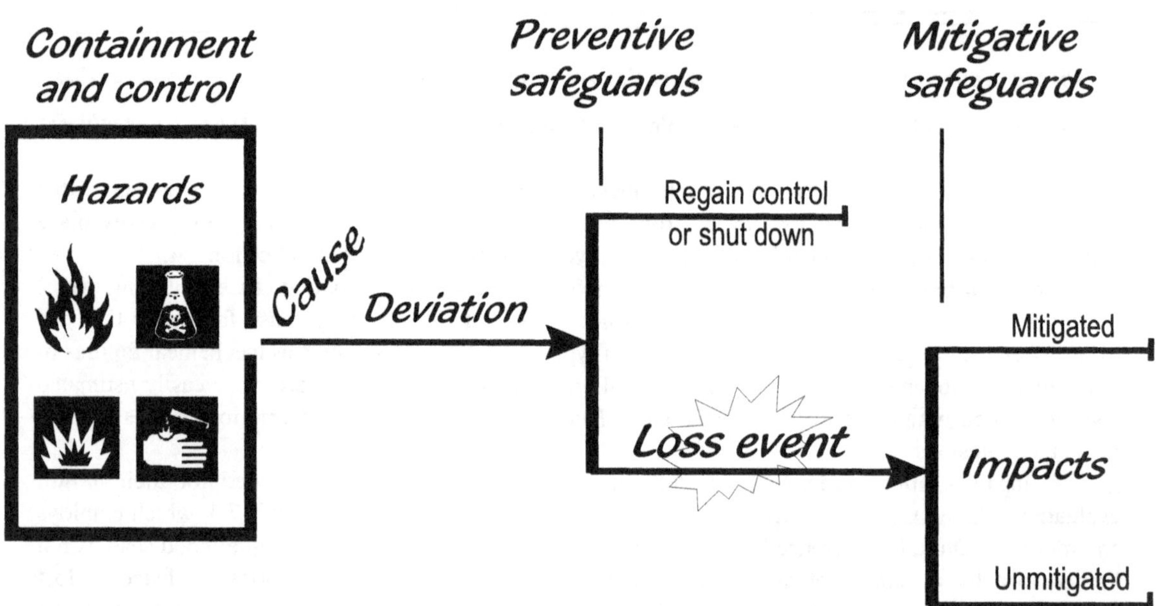

Figure 7.2 Preventive and mitigative safeguards

Table 7.3 Example initiating cause frequency scale (order-of-magnitude basis)

Magnitude 10^x/yr	Equivalent cause likelihood	Comparison with experience
0	Once a year	Unpredictable as to when it will occur, but within realm of most employees' experience
-1	1 in 10 (10% likelihood) per yr of operation	Outside of some employees' experience; within realm of process' experience
-2	1 in 100 (1% likelihood) per yr of operation	Outside of almost all employees' experience; within realm of plant-wide experience
-3	1 in 1,000 per yr of operation	Outside of almost all process experience; may be within realm of company-wide experience
-4	1 in 10,000 per yr of operation	Outside of most companies' experience; within realm of industry-wide experience
-5	1 in 100,000 per yr of operation	May be outside the realm of industry-wide experience, except for common types of facilities and operations

For a given cause-consequence pair, if the loss event would be realized every time the initiating cause occurs, the failure probability of the preventive safeguards is equal to one. In this case, the overall scenario frequency will be equal to the initiating cause frequency. A typical example is a scenario involving a loss-of-containment cause and a hazardous material release loss event. As soon as loss of containment occurs (e.g., due to a flange failure in a pipeline continuously in use transporting a hazardous fluid), the hazardous material release is realized. (*Mitigative* safeguards would come into play at this point, determining the impact of the hazardous material release.) This example illustrates that for many loss-of-containment causes, containment integrity and mitigative safeguards are more important than preventive safeguards.

To take this example one step further, a second consequence of concern might be a fire of some kind if the hazardous material is flammable or is combustible and heated above its flash point temperature. This scenario may have the same flange failure initiating cause as before, but now has a different loss event. One preventive safeguard may now intervene between the loss-of-containment cause and the fire loss event; namely, the necessity for a sufficiently energetic ignition source to also be present in the same location at the same time as the flammable vapor/air mixture. This preventive safeguard will be called "ignition source control" for purposes of this example. If a fire will not likely ensue every time the flange failure occurs, then the ignition source control safeguard has some degree of effectiveness associated with it. A hazard evaluation team could document "ignition source control" as a safeguard when analyzing this scenario, and then make an assessment as to the effectiveness of this safeguard. If the entire area in which a flammable vapor/air mixture would exist has properly maintained electrically classified equipment, and the most likely identified ignition source is infrequent plant vehicle traffic in the vicinity, then the review team might assess that the probability of a fire being ignited upon flange failure is low, or, quantitatively, that in this case there would be on the order of a 10% probability of vehicle traffic passing through the area between (a) the time that the flange failure occurred and a flammable vapor/air mixture was formed and (b) the time the flammable vapors were dissipated or controlled or traffic to the area was blocked. (In many other situations, the probability of an ignition source being present may be closer to 100% than 10%, but 10% is used for purposes of this example.)

It can now be seen how this preventive safeguard probability of 10% (0.1) can be used in assessing the scenario frequency. If the initiating cause (flange failure) frequency is determined by some means to be on the order of one chance in 100 per year of operation (0.01/year), then the overall scenario frequency is equal to the initiating cause frequency times the failure probability of the safeguards, or 0.01/year x 0.1 = 0.001/year likelihood of a fire from this cause.

When more than one preventive safeguard applies to a cause-consequence pair, then an overall failure probability of the preventive safeguards must be determined (qualitatively or quantitatively) to judge or calculate the scenario risk. The key question to ask when reviewing multiple preventive safeguards is whether the safeguards are independent of each other ("independent protection layers," or IPLs). If independent, then their individual failure probabilities can be multiplied together to obtain the overall failure probability of the safeguards.

Besides ignition source control, other typical preventive safeguards include check valves, operator response to alarms, instrumented protective systems, and last-resort preventive safety systems such as reactor dump and quench systems and emergency venting and relief systems. Example preventive safeguard failure probabilities are shown in Table 7.4. As for initiating cause frequencies, partial orders of magnitude are allowed by some companies when conducting such hazard evaluations. A more formalized approach to determining the overall failure probability of preventive safeguards, known as Layer of Protection Analysis (LOPA), is discussed in Section 7.6. See also the discussion in Section 7.2 on source-mitigative safeguards for layers of protection that act after the loss event occurs but significantly affect the "source term" (e.g., rate and/or duration of material or energy release) for determining the loss event consequences and impacts.

7.5 Risk Estimation using Risk Matrix or Direct Calculation

As soon as a hazard evaluation team or process risk analyst has an estimate of the scenario impact (consequence severity) and scenario frequency (initiating cause frequency times the failure probability of the preventive safeguards), a scenario risk estimate can be made. This scenario risk estimate can then be used to judge whether the existing safeguards are adequate to control the scenario risk. This must, of course, be repeated for every scenario with consequences of concern. The next steps, development and prioritization of recommendations, are discussed in Chapter 8.

Hazard evaluation procedures such as What-If Analyses and HAZOP Studies were originally developed as purely qualitative methodologies. For each identified potential incident scenario, deciding on whether or not safeguards were adequate typically relied on a qualitative judgment of the scenario risk, which was based on a team discussion of the scenario severity and likelihood.

Risk matrices have now become widespread tools for relating scenario likelihood and severity to defined risk boundaries and risk reduction requirements. A typical risk matrix of this type is shown in Figure 8.1.

Table 7.4 Example preventive safeguard failure probabilities

10^{-x}	Safeguard failure probability	Typical examples
0	~ 1	Next event in incident sequence will occur essentially every time the previous event occurs (e.g., if there are no other preventive safeguards, the loss event can be expected to ensue every time the initiating cause occurs)
0	~ 1	Operator response to non-independent alarm or with minimal time to respond*
0	~ 1	Software or hardware automatic safety system that is not functionally tested; e.g., typical single check valve
0	~ 1	Ineffective ignition source control or continuous ignition source nearby; high probability of released vapors/mist/dust cloud igniting
1	~ 0.1	Control action implemented in BPCS with sensor independent of the initiating cause, designed and managed to achieve a PFD ≤ 0.1, running in automatic mode during all operating phases where the incident scenario could occur
1	~ 0.1	Operator response to independent alarm with adequate time to respond and with action included in procedures and training**
1	~ 0.1	Ignition source control sufficiently effective to avoid lighting off ignitable vapors/mist/dust cloud on the order of nine times out of ten events
1	0.01 to 0.1	Safety instrumented function with Safety Integrity Level of SIL1†
2	~ 0.01	Operator response to independent alarm with 100% operator coverage, ample time to diagnose and respond, straightforward diagnosis and action included in procedures and training, and reliable means of regaining control or bringing the process to a safe state***
2	~ 0.01	Relief valve or rupture disk in clean service, routinely inspected or replaced, with entire emergency relief system designed and sized to protect against the particular overpressure event being analyzed‡
2	0.001 to 0.01	Safety instrumented function with Safety Integrity Level of SIL2†
3	0.0001 to 0.001	Safety instrumented function with Safety Integrity Level of SIL3†

* E.g., less than 10 min to respond, with troubleshooting or diagnostics required to determine the appropriate action.

** E.g., 10 to 30 min to respond and no troubleshooting or diagnostics required to determine the appropriate action, or 40+ min to respond with minor troubleshooting or diagnostics required. For these examples, alarm may be implemented either within the basic process control system or independent of the BPCS, but must have a sensor independent of other protective layers.

*** E.g., Operator response to alarm with more than 24 hours response time, with multiple operators having opportunity to detect the alarm(s) and take action; alarm automatically repeated at an interval necessary to ensure that each shift is notified of the process condition; minor troubleshooting or diagnostics allowed if needed before taking the action; alarm system is independent of the BPCS and independent of other layers of protection.

† Most companies take credit for only one order of magnitude protection for a SIL1 SIS, even though the defined order-of-magnitude range is >1 to 2. Similarly, most companies only take two orders of magnitude risk reduction credit for a SIL2 SIS and three orders of magnitude credit for a SIL3 SIS. A midpoint value (e.g., **1.5**) is sometimes used if the SIS is known to have an intermediate PFD (e.g., 0.003). Other requirements such as functional testing must be satisfied to take credit for SILs.

‡ From *Layer of Protection Analysis: Simplified Process Risk Assessment* [4] as an example screening value. PFD ranges from literature and industry are given as 0.1 to 0.00001. Quantitative analysis may justify a PFD value lower than 0.01 depending on the particular service (including both upstream and downstream of device), inspection/replacement interval, and effectiveness of quality control and maintenance procedures to prevent wrong valve/disk from being installed and to avoid improper installation.

The combining of factors on the two axes of a matrix is the same as multiplying the two factors together. Thus a risk matrix, even with qualitative scales of increasing likelihood and severity on the two axes, is performing the same function as the risk equation at the beginning of this chapter:

$$\textit{Scenario Frequency} \times \textit{Scenario Impact} = \textit{Scenario Risk}$$

Quantitative risk matrices are also employed in hazard evaluations. Figure 7.3 is an example of a risk matrix that combines scenario frequency magnitudes (exponents of the order-of-magnitude annual frequencies) on one axis with the severity categories of Table 7.2 on the other axis. Since adding the exponents of two numbers is the same as multiplying the two numbers, the mathematics of determining a risk magnitude is greatly simplified in this example (i.e., risk magnitude = scenario frequency magnitude + consequence severity magnitude).

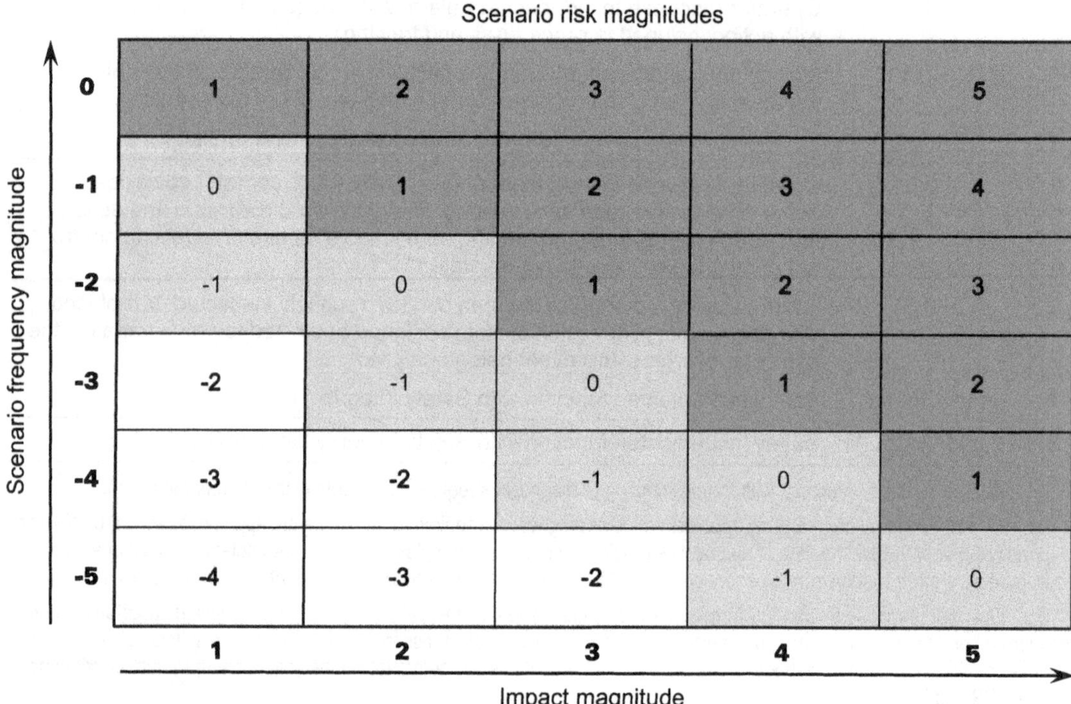

Figure 7.3 Example risk matrix using order-of-magnitude frequency and severity categories

Risk increases toward upper right (dark), where prompt risk-reduction actions are typically required. Risk decreases toward lower left (white), where risk becomes broadly acceptable and no action is typically required. Intermediate range may require closer analysis, and near-term risk reduction would generally be warranted unless risk is shown to be as low as reasonably practicable.

Chemical Process Quantitative Risk Analyses (CPQRAs) have been conducted for many years, using established methods such as those in Reference 4. These quantitative risk analyses explicitly calculate risk estimates for every scenario, albeit in different ways depending on the specific method employed. For example, a fault tree might be developed and resolved into its minimal cut sets (scenarios). The overall frequency for each minimal cut set can then be calculated by multiplying the initiating cause frequency by the preventive safeguard failure probabilities and also by any enabling conditions that are not failures *per se* (e.g., the fraction of time the ambient temperature is within a certain range, if that is pertinent to the scenario). The total "Top event" frequency can be determined by adding together the individual scenario frequencies, and then combined with an estimate of the severity of consequences to determine the total risk.

A word of caution: It must be recognized that the risk-based approach to evaluating the adequacy of safeguards requires many subjective decisions and thus, although is a structured and quantitative approach, it nevertheless will always have an element of subjectivity. This means that two people well-versed in risk assessment may come up with different risk values for the same scenario evaluation, which may result in different views on whether safeguards are adequate. Proceeding through the risk determinations with a sound methodology and based on the consensus decisions of a knowledgeable evaluation team will partly offset the subjectivity and result in meaningful conclusions as to where further risk reduction is warranted.

7.6 Layer of Protection Analysis

Layer of Protection Analysis (LOPA) is a simplified form of quantitative risk analysis, using order of magnitude categories for initiating cause frequency, consequence severity, and the likelihood of failure of independent protection layers (IPLs) to analyze and assess the risk of one or more scenarios. LOPA can be useful in the process development, process design, operational, maintenance, modification, and decommissioning life cycle phases. The LOPA methodology as described in the CCPS Concept Book *Layer of Protection Analysis: Simplified Process Risk Assessment*[5] is summarized in this section with minor updates.

Purpose

LOPA was developed around the need to answer questions such as:

- *How safe is safe enough?*
- *How many protection layers are needed?*
- *How much risk reduction should each layer provide?*

LOPA can help answer these questions without having to conduct a complete, detailed quantitative risk analysis (QRA). LOPA is an order-of-magnitude quantitative method (sometimes termed "semi-quantitative") that builds on qualitative hazard evaluations such as HAZOP Studies. By analyzing selected scenarios in detail, effective application of LOPA can determine whether the risk posed by each analyzed scenario has been reduced to be within a tolerable risk range. However, if the analyst or team can make a reasonable risk decision using only qualitative methods, then LOPA may be overkill.[5] Qualitative hazard evaluation methods such as HAZOP Studies are intended to identify a comprehensive set of incident scenarios and qualitatively analyze those scenario for the adequacy of safeguards. By contrast, LOPA might be used to analyze 10 to 30% of the incident scenarios, and Chemical Process Quantitative Risk Analysis (CPQRA) might be employed to study only 1% of the scenarios in detail.

Benefits of LOPA. Many benefits of employing the LOPA methodology can be cited:

- LOPA requires less time than quantitative risk analysis. This benefit applies particularly to scenarios that are too complex for qualitative assessment of risk.
- LOPA helps resolve conflicts in decision making by providing a consistent, simplified framework for estimating the risk of a scenario and by providing a common language for discussing risk. LOPA provides a better risk decision basis, as compared to subjective judgments with respect to the adequacy of safeguards.
- LOPA can improve the efficiency of hazard evaluation meetings by providing a tool to help reach risk judgments more quickly.
- LOPA facilitates the determination of more precise cause-consequence pairs, and therefore improves scenario identification.
- LOPA provides a means of comparing risk from unit to unit or plant to plant, if the same approach is used throughout the company.
- LOPA provides more defensible comparative risk judgments than qualitative methods, due to the more rigorous documentation and the specific values assigned to frequency and consequence aspects of the scenario.
- LOPA can be used to help an organization decide if a risk is "as low as reasonably practicable" (ALARP), which may also serve to meet specific regulatory requirements.
- LOPA helps identify operations and practices that were previously thought to have sufficient safeguards, but, upon more detailed analysis (facilitated by LOPA), the safeguards may be found to be insufficient to reduce the risk to a tolerable level.
- LOPA provides a basis for a clear, functional specification of independent protection layers (IPLs).
- Information from LOPA helps an organization focus its mechanical integrity and training programs on IPLs identified during the study. Therefore, LOPA is a tool that supports an effective mechanical integrity or risk-based maintenance system by aiding in the identification of safety-critical features and tasks.

Limitations of LOPA. Despite all its benefits, it should be remembered that LOPA is just another tool that must be used correctly by those employing it. The limitations imposed on LOPA result in a work process that is much less complex than quantitative risk analysis, while generating useful estimates of risk. LOPA adds simplifying assumptions that are intended to be conservative if compared to the results of a fully quantitative analysis. LOPA is subject to the following limitations:

- LOPA is not intended to be a scenario identification tool. LOPA depends on qualitative hazard evaluation methods to identify incident scenarios including initiating causes and pertinent safeguards.

- LOPA is not intended to be used as a replacement for detailed quantitative analysis. QRA may be warranted to analyze complex scenarios.
- LOPA requires more time to reach a risk-based decision than qualitative methods such as HAZOP Studies and What-If Analyses. The amount of effort required to implement LOPA may be excessive for some risk-based decisions and is overly simplistic for other decisions. For simple decisions, the value of LOPA is minimal.
- Risk comparisons of scenarios are valid only if (a) the same methods are used for selecting failure data and (b) comparisons are based on the same risk tolerance criteria.
- LOPA results cannot generally be compared between organizations due to differences in risk tolerance criteria and in LOPA implementation.

Types of Results

Implementation of LOPA results in a set of order-of-magnitude risk estimates for a range of scenarios, which may or may not be grouped by and summed over the various consequence categories being considered. LOPA results include:

- An identification of initiating causes for scenarios involving process hazards.
- Order-of-magnitude estimates for initiating cause frequency, consequence severity, and probabilities of failure on demand for preventive safeguards intended to keep the initiating cause from leading to the loss event.
- A formal designation of preventive safeguards as *independent protection layers* (IPLs) based on their independence, effectiveness, and auditability.
- An assessment of the adequacy of the IPLs for each scenario, in the form of risk-based decisions, as well as recommendations and specifications for additional IPLs if deemed necessary.

Resource Requirements

Prerequisites for LOPA implementation fall into several categories:

- Corporate experience and culture. Corporate standards requiring hazard analysis, safety reviews, reliability analysis, root cause/failure analyses, and design checks set the stage for successful LOPA implementation. Corporate experience with quantitative hazard evaluation methods is preferred, since LOPA bridges the gap between qualitative and quantitative methods.
- Data requirements. Incident consequences, equipment failure rates, operator error rates, and safeguard performance are areas of knowledge that the company must develop, either in-house or from literature sources.
- IPL retention. An organization must establish a system to periodically assess (audit) the components and human interventions identified as IPLs to ensure that the IPLs remain in service at the anticipated safeguard effectiveness.

- Risk tolerance criteria. To achieve consistent results, it is strongly advised that organizations define risk tolerance criteria before implementing LOPA. Such risk criteria can be useful as the basis for other risk-based decisions within the organization, in addition to being needed to conduct LOPAs.

Significant time and personnel resources are required to complete and document a LOPA. Once a company has established its LOPA protocol and begun using LOPA on a regular basis, the number of hours required for preparation, team meetings and documentation will be nearly a linear function of the number of scenarios analyzed using the LOPA methodology.

Description

LOPA is typically applied after, and builds upon, the information gathered in a qualitative hazard evaluation, but can be applied to scenarios gathered from any source, such as an incident investigation. It, in turn, can be used as a screening tool for scenarios prior to application of a quantitative risk analysis. LOPA can also be implemented in conjunction with a qualitative scenario-based hazard evaluation method such as What-If Analysis or HAZOP Study.

After the scope of the study is defined, LOPA consists of the six steps summarized below. The LOPA results can then be used to make risk-based decisions. It should be noted that these steps do not necessarily represent the order in which they may be performed. A common alternative sequence is to select a scenario that meets given screening criteria for consequences of concern, determine a maximum tolerable scenario frequency based on established risk criteria and the loss event impacts of the scenario, estimate the initiating cause frequency, determine the pertinent IPLs and their probabilities of failure on demand (PFDs), then compare the calculated scenario frequency to the maximum tolerable frequency to determine whether further risk reduction is needed.

Step 1: Identify the consequence to screen the scenarios. Since LOPA typically evaluates scenarios that have been developed in a prior study, a first step by the LOPA analyst or team is to screen these scenarios, and the most common screening method is based on consequence.

LOPA consequence severity estimates and consequence screening thresholds may be defined in a number of ways, each having strengths and weaknesses:

- Method 1: Category approach without direct reference to human harm – By this method, consequences are categorized in terms of the type and magnitude of a release or other consequence characteristic, avoiding any overt appearance that injuries and fatalities are tolerable, and avoiding the difficulty of estimating how many and how severe the injuries from a particular release might be.

- Method 2: Qualitative estimates with human harm – By this method, human impacts are overtly considered, usually allowing direct comparison with organizational guidelines, but estimates of impact magnitudes are arrived at using qualitative judgment.

- Method 3: Qualitative estimates with human harm, with adjustments for post-release probabilities – This method replaces the qualitative judgment in Method 2 with additional probabilities, giving a better estimate of the risk of human harm.

- Method 4: Quantitative estimates of human harm – This method is similar to Method 3 but uses detailed analyses in determining the effects of a release and its effects upon individuals and equipment. Tools associated with full CPQRAs may be used here, including dispersion and blast effects analysis, adding complexity, need for expertise, time, etc. The level of sophistication required for these consequence analyses may be disproportionate to the order-of-magnitude frequency estimates employed within LOPA.

Step 2: Select an incident scenario. LOPA is applied to one scenario at a time. The scenario can come from other analyses, such as qualitative hazard evaluations, but each scenario must describe a single initiating cause-loss event ("cause-consequence") pair.

When scenarios are selected from a qualitative hazard evaluation, such as from a HAZOP Study, they may need to be separated into multiple scenarios for evaluation. For example, where a scenario involves an emergency relief system, the LOPA could be applied both for the case in which the emergency relief does not function properly (greater severity but lesser likelihood) and the case in which it does function properly (lesser severity but greater likelihood).

Step 3: Identify the initiating cause of the scenario and determine its frequency. The initiating cause must lead to the loss event, given failure of all of the preventive safeguards. The frequency must account for background aspects of the scenario, such as the frequency of the mode of operation for which the scenario is valid. Most companies provide guidance on estimating the frequency to achieve consistency in LOPA results; suggested frequencies are also given by CCPS.[5] The team should determine whether the suggested value is appropriate, based on plant historical performance and/or experience with the initiating cause under similar plant conditions.

Background aspects, such as the probability that the process is in a certain mode of operation at the time another failure occurs, or the frequency with which a certain procedural step is performed during which another failure may occur, are not initiating causes, but *enabling events or conditions*. Under LOPA, either (a) their probabilities modify the initiating cause frequency, or (b) their frequency is used as the beginning of the event sequence and is modified by the probability that the initial failure will occur. For example, if the incident scenario involves an operational error during an unloading operation resulting in a tank overflow, the operational error is the initiating cause, but the initiating cause frequency is a combination of the frequency at which the unloading operation is performed (the enabling event or condition) and the probability of operator error per unloading operation.

Step 4: Identify the IPLs and estimate the PFD of each IPL. Recognizing the existing preventive safeguards that meet the requirements of *independent protection layers* (IPLs) for a given scenario is the heart of the LOPA methodology. An IPL is a device, system, or action that is capable of keeping a scenario from proceeding to the undesired loss event, independent of the initiating cause or the action of any other layer of protection associated with the scenario. Preventive safeguards are known as IPLs when they are designed and managed to achieve the following seven core attributes:

- Independence—the performance of a protection layer not being affected by the initiating cause or by the failure of other protection layers
- Functionality—the required operation of the protection layer in response to a specific off-normal condition
- Integrity—the risk reduction that can reasonably be expected given the protection layer's design and management

- Reliability—the probability that a protection layer will operate as intended under stated conditions for a specified time period
- Auditability—the ability to inspect information, documents, and procedures that demonstrate the adequacy of and adherence to the design, inspection, maintenance, testing, and operation practices used to achieve the other core attributes
- Access security—the use of administrative controls and physical means to reduce the potential for unintentional or unauthorized changes
- Management of change—the formal process used to review, document, and approve modifications to equipment, procedures, raw materials, processing conditions, etc., other than "replacement in kind," prior to implementation of the modifications.

IPLs can be viewed as "lines of defense" against potential incident scenarios. The independence of the IPL from the initiating cause and from other IPLs is very important. The LOPA team must assess the independence of each IPL and estimate its probability of failure based on the IPL design and management. All IPL equipment should be included in the mechanical integrity program and be subjected to inspection and proof tests as necessary to maintain the IPL in the "as good as new" condition. IPLs depending on operating personnel should be covered by written procedures, with identified personnel being trained and tested on the procedures. Access to IPL equipment should be controlled and proposed changes should undergo management of change review prior to implementation.

As an example of one type of IPL, an alarm with operator response may be considered as an IPL, but should meet the following conditions:[6]

- The alarm and response field devices are independent of the initiating cause and any other IPLs.
- There is sufficient time for the operator to be effective in making the response.
- The action taken brings the system to a safe state and minimizes the risk to the responder.
- There is a simple, well-documented procedure with clear and reliable indication that a specified action is required.
- The operator is trained and tested on the procedure at an interval not to exceed three years. More frequent training may be necessary to achieve the risk reduction.
- The alarm set point is protected and under management of change.
- The devices are proof tested as appropriate for the allocated risk reduction.

Only one alarm is generally credited as an "Alarm IPL" for each identified incident scenario, unless the alarms occur at distinctly separate times (an hour or more apart) and utilize separate field instrumentation. Multiple alarms that rely on response from the same operator should not be listed as separate Alarm IPLs, due to the potential for common mode errors. The design and test requirements must follow good engineering practices.

Another type of IPL is the safety instrumented system (SIS). CCPS has published guidelines covering the life cycle of safety instrumented systems and other instrumented protective systems.[6]

The integrity of each IPL is quantified in terms of its probability of failure on demand (PFD), which is a dimensionless number between 0 and 1. The PFD value for an IPL is the probability that, when needed for the scenario in question, the IPL will not perform the required task. Most companies provide a predetermined set of PFD values for use by the LOPA analyst, so the analyst may pick the value that best fits the scenario being analyzed.

Step 5: Calculate the scenario frequency. The overall frequency of the scenario is estimated by mathematically combining the initiating cause frequency and the IPL PFDs. Combining methods include arithmetic formulae and graphical approaches. Regardless of the method, most organizations provide a standard form for documenting LOPA intermediate and final results. The following mathematical approach is applicable for low-demand situations (f_i^I less than twice the test frequency for the first IPL).

$$f_i^C = F_i^I \cdot \prod PFD_{ij}$$

$$= F_i^I \cdot PFD_{i1} \cdot PFD_{i2} \cdot \ldots \cdot PFD_{ij}$$

where

f_i^C is the frequency for consequence C for initiating cause i

F_i^I is the initiating cause frequency for consequence C for initiating cause i

PFD_{ij} is the probability of failure on demand (fail-dangerous failure mode) of the jth IPL that protects against consequence C for initiating cause i.

For example, if the frequency for the first initiating cause (i =1) is estimated to be once every ten years ($F_1^I = 10^{-1}$/year), and two IPLs each having a PFD of 0.01 ($PFD_{11} = 10^{-2}$ and $PFD_{12} = 10^{-2}$) are protecting against this particular initiating cause and will prevent consequence C from being realized if either IPL works successfully in response to initiating cause 1, then the calculated frequency for consequence C for initiating cause 1 (f_1^C) is equal to 10^{-1}/year x 10^{-2} x 10^{-2} = 10^{-5}/year.

Other factors may be included in the scenario frequency calculation, depending on the definition of the consequence. For example, when the company is estimating the frequency of direct human harm, the analysis may include additional, modifying probabilities in the scenario frequency, such as those for ignition or presence of personnel near a release. Those probabilities are then included as additional factors in the frequency equation, as in

$$f_i^C = f_i^I * \prod PFD_{ij} * P^{ignition}$$

for the frequency of ignition, and

$$f_i^C = f_i^I * \prod PFD_{ij} * P^{ignition} * P^{person\_present}$$

for the frequency that a person might be present near the fire, and so on.

Step 6: Evaluate the risk to reach a decision concerning the scenario. Where a risk index is the desired outcome, the frequency of the outcome of interest is multiplied by a factor related to the magnitude of the consequences

$$R_k^C = f_k^C * C_k$$

where

R_k^C is the risk index of incident outcome of interest k, expressed as a magnitude of consequences per unit time (unit will vary according to the type of risk being estimated).

f_k^C is the frequency of the incident outcome of interest k, in inverse time units

C_k is a specific measurement of the severity of consequences (impact) of the incident outcome of interest k (e.g., fatalities, dollars of economic loss, etc.).

Where a company has individual or geographic risk criteria, the frequencies and risks of scenarios having the same consequence can be summed to enable the LOPA results to be compared against those criteria.

The scenario risk or frequency may also be determined qualitatively using lookup tables. Typically, such matrices also include a target (or required) number of IPLs for different risk categories. Some matrices may include the frequency of the consequence. Calculations from the equations in this chapter and the risk tolerance criteria are embedded in the look-up table.

Using LOPA to make risk decisions. Once LOPA has been applied to yield order-of-magnitude risk estimates for a group of scenarios or consequences, risk decisions can be made. This evaluation is normally in relation to an organization's risk tolerance criteria. If the calculated risk exceeds the tolerable risk level, the proportion by which the calculated risk exceeds the tolerable risk level indicates by how much the risk must be reduced. Risk reduction can be achieved by various means, including eliminating scenarios by inherent safety approaches, reducing initiating cause frequencies, increasing integrity of IPLs, adding new IPLs, and/or reducing loss event impacts. Many other uses have been found for LOPA results, including adjusting the mechanical integrity program to emphasize oversight of particular equipment components.

Anticipated Work Product

Implementation of LOPA results in a set of order-of-magnitude risk estimates for a range of scenarios that may or may not be grouped by and summed over the various consequence categories being considered. LOPA also includes an assessment of the adequacy of the independent protection layers for each scenario in the form of risk-based decisions, and, if deemed necessary, recommendations and specifications for additional IPLs. An example LOPA worksheet that illustrates these results for a single scenario is shown in Table 7.5. Reference 5 should be consulted to understand the details of this example.

Table 7.5 Example LOPA worksheet from Table B.2 of Reference 5

<table>
<tr><td>Scenario number 1</td><td>Equipment number</td><td colspan="3">Scenario title: Cooling water failure with runaway reaction and potential for reactor overpressure, leakage, rupture, injuries and fatalities. Agitation assumed.</td></tr>
<tr><td colspan="2">Date: ##/##/####</td><td>Description</td><td>Probability</td><td>Frequency (per year)</td></tr>
<tr><td colspan="2">Consequence description/category</td><td>Runaway reaction and potential for reactor overpressure, leakage, rupture, injuries, and fatalities Category 5</td><td></td><td></td></tr>
<tr><td colspan="2">Risk tolerance criteria (category or frequency)</td><td>Unacceptable (Greater than)
Tolerable (Less than or equal to)</td><td></td><td>1×10^{-4}
1×10^{-6}</td></tr>
<tr><td colspan="2">Initiating cause (typically a frequency)</td><td>Loss of cooling water</td><td></td><td>1×10^{-1}</td></tr>
<tr><td colspan="2">Enabling event or condition</td><td>Probability that reactor in condition where runaway reaction can occur on loss of cooling (annual basis)</td><td>0.5 (per reactor)</td><td></td></tr>
<tr><td colspan="2" rowspan="4">Conditional modifiers (if applicable)</td><td>Probability of ignition</td><td>N/A</td><td></td></tr>
<tr><td>Probability of personnel in affected area</td><td>N/A</td><td></td></tr>
<tr><td>Probability of fatal injury</td><td>N/A</td><td></td></tr>
<tr><td>Others</td><td>N/A</td><td></td></tr>
<tr><td colspan="3">Frequency of unmitigated consequence</td><td></td><td>5×10^{-2}</td></tr>
<tr><td colspan="3">Independent protection layers</td><td></td><td></td></tr>
<tr><td colspan="2">BPCS alarm and human action</td><td>Shortstop addition on BPCS loop high reactor temperature alarm</td><td>1×10^{-1}</td><td></td></tr>
<tr><td colspan="2">Pressure relief valves</td><td>With required modifications to system (see Actions)
(PFD may be conservative if modifications added)</td><td>1×10^{-2}</td><td></td></tr>
<tr><td colspan="2">SIF (Required PFD = 1×10^{-3}) (Part of SIS for all 3 reactors)</td><td>SIF to open vent valves (see Actions for design details)
Required PFD set by Scenario 5
TO BE ADDED — see Actions/Notes</td><td>1×10^{-3}</td><td></td></tr>
<tr><td colspan="2">Safeguards (non-IPLs)</td><td colspan="3">Operator action. Other operator actions not independent of the same operator already credited.
Emergency cooling system (steam turbine). Not credited as an IPL as too many common elements (piping, valves, jacket, etc.) that could have initiated initial cooling water failure.</td></tr>
<tr><td colspan="3">Total PFD for all IPLs</td><td>1×10^{-6}</td><td></td></tr>
<tr><td colspan="3">Frequency of mitigated consequence</td><td></td><td>5×10^{-8}</td></tr>
<tr><td colspan="5">Risk tolerance criteria met? (Yes/No): Yes with added SIF</td></tr>
<tr><td colspan="2">Actions required to meet risk tolerance criteria</td><td colspan="3">Add SIS for all 3 reactors. Install SIF with minimum PFD = 1×10^{-3} for opening vent valves on high temperature. Separate nozzles and piping for each vent valve. Install separate nozzle and vent lines for each PSV to minimize blockage and common cause.
Consider nitrogen purges under all vent valves / PSVs.
Responsible group / person / date: Plant Technical / J.Doe / January ####</td></tr>
<tr><td colspan="2">Notes</td><td colspan="3">Ensure operator response to high temperature meets requirements for IPL.
Ensure RV design, installation, maintenance meet requirements for PFD 1×10^{-2} as a minimum. If determined to be better, consider PFD for Vent Valve SIF PFD.</td></tr>
</table>

Chapter 7 References

1. R. W. Johnson, "Risk Management by Risk Magnitudes," *Chemical Health and Safety* 5(5), September/October 1998, 9-13.
2. "HAZOP Studies and other PHA Techniques for Process Safety and Risk Management," AIChE/ASME Professional and Technical Training Courses, American Society of Mechanical Engineers, New York, 2007.
3. Center for Chemical Process Safety, *Guidelines for Consequence Analysis of Chemical Releases*, ISBN 0-8169-0786-2, American Institute of Chemical Engineers, New York, 1999.
4. Center for Chemical Process Safety, *Guidelines for Chemical Process Quantitative Risk Analysis, 2nd Ed.*, ISBN 0-8169-0720-X, American Institute of Chemical Engineers, New York, 1999.
5. Center for Chemical Process Safety, *Layer of Protection Analysis: Simplified Process Risk Assessment,* ISBN 0-8169-0811-7, American Institute of Chemical Engineers, New York, 2001.
6. Center for Chemical Process Safety, *Guidelines for Safe and Reliable Instrumented Protective Systems*, ISBN 978-0-471-97940-1, American Institute of Chemical Engineers, New York, 2007.

8
Analysis Follow-Up Considerations

Hazard evaluations can greatly benefit the organizations that fully use them. As a result of hazard evaluations, team members acquire insights into the hazards that exist in a process or facility and make recommendations that might help reduce the risk of hazardous situations. In addition, the facility can achieve higher quality, incur lower operating costs, and give managers greater confidence that hazards have been identified and are being controlled. But if a hazard evaluation is not properly documented, communicated to management in the form of meaningful recommendations, and then followed up, the facility will not see the study's benefits, nor will the hazard evaluation activity be complete. Then, the resources applied to the effort may have been wasted. In addition, failure to effectively follow up a hazard evaluation may present a legal liability for the sponsoring organization.

Organizations should consider the following activities to help make the most effective use of the products of a hazard evaluation: (1) develop meaningful recommendations, (2) prioritize the analysis results, (3) document the hazard evaluation, (4) develop a management response to the study, and (5) resolve the actions resulting from the risk management decision-making process in a timely manner. As the selected design and operating improvements are implemented, these changes should be communicated to affected employees and others who can benefit from the results.

8.1 Development of Recommendations

Most hazard evaluations result in recommendations or action items to improve the safety of a process. Well-thought-out, clearly written and relevant recommendations are necessary to effectively communicate the findings from a hazard evaluation. For management to understand and act on the results of a hazard evaluation, recommendations must be associated with improving the safety of an operation and must be tied directly to the hazard evaluation. Clearly written and pertinent recommendations facilitate a positive management response to a hazard evaluation.

Some hazard evaluations may identify deficiencies by comparison with safety-related codes, standards, practices or regulations. Recommendations arising from experience-based techniques such as Checklist Analysis will generally address identified areas of non-compliance with relevant safety codes, standards, practices and regulations. For example, if the pertinent fire code requires sprinklers of a certain design in a warehouse handling flammable liquids, a recommendation at the design stage would be to adhere to the particular fire code sprinkler system design requirements. A Checklist Analysis of an existing facility will compare the current design and operation to the relevant experience base and point out areas of noncompliance. A hazard evaluation team might also make a recommendation to investigate an area of uncertainty such as whether a recently updated code contains any new requirements that are relevant to the facility.

Scenario-based hazard evaluations using the methods described in Chapter 5 identify and analyze potential incident scenarios. Where safeguards are judged to be inadequate or questionable, recommendations are developed by the person or team conducting the hazard evaluation. To be effective, recommendations should be developed that either reduce the likelihood or potential severity of possible incident scenarios or, if possible, completely eliminate the scenarios. For example, if in a HAZOP Study it is discovered that a high temperature deviation could result in a runaway reaction and a vessel rupture explosion, a recommendation might be made to install an emergency cooling safety system to avoid the vessel rupture loss event. A more effective recommendation could be to limit the heat transfer system maximum temperature to well below the temperature at which a runaway reaction could be initiated in the particular reactor configuration, in order to altogether avoid the possibility of a runaway reaction initiated by excess heat input from the heat transfer system.

The preceding example illustrates a hierarchy of strategies or approaches for hazard evaluation recommendations. *Inherently safer strategies* eliminate or reduce the underlying process hazard by changing the nature or magnitude of the hazard itself, such as by substituting a material that is less hazardous or reducing the quantity of the hazardous material. Recommendations that are inherent in nature can have a greater positive impact than those that are protective in nature. Inherently safer approaches are generally the most effective in avoiding a loss event. For example, it is better to eliminate the use of a flammable liquid in a process (if feasible) than to install fire protection systems but still have the fire hazard be present. However, as mentioned in Section 3.7, implementing inherently safer strategies warrants careful examination, since some strategies can reduce hazards but actually increase the overall risk.

Passive approaches to safety improvement involve systems that do not require an action to occur for protection to be realized. Examples are robust pressure vessel design, dikes and berms, and engineered blast protection. Passive approaches can be less reliable than inherently safer measures (e.g., if a passive system is not properly maintained) but nevertheless be highly reliable.

Active safeguards require a physical action to occur in response to a process event. For example, a fusible link or other engineered device must function to successfully trip a fire protection system. Another example is a safety instrumented system, where a sensor, logic solver and final control element must all work properly for the safety function to be effective. Active safeguards are generally less reliable than passive approaches; nevertheless, hazard evaluation recommendations very commonly address either the addition or the improvement of active safeguards.

Overall, recommendations that are usually least reliable involve *administrative approaches* that require human action and introduce the likelihood of human error. Examples of administrative approaches are procedural checks, operator actions in response to process deviations, and emergency response following a loss event.

Chemical processes require all types of strategies and approaches to keep the process risk at a tolerable level. The hazard evaluation team should consider recommendations related to inherently safer strategies, passive approaches, active safeguards and administrative approaches as appropriate to reduce the risk. By using diverse risk reduction measures, the chance of common safeguard system weaknesses can be minimized and preparedness for loss events can be enhanced.

8.2 Prioritization of Hazard Evaluation Results

It is sometimes difficult to rank the safety improvement recommendations from hazard evaluations because hazard evaluation techniques generally do not provide definitive, quantitative characteristics useful for ranking purposes. This section describes strategies for prioritizing hazard evaluation recommendations and discusses specific hazard evaluation techniques, the results of which can be easily ranked. It also describes other hazard evaluation techniques that make prioritization easier. Three basic questions must be addressed when discussing prioritization of the hazard evaluation results: (1) What are the primary results of a hazard evaluation? The results should be put in the form of recommendations. (2) What are appropriate criteria for a hazard evaluation team to use in prioritizing the results? (3) What approaches exist for performing the prioritization?[1]

Types of Results

There are three basic types of hazard evaluation results: (1) lists of identified hazards, perceived problems, or potential incident scenarios; (2) descriptions of the significance of these problems or incidents (e.g., risk posed by each scenario); and (3) recommendations for reducing or eliminating the hazards, coming into compliance with codes or standards, or reducing the risks associated with the incident scenarios. Not all hazard evaluations will contain all three types of results; in fact, some hazard evaluation objectives may be accomplished by simply identifying potential problems. For example, a hazard evaluation team may be requested to review the conceptual design of a process to see if there are any types of incidents that have been overlooked when specifying protective systems. Other, more ambitious objectives may require that the hazard evaluation team develop a prioritized list of potential incidents that can be used to aid emergency response planning activities. Most hazard evaluations of existing facilities using detailed hazard evaluation techniques will contain all three types of results.

The type of results available from a hazard evaluation is influenced by several factors, one of which is the technique the hazard analysts use to perform the work. Some methods, such as Checklist Analysis or Safety Review, will provide a general list of problem areas associated with the subject process. Other techniques, such as HAZOP Study, Fault Tree Analysis, or Event Tree Analysis can give detailed lists of potential incidents.

Some hazard evaluation techniques encourage the analyst(s) to assess the significance of the potential incidents in a very systematic and detailed fashion. To help assess this significance, analysts can examine the potential causes and effects of an incident. While some hazard evaluation techniques analyze the effects of an incident, they may not provide detailed insight into the potential causes (e.g., Relative Ranking, Checklist Analysis, and Safety Review). But other techniques focus mostly on the causes of incidents (e.g., Fault Tree Analysis, Human Reliability Analysis). Finally, some techniques encourage hazard evaluation teams to consider both the potential causes and effects of incidents (e.g., HAZOP Study, FMEA, and Event Tree Analysis). Methods that focus on both the causes and effects of incidents generally provide the most balanced insights into potentially unsafe situations and ways to correct them. Results of this type of study are most useful when hazard analysts must prioritize the solutions to safety problems developed during a hazard evaluation, possibly through the application of more detailed quantitative analysis techniques.

Most hazard evaluations result in lists of recommendations for reducing the risk of the subject process. The hazard evaluation team should use their judgment to define the significance of potential incidents in order to determine whether the problem warrants a recommendation to management.

Basis for Prioritization

A frequent problem facing the users of hazard evaluation results is that the hazard evaluation team creates a long list of recommendations for management to consider implementing in order to improve safety. In cases where this occurs, decision makers can rightfully wonder whether these results are of any practical use to them. They may ask, "Where do we start?" or "Which are the most important recommendations?"

To help management make these decisions, an efficient hazard evaluation team should give them as much information as possible. One way to do this is to rank the results of the hazard evaluation. Ranking the safety improvement recommendations from a hazard evaluation allows management to prioritize the immediate efforts of resolution and follow-up or, if resources are scarce, to influence which potential improvements are actually implemented. Although all of the recommendations of a well-structured and well-implemented hazard evaluation should receive serious management consideration, it is often clear that it is more important to implement some items than others. Prioritizing recommendations clearly shows how the hazard evaluation team judges the importance of each item. ***While it is possible to simply rank the significance of problems identified by a hazard evaluation, it is often more useful and efficient to rank the proposed lists of recommendations or action items for safety improvements.***

The five most common criteria for ranking the safety improvement recommendations of a hazard evaluation are:

- The identification of any situation that violates a regulatory requirement, corporate or site standard, or applicable recognized and generally accepted good engineering practice
- The identification of any factual error or discrepancy, such as a mislabeled piece of equipment, an omission from a piping and instrumentation diagram, or steps in a standard operating procedure that are out of sequence
- The analysts' understanding of the ***risk posed*** by the potential incidents addressed by each recommendation
- The analysts' perception of the risk reduction that would be gained by implementing the specific recommendation.
- The risk reduction gained by implementing the recommendation in comparison to the resources required to implement it (if such information is available at the time).

Identifying situations falling into the first two categories above are generally put at the top of the priority list for starting corrective action. The third criterion ranks items based on their associated level of risk - it makes sense to resolve the most important problems first. The last criterion ranks proposed improvements by how much they will benefit the facility, not necessarily on how serious the problem is.

Hazard analysts should naturally focus their attention on the basic concepts of risk (i.e., likelihood, consequence - presented in Chapter 1) when determining how to rank the results of hazard evaluations.

Even their qualitative insights on the significance of potential incidents will greatly benefit the manager who must make important resource decisions.

There are many other possible criteria that could be used to prioritize the results of a hazard evaluation (e.g., cost, implementation schedule, competing production priorities, uncertainty concerning technical feasibility), but these are most appropriately used by an organization's management when they are choosing a specific method of risk reduction. However, it is important for hazard evaluation teams to keep these factors in mind when developing their recommendation lists so that they propose practical solutions to management. On the other hand, it would not be appropriate to have a team's process safety judgment overshadowed by these other concerns. Hazard analysts should practice a "call them like you see them" approach, tempered by the pragmatic understanding that safety improvement resources are limited. The following section discusses how an organization can use these results in making risk management decisions.

Ultimately, the qualitative ranking of recommendations from a hazard evaluation may not satisfy the manager's information needs. In these cases, managers should use more detailed hazard evaluation techniques or sponsor CPQRA studies to gain the necessary insights. In addition, it should be noted that the prioritization of recommendations is dynamic. As some recommendations are implemented, the priority of remaining ones may change.

Approaches for Prioritizing Study Recommendations

Hazard evaluations typically result in a list of recommendations or action items. These safety improvement alternatives can be ranked in two fashions: (1) with a detailed assessment of the significance of causes and effects of potential incidents or (2) without the benefit of this detailed incident cause and effect information. Table 8.1 lists the hazard evaluation techniques covered in these *Guidelines* and classifies them according to whether they normally provide detailed information concerning the causes and effects of potential incidents. Highly experienced analysts can sometimes use the more general approaches listed in the "may" column to generate lists of specific incident scenarios for ranking purposes.

Table 8.1 Classification of hazard evaluation techniques for the purpose of ranking action items

Technique provides specific incident scenario information useful for prioritizing action items:

USUALLY DOES NOT	MAY	USUALLY DOES
Safety Review	What-If Analysis	HAZOP Study
Checklist Analysis	What-If/Checklist Analysis	Fault Tree Analysis
Relative Ranking	Failure Modes and Effects Analysis	Event Tree Analysis
Preliminary Hazard Analysis	Human Reliability Analysis	Cause-Consequence Analysis

An experienced team's ability to formally rank safety improvement recommendations will depend largely on which hazard evaluation method they used to perform the study. For example, if a hazard evaluation was performed using the traditional Checklist Analysis method, then normally the study's only result will be a list of deficiencies that were identified when the actual system was compared to the desired system attributes. Since a list of incident causes and effects is not customarily the result of a traditional Checklist Analysis, analysts must use their own judgment to assess the significance of the deficiencies and prioritize any resulting action items.

In cases when no specific incident scenario information is available, the hazard evaluation team or the user of the study results may simply choose to rank the individual recommendations according to some broad classification. Table 8.2 lists some ways that hazard analysts have prioritized the action items from a hazard evaluation when detailed incident scenario information is not available. At the very least, a hazard evaluation team can simply use their judgment to informally place all of their recommendations in decreasing order of importance. Table 8.3 is an example of safety improvement recommendations that resulted from a What-If Analysis of a process unit - the items are arranged in three qualitative categories of decreasing importance. Many times, even this effort will be enough to help the ultimate decision maker use the results of an inexpensive study to satisfy an organization's risk management needs.

The analyst can also prioritize the recommendations from a hazard evaluation by using more detailed information on the causes and effects of potential incidents. Sometimes this information is provided through the use of the standard hazard evaluation techniques. For example, the standard results of an FMEA include qualitative descriptions of the failure effects of single pieces of equipment. These individual failures may result in a spectrum of consequences, with the severe consequence failures indicating that safety improvement is needed. The hazard analyst(s) performing this FMEA can use this information to perform a consequence-based ranking of the recommendations that result from the study.

Another way to prioritize recommendations is to use the incident "cause" information available from some hazard evaluation techniques. Hazard analysts can perform a likelihood-based ranking of the recommendations using this information. For example, the Fault Tree Analysis technique results in a list of combinations of failures, called minimal cut sets that can cause the Top event - an incident with some specific consequence. The list of minimal cut sets can be ranked in terms of their structural importance, which means the cut set is categorized by the number and types of failures in it.

Table 8.2 Typical ways of ranking recommendations from hazard evaluations

- Divide the list of safety improvement alternatives into two or more categories on the basis of practicality, such as high and low priority; or high, medium, and low priority
- Classify the list according to the perceived urgency with which the items should be considered for implementation; for example, use three categories: begin correction immediately, implement within 90 days, and implement within one year
- Have the hazard evaluation team pick the top one, five, or ten action items, and document the team's justification for each of them

Table 8.3 Example ranking of recommendations in qualitative categories of urgency

Recommendations requiring immediate action

- Obtain an SCBA unit to be stationed on the west side of the unit
- Survey all relief valves that have manual isolation valves and ensure that the isolation valves are locked open

Recommendations that should be planned and scheduled as soon as practical

- Develop an emergency training session for the fire brigade involving DOT Chemical 2108
- Consider adding an independent high temperature alarm for the scrubber outlet flow to help detect a reaction in the scrubber
- Extend the handwheels on the manual isolation valves for the compressors through the building walls so the valves can be closed without entering the compressor building

Recommendations requiring further evaluation to determine priority

- Consider providing a compressor shutdown system that includes the capability to remotely stop and isolate the compressors
- Consider providing improved ventilation that is capable of preventing the buildup of a flammable environment in the compressor building

The concept of structural importance recognizes the fact that the more failures required to produce a certain incident, the less likely the incident is to occur.[2,3] For example, an incident caused by a single failure is usually more likely than an incident caused by multiple failures. Also, structural importance recognizes that some types of failures are more likely to occur than others. The following failure types are listed in order of decreasing likelihood:

- Human errors of omission (e.g., operator fails to check equipment as required)
- Human errors of commission (e.g., operator inadvertently trips system)
- Active equipment failures (e.g., pump stops, valve fails to close)
- Passive equipment failures (e.g., pipe ruptures, structural failures)

Obviously, there are circumstances that may alter the ranking of such failures (especially those involving human error); however, using such a list for general ranking can produce prioritizations that are adequate for many analysis objectives. For example, consider the failure scenarios listed in Table 8.4 that could result in release of toxic material. A hazard evaluation team that makes recommendations for reducing the risk associated with scenarios B and E might be expected to assign a "B recommendation" a higher priority that an "E recommendation." Analysts often use this likelihood-based ranking to rank risk reduction options even when no explicit information is known about the cause of an incident.

Table 8.4 Example of structural importance ranking

Scenario	Failures involved	Number and type of failures involved
One-event cut sets		
A	Pipe containing toxic material ruptures	One - passive hardware
B	Operator fails to re-close drain valve after draining water, allowing toxic material to escape	One - human error of omission
C	Vent valve fails to re-close after venting noncondensables, allowing venting of toxic material	One - active hardware
Two-event cut sets		
D	Tank outlet pump transfers off, level control safety system fails to stop feed to tank (tank overflows, releasing toxic material)	Two - both active hardware failures
E	Tube rupture overpressures exchanger shell (i.e., the exchanger shell opens); relief valve transfers open, releasing toxic material to flare, *and* flare system not operable	Two - one passive hardware and one active hardware (flare system not analyzed, assumes not redundant)

A few hazard evaluation techniques can provide explicit cause and effect information that a hazard evaluation team can use to prioritize their list of recommendations. Table 8.5 lists the various hazard evaluation techniques described in these Guidelines and indicates whether their results can be used to prioritize study recommendations. The table shows the techniques as they are normally applied, along with some extensions of the techniques that can easily be used to rank hazard evaluation results. For example, hazard analysts sometimes assign a "criticality" attribute to the severity associated with a component failure in an FMEA. This extension of the FMEA technique is formally known as the Failure Modes, Effects, and Criticality Analysis (FMECA) method.[4-6]

Table 8.6 is an example of a FMECA table. Similar extensions can be made to any technique that provides information about incident causes and effects (e.g., HAZOP Study).

Using either standard or extended hazard evaluation methods, analysts often simplify their descriptions of the likelihood and impact of a potential incident. Frequently, qualitative or semi-quantitative scales will be used; Tables 8.7 and 8.8 are examples of such scales.[7,8] Typically, hazard evaluation teams discuss the potential causes and effects of particular incident scenarios and assign them to likelihood and impact categories.

Table 8.5 Prioritization attributes of hazard evaluation techniques

Technique	Provides incident scenario information?	Provides frequency information?	Provides consequence information?	Event ranking possible? (with typical results)	Comments
Checklists	No, specific scenarios usually not identified	No	No	No	
Safety Review	No, specific scenarios usually not identified	No	No	No	
Dow and Mond Indexes	Yes, on a unit or a major system basis	No	Yes	Consequence ranking	
Preliminary Hazard Analysis	No, specific scenarios usually not identified	No	Yes	Yes	
What-If Analysis, What-If/Checklist Analysis	Yes, if What-If condition is an initiating cause	Yes, if What-If condition is an initiating cause	Yes	Consequence or risk ranking	When full scenarios are developed, simple risk ranking is possible
HAZOP Study	Yes	Yes	Yes	Consequence or risk ranking	Since detailed causes and consequences are identified, simple risk ranking is possible
FMEA	Yes, if failure mode is an initiating cause	Yes, if failure mode is an initiating cause	Yes	Consequence or risk ranking	When full scenarios are developed, simple risk ranking is possible
FMECA	Yes, if failure mode is an initiating cause	Yes, if failure mode is an initiating cause	Yes	Consequence or risk ranking	When full scenarios are developed, simple risk ranking is possible
Fault Tree Analysis	Yes, if minimal cut sets are determined	Yes, based on size and number of cut sets and type of failures involved	No	Frequency ranking based on structural importance	Quantitative FTA techniques are available to estimate Top event frequencies
Event Tree Analysis	Yes	Yes, based on number of incident scenarios and number and type of failures involved	Yes, consequence categories can be assigned to each scenario	Yes	Quantitative Event Tree Analysis techniques are available to estimate incident scenario frequencies
Cause-Consequence Analysis	Yes	Yes, based on number of incident scenarios and number and type of failures involved	Yes, consequence categories can be assigned to each scenario	Yes	Quantitative Cause-Consequence Analysis techniques are available to estimate incident scenario frequencies
Human Reliability Analysis	Yes	Yes, based on number and length of scenarios and type of human errors involved	No	Frequency ranking	Quantitative Human Reliability Analysis techniques are available to estimate human error probabilities

Table 8.6 Example of a Failure Modes, Effects and Criticality Analysis table

				Criticality category assignments				
Item no.	Component description	Failure mode	Effect	Worker safety	Production cost	Equipment cost	Frequency category	Actions
1	Feed gas line from high pressure amine contactor to feed gas knockout drum	Leak	Release of feed gas. Fire likely. Plant 8 shut down to isolate rupture	2	2	1	3	—
2	Feed gas line from high pressure amine contactor to feed gas knockout drum	Rupture	Release of feed gas. Explosion and/or fire likely. Plant 8 shut down to isolate rupture. Major damage to pipe racks (primarily electrical and instrumentation cables)	3	4	3	1	1, 2
3	Feed gas knockout drum	Leak	Release of feed gas. Fire likely. Some damage to Plant 7 pipe rack	1	1	1	3	—
4	Feed gas knockout drum	Rupture	Release of feed gas. Major explosion and/or fire likely. Damage to Plant 7 pipe rack and nearby equipment	3	4	2	1	3
5	Feed gas knockout drum	Level Controller Fails High	Potential carry-over of amine to downstream equipment. Minor upset	1	1	1	4	—
6	Feed gas knockout drum	Level Controller Fails Low	Blowdown of gas to vessels in liquid service. Damage to vessel not designed for gas flow	3	4	3	4	4
7	Outlet line from feed gas knockout drum to feed gas caustic	Leak	Release of feed gas. Fire likely. Some damage to Plant 7 pipe rack	2	1	1	3	—

After hazard analysts have assigned each incident scenario to likelihood and impact categories, a risk matrix (Figure 8.1) can be used to prioritize the action items associated with each potential incident.[6,8-10] The size of the matrix and category definitions should be defined to meet the needs of the organization. Typically, analysts will place in each matrix cell the total number of incidents that were assigned to a particular likelihood and impact category. This number may indicate the overall significance of incidents that have the perceived likelihood and impact. Some analysts include the reference number for each specific incident scenario so the team can tell which incidents are involved.[8] Table 8.9 lists descriptions of how some organizations use a risk matrix to prioritize their response to safety improvement recommendations.[6] The numbers in each risk matrix cell in Figure 8.1 correspond to the category numbers in Table 8.9. For example, incidents that are closer to the upper right corner of the risk matrix (Category I) are individually considered to be higher risk events (i.e., they have higher likelihoods and higher consequences) than events below and to their left (Category II, III, or IV). If the likelihood and consequence categories are selected consistently (i.e., if the same ratio between adjacent categories is used for both the likelihood and consequence categories), events on a diagonal from the upper left to lower right are of equivalent risk. This is because risk is the combination of likelihood and consequence (e.g., often expressed as a simple mathematical product); therefore, the apparent decrease in risk from being a likelihood category lower (i.e., one horizontal row down) is offset by being a consequence category higher (i.e., one vertical row to the right). In this example, a hazard evaluation team would rank incident scenarios in the upper right region as presenting greater risk than ones toward the lower left. Likewise, any recommendation the team makes for reducing the risk of these scenarios would be ranked accordingly.

Table 8.7 Example criticality (impact) categories

Public safety		Employee safety	
Category	Description	Category	Description
1	No injury or health effects	1	No injury or occupational safety impact
2	Minor injury or minor health effects	2	Minor injury or minor occupational illness
3	Injury or moderate health effects	3	Injury or moderate occupational illness
4	Death or severe health effects	4	Death or severe occupational illness
Production loss		**Facility/equipment damage (millions of dollars)**	
Category	Description	Category	Description
1	Less than one week	1	Less than 0.1
2	Between one week and one month	2	Between 0.1 and 1
3	Between one and six months	3	Between 1 and 10
4	More than six months	4	Above 10

Source: ABS Consulting, Knoxville, Tennessee

Table 8.8 Example frequency categories

Category	Description
1	Not expected to occur during the facility lifetime
2	Expected to occur no more than once during the facility lifetime
3	Expected to occur several times during the facility lifetime
4	Expected to occur more than once in a year

Source: ABS Consulting, Knoxville, Tennessee

If a hazard evaluation provides information about specific incident scenarios as well as recommendations for improving safety, the recommendations can be ranked by the scenarios they address and the extent to which the team believes the recommendation will reduce the risk. These risk-based rankings represent the most balanced view possible in that the team considers both likelihood and consequence in determining the importance of their recommendations. Table 8.9 lists examples of how a hazard evaluation team might categorize recommendations from a hazard evaluation using the risk matrix in Figure 8.1.

The techniques presented here for prioritizing hazard evaluation results are applicable to most hazard evaluations. However, more detailed analyses may be necessary if (1) the evaluation identifies complex failure scenarios (e.g., involving many failures), (2) the potential changes considered are expensive, or (3) there are important public or regulatory issues involved. In many cases, these more detailed evaluations will require quantitative risk analysis. The *Manager's Guide to Quantitative Risk Assessment*, published by the American Chemistry Council, provides guidance on when CPQRA is an appropriate tool to use, and the CCPS' *Guidelines for Chemical Process Quantitative Risk Analysis* provides detailed information on conducting such analyses.[1,11]

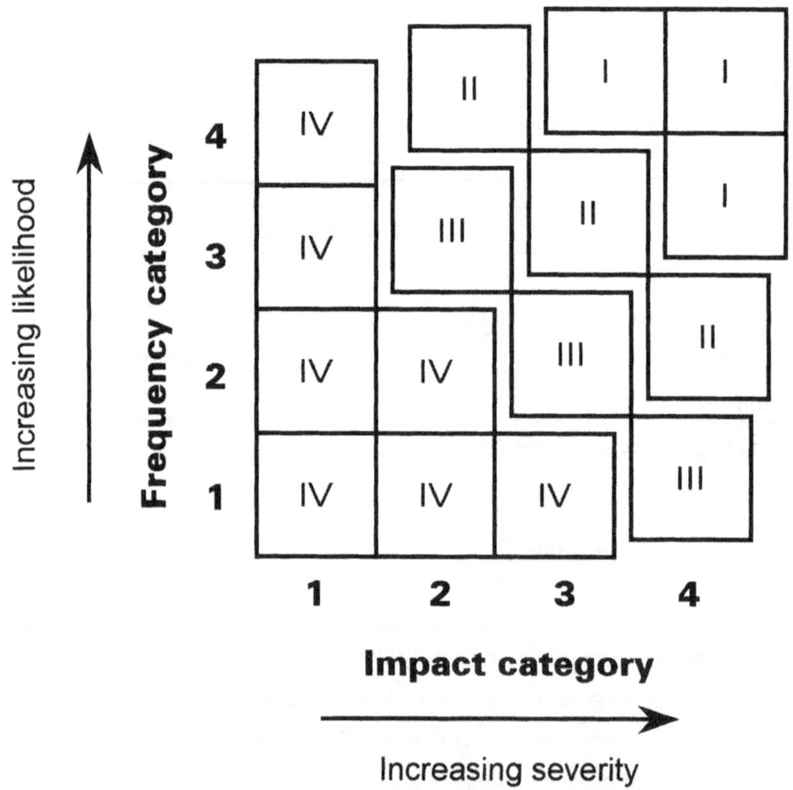

Figure 8.1 Example risk matrix

Table 8.9 Example risk ranking categories

Number	Category	Description
I	Unacceptable	Should be mitigated with engineering and/or administrative controls to a risk ranking of III or less within a specified time period such as 6 months
II	Undesirable	Should be mitigated with engineering and/or administrative controls to a risk ranking of III or less within a specified time period such as 12 months
III	Acceptable with controls	Should be verified that procedures or controls are in place
IV	Acceptable as is	No mitigation required

Source: Stone and Webster Engineering Corporation, Houston, Texas

8.3 Documentation of Hazard Evaluations

After a hazard evaluation is completed, hazard analysts should communicate their findings to the sponsor of the study so that appropriate risk management decisions can be made. Some of the recommendations for improving safety will be accepted and implemented; other recommendations may be modified or rejected. In all cases the rationale for the decisions should be formally documented.

The importance of properly documenting the results of a hazard evaluation is often understated or given too little consideration. Without appropriate documentation, it is more difficult for an organization to achieve the desired objectives from a hazard evaluation. Moreover, studies that are not appropriately documented can represent legal liabilities to the organization that sponsors the work.

There are three additional purposes for documenting the results of a hazard evaluation:

1. To consolidate and preserve the primary results of the study for future use
2. To provide evidence that the hazard evaluation has been performed according to sound engineering practices
3. To support other PSM program activities

The higher the quality of the study documentation, the easier it will be to achieve these three objectives.

Documentation needs should be defined before the hazard evaluation begins (or as soon as possible thereafter) since these needs may influence how the analysis is actually carried out. For example, consider a company that wants to rapidly perform a hazard evaluation of a process that is being considered for acquisition. The hazard analysts choose the HAZOP Study approach, but to save time, decide only to record any action items or unresolved questions that arise during the meetings. Later, after purchasing the facility, the company wants to use the previously completed HAZOP Study to help provide plant operators with training in emergency responses to potential process upsets. Because this information was not documented, the company had to spend additional effort to generate the information needed to support the training activities. Within reason, it is always more prudent to provide ample documentation.

When determining the extent and level of detail of hazard evaluation documentation, hazard analysts should consider the following questions: What is the purpose of the hazard evaluation and who is the audience? First, a hazard evaluation report should be designed to meet the stated primary purpose and anticipated future needs for the study. There is no sense in writing a 300-page report if the objectives of the study can be met by a 30-page document. For example, if the hazard evaluation was used to analyze the conceptual design of a process and to determine whether any safer design features could be used to eliminate hazards, there is no sense in creating a report that will be out of date once the detailed design effort begins.

On the other hand, if an organization wants to use a hazard evaluation report to more easily perform periodic process reviews, then it makes sense to put together a report that documents as much of the hazard evaluation team's analysis and judgments as possible. Then, this information will be retrievable when the next periodic review of the process is performed.

Second, preparers of hazard evaluation reports should also consider who will be using the document. For example, if the immediate audience will be made up of people who are very knowledgeable about the process design project (e.g., the project management and engineering staff), then less background and detail will be needed to explain the results of the work. On the other hand, if a report is likely to be reviewed by people (either now or in the distant future) who are unfamiliar with the process, then considerably more information should be provided to help them understand the results of the study. Table 8.10 lists some other questions that hazard analysts should consider when determining the documentation needed for a particular study.

Other considerations can also influence the design of a hazard evaluation report.

- The report should contain sufficient information to allow adequate scrutiny of the technical quality of the analysis.
- The report should carefully define its scope and limitations. For example, if a review focuses only on events that could impact the public, the report should state that it does not consider all the ways plant personnel could be injured by those same events or ones that were not included because their effects could not reach beyond the plant boundaries.
- The team should consider documenting any action items or recommendations that were identified and resolved immediately. That is, if based on the team's review some potentially unsafe situations were recognized and corrected immediately, these should be documented. In this way, similar hazards may be reduced if other locations in the company try to use the results of this study.

Table 8.10 Some issues that influence the contents of hazard evaluation reports

- What level of knowledge and technical expertise does the primary audience of the report have?
- Will other locations of the organization be able to use the report to help them examine their similar processes?
- Will the report be made available to the public?
- Will it be released for third-party review as part of a permitting, regulatory, or legal matter?
- Will the hazard evaluation team be involved in helping the organization's managers decide which recommendations to implement?

- The report should provide the basis for following up the unresolved action items. In some cases the report can document the assignment of responsibilities and dates for actions to be completed; in other cases it is more appropriate to simply provide a list of the unresolved action items that management needs to track.

A hazard evaluation report can contain a variety of elements, depending upon the needs defined by the sponsoring organization. The hazard evaluation team leader should develop a preliminary outline of the proposed hazard evaluation report early in the project. This way the appropriate information can be saved during the study to aid in producing the final report. Table 8.11 lists various types of information that can be included in a hazard evaluation report. Not all of these types of information are needed for every report. The appropriate elements of a report should be selected based on all of the preceding considerations.

Table 8.11 Items to consider including in hazard evaluation reports

Item	Description
Study objectives	Primary purpose and ancillary needs
Physical and analytical scope	Specific process areas, systems, and equipment covered. Hazards analyzed along with situations beyond the study's scope
Analysis team composition	Participants in the review (categorized by organization and job/experience)
Meeting dates and duration	When, where, and for how long the team met
Hazard evaluation technique description	Rationale for selection of the hazard evaluation technique(s). Description of the technique(s) employed
Process/activity description	Description of the system analyzed. Explanation of the procedural steps used in batch operations. Listing of important process conditions. Descriptions of safety systems. Listing of materials and hazards
Drawing, specification, and procedures list	Documents (with dates and revision numbers) used as the basis for the analysis
Action items/suggestions	Recommendations made to management for risk reduction
Detailed documentation of any review meetings; technique-specific lists, forms, or graphical models	Minutes of meetings, summaries of topics covered, attendance lists Checklists, lists of What-If questions, worksheets, FMEA or HAZOP tables, fault trees, event trees, cause-consequence diagrams, HRA event trees

Even though the specific contents of each hazard evaluation report may differ due to the needs of the study, there are essential features that every report should contain. The following questions should be addressed by every hazard evaluation report, no matter how simple the report is or how minimal the objectives of the study.

- Who did the study?
- On what process area or operation was the study performed?
- Which technique was used?
- What are the important results and conclusions?

Generally, the more an organization wants to maximize the benefits derived from a hazard evaluation, the more extensive the documentation needs to be. Following is a minimum set of information that could be used to document the results of a HAZOP Study of an existing process:

- Each hazard evaluation team member's name, function/expertise and organization
- A list of drawings and procedures used (along with their revision dates)
- A simple statement indicating how the HAZOP Study technique was used
- A list of the action items that resulted from the team's deliberations

In this example, not even the HAZOP table that may have been produced in the meetings would be a ***necessary*** part of the documentation, as long as there was no need to (1) use the results in later years, (2) "prove" that an adequate HAZOP Study was done, or (3) perform quality assurance of the study beyond that which was done during the actual team meetings.

On the other hand, a much more detailed report would be appropriate if the organization had the following objectives:

- To use the results of the study to support more efficient periodic reviews of the process
- To use the study at other locations to improve the design/operation of similar processes
- To support a technical position in a permitting or regulatory matter
- To support other PSM activities, such as developing procedures, management of change reviews, performing CPQRA studies, improving training, or testing emergency response readiness

In this situation, the contents of the hazard evaluation report would be much more extensive and could include the following elements:

- A section describing the specific details of the hazard evaluation technique(s) used in the study. Any extension or nonstandard use of the method(s) should be pointed out.

- A description of the process, including simplified schematics. This would be useful if people who are unfamiliar with the process will have access to the report.

- Justifications for each of the action items. Beyond the simple list of one-sentence descriptions, this section would summarize why the hazard evaluation team believed the potential incident situation was significant enough to warrant the proposed action.

- Typed versions of the hazard evaluation team meeting notes that have been thoroughly reviewed for consistency and accuracy.

- Copies of any technique-specific models or worksheets that may have been used to explain why the hazard analysts concluded particular events were significant.

Analysts should prepare hazard evaluation reports with the same care and diligence as they would any important technical document. These reports should have concise summaries with the detailed background material placed in easy-to-use appendixes. Many companies are using personal computers to prepare and archive their hazard evaluation reports on electronic media. This can help reduce the sometimes overwhelming amount of paper that results from these studies.

Regardless of the format of these reports, a prudent organization will retain them throughout the life of the process or operation. Only when a process has been decommissioned can an organization really be sure that they will not need the information. Even then, companies may wish to summarize the lessons they learned from studies performed over a process's lifetime so that this feedback can be given to the process development, design, and operating departments to help build new facilities in the future. Sometimes, legal concerns may cause organizations to retain hazard evaluation reports long after a process has been shut down.

8.4 Development of a Management Response to a Hazard Evaluation

Once the results of a hazard evaluation have been documented, the report must be presented to management so they can make appropriate decisions. Sometimes a team may need to identify specific options for safety improvement before a decision can be made. For example, the list of action items may include several areas that need additional design review to help determine if there are any alternative ways to solve the problem. After specific alternatives are proposed, an organization's management can make a decision concerning which changes to implement.

The decision to accept or reject the recommendations of a hazard evaluation team may be very difficult. The hazard evaluation team's criteria for identifying solutions to hazard problems is limited compared to the criteria that the managers use. Table 8.12 lists some examples of the considerations that managers typically deal with when deciding how to allocate their organization's resources to get the greatest overall benefit. Usually the first two considerations, "how serious is the safety problem" and "how beneficial is the solution," are used by the hazard evaluation team when prioritizing action items. The managers' jobs are much more difficult because they must factor in the other competing concerns listed in the table.

Table 8.12 Examples of risk management considerations

- How serious is the safety problem?
- How beneficial is the potential safety solution?
- Does the proposed solution create adverse effects elsewhere?
- How expensive is the proposed solution?
- Does the solution require capital resource allocation or adding more personnel, or can it be done under the current operating budget?
- Can the work be done using company personnel only, or will contractors or vendors be required?
- What is the effectiveness of the proposed improvement? How much benefit is gained per dollar spent?
- How long will it take to implement the change? Can it be done while the unit is operating, or will it require an extended shutdown?
- How will the improvement affect other aspects of the facility, such as labor relations?
- Does the change represent an exception to corporate policy? Will company standards have to be modified? Will other facilities be affected?
- What regulatory implications are involved?

Once recommendations from a hazard evaluation team have been fully evaluated and alternatives have been identified, an organization must formally respond to the study. Sometimes it is necessary for a follow-up technical review by a team different from the hazard evaluation team to demonstrate the recommendation is valid and acceptable to local management. Management of Change reviews are often appropriate to ensure that new hazards are not introduced by the change and that the proper documentation and training is done. Documenting this management response to the study is very important and is required by OSHA's PSM standard (29 CFR 1910.119). Some organizations append their formal responses to the original hazard evaluation report; others create a separate document. Either approach is acceptable as long as there is an auditable link between the action items and the responses.

It is quite appropriate for management to decide not to implement specific recommendations; however, the reasons for those decisions need to be justified and clearly stated. These reasons may include examples of other safety improvements that make implementing the recommendation unnecessary. Sometimes a hazard evaluation team should reconvene to consider the hazards of other safety improvements that are developed during the decision making process. Table 8.13 lists typical reasons an organization might have for not implementing a particular recommendation suggested by a hazard evaluation team. Depending on the facility's particular regulatory and corporate requirements, not all of these reasons may be allowable.

When a decision is made to implement an improvement, management should be told how long it will take to implement the change. Based on how they perceive the risk associated with a potential incident situation, managers must either take other special measures (e.g., operate at reduced throughput) until the change is implemented or operate as usual until the change is made. This is especially important if managers must wait until the next major outage to implement the change. Again, managers should assign responsibility for implementing the change to someone with sufficient authority to ensure adequate follow-up.

Table 8.13 Typical reasons why rejecting a hazard evaluation team recommendation might be justified

- A detailed engineering analysis following the hazard evaluation indicates the suggestion was not a good idea [or not feasible] for the following specific reasons...
- Other information, which was not available to the hazard evaluation team, indicates that the hazard is not as significant a problem as the team thought...
- A review of the data upon which the team based its finding revealed an error in the data, which, when corrected, indicates the finding is no longer warranted...
- The situation has changed; the recommendation is no longer valid because...
- Implementation of other action items from the hazard evaluation causes this recommendation to no longer be necessary...
- The recommendation, although it would provide some improvement in safety, does not provide as effective a level of risk reduction as do other safety improvement alternatives...
- Although a good idea for possibly improving operability of the process, the recommendation is not related to protecting the safety of workers or the public or protecting the environment, so it will be forwarded to the Operations Manager for consideration but not tracked as a process safety item...
- The cost of implementing the change is not justified in light of the relatively minor risk reduction afforded by it compared to other recommendations...
- A detailed review of the scenario risk, the risk reduction gained by identified potential risk reduction options, and the costs involved in implementation of the options shows that the scenario risk is now as low as reasonably practicable (ALARP), with the costs associated with implementing additional risk reduction options being grossly disproportionate to any further risk reduction to be gained...

8.5 Resolution of Action Items

Organizations should establish a management system to track progress on action items from the time the items are recommended until they are resolved. The follow-up documentation to a hazard evaluation should identify the individual(s) who is evaluating a solution to an issue, or an organization responsible for deciding whether a specific recommendation is to be implemented. This documentation should include the specific date by which a response or resolution is required. If the date is not met, then the tracking system should require that a new date and an interim response (which may be a temporary risk control measure) be provided. Table 8.14 provides an example of a table used for documenting the current status of action items.

This tracking process may take a long time—possibly years—if the implementation of a recommendation requires a major outage to allow the change to be made, or requires design and construction of new facilities. Because of the differences in time that it takes to analyze and implement changes, the follow-up system will necessarily be tracking recommendations at many different stages in the review, approval, and implementation process. When issues are divided for evaluation and implementation, there is a tendency to consider the entire issue closed when any part of the change is completed. Thus, this tracking process should be reviewed periodically to help ensure that the full intent of all authorized improvements is satisfied. The tracking system should allow for interim actions if full implementation of a recommendation cannot be accomplished in one year. If any imminent risks are identified, interim risk control measures may need to be implemented immediately and kept in place until more permanent solutions are found. For example, if an undersized relief valve is found that cannot be replaced at once, some manual valves may need to be locked open to ensure that relief valves on other equipment can provide the difference between the existing and the required relief capacity.

Table 8.14 Example action item tracking log

Item	Description	Priority	Responsibility	Status and response date*
1	Obtain an SCBA unit to be stationed on the west side of the unit near the tank farm	I	Fire Dept.	Completed ##/##/##
2	Survey all relief valves with manual isolation valves and ensure that the isolation valves are locked open	I	Unit Operations Superintendent	Completed ##/##/##
3	Develop an emergency training session for the fire brigade, presenting the special hazards of fires involving DOT Chemical 2108	II	Fire Dept.	Under development — planned for second quarter training
4	Consider adding an independent high temperature alarm for the scrubber outlet flow to help detect when a reaction is occurring in the scrubber	II	Process Engineering	Rejected — other improvements obviate the need for this change
5	Provide handwheel extensions to the compressor manual isolation valves so the valves can be closed without entering the compressor building when a leak occurs	II	Design Engineering	In detailed design — planned for turnaround in June ####
6	Consider providing a compressor shutdown system that includes the capability to remotely stop and isolate the compressors	III	Design Engineering	Under evaluation — ##/##/##
7	Consider providing an improved ventilation system that is capable of preventing the buildup of a flammable environment in the compressor building	III	Design Engineering	Under evaluation — ##/##/##

* If the action item is not completed, the date shown is when the next update is required from the responsible organization.

After managers reach a decision about which hazard evaluation recommendations are to be implemented, the tracking system should focus on the status of the implementation for each action item. Before implementing the recommendations, each specific engineered and procedural solution should be reviewed for hazards that it may create. Some recommendations may be very simple and can be handled through a facility's normal management of change program. For other recommendations, particularly those involving major design changes, it may be necessary to conduct a formal hazard review of the detailed plans for the change before it is implemented. Often the more complex recommendations are returned to the original hazard evaluation team for analysis; however, such a policy may be impractical if the particular recommendation requires a long period of time for evaluation. In any event, all changes must be analyzed to avoid introducing new hazardous situations into the process or operation.

The person who is responsible for resolving the hazard evaluation recommendations should have the authority to ensure that actions are resolved in a timely manner. If the individual who actually does the tracking does not have that type of authority, the current status of all hazard evaluation recommendations should be reviewed regularly with the facility's senior management (e.g., at a plant manager's monthly safety meeting or a periodic safety committee meeting).

Resolving issues raised or recommendations made by a hazard evaluation team is a critical element in an effective process safety management (PSM) program. Timely resolution of recommendations from a hazard evaluation is required by the U.S. OSHA PSM Standard. Moreover, failing to resolve such items leaves the facility with inadequate safeguards and may increase the liability a company faces if an incident occurs. From a liability standpoint, not addressing an issue that has been identified may be worse than not realizing that the issue existed. This does not mean that all items from a hazard evaluation must be implemented. It may be acceptable to reject a specific recommendation (see Table 8.13 and the accompanying text); however, someone must document the resolution no matter what decision is made.

8.6 Communication of Special Findings / Sharing of Information

Sometimes a hazard evaluation may discover findings or information that should be communicated beyond the plant or process evaluated by the study. For example, a hazard evaluation may be done on new technology that has the potential to be used at other locations in the company. It would be prudent to share the results of such a study with the other locations so they can benefit from the hazard evaluation results. Similarly, the results of hazard evaluations being done in plants with the same technology may need to be shared with sister plants to ensure that important issues in their hazard evaluations had not been missed. The results of some hazard evaluations that may have industry wide implications may be shared with industry groups or presented at national technical conferences. Because hazard evaluations can require significant resources and their results can prevent major incidents, the results of all hazard evaluations should be communicated to those who can benefit from the knowledge.

8.7 Use of Hazard Evaluation Results over the Plant Lifetime

As previously stated in Section 6.6, process hazards exist from the onset of a project through its end or assimilation into another project. In the course of the necessary hazard evaluations that occur over the lifetime of the project, the process safety features are documented along with some form of qualitative or quantitative assessment of process risks. The documentation represents the combined knowledge and experience of those most familiar with the specific hazards of the process.

The following plant lifetime stages have been delineated by CCPS:

1. Process development (includes research & development and project implementation)
2. Detailed design
3. Construction and start-up
4. Operating lifetime (includes production operations and cyclic reviews)
5. Extended shutdowns
6. Decommissioning.

Each stage makes use of the hazard evaluation information in many ways. Table 2.1 in Section 2.2 listed typical hazard evaluation objectives at different life cycle stages of a process. Similarly, Table 8.15 identifies some of the ways the hazard evaluation results developed in each stage are used during the life of a project. Other examples of how the results can be used can be found in Section 8.3 on Documentation of Hazard evaluations.

Chapter 8 References

1. J. S. Arendt et al., *Evaluating Process Safety in the Chemical Industry—A Manager's Guide to Quantitative Risk Assessment,* American Chemistry Council, Arlington, VA, 1989.
2. W. E. Vesely et al., *Fault Tree Handbook*, NUREG-0492, U. S. Nuclear Regulatory Commission, Washington, DC, 1981.
3. J. B. Fussell, "Fault Tree Analysis: Concepts and Techniques," *Generic Techniques in System Reliability Assessment*, NATO, Advanced Study Institute, Brussels, 1976.
4. W. E. Jordan, *Failure Modes, Effects and Criticality Analysis*, proceedings of the Annual Reliability and Maintainability Symposium, San Francisco, Institute of Electrical and Electronics Engineers, New York, 1982.
5. *Procedures for Performing a Failure Mode, Effects and Criticality Analysis*, MIL-STD-1629A, Department of Defense, Washington, DC, 1980.
6. H. R. Greenberg and J. J. Cramer (eds.), *Risk Assessment and Risk Management for the Chemical Process Industry*, ISBN 0-442-23438-4, Van Nostrand Reinhold, New York, 1991.
7. J. S. Arendt et al., *Qualitative Risk Assessment of Engineered Systems*, AIChE 71st Annual Meeting, Miami, November 1978, American Institute of Chemical Engineers, New York, 1978.
8. M. L. Casada et al., "Facility Risk Review as an Approach to Prioritizing Loss Prevention Efforts," *Chemical Engineering Progress*, October 1990.
9. P. L. Clemens, "System Safety Principles Applied in Industrial Hygiene Practice," *Hazard Prevention*, November/December 1987.
10. J. A. Alderman, "PHA—A Screening Tool," presented at the AIChE Spring National Meeting, Orlando, March 1990.
11. Center for Chemical Process Safety, *Guidelines for Chemical Process Quantitative Risk Analysis, 2nd Ed.*, American Institute of Chemical Engineers, New York, 1999.

Table 8.15 Some uses for hazard evaluation results over the life of a project

Project phase	Uses for hazard evaluation results
Process development	▪ Determine whether any inherently safer design features could be used to eliminate hazards
	▪ Justify many design specifications for safety systems in preparation for the detailed design
Detailed design	▪ Identify similar hazards/risks that may be reduced if other locations in the company were to use the results of the study
	▪ Provide evidence that an adequate hazard evaluation was done
	▪ Support a technical position in permitting or regulatory matters
Construction and start-up	▪ Identify reasons for specific construction requirements, such as gauge, valve, and instrument locations, vessel orientation, etc.
Operating lifetime	▪ Use a hazard evaluation report to more easily perform periodic process reviews and as the basis for revalidations
	▪ Share lessons learned or discovered with other operating plants
	▪ Support other process safety activities, such as revising procedures, reviewing proposed changes, performing CPQRA studies, improving training, and testing emergency response readiness
Extended shutdowns	▪ Use a previously completed hazard evaluation on the process to help identify risky restart issues and better prepare for resumed operations
Decommissioning	▪ Summarize the lessons learned from studies performed over the lifetime of the process so that feedback can be given to the process development, design, and operating departments when building future new facilities
	▪ Legal and liability concerns may require organizations to retain hazard evaluation reports long after a process has been shut down
Mergers and acquisitions	▪ Use a hazard evaluation as part of the due diligence process to help assess process safety issues related to the acquisition
	▪ As a tool in the merger process, use the hazard evaluation to understand and control risk during the merger transition

9

Extensions and Special Applications

This chapter gives further information on special topics that are related to hazard evaluation procedures. These include hazard evaluation of procedure-based operations, processes controlled by programmable systems, and chemical reactivity hazards; consideration of human factors and Human Reliability Analysis; consideration of facility siting, and the combination of tools such as Hazard and Operability Studies with Layer of Protection Analysis.

9.1 Hazard Evaluation of Procedure-Based Operations

During the period 1970 to 1989, 60 to 75% of major incidents in continuous processes occurred during "non-routine" modes of operation; i.e., in operating phases other than the continuous operation of the process after start-up.[1] As discussed throughout these *Guidelines,* hazard evaluations are employed to help determine where existing safeguards may not be adequate. However, many facilities have applied hazard evaluation techniques only to the continuous operation of a process and not to procedure-based aspects such as start-up, shutdown, emergency operations, sampling, and catalyst change-out. Likewise, even the evaluation of explicitly procedure-based operations such as batch chemical reactions has often used only the piping and instrumentation diagram (P&ID) as the basis of the analysis, without explicitly looking at the batch procedure.

This section describes approaches that have been employed to identify and evaluate incident scenarios associated with procedure-based modes of operation, including the capability to more fully address the human factors aspects that can make operations during procedure-based modes more hazardous.[2,3] (See Section 9.5 for discussion of Human Reliability Analysis and other human factors considerations.) Personnel may have less operating experience with procedure-based operations that are heavily dependent on task performance and operator decision-making. In addition, safeguards may be bypassed or not fully functional during some modes of operation such as at start-up of a continuous process. Performing a hazard evaluation of procedures can identify steps where the operator is most vulnerable and point to means of reducing the risk of an incident, such as by adding engineered safeguards and improving administrative controls.

The approach outlined in this section applies equally well to the hazard evaluation of any situation where the steps for a procedure-based operation are well-defined and in written form, including the hazard evaluation of existing processes, hazard evaluations during preliminary and detailed design phases of projects (for new or revised processes), and management of change hazard reviews. A case study illustrates the analysis approach.

Overview of Methodology

The hazard evaluation of procedure-based operations involves reviewing procedures using a HAZOP Study, simplified HAZOP Study, or What-If Analysis to uncover potential incident scenarios associated with procedure-based modes associated with both continuous and batch operations. By analyzing procedural steps where human error is more likely, and where human error or component failure could lead to a consequence of concern, risk can be reduced. The objective of the hazard evaluation team studying a procedure-based operation is to evaluate the adequacy of safeguards associated with incident scenarios that have as their initiating causes either the skipping of a procedural step or the performing of a step incorrectly.

Preliminary Hazard Analysis, Checklist Analysis, Failure Modes and Effects Analysis, and other hazard evaluation methods are not useful or applicable for accomplishing a detailed hazard evaluation of procedure-based operations, although a checklist of human factors issues or procedure format/ presentation rules can be useful to supplement an evaluation of the risk of skipping steps and performing steps incorrectly. As with scenarios uncovered during the evaluation of continuous mode of operation, more detailed analyses such as Layer of Protection Analysis or Human Reliability Analysis may need to be performed to more fully address any unresolved or complex issues raised in the hazard evaluation of procedure-based operations.

Purpose

The purpose of evaluating the hazards of a procedure-based operation is to ensure a facility has sufficient safeguards for the inevitable situation when a procedural step is skipped, performed out of sequence, or performed incorrectly. Human factors can be considered in hazard evaluations by asking why an operator might make an error that leads to a process deviation. However, this approach only addresses a small fraction of the potential human errors that can affect process safety. Many analysts attempt to find incident scenarios in procedure-based operations by adding generic guide words such as "deviations during start-up" and "deviations during maintenance/sampling" to the hazard evaluation of equipment nodes or subsystems. This has been found to be insufficiently rigorous, as the hazard evaluation team is focused primarily on the equipment and process parameters and on normal-operation safeguards rather than the details of the start-up or maintenance/sampling procedures. Experience has shown that the step-by-step analysis of procedures has revealed many more potential incident scenarios than trying to address procedure-based modes of operation during P&ID-driven hazard evaluations.

While a company may not currently conduct hazard evaluations of procedures, it is likely to perform some type of Job Safety Analysis (JSA).[4] The JSA can be an excellent starting point for an evaluation of procedures, since a JSA identifies the tasks that workers must perform and the equipment required to protect workers from typical industrial hazards (slips, falls, cuts, burns, fumes, etc.). For a few procedure-based tasks such as simple sampling operations, a JSA might be considered adequate, but a typical JSA does not address detailed process safety or human factors issues. For example, from a JSA perspective, it may be perfectly safe for an operator to open a steam valve before opening a feed valve; however, from a process safety perspective, the operator may need to open the feed valve before the steam valve to avoid the potential for overheating the reactor and initiating an exothermic decomposition. The primary purposes of JSAs and other traditional procedure review methods have been to ensure the procedures are accurate and complete, the hazards of performing the tasks are recognized, and precautions such as the use of personal protective equipment are taken.

Even the best procedure may not be followed for any number of reasons, and these failures to follow the prescribed instructions can and do result in incidents. The chance of a worker making an error in following a procedure as a whole is generally considered to be greater than 1%. When factoring in common human factor deficiencies that accompany nonroutine operations, such as fatigue, lack of practice, and the rush to restart and return to full production, the probability of error can climb to 10% per task, with a "task" typically involving anywhere from one to ten detailed steps. Some operational errors are obviously more critical than others. Measures such as cross-checking and the use of checklists can reduce the chances of error, but only to a limited extent. Industry has found that a HAZOP Study or What-If Analysis, structured to address procedures, can be used effectively for finding the great majority of incident scenarios that can occur during procedure-based modes of operation.[5,6]

HAZOP Method for Analyzing Deviations of Procedural Steps

The Hazard and Operability (HAZOP) Study method can be considered to have two major variations:

1. For continuous mode of operation, the team brainstorms what would happen if the process deviates from normal operating parameters.
2. For procedure-based operations, the team brainstorms what would happen if the steps of a procedure are not followed correctly.

The second variation can be used when analyzing batch or other procedure-based operations where there are several steps generally involving manual operations. It is an expansion of a hazard evaluation method based strictly on asking:

- ***What happens if the step is skipped?***
- ***What happens if the step is performed incorrectly?***

This method of brainstorming incident scenarios is based on the understanding that human errors occur by someone not doing a step (errors of *omission*) or by doing a step incorrectly (errors of *commission*). Hence, simply asking what would happen if the operator omitted a step or performed a step incorrectly is one way to structure the hazard evaluation of a step-by-step procedure. To be more thorough, these two types of errors can be broken out into subparts using the original seven HAZOP Study guide words:

Omission:	NO (Skip a Step, or Step Missing)
	PART OF
Commission:	MORE
	LESS
	AS WELL AS
	REVERSE (Out of Sequence)
	OTHER THAN

The guide word NO can be augmented by adding the option of discussing whether there are any steps ***missing*** from the procedure.

To apply the HAZOP Study method to procedural steps for start-up, shutdown, loading/unloading, maintenance, and other procedure-based modes of operation, the facilitator or team must first divide the procedure into individual actions. This is already done if there is only one action per step. The set of guide words or questions is then systematically applied to each action of the procedure, resulting in procedural deviations or what-if questions. The guide words shown in Table 9.1 were derived from HAZOP guide words commonly used for analysis of batch processes. The definition of each guide word is carefully chosen to allow universal and thorough application to both routine batch and nonroutine continuous and batch procedures. The actual review team structure and meeting progression are nearly identical to those of a process equipment HAZOP Study or What-If Analysis, except that active participation of one or more operators is even more important.

For each deviation from the intent of the process step, identified by applying these guide words to the process step or action, the team should look beyond the obvious cause of "operator error" to identify root causes associated with human error such as "inadequate emphasis on this step during training," "responsible for performing two tasks simultaneously," "inadequate labeling of valves" or "instrument display confusing or not readable." The guide word NO might elicit causes such as "no written procedural step," "no formal training to obtain a hot work permit before this step" or "no written procedural step or formal training to open the discharge valve before starting the pump." If not explicitly documented, these underlying causes should at least be discussed by the team when assessing the likelihood of the operational error. The team should also discuss systematic reasons that raise the likelihood of skipping steps, such as fatigue, miscommunication or misunderstanding of responsibility.

Table 9.1 Definitions of guide words (with alternatives) for HAZOP Study of procedure-based operations

Guide word	Meaning when applied to a step
MISSING*	A step or precaution is missing from the written procedure prior to this step
NO, NOT or SKIP	The step is completely skipped or its specified intent is not performed
PART OF	Only part of the full intent is performed (usually applies to a task that involves two or more nearly simultaneous actions, such as "Open valves A, B, & C")
MORE or MORE OF	Too much of the specified intent is done (too much quantity added, performed for more time than intended, etc.) or the step is performed too rapidly**
LESS or LESS OF	Too little of the specified intent is done or the step is performed too slowly**
AS WELL AS or MORE THAN	Something happens, or the operator does another action, in addition to the specified step being done correctly
REVERSE or OUT OF SEQUENCE	This step is performed too early in the sequence, or a later step is performed at this time instead of the intended step
OTHER THAN	The wrong material is selected or added or the wrong device is selected, read, operated, etc. in a way other than intended

*optional guide word ** MORE and LESS are not applicable to simple on/off or open/close functions

Two Guide Word Method for Analyzing Deviations of Procedural Steps

A streamlined approach known as Two Guide Word Analysis has proven useful for analyzing procedures related to less hazardous operations and tasks. It might also be used by a team leader having extensive experience in the use of the guide words mentioned previously and capable of effectively using a more streamlined approach with comparable results. The two guide words (or guide phrases) for this approach, as defined in Table 9.2, encompass the basic human error categories of errors of omission and errors of commission. These guide words are used in an identical way to the guide words introduced earlier. Essentially, OMIT or "Step Skipped" includes the errors of omission related to the guide words MISSING, NO and PART OF mentioned earlier. The guide word INCORRECT or "Step Performed Incorrectly" incorporates the errors of commission related to the guide words MORE, LESS, AS WELL AS, REVERSE, and OTHER THAN. The analysis of an emergency shutdown procedure in Table 9.3 illustrates this approach.

What-If Method for Analyzing Deviations of Procedural Steps

The What-If Analysis method for analyzing procedure-based modes of operations uses the same basic brainstorming approach as for a continuous operation. The hazard evaluation team reads the procedure and then answers the question, "What errors might lead to the consequences of concern?" The team lists these operational errors, and then brainstorms the full causes, consequences, and existing safeguards as is done for the guide word approaches. This type of what-if brainstorming often attempts to cover the entire task (procedure) at one time. However, some companies use the What-If Analysis or What-If/Checklist Analysis approach on either individual steps or groups of steps rather than on the entire procedure.

Table 9.2 Guide words for Two Guide Word Analysis of procedure-based operations

Guide word	Meaning when applied to a step
OMIT (Step Skipped)	The step is not done or part of the step is not done. Some possible reasons include the operator forgot to do the step, did not understand the importance of the step, or the procedure did not include this step
INCORRECT (Step Performed Incorrectly)	The operator's intent was to perform the step (not omit the step); however, the step is not performed as intended. Some possible reasons include the operator does too much or too little of the stated task, the operator manipulates the wrong process component, or the operator reverses the order of the steps.

Table 9.3 Example Two Guide Word Analysis documentation for part of an emergency shutdown procedure

Item	Deviation	Causes	Consequences	Safeguards	Recommendation
Drawing or procedure: SOP-03-002, Cooling Water Failure		**Unit:** HF Alkylation		**Method:** Two Guide Word Analysis	**Documentation type:** Cause by cause
Node: 23		**Description:** STEP 2. Block in olefin feed to each of the two reactors by blocking in feed at flow control valves.			
23.1	Step Skipped	Operator fails to block in one of the reactors, such as due to miscommunication between control room operator and field operator, or control valve sticking open or leaking through	High pressure due to possible runaway reaction (because cooling is already lost) due to continued feeding of olefin (link to 11.7 - High Rxn Rate; HF Alky Reactor #1/#2) High pressure due to high level in the reactor, due to continued feeding of olefin (link to 11.1 - High Level; HF Alky Reactor #1/#2)	High temperature and pressure alarms on reactor Field operator may notice sound of fluid flow across valve Flow indication (in olefin charge line to reactor that is inadvertently NOT shut down) Level indicator, high level alarm, and independent high-high level switch/alarm	
		Operator fails to make sure bypass valve is also closed, since this precaution is not listed in the written procedure; or the bypass valve leaks through	High pressure due to possible runaway reaction (because cooling is already lost), because of continued feeding of olefin (link to 11.7 - High Rxn Rate; HF Alky Reactor #1/#2) High pressure due to high level in the reactor, because of continued feeding of olefin (link to 11.1 - High Level; HF Alky Reactor #1/#2)	High temperature and pressure alarms on reactor Field operator skill training requires always checking bypasses are closed (where applicable) when blocking in control valves Field operator may notice sound of fluid flow across valve Flow indication in olefin charge line (but likely not sensitive enough for small flows) Level indicator, high level alarm, and independent high-high level switch/alarm	
		Operator fails to close flow control valve manually from the DCS because the phrase "block in" is used instead of the word "close"	Valve possibly opens full at restart, allowing too much flow to reactor at restart, resulting in poor quality at start-up and/or possibly resulting in runaway reaction and vessel rupture	Control room skill training requires always manually commanding automatic valves closed before telling field operator to block in control valve	37. Implement best-practice rules for procedure writing, including the use of common terminology
23.2	Step Performed Incorrectly	Operator closes the olefin charge flow control valves before shutting down the charge pump, primarily because the steps are written out of the proper sequence	Charge pump dead-headed, leading to possible pump seal damage/failure and/or other leak, resulting in a fire hazard affecting a small area	Step 3 of procedure that says to shut down charge pump The step to shut down the charge pump (Step 3) is typically accomplished before Step 2 (in practice)	41. Move Step 3 ahead of Step 2
		Field operator closes both upstream and downstream block valves	Possible trapping of liquid between block valve and control valve, leading to possible valve damage from thermal expansion	Field operator skill training stresses that only one block valve should be closed	

Choosing the Right Method for Analysis of Procedure-Based Operations

The What-If Analysis approach that covers an entire task (procedure) at one time can be expected to take less time than the Two Guide Word method, which in turn can be expected to take less time than applying a full set of seven or eight HAZOP Study guide words to each procedural step. Experience has shown that hazard evaluation facilitators, newly trained in these three techniques, tend to overwork the hazard evaluation of procedures, so a tiered approach is best.

In this tiered approach, the first step in choosing a method to analyze procedure-based operations would be to screen the procedures and select only those procedures expected to involve major hazards. These procedures should be subjected to a detailed procedure-based HAZOP Study using the seven or eight guide words in Table 9.1. The Two Guide Word set can be effectively used for more moderate hazards or less complex tasks. The What-If Analysis method should be reserved for low-hazard, low-complexity, or very well-understood tasks.

Figure 9.1 shows a rough breakdown of the usage of the three methods described above for a set of typical operating procedures within a complex chemical plant or refinery. Most of the procedures may be simple enough, or have sufficiently low hazards, to warrant using the What-If Analysis method. The full seven or eight guide words with the HAZOP Study method are used less frequently, since most tasks do not require that level of scrutiny to find the incident scenarios associated with procedure-based operations.

As mentioned in Chapter 6, Selection of Hazard Evaluation Techniques, the experience of the leader or the team plays a major part in selecting the method to use for each task/procedures to be analyzed. However the first decision will always be "Are these procedures ready to be risk-reviewed?" If the procedures are up to date, complete, clear, and used by operators, then the best approach for accomplishing a complete hazard evaluation of all modes of operation, including procedure-based modes of operation, is shown in Table 9.4. For continuous processes, if the procedures are not ready for review, the best approach is to develop accurate and up-to-date procedures as quickly as possible; in the meantime, the HAZOP Study or What-If Analysis of the continuous operation can proceed.

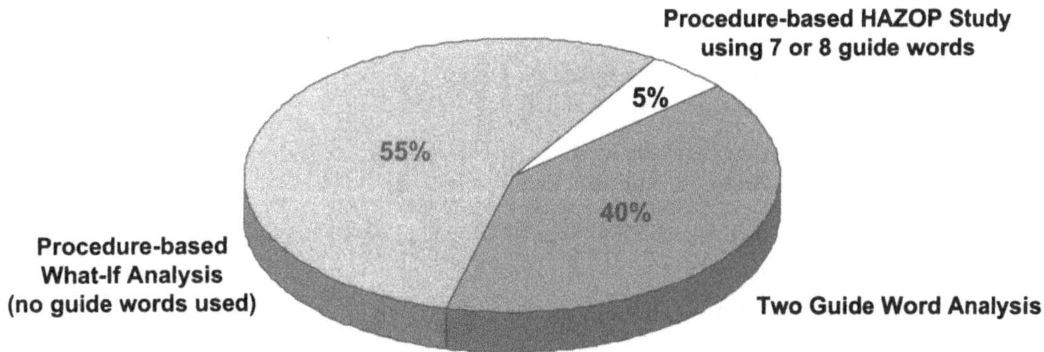

Figure 9.1 Typical usage of procedure-based techniques at some facilities

Table 9.4 Example choice of methods for hazard evaluation of all modes of operations

Continuous mode (if applicable)	Procedure-based modes (batch, start-up, shutdown, maintenance, emergency shutdowns, unloading, etc.)
▪ HAZOP Study of parameters ▪ What-If Analysis of simple sub-systems	▪ HAZOP Study of procedural steps - Full seven or eight guide words - Two Guide Word Analysis ▪ What-If Analysis of simple tasks

Any procedure can be analyzed using these techniques, including procedures that may be incorporated into operator check sheets or programmable control. Review of routine procedures is important; review of nonroutine procedures can be even more important. The nature of nonroutine procedures means that operators have much less experience performing them; in addition, the procedures may not receive as much scrutiny to ensure their accuracy and completeness.

Using these approaches, a company doing a complete hazard evaluation of an existing continuous process will invest roughly 60% of its time evaluating normal (i.e., continuous-mode) operations and the remaining time evaluating the associated procedure-based operations. The following are general guidelines for analyzing procedure-based operations.

- Define the assumptions about the system's initial status. "What are assumed to be the starting conditions when the user of the procedure looks at Step 1 of the procedure?"
- Define the complete design intention for each step. "Is the step actually three or five actions instead of one? If so, what are the individual actions to accomplish this task?"
- Do not analyze safeguard steps that start with Ensure, Check, Verify, Inspect, etc., or where the consequence of Skip is "Loss of one level of safeguard/protection against..." These steps need not be analyzed if failure to perform them means only that there will be less protection against deviations at other steps. This approach is similar to not analyzing inadequate flow through a relief valve during a continuous-mode HAZOP Study; instead, the effectiveness of the relief system is evaluated as a safeguard against overpressure deviations initiated by other causes.
- Before the team review meeting begins, work with an experienced operator to identify the sections of the procedures that warrant use of:
    - HAZOP Study using all guide words, where deviations can lead to major impacts or where the process or operation is highly complex
    - Two Guide Word Analysis, for systems involving moderate complexity or moderately severe impacts
    - What-If Analysis, for simpler or lower-hazard systems.

- For overly complex procedural steps, break down each written step into a sequence of actions beginning with active verbs.
- Apply guide words directly to the intentions of each action (except when using What-If Analysis).
- When analyzing batch chemical reactions and many other types of procedure-based operations, one step in the sequence is often very similar to a continuous process—more time is required, and process variables are closer to being steady-state. This is usually the reaction step for batch chemical processes, where e.g. heating with agitation is required for a predetermined time or until a specified endpoint is reached. For loading or unloading operations, this is the transfer time after hookup is completed and flow is started. Similar timeframes can be identified for other procedure-based operations such as batch distillation. For such operations, one approach that has been found useful is to use a procedure-based analysis method to study the batch sequence up to the point of this more "continuous" operation, then study the equipment and controls using the P&ID at this point (e.g., using the HAZOP Study methodology as for a continuous process). When this is complete, return to the procedure-based review for the remaining steps of reactor discharge and cleanout, transfer line blow-through and disconnecting, etc. This approach can also be employed for continuous operations by first examining the start-up procedure, followed by the continuous operation, and finally the shutdown procedure and any other pertinent procedures.

The following preparation steps may also be needed:

- Review the procedure with one or more operators to observe the work situation and verify the accuracy of the written procedure.
- Determine if the procedure follows best practices for "presentation" of the content; these best practices will limit the probability of human error.
- Discuss generic issues related to standard operating procedures, such as:
  - staffing (normal and temporary)
  - human-machine interface
  - worker training, certification, etc.
  - management of change
  - policy enforcement.
- Review other related procedures such as lockout/tagout and hot work permitting.

Case Study

A company had traditionally performed Checklist Analyses of its process systems and Job Safety Analyses (JSAs) of its procedures. After an explosion that resulted in fatalities, the company embarked on an aggressive program of performing hazard evaluations of its process equipment and procedures using primarily the HAZOP Study technique. The following results were taken from analysis of a toxic material unloading system, as illustrated in Figure 9.2.

The HAZOP Study of the unloading procedure considered the guide word LESS as it applied to the step "Pull vacuum in the unloading line before starting the unloading process." To complete this step, the operator had to align several valves and start a steam ejector system in an adjacent building. The review team realized that reading a vacuum gauge at the steam ejector did not ensure that a vacuum had been pulled in the unloading line out to the unloading rack. If the unloading line was not evacuated, leftover material in the line could contaminate other storage tanks, reducing product quality, and could also cause very rapid corrosion in other downstream equipment, likely resulting in loss of containment. The team recommended installing a vacuum gauge at the unloading rack so the operator could verify that vacuum had been achieved and maintained at that location before starting to unload the truck.

During the HAZOP Study, the team discovered that procedures had not been updated in a timely fashion. The operators had made several modifications (mostly improvements) to the procedures that had not been documented, and management was unaware of these changes. In addition, the procedures had not been reviewed for accuracy in over two years. Interviews revealed the existence of administrative requirements for (1) conducting periodic reviews of operating procedures and (2) implementing changes in design and/or operations documents, but management had not taken steps to ensure compliance with these administrative controls. One remedy suggested by the team was to have the document control clerk issue a schedule and audit the status of procedure reviews. Also, it was suggested that the procedure review team include both operators and engineers and that any procedural changes be subjected to a HAZOP Study by an independent team of similar composition.

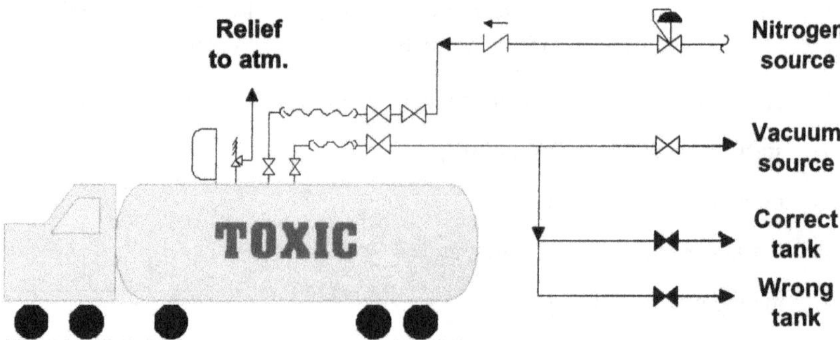

Figure 9.2 Illustration for case study

Linkage to Detailed Human Reliability Analysis

In the HAZOP Study, the team identified several operational errors that could cause a toxic release. As discussed above, improvements were made regarding some of the specific errors identified. However, company management felt that an additional, more detailed analysis should be conducted. To accomplish this, a human reliability expert observed operators (on various shifts, with varying degrees of experience) performing routine operations. A qualitative analysis revealed several additional recommendations, including the following:

- Improve the outdoor lighting near the unloading rack.
- Unload only during daylight hours, when leaks are easier to see and emergency response personnel are more readily available.
- Lock closed the crossover valve to reduce the chance of material being unloaded into the wrong tank.
- Provide a local indicator of storage tank pressure at the unloading rack. (Operators do not check the pressure frequently, because the existing indicator is on top of the storage tank.)

The detailed analysis was stopped at this point, since quantitative results were not necessary to reach a decision to implement the recommended changes.

Section 9.1 References

1. B. Rasmussen, "Chemical Process Hazard Identification," *Reliability Engineering and System Safety 24,* Elsevier Science Publishers Ltd., UK, 1989.
2. W. G. Bridges et. al., "Addressing Human Error During Process Hazard Analyses," *Chemical Engineering Progress,* May 1994.
3. F. L. Leverenz and R. W. Johnson, "HAZOP Studies and Other PHA Techniques for Process Safety and Risk Management," AIChE/ASME Continuing Education Course CH157.
4. OSHA Publication 3071, Job Hazard Analysis, U.S. Department of Labor, Occupational Safety and Health Administration, Washington, DC, 2002, http://www.osha.gov/Publications/osha3071.pdf.
5. A similar approach for analysis of procedures is discussed in Chapter 9 of W. Hammer and D. Price, *Occupational Safety Management and Engineering, 5th Ed.,* Prentice Hall, Upper Saddle River, New Jersey, 2000.
6. A HAZOP Study approach to analyzing procedural steps is demonstrated in H. E. Kongso, "Application of a Guide to Analysis of Occupational Hazards in the Danish Iron and Chemical Industry," *International Conference on Hazard Identification and Risk Analysis, Human Factors and Human Reliability in Process Safety,* Center for Chemical Process Safety, AIChE, New York, 1992.

9.2 Hazard Evaluation of Processes Controlled by Programmable Systems

A *programmable controller* (sometimes termed *programmable logic controller* or *PLC*) is a computer, hardened for an industrial environment, for implementing specific functions such as logic, sequencing, timing, counting, and control.[1] Programmable controllers are one category of *programmable systems,* which are computer-based systems connected to sensors and final control elements for the purposes of control, protection, or monitoring. Programmable controllers are implemented in chemical process operations to perform the functions of the basic process control system (BPCS); i.e., the control equipment and system installed to regulate normal production functions.[1]

Safety instrumented systems (SISs) may also be implemented in programmable systems. A safety instrumented system is designed to achieve or maintain a safe state of the process when unacceptable process conditions are detected. The instrumentation and controls associated with the SIS operate independently of the BPCS. See Step 4 in Section 7.6 on Layer of Protection Analysis (LOPA) for description of an approach to addressing programmable systems in the context of hazard evaluations.

An analysis of 34 incidents involving control system failures[2] revealed that 44% had inadequate specification as their primary cause, 20% involved changes after commissioning, 15% design and implementation, 15% operation and maintenance, and 6% installation and commissioning. The study concluded that approximately three-fifths of all control system failures are built in before operation commences. Although the control systems in this analysis were not limited to programmable systems, the findings are nevertheless indicative of primary causes that are likely to apply to programmable systems.

This section addresses the unique considerations and challenges that apply to evaluating hazards and risks in chemical processes where a programmable system is employed for process control. Questions that need to be addressed include:

3. What are the failure modes of programmable systems?
4. How likely are different failure modes to occur in such a system?
5. What are some different approaches to systematically identifying and evaluating scenarios when analyzing processes involving programmable systems?

A programmable system failure can be either a software failure or a hardware failure. The following discussion of programmable system failure modes, particularly as they apply to the basic process control system, is extracted from information in Reference 5.

Software Failures in Programmable Systems

In the context of this section, the term *software* refers to both the code and the data residing in the programmable system. *Data* refers to current conditions, as reported from sensors through the input card(s) to the controller, as well as alarm setpoints, while the application software *(program)* is the logic developed by the user to assess the data.

Software failures in programmable systems occur when an incorrect signal is sent to a final control element, or when a necessary signal is *not* sent, despite all hardware functioning properly, including sensors and final control elements. Such errors can be rooted in the program itself (i.e., a fault in the program logic as originally configured), or in human error on the part of the operator or other person with sufficient access to the controller to alter either the application software or data or both.

Thus, the hazard evaluation team must consider the possibility that BPCS performance can be degraded by human error, especially when taking credit for more than one function of a single BPCS within a particular incident scenario.

Hardware Failures in Programmable Systems

Programmable system hardware failures can include failure of an input or output card, main processor, or a connection between components. The input and output cards used for transferring information into and out of the main processor are components that may fail at a higher rate than the logic solver itself. A more complete list of programmable system hardware failure modes and their possible effects can be found in Appendix G of Reference 1. A taxonomy of common cause failures and methods for their analysis are given in Appendix D of Reference 8.

Normally, programmable system hardware failures occur at a significantly lower rate than do failures in either the sensors that input information to the controllers or the final control elements to which controllers send outputs. Thus, in considering how a control loop might initiate (be the Cause for) an incident scenario, hazard evaluation teams will generally focus on the failure modes and frequencies of the loop sensor (e.g., level transmitter or mass flowmeter), final control element (e.g., fail-closed control valve), and ways in which personnel might affect the control parameters (e.g., not switch control from MANUAL to AUTO, or input the wrong loop setpoint).

The basic process control system, including normal manual controls, is generally the first layer of protection during normal operation. Using the Layer of Protection Analysis (LOPA) approach, the BPCS may be credited as an independent protection layer (IPL) if it meets certain criteria, despite the fact that its failure can be an initiating cause in some scenarios. The quantity of interest in evaluating the effectiveness of the BPCS as a protective system is the system's overall probability of failure on demand. CCPS[8] gives an explanation of how the BPCS design and management affect its performance.

Systematic Hazard Evaluation of Processes with Programmable Systems

Programmable systems are at one end of a spectrum that has full manual control at the opposite end. Table 9.5 highlights important differences between programmable control and manual operations.

The manual control aspects are generally less important for continuous processes that use loop controllers (even if not programmable devices), as opposed to many batch operations. However, it should be noted that for continuous processes, many automated control systems are placed in manual mode by the operator when process upsets or transitions occur. Also, control systems are often not tuned well enough to operate in automatic mode as designed, and hence these control loops are operated in manual mode the majority of the time.

For batch or other procedure-based operations, including start-up and shutdown of continuous processes, the operation is likely to be either fully automated (except for batch initiation) or fully manual (except for specific steps such as automatic weighing of a raw material once a totalizer is set and flow is started). It is even more important to consider the possible failure modes of the programmable controller for computer-controlled batch processes than for computer-controlled continuous processes, since additional parameters of timing, quantity and sequencing are involved and the position of valves is often being switched.

Table 9.5 Programmable versus manual control

Programmable system	Operator
Does exactly as instructed: no more, no less. Cannot "think"; hence, will not respond to additional inputs such as noise/vibration (or lack thereof), unexpected (hence unprogrammed) signals, etc.	Operator can consider all inputs received. Makes judgments, etc. Can be a preventive safeguard based on "Operator Awareness."
Electronic equipment highly reliable, but sensitive to environment. Hence, computer will follow instructions explicitly with high reliability. This means that when things go wrong, the programmable system will continue, unless programmed to stop or is interrupted. This also means that "skipping a step" is very unlikely. It is more likely that the devices being controlled by the programmable system will fail rather than the programmable system itself.	Operator is subject to errors such as skipping steps and reversing steps. These errors may be more likely than failure of the devices being controlled. In addition, they are more likely than similar programmable system failure modes.
Unless programmed otherwise, accepts all inputs as true. That is, does not know whether or not the input from a device is reasonable, unless programmed to diagnose.	Operator has a general sense of what is a reasonable response from the instrumentation, etc.
Executes quickly. In essence, sequential instructions are executed almost instantaneously. Only waits if programmed to wait for a response, or for a fixed time.	Limited by ability or time required to perform actions. Also will need breaks, which take time.
Instruction set can be dramatically affected by someone changing the program or making an error in the process of making a change (e.g., intending to change set point of instrument, but changing something else instead).	Operator not prone to this type of error; however, if a procedure is changed, habit tends to cause errors of doing the old procedure.

The development of software for a programmable system goes through four basic stages:

1. Operational concept (e.g., functional description of "Transfer 1000 kg ethanol to the reactor")
2. Program description (e.g., "1. Store initial weight for source vessel. 2. Check whether source vessel inventory is less than required amount. 3a. YES: Alarm Low Level. Go to ABANDON. 3b. NO: Close & check route and off-route valves." Etc.)
3. Computer programming (e.g., READ TK1LVL / STO TEMP1 / IF(TEMP1< AMOUNT) GO TO ALARM A / STO VOVPOS1 = 0 / etc.)
4. Machine language (binary/hex).

The focus of a hazard evaluation of this type of computer-controlled operation will usually be on the program description (stage 2 above). The program description, which clearly illustrates the process control strategy for the software to be developed, may be in the form of a stepwise instruction similar to a standard operating procedure, or in the form of a flow chart.

The same general approach to scenario development as described in Chapter 5 applies to the analysis of computer controlled processes. Once an initiating cause is identified, whether by What-If questioning, HAZOP Study, etc., consider the continuing programmable system execution until a loss event occurs. Anything that would keep the loss event from occurring would then be a preventive safeguard. Some companies limit the number of safeguards that would need to be breached, such as stopping when more than three protective layers must fail for the loss event to occur. Other companies make this number-of-safeguards limit a variable that is based on the anticipated severity of the incident to be prevented.

Table 9.6 shows the types of deviations that might be identified by HAZOP Study guide words, along with some possible accompanying causes. Additional considerations for evaluating programmable systems are as follows:

- Consider all transfers to routines such as a "hold" or a "shutdown" given the postulated cause and what those routines do.
- If correct software execution is important to safety, ensure the actual software is reviewed versus requirements and simulated off-line (where possible) before being installed on the plant process computer.
- The reviewer should be someone (other than programmer) who clearly understands the process operation and the needed program operation.
- Consider "serious" program skips to other routines/operations; ensure program checks for proper entry.
- All software changes should be part of the facility's management of change process.

Table 9.6 Example deviations and causes with programmable control

Guide word	Deviation	Possible causes
NONE	Commanded device does not respond	Device has failed Computer D/A converter failed
MORE OF	Quantitative command exceeded	Device exceeds commanded value
	Device is sampled at "high" value	Device provides high reading Computer A/D converter fails
LESS OF	Quantitative command not achieved	Device does not reach commanded value
	Device is sampled at "low" value	Device provides low reading Computer A/D converter fails
REVERSE	Required state is opposite of indicated state	Position indicator shows opposite state On/off indicator shows opposite state
OTHER THAN	Interrupt of normal programmed flow	Power transient Operator interrupt Programmed interrupt

Kletz[5] describes several incidents that have occurred in computer-controlled process plants, and gives further guidance on applying the HAZOP Study method to computer-controlled plants. Redmill et al.[6] also address software HAZOP Studies. The HAZOP Study method as applied to programmable systems is also known as a "computer HAZOP" or CHAZOP. In this approach, a checklist is used by a study team in the early design stages to define the functionality scope of the programmable system; later, HAZOP-type guide words (more of, less of, none of, other than, sooner/later than, corruption of, what else, reverse of) are applied to data transfers between major components of the system.[7]

Section 9.2 References

1. Center for Chemical Process Safety, *Guidelines for Safe Automation of Chemical Processes*, American Institute of Chemical Engineers, New York, 1993.
2. UK Health & Safety Executive, *Out of Control: Why Control Systems Go Wrong and How To Prevent Failure, 2nd Edition,* ISBN 0717621928, HSE Books, 2003.
3. Center for Chemical Process Safety, *Layer of Protection Analysis: Simplified Process Risk Assessment*, American Institute of Chemical Engineers, New York, 2001.
4. F. L. Leverenz and R. W. Johnson, "HAZOP Studies and Other PHA Techniques for Process Safety and Risk Management," AIChE/ASME Continuing Education Course CH157.
5. T. A. Kletz, *Computer Control and Human Error,* Institution of Chemical Engineers, Rugby, UK, 1995.
6. F. Redmill, M. Chudleigh, and J. Catmur, *System Safety: HAZOP and Software HAZOP,* Wiley, New York, 1999.
7. D. Macdonald, *Practical Industrial Safety, Risk Assessment and Shutdown Systems,* Elsevier, Oxford, 2004.
8. AIChE Center for Chemical Process Safety, *Guidelines for Safe and Reliable Instrumented Protective Systems*, ISBN 978-0-471-97940-1, American Institute of Chemical Engineers, New York, 2007.

9.3 Hazard Evaluation of Chemical Reactivity Hazards

Chemical reactivity hazard evaluations, like all hazard evaluations, should be done throughout the life cycle of a facility. The most effective reactivity hazard evaluations are done during the process development stage when changes in chemistry are easier to implement.

Chemical reactivity hazards are best evaluated using a team approach. The team must have expertise in chemistry and reactive chemicals testing and may include:

- A person familiar with the hazard evaluation technique being used
- A person familiar with the chemicals and chemistry
- A person familiar with the operation of the facility
- A person familiar with maintenance and equipment
- A person familiar with the design and design basis of the facility and equipment
- Other people with specialized knowledge as necessary

Because of the specialized knowledge in chemistry and reactive chemicals testing required do a good evaluation, reactivity hazard evaluations are often done separately from other hazard evaluations.

Most of the hazard evaluation techniques in Chapters 4 and 5 can be used for evaluating reactivity hazards. Some of the initiating causes involving reactive chemistry that deserve attention include:

- Not charging any of a particular ingredient
- Not charging enough or charging too slowly
- Charging too much or too fast, including double charging
- Excessive or delayed catalyst addition
- Starting temperature too high or too low
- Loss of heating or cooling during reaction
- Excessive or prolonged heating or cooing during a reaction
- Loss of vacuum if the reaction is done under vacuum
- Inadequate venting of off gas
- Loss of pH or level control
- Inadequate or excessive agitation
- Contamination
- Adding a material out of sequence
- Adding the wrong material
- Delayed discharge of material from vessel

Once the scenarios have been identified, the hazard evaluation team can then make recommendations to control the process hazards and reduce the risk where warranted. Inherently safer alternatives can be used or passive, active or procedural controls can be implemented. Protective and mitigative measures such as emergency cooling, secondary containment, fire protection, emergency response, and spill response are often necessary. It is important that safe operating limits be established and maintained, such as for temperature, pressure, reaction time, charge rate and quantity, and concentration.

The evaluation of chemical reactivity hazards is just one example of a specialized hazard evaluation. More information on the evaluation of chemical reactivity hazards can be found in References 1-6. An example of a checklist that can be used to help identify chemical reactivity hazards and scenarios is shown in Appendix A3, and a reactivity-group chemical compatibility chart is presented, along with known exceptions, in Appendix E.

Section 9.3 References

1. J. Barton and R. Rogers (eds.), *Chemical Reaction Hazards: A Guide to Safety, 2nd Ed.,* ISBN 0-88415-274-X, Gulf Publishing Company, Houston, 1997.
2. Center for Chemical Process Safety, *Guidelines for Chemical Reactivity Evaluation and Application to Process Design,* American Institute of Chemical Engineers, New York, 1995.
3. Center for Chemical Process Safety, *Guidelines for Safe Storage and Handling of Reactive Materials,* American Institute of Chemical Engineers, New York, 1995.
4. D. Mosley, A. Ness and D. Hendershot, "Screen Reactive Chemical Hazards Early in Process Development," *Chemical Engineering Progress (96)*11: 51-65, November 2000.
5. R. W. Johnson, S. W. Rudy and S. D. Unwin, *Essential Practices for Managing Chemical Reactivity Hazards,* American Institute of Chemical Engineers, New York, 2003.
6. R. W. Johnson and S. D. Unwin, "Addressing Chemical Reactivity Hazards in Process Hazard Analysis," *Proceedings of the International Conference and Workshop on Managing Chemical Reactivity Hazards and High Energy Release Events,* American Institute of Chemical Engineers, New York, 2003.

9.4 Combinations of Tools

As discussed in Chapter 6, the analysis of a particular process may involve using more than one hazard evaluation technique. This generally is done by applying different methods to discrete parts of the facility under study, with each part being analyzed using the most appropriate technique depending on the characteristics of the process.

However, the use of different hazard evaluation techniques can also be combined in the analysis of a single process or part of the process, such that their combined use gives a more effective result than the employing of a single methodology. One approach is to use both an inductive method such as What-If Analysis or FMEA and a deductive method such as Fault Tree Analysis to develop a comprehensive set of incident scenarios. Although this would be more time-consuming than use of a single technique, for critical applications it would give a greater assurance of identifying all important scenarios.

Another approach is to use the results of one hazard evaluation technique as the starting point for further analysis using a second methodology. Incident scenarios identified by any of the scenario-based hazard evaluation methods in Chapter 5 can be used as the starting point for a Layer of Protection Analysis, as discussed in Section 7.6. Notes and details captured during the initial hazard evaluation can also provide key information that is useful in the subsequent LOPA study.

What-If/LOPA and HAZOP/LOPA

A third approach is to amalgamate two methods such that the resulting technique hopefully takes advantage of the strengths of both methods and mitigates the weaknesses of each. The Bow-Tie Analysis technique is an example of a method that has evolved by realizing the synergism between using Fault Tree Analysis to study what can lead up to a loss event and Event Tree Analysis to study the event outcomes.

Another strategy that has proven useful is the combination of scenario-based hazard evaluation techniques with Layer of Protection Analysis (LOPA). Any qualitative hazard evaluation method that identifies scenarios in terms of their initiating causes, event sequences, consequences and safeguards can be readily extended to record much of the results of a LOPA study (Section 7.6), whenever the analyst or analysis team deems a scenario to require such analysis. The What-If Analysis and HAZOP Study methods are thus well-suited to combination with LOPA.

In a tabular scenario presentation, for example, columns may be added in which to record the order-of-magnitude frequency and probability estimates involved in LOPA. Text that identifies safeguards must differentiate between independent protection layer (IPL) and non-IPL safeguards. Some of the hazard evaluation software listed in Appendix D already implements order-of-magnitude scoring on cause frequency, safeguard effectiveness (or failure-on-demand probabilities), and consequence severity. In parallel to properly formulated text descriptions, such scoring systems, once in place, need to be extended to separate the aggregate probabilities of failure on demand (PFDs) for IPL safeguards and non-IPL safeguards, and to indicate the integrity of each IPL when multiple IPLs protect against the same scenario.

To get the most benefit of an integrated approach for recording both the scenario-based method (HAZOP Study or What-If Analysis) and LOPA, the team or analyst should take time up front to *initially* formulate scenarios in a LOPA-compatible manner (as discussed previously in Section 4.8; refer to Steps 3 and 4); specifically, proper identification of the true initiating cause in combination with any enabling events, and identification of a single loss event for the initiating cause. Depending on the software interface, this may require re-entering the initiating cause and then the event sequence leading to alternative consequences. If such structured thinking is not imposed, then the potential benefit of combining methods may not be derived.

The benefit of these combined methods, given that LOPA results satisfy the objectives of the analysis, is that the analysis team may not need to reconvene for a follow-up LOPA study. The team can instead achieve scenario screening and provide the consistent assessment of the need for recommendations as offered by LOPA in a single study. IPL design bases and other supporting documentation will still need to be retained in the same manner as a standalone implementation of LOPA.

The analysis procedures for What-If/LOPA and HAZOP/LOPA are thus identical to the root scenario development method, with some adjustments, additional analyses and documentation:

- Initiating causes, enabling events, and consequences (unmitigated) are assigned order-of-magnitude frequencies, probabilities and severities according to a pre-established protocol.
- Risk-based scenario screening criteria can be applied, again based on an organization's risk tolerance or other criteria.
- If a scenario's risk is above a certain threshold, then protective and source-mitigative safeguards are identified and classified as to whether they are independent protection layers (IPLs). Refer to Section 7.6, Layer of Protection Analysis, for discussion of criteria that qualify a safeguard as an IPL.
- IPLs are fully documented with respect to instrument and device tag numbers, standard operating procedure (SOP) number, mechanical integrity plan, etc. A functional description of how the IPL works should also be provided to assist the team in determining the amount of risk reduction provided by the IPL.
- For scenarios with low unmitigated risk, safeguards are assessed per the root hazard evaluation methodology. For higher risk scenarios, the mitigated risk is determined, based on the expected performance of identified IPLs.
- The need for further risk reduction is determined and documented, including the target frequency reduction. Again, risk tolerance criteria and other guidelines come into play. As with stand-alone What-If Analysis or HAZOP Study methods, additional IPLs should be fully described, whether they would be new installations or existing safeguards modified to qualify as IPLs. The analyst(s) must ensure that the added IPLs and/or safeguards will provide the appropriate level of risk reduction.

9.5 Human Factors and Human Reliability Analysis

"Human factors" involve designing machines, operations and work environments so that they match human capabilities, limitations and needs. It is based on the study of operators, managers, maintenance staff, and other people in the work environment, and of factors that generally influence humans in their relationship with a technical installation.[16] It is now recognized that such factors go well beyond basic ergonomics and operator-machine interface considerations and include aspects of a safety culture such as management leadership and commitment, clear communication of expectations, and operating discipline.

Although modern control systems achieve a high degree of automation, the process operator still has the overall immediate responsibility for safe and economic operation of the process. Opinions differ as to the extent to which the function of safety shutdown or other response to abnormal situations should be removed from the operator and assigned to an instrumented protective system. In general, the greater the hazards are, the stronger is the argument for protective instrumentation. Whatever approach is adopted, the operator still has the vital function of running the plant so that control is maintained when possible and operator action is taken when needed to avoid a loss event. The job of the process operator is therefore a crucial one, and therefore should be considered when conducting a hazard evaluation.

In today's world, there are few tools to allow a thorough review of human factors as they relate to the process being studied. Most often, facilitators feel that the "human" aspect of process safety will be covered throughout the study as they go through their nodes, deviations and guide words and therefore feel there is no reason to review this topic separately. This may be the case in very detailed hazard evaluations done on procedure-based operations and batch process operating procedures; however, in a hazard evaluation on a continuous process, human factors rarely get the level of review needed.

Hence, a separate, global review of human factors considerations is often warranted as part of an overall hazard evaluation. Such a review can be conducted either at the beginning or the end of the team review meetings. Human factors reviews, ergonomic studies, and targeted Human Reliability Analyses may also be commissioned apart from and independent of hazard evaluations.

One approach that has been found to be effective is to make an assessment in various categories, such as those listed in Table 9.7, as to where the facility currently stands with respect to having "positive" or "negative" human factors. Positive human factors tend to decrease the likelihood of serious operational or maintenance errors or inadequate responses to abnormal situations. Negative human factors tend to increase the likelihood of errors or inadequate responses.

Note that generically positive human factors may be beneficial in most situations but impractical or even undesirable in other situations. Also, what may be a negative factor in one sense may nevertheless be necessary or beneficial in another sense. For example, some personal protective equipment (PPE) may be cumbersome for performing tasks but essential for personal protection. Compensating provisions may be warranted in such cases.

Standard practice for a given type of operation often falls somewhere between the descriptions for positive and negative in Table 9.7. Where a negative human factor is found, a recommendation may be warranted to address the negative factor. For example, if certain standard operating procedures are found to be written in a wordy, inconsistent style (a negative human factor, tending to increase the likelihood of an error being made when using the procedure), the hazard evaluation team may document a finding to that effect, and the facility may respond by rewriting the procedures in concise, imperative language that gives clear, step-by-step instruction.

Table 9.7 Example positive and negative human factors

Generic positive factors	Generic negative factors
1 Operating and maintenance instructions	
1.1 Procedures are at the right level of detail	too general or too detailed
1.2 Procedures are written in concise, imperative language	wordy, inconsistent style
1.3 Notes, cautions and warnings set off from procedural steps (e.g., in text boxes placed immediately before applicable steps)	mixed together with text of procedural steps
1.4 Clear identification of equipment in procedure	ambiguous identification/location
1.5 Labeling consistent between procedures, P&IDs, equipment	inconsistent, contradictory
1.6 Diagrams, photographs, tables, checklists as appropriate	all text; word descriptions only
1.7 Procedures/checklists used in the performance of tasks	task sequence done by memory
1.8 Procedures include appropriate supervisory checks	no cross-checking or verification
2 Personnel / equipment interface	
2.1 Process equipment is easily accessible for operations	difficult or dangerous to access
2.2 Process equipment is easily accessible for maintenance	difficult or dangerous to access
2.3 Layout of process equipment is logical and well-planned	layout is confusing, inconsistent
2.4 Process equipment is clearly and uniformly labeled	mislabeled or not labeled
2.5 Sequences are as would be expected	unexpected (e.g., 1, 2, 4, 3, 5)
2.6 Components are easily distinguished from one another	several similar components in same area or grouped together
2.7 Process equipment is easily operated (manual valves, etc.)	difficult to operate, excess leverage required
3 Operator / process controls interface (BPCS = basic process control system)	
3.1 Process controls are clearly and unambiguously labeled	mislabeled or not labeled
3.2 Process control interface is clear, simple, representational	unclear, complex, misleading
3.3 BPCS is fully automatic, well-tuned	manual, many steps
3.4 BPCS gives immediate, unambiguous feedback to operator	none, delayed, misleading
3.5 Readings, indicators, and gauges are reliable	unreliable, misleading, don't work
3.6 Readings, indicators, and gauges are easy to read	difficult to read, easy to misread
3.7 Units of measure used onsite are consistent, understood	inconsistent (e.g., °F and °C), confusing, unusual
4 Preventive safeguards involving operator action	
4.1 100% operator coverage with backup	unattended operation, operator not always present, inadequate coverage for abnormal situations
4.2 Operator continually involved and interacting with process	operator detached, time needed to figure out status of process
4.3 Early detection of deviation onsets	undetectable causes, deviations; long detection time
4.4 Prioritized, first-out, safety-critical alarm presentation	similar, simultaneous alarms
4.5 Minimal false or spurious alarms	many false or spurious alarms, alarms ignored or disabled
4.6 Simple action(s) to be taken to control abnormal situations	complex process, diagnosis required, many possible actions

Table 9.7 Example positive and negative human factors (continued)

Generic positive factors	Generic negative factors
4.7 No time pressure for response to abnormal situations	consequence occurs too rapidly for timely operator response
4.8 Safety shutdown systems known to be reliable, effective	unreliable, ineffective, untested
4.9 Safety shutdown systems never bypassed	frequent trouble or maintenance bypassing of safety systems
4.10 Last-resort shutdown not discouraged	shutdown discouraged/unsafe
5 Emergency operations	
5.1 Early detection of releases	long detection time; sensors too few, unreliable, misplaced
5.2 Control room is temp safe haven to allow orderly shutdown	unsafe location, immediate control room evacuation likely
5.3 HVAC shutoff quickly accessible, easy to actuate	location unknown, difficult to access or actuate
5.4 Readily available PPE for escape, emergency operations	insufficient, unavailable, locked, ineffective, untested
5.5 Clearly visible, understandable exit route signs, instructions	no signage, obscured, unclear, confusing, ambiguous
5.6 Protected, well-lit emergency exit, evacuation routes	exit goes through dim or dangerous route
5.7 All emergency isolation valves remotely actuated or safely accessible; quick-closing	not at ground level, inaccessible during emergency, inoperable
5.8 Emergency procedures readily accessible, clear, simple	inaccessible, complex, not suitable for use in emergency
6 Operations, maintenance, supervisory and emergency communications	
6.1 Formal communication and turnover log for shift changes	inadequate communication between shifts
6.2 Constant communication between control & field operators	no communication with field
6.3 Reliable control/field communication equipment (two-way radios, telephone, etc.) with alternative means	unreliable, no alternative, may not work in emergency situation
6.4 Clear, unambiguous site-wide emergency warning system	no distinction between areas or events, some onsite locations do not hear, unreliable, untested
6.5 Good communication between operations and maintenance	no or inadequate coordination
6.6 Frequent supervisory communications	little or no supervisory checks
6.7 Expectations communicated, rules consistently enforced	confusion of expectations, lack of consistency
7 Operational environment	
7.1 Noise level low enough to not hinder communications	hearing protection required
7.2 Sheltered operations/maintenance environment	precipitation, wind
7.3 Climate-controlled operations/maintenance environment	temperature/humidity extremes
7.4 Clear or enhanced visibility	fog, smoke, obscuration
7.5 Lighting matched to tasks	inadequate for tasks
7.6 Required PPE does not affect performance of tasks	cumbersome, impairing PPE
8 Task scheduling and staffing	
8.1 Overtime limited and reasonable	extreme, affecting performance
8.2 Permanent shift assignments	disruptive shift rotations
8.3 Number of tasks well-matched to work force	required tasks exceed resources

Table 9.7 Example positive and negative human factors (continued)

Generic positive factors	Generic negative factors
8.4 Pace of operations is normal	different tasks, rapid succession
8.5 Required tasks are regularly performed	infrequent or constantly changing
8.6 Turnover of operations/maintenance personnel is minimal	high turnover, less experience
9 Personnel training and qualifications	
9.1 Hiring qualifications consistent with task requirements	unqualified staff or contractors, language or similar problems
9.2 Consistent, thorough substance abuse policies, screening	substance abuse problems
9.3 Well-trained operations personnel for all normal operations	untrained, inexperienced
9.4 Training, drills, simulations for abnormal situations	unprepared for abnormal situations, emergencies
9.5 Well-trained maintenance staff including safe work practices	untrained, inexperienced
9.6 Well-organized training program including refresher training	haphazard, incomplete, behind schedule, undocumented
9.7 Verification of training includes both tests and observations	no or inadequate verification

Source: Unwin Company, Columbus, Ohio

Another approach for evaluating human factors is the use of a standardized checklist of questions. Examples of human factors checklists of this nature can be found in Appendix G of the CCPS Concept Book *Revalidating Process Hazard Analyses*[17] and in the Appendix to another CCPS Concept Book *Human Factors Methods for Improving Performance in the Process Industries*.[18] A similar "Latent Conditions Checklist" (or an approved alternative) is required to be used by certain facilities in Contra Costa County, California.[20] In addition to being analyzed as a global review, human factors should also be considered when evaluating the likelihood of specific initiating causes and the effectiveness of particular safeguards (e.g., whether an operator has time to respond to a deviation).

It should be noted that many features and activities that are foundational to effectively addressing human factors are covered by other elements of process safety management. These include the content, completeness, and accessibility of operating and maintenance procedures, as well as training content, contractor safety, safe work practices, emergency planning and response, and management commitment.

Human Reliability Analysis

A Human Reliability Analysis (HRA) is a systematic evaluation of the factors that influence the performance of operators, maintenance staff, technicians, and other plant personnel. It involves one of several types of task analyses; these types of analyses describe a task's physical and environmental characteristics, along with the skills, knowledge, and capabilities required of those who perform the tasks. A Human Reliability Analysis will identify error-likely situations that can cause or lead to incidents. A Human Reliability Analysis can also be used to trace the causes of human errors. Human Reliability Analysis is usually performed in conjunction with other hazard evaluation techniques.

Purpose

The purpose of Human Reliability Analysis is to identify potential human errors and their effects, or to identify the underlying causes of human errors.

Types of Results

A Human Reliability Analysis systematically lists the errors likely to be encountered during normal or emergency operation, factors contributing to such errors, and proposed system modifications to reduce the likelihood of such errors. The results are qualitative, but may be quantified. The analysis includes identifying system interfaces affected by particular errors, and ranking these errors in relation to the others, based on probability of occurrence or severity of consequences. The results are easily updated for design changes or system, plant, or training modifications. A worked example showing Human Reliability Analysis results can be found in Part II, Chapter 21.

Resource Requirements

Using Human Reliability Analysis requires the following data and information sources: plant procedures; information from interviews of plant personnel; knowledge of plant layout, function, or task allocation; control panel layout; and alarm system layout.

Staffing requirements vary based on the scope of the analysis. Generally, one or two analysts with human factors training should be able to perform an HRA for a facility. The analyst(s) should be familiar with interviewing techniques and should have access to plant personnel; to pertinent information, such as procedures and schematic drawings; and to the facility. The analyst should be familiar with (or know someone who is familiar with) the plant response or consequences caused by various human errors.

The time and cost for this type of analysis are proportional to the size and number of tasks, systems, or errors being analyzed. As little as an hour should be sufficient to conduct a rough HRA of the tasks associated with a simple plant procedure. The time required to identify likely sources of a given type of error will vary with the complexity of the tasks involved, but this analysis could also be completed in as little as an hour. If the results of a single task analysis were used to investigate several sources of potential human error, the time requirement per source of error would be significantly decreased. Identifying potential modifications to reduce the incidence of human errors would not add materially to the time required for a Human Reliability Analysis. Table 9.8 lists estimates of the time needed to perform a hazard evaluation using the HRA technique.

Table 9.8 Time estimates for using the Human Reliability Analysis technique

Scope	Preparation	Model Construction	Qualitative Evaluation	Documentation
Small system	4 to 8 hours	1 to 3 days	1 to 2 days	3 to 5 days
Large process	1 to 3 days	1 to 2 weeks	1 to 2 weeks	1 to 3 weeks

Technical Approach

Reliable human performance is necessary for the success of human-machine systems and is influenced by many factors.[1-5] These performance shaping factors (PSFs) may be internal attributes such as stress, emotional state, training, and experience, or external factors such as work hours, environment, actions by supervisors, procedures, and hardware interfaces. The number of PSFs that affect human performance is almost infinite. While some PSFs cannot be controlled, many can be, thus significantly influencing the success or failure of a process or operation.

Many of the hazard evaluation techniques illustrated in the *Guidelines* can be used to pinpoint the potential for human-failure-caused incidents. For example, an analyst may include operator errors in fault tree models. A What-If/Checklist Analysis may consider issues concerning the potential for an operator to open a valve instead of closing it during an upset condition. Typical HAZOP Studies frequently include general operator errors as causes of process deviations. Even though these hazard evaluation techniques can be used to address general human errors, they tend to focus mainly on the hardware aspects of potential incidents. When process operations include many manual activities, or when the complexity of the human-machine interface makes it difficult for standard hazard evaluation techniques to evaluate the significance of potential human errors, then more specific approaches are needed to evaluate these human factors.

Human factors is a discipline concerned with designing machines, operations, and work environments so they match human capabilities, limitations, and needs. There are many different techniques used by human factors specialists to evaluate work situations.[6,7] One common technique is called job safety analysis (JSA);[19] however, its focus is usually the safety of individual workers performing the required tasks. A job safety analysis is an excellent starting point, but for process safety analyses, a technique known as Human Reliability Analysis (HRA) is more useful.[8-10] Human Reliability Analysis techniques can be used to identify and improve PSFs, thereby reducing the likelihood of human errors. These techniques analyze the characteristics of systems, procedures, and operators to identify likely sources of errors.

Performing an HRA outside the context of an entire system analysis may place too much emphasis on human performance and not enough on equipment characteristics. This could be appropriate if the system in question were one known to be prone mainly to human-error-initiated incidents. In most cases, however, it is recommended that the HRA be performed in conjunction with other hazard evaluation techniques. This can be done by performing all the hazard evaluations at once or by performing them separately and integrating the results into the overall hazard evaluation. Typically, HRA is performed after other hazard evaluation techniques, such as HAZOP, FMEA, or FTA, have identified specific human errors with severe consequences.

Specific methods for HRA are part of a specialized sub-discipline of hazard evaluation and, as such, are only briefly described.[1,2,11-16] More information on these types of evaluations is available from several specialized sources (see ***References*** at the end of this section).

Analysis Procedure

Human Reliability Analysis involves the following steps: (1) describing the characteristics of the personnel, the work environment, and the tasks that are performed; (2) evaluating (through HRA) the human-machine interfaces; (3) performing a task analysis of intended operator functions; (4) performing a human error analysis of intended operator functions, and (5) documenting the results.[1,2,8,11] An HRA of a specific situation may not require completion of all of these steps. In fact, the goals and the scope of the study should be clearly defined to include only the level of detail in the analysis necessary to satisfy the analysis objectives. However, each of these steps should be performed in order, until a sufficient information is obtained. For instance, if the goal of a certain study is to improve operator efficiency in a control room, the study may be stopped after the task analysis step. Chapter 2 provides general advice on preparing for a hazard evaluation.

Describing the characteristics of the personnel, the work environment and the tasks that are performed. The first step in HRA is to develop an understanding of the human-machine system. Information describing the personnel, the work environment, and the tasks that are performed is gathered from several sources. Typical information sources include:

- Employee demographics describing pertinent characteristics such as language, education levels, and physical aspects
- Written and graphical descriptions of work environments
- Written operating procedures
- Analyst visits to the work area
- Analyst interviews with operators who actually perform operations (which they may or may not perform according to written procedures)
- Other previous "human error" incidents or studies that may provide insight into error-likely situations

Evaluating the human-machine interfaces. Human factors engineering (ergonomics) analyzes the compatibility between human needs and the limitations of machine designs and layouts, process operations, and work environments. Human factors engineering systematically evaluates each aspect of human-machine interfaces to ensure that demands on the operator are appropriate. For instance, human factors engineering techniques might identify problems such as:

- Process measurement indicators with irregular or oddly defined scales that require fine adjustments
- Placement of indicators that makes readings difficult or that makes an operator reluctant to take a reading
- Manual valves that cannot be operated by someone of below-average strength or height
- Lack of an effective means for operators to communicate batch status between shifts

The goal of human factors analysis is to identify any general deficiencies associated with the human-machine system.

Performing a task analysis of intended operator functions. After identifying and correcting the general deficiencies of a human-machine system, a detailed study of the specific actions required of employees may be undertaken. This process, called task analysis, separates a specific operator function or goal into constituent tasks. Each task represents a specific action that an operator must perform to complete a function.

Task 1 + Task 2 + Task 3 + Task 4 + Task 5 ... = Operator Function

Every task represents an opportunity for human error; that is, each of these steps must be satisfactorily performed. After listing the tasks, the analyst will evaluate each one to identify any error-likely situations that may cause the operator to fail to successfully complete one or more of the tasks. This process is a knowledge-based approach, similar to a checklist, that identifies conditions that have been causes of human errors in the past. Table 9.9 lists some contributors to error-likely situations. Hardware, procedural, and policy changes may be necessary to alleviate the causes of error-likely situations.

Performing a human error analysis of intended operator functions. The team analyzes these tasks by creating event trees for each one. The logic may be represented using a traditional event tree or may be shown using a special HRA event tree format (Figure 9.3). The event tree conveys the task analysis information and suggests a scheme to quantitatively assess the combination of failures as a part of a CPQRA. The tree may include hardware failures, as appropriate, in addition to human errors. Other human error techniques are available, and some are listed in the ***References***. Many of these other techniques are applicable only to specific situations or conditions.

Documenting the results. The final step in performing a Human Reliability Analysis is to document the results of the study. The hazard analyst should provide a description of the human-machine system analyzed; a discussion of the procedures analyzed; a list of assumptions; any HRA event tree model(s) that were developed; lists of error-likely situations, along with a discussion of the specific PSFs that contributed to the potential operator errors; a discussion of the consequences of the various incident sequences; and an evaluation of the significance of the human error situations. In addition, any recommendations that arise from the HRA should be presented.

Table 9.9 Contributors to error-likely situations

- Deficient procedures
- Inadequate, inoperative, or misleading instrumentation
- Insufficient knowledge
- Conflicting priorities
- Inadequate labeling
- Inadequate feedback
- Policy/practice discrepancies
- Disabled equipment
- Poor communication
- Poor layout
- Violations of populational stereotypes (e.g., left-hand threads)
- Overly sensitive controls
- Excessive mental tasks
- Excessive opportunities for errors
- Inadequate tools
- Sloppy housekeeping
- Extended, uneventful vigilance
- Inadequate practice with backup control systems
- Inadequate physical restrictions (e.g., acid and caustic connections of the same size)
- Appearance at the expense of functionality (e.g., prohibition of tape, marks, etc. that would help operators)

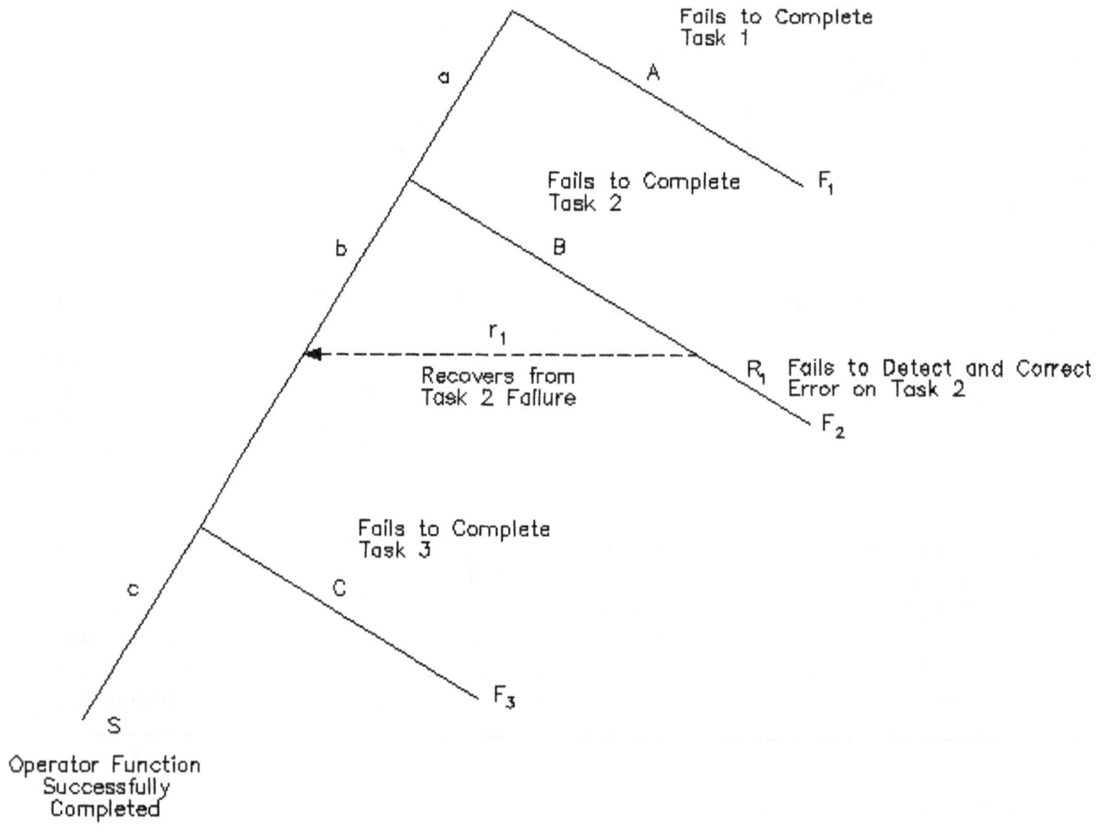

Figure 9.3 Example HRA event tree structure

Anticipated Work Product

Different levels of Human Reliability Analysis produce different work products. The analysis should produce a list of corrective actions that will reduce the likelihood of specific types of human errors. The analysis will also produce a detailed description of operator tasks, in the form of a list or graph, which can be used to establish procedures, training, or policies. Additionally, the analysis may generate a set of combinations of human errors that may be ranked using human error data. Evaluation of these results can lead a hazard analyst to suggest both hardware and procedural improvements. Chapter 8 provides additional information concerning analysis follow-up considerations.

Example

Consider an operator in a control room with a manually activated emergency shutdown feature. If one of five critical alarms annunciates, the operator must switch on the emergency shutdown system. (Other noncritical alarms also annunciate in the control room, but do not require emergency shutdown of the process.)

A Human Reliability Analysis identified that:

- Excessive noise in the control room made the audible alarms difficult to hear.
- The visual indicators for the five critical alarms that require emergency shutdown system activation were in several different locations on a large, crowded control board, making identification difficult.
- The switch for the emergency shutdown system was in a protected location to prevent inadvertent operation, but this made locating and operating the switch difficult.
- The switch for the emergency shutdown system did not have an indicator to inform the operator that the system had been successfully initiated.

Subsequently, a task analysis divided the operator emergency function into the following tasks:

1. Recognize the annunciating alarm.
2. Identify the type of alarm.
3. Recall the appropriate corrective action.
4. Locate the emergency shutdown switch.
5. Operate the emergency shutdown switch.

The task analysis, supported by a thorough understanding of the actual control room operations and conditions, allowed the following error-likely situations to be identified:

- ***Policy/practice discrepancies*** - Although the written procedures called for immediate shutdown, operating practice encouraged operators to make "last ditch" efforts to control the process before initiating the shutdown system.
- ***Poor layout*** - The control board had been expanded with recent process modifications, and controls/instrumentation were placed where space was available, not necessarily in a logical or ergonomically desirable location.
- ***Extended, uneventful vigilance*** - The control room is typically occupied by only one operator whose only duties, other than observation, occur once every hour. Occasionally, operators have been known to take short naps during night shifts.

The HRA event tree for this task analysis is shown in Figure 9.4. The incident sequences for the event tree are shown in Table 9.10. They are identified by the path of successes and failures to each failure term. For example, consider failure F_5. This would only occur if the alarm works (δ), the operator detects (a) and identifies (b) the alarm, the operator attempts to initiate corrective action (c), but fails to locate or operate the switch (D), and then the supervisor also fails to recover from the operator error by locating or operating the switch (R_1). Subsequent evaluation of these results helped the hazard analysts suggest specific equipment and procedural changes to correct the noted deficiencies.

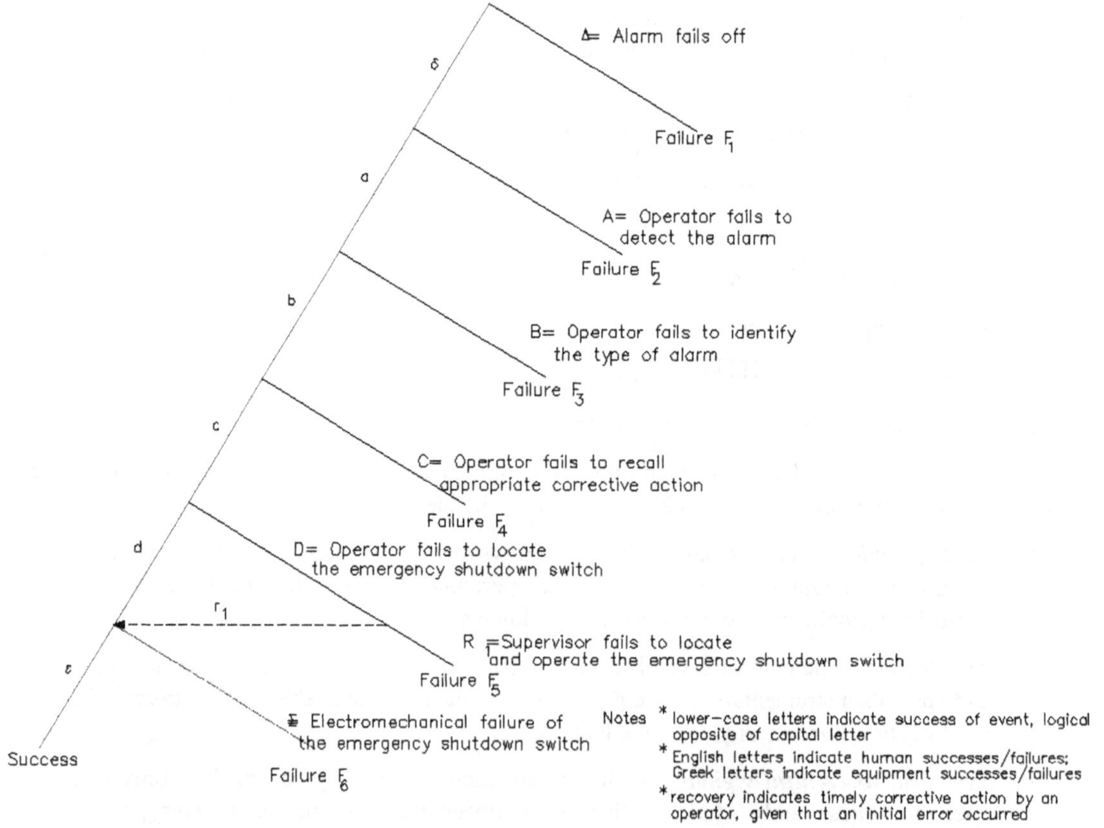

Figure 9.4 HRA event tree for the operator response to an alarm example

Table 9.10 HRA event tree incident sequences for the operator response to an alarm example

F_1	$= \Delta$
F_2	$= \delta A$
F_3	$= \delta aB$
F_4	$= \delta abC$
F_5	$= \delta abcDR_1$
F_6	$= \delta abc(d+Dr_1)\Sigma$
F_{Total}	$= F_1+F_2+F_3+F_4+F_5+F_6$

Section 9.5 References

1. Center for Chemical Process Safety, *Guidelines for Preventing Human Error in Process Safety*, American Institute of Chemical Engineers, New York, 1994.
2. D. K. Lorenzo, *A Manager's Guide to Reducing Human Errors—Improving Human Performance in the Chemical Industry*, American Chemistry Council, Arlington, Virginia, July 1990.
3. H. R. Greenberg and J. J. Cramer (eds.), *Risk Assessment and Risk Management for the Chemical Process Industry*, ISBN 0-442-23438-4, Van Nostrand Reinhold, New York, 1991.
4. J. Stephenson, *System Safety 2000—A Practical Guide for Planning, Managing, and Conducting System Safety Programs*, ISBN 0-0442-23840-1, Van Nostrand Reinhold, New York, 1991.
5. A. E. Green, ed., *High Risk Safety Technology*, ISBN 471-10153-2, John Wiley & Sons Ltd., New York, 1982.
6. E. J. McCormick and M. S. Sanders, *Human Factors in Engineering Design*, 5th ed., ISBN 0-07-044902-3, McGraw-Hill Company, New York, 1982.
7. B. S. Dhillon, *Human Reliability with Human Factors*, ISBN 0-08-032774-5, Pergamon Press, New York, 1986.
8. E. M. Dougherty, Jr., and J. R. Fragola, *Human Reliability Analysis*, John Wiley & Sons, New York, 1988.
9. K. S. Park, *Human Reliability—Analysis, Prediction, and Prevention of Human Errors*, Elsevier Science Publishing Company, Inc., New York, 1987.
10. B. J. Bell, *Evaluation of the Contribution of Human Errors to Accidents*, ISBN 0-8169-0411-1, presented at the International Symposium on Preventing Major Chemical Accidents, held in Washington, DC, February 1987, American Institute of Chemical Engineers, New York.
11. A. D. Swain and H. E. Guttmann, *Handbook of Human Reliability Analysis with Emphasis on Nuclear Power Plant Applications,* NUREG/CR-1278, U.S. Nuclear Regulatory Commission, Washington, DC, August 1983, with *Addendum #1 to NUREG/CR-1278, August 1983,* September 1, 1985, by A. D. Swain.
12. B. J. Bell and A. D. Swain, *A Procedure for Conducting a Human Reliability Analysis for Nuclear Power Plants*, NUREG/CR-2254 and SAND81-1655 RX (Dec). U.S. Nuclear Regulatory Commission, Sandia National Laboratories, Albuquerque, NM, 1981.
13. J. Williams, *HEART: A Proposed Method for Assessing and Reducing Human Error*, Advances in Reliability Technology Symposium:B3/R/1-13, held at National Centre of Systems Reliability, Warrington, UK, 1986.

14. G. W. Hannaman and A. J. Spurgin, *Systematic Human Action Reliability Procedure (SHARP)*, EPRI NP-3583, Electric Power Research Institute, Palo Alto, CA, 1984.
15. K. Webley and P. Ackroyd, *The THERP Model,* Human Reliability Assessment Course, IBC Technical Services, UK Atomic Energy Authority, Culcheth, UK, 1986. (Republished in 1987 as *The Human Reliability Assessor's Guide.*)
16. "OECD Guiding Principles for Chemical Accident Prevention, Preparedness and Response," 2nd ed., Organisation for Economic Co-operation and Development, Paris, 2003.
17. W. L. Frank and D. K. Whittle, *Revalidating Process Hazard Analyses*, ISBN 0-8169-0830-3, American Institute of Chemical Engineers, New York, 2001.
18. D. A. Crowl (ed.), *Human Factors Methods for Improving Performance in the Process Industries*, ISBN 978-0470117545, American Institute of Chemical Engineers, New York, 2007.
19. OSHA Publication 3071, Job Hazard Analysis, U.S. Department of Labor, Occupational Safety and Health Administration, Washington, DC, 2002, http://www.osha.gov/Publications/osha3071.pdf.
20. Contra Costa County Human Factors Program Guidance, "Section B: Human Factors Program" and "Attachment A: Latent Conditions," Contra Costa County, California, December, 1999.

9.6 Facility Siting

Akin to human factors considerations (Section 9.5), facility siting is a cross-cutting topic that is often considered in addition to the hazard evaluation procedures detailed in Chapters 4 and 5.

"Facility siting" can be considered to have three overlapping aspects:

- Facility location with respect to its surroundings
- Facility layout and spacing of equipment and buildings
- Personnel protection in occupied buildings.

Each of these is discussed as it relates to hazard evaluations. Although some facility siting considerations such as initial site selection are one-time decisions, many facility location/layout and personnel protection factors can change significantly over time as e.g. surrounding populations encroach, plant capacity increases, and new materials are brought on-site. Hence, addressing facility siting considerations should not be a one-time exercise, but should rather be included as an important part of each successive hazard evaluation. In addition, available tools and methods for addressing facility siting considerations continue to improve over time. Facility siting implications have also been included as part of management of change reviews.

Facility Location

"Facility siting" (also termed "stationary source siting") immediately brings to mind considerations of where to site, or locate, a process facility. For existing facilities, this is obviously a moot point, since the site has already been selected based on any number of historic, economic, business, and legacy reasons. However, when a determination needs to be made where to site a new facility, either as a greenfield project or as an addition to an existing company plant site with one or more alternative locations, process hazards and risks should play a primary role in the decision making process. Features that can vary significantly between potential sites and that can have major implications for facility risks include:

- Dominant wind direction; downwind/surrounding populations
- Sensitive population receptors nearby (schools, nursing homes, etc.)
- Capability of surrounding populations to shelter in place
- Site topography; flooding potential
- Sensitive environmental receptors nearby
- Climate and weather extremes; windstorms
- Settling potential; seismic events
- Air traffic patterns and frequency; proximity to landing sites
- Neighboring facilities that could affect new facility
- Reliability of electrical power and water supplies
- Distance to and capability of offsite emergency support services
- Potential for obscuring cloud drifting across roadway (e.g., cooling tower plume or dense cloud obscuring motorists' vision)
- Availability of capable workers and contractors to support operations
- Language and cultural considerations relative to the available workforce
- Mode, route, and distance to transport hazardous materials
- Ability to secure the site against malevolent human actions

Appendix B of CCPS' *Guidelines for Facility Siting and Layout*[1] contains a more detailed requirement list for site selection considerations.

Detailed hazard evaluations using the methodologies in Chapters 4 and 5 are not generally conducted at the time a suitable site is being selected. However, hazards and risks should obviously be considered at this stage, such as by determining and documenting comparative site features as input to the final site selection. Some countries and locations even require a quantitative risk analysis to be performed[2] and the calculated risks must meet pre-established risk targets before a new facility plan will be approved by the governing authorities. Risk targets or planning criteria[3] are often more stringent for new facilities than for existing operations.

Facility Layout

Facility layout refers to the location of equipment and buildings within a facility complex. Historically, facility layout has primarily focused on plant protection from fires and explosions, including location of control rooms and other occupied structures, as well as considerations of operating efficiency.

Two basic approaches to addressing facility layout are the use of standard spacing distances and the estimation of actual physical effect distances (e.g., distance to a given blast overpressure or impulse or fire thermal radiation flux). Standard spacing distances have been provided by insurance companies,[5] with the primary focus being adequate spacing to reduce the likelihood of a fire in one unit or storage tank from further involving adjacent facilities. A newer set of tables showing typical spacing distances, which were developed with broad industry input, can be found in Appendix A of *Guidelines for Facility Siting and Layout*.[1]

Estimating actual effects has the benefit of considering explosion and toxic release effects, as well as fire exposures. The Dow Fire/Explosion Index,[6] discussed in Section 3.6, is one method that has been widely used in predicting a fire/explosion effect radius within which property damage can be expected. It relies on an empirical correlation of many factors that have been found to influence the extent of effects. A more rigorous treatment would require actual calculations of source terms and building/equipment impacts.[7,8]

Other considerations for facility layout include a careful look at the potential for a fire/explosion incident affecting critical utilities such as electric power and firewater mains, as well as an evaluation of the potential impact of adjacent equipment affecting the hazardous process of concern (e.g., a boiler steam drum explosion having blast effects that could cause loss of containment of a hazardous material). A fuller discussion of facility layout as it pertains to process hazards has been presented elsewhere.[1]

Personnel Protection

The primary thrust of facility siting as it relates to hazard evaluations is to ensure people are adequately protected against major incidents involving sudden, unplanned releases of process material and/or energy. One approach that exemplifies this thrust is the American Petroleum Institute's recommended practice RP 752.[9] The focus of RP 752 is protection of personnel in on-site occupied buildings, particularly from explosion events such as vapor cloud explosions. It also addresses fire and toxic release considerations. The risk-based approach of RP 752 examines potential incident sources and magnitudes, estimates the incident likelihoods (by major unit operation, using historical data where available), identifies potentially affected occupied structures, estimates the potential effects on each structure and its occupants from each incident source, and compares the combined information with risk tolerability criteria to determine where further protection may be warranted. Alternative approaches are to address occupied structures based on potential consequences or minimum distance rather than based on risk. Industry guidance for protection of personnel in portable buildings has also been developed.[11]

Other approaches address personnel protection in a more qualitative manner, but with the same general consideration of hazard sources, personnel locations, and plant features such as distances and building construction that affect whether people are adequately protected. Table 9.11 is an example of facility siting considerations that fall into two broad categories: those associated with *preventing* major incidents, such as the location and guarding of vulnerable piping and equipment relative to on-site traffic patterns, and those associated with *mitigating* major incidents, or reducing the severity of consequences if a fire, explosion, or hazardous material release does occur. After identifying hazard sources and receptor locations, these items can be used as part of a checklist-type review to facilitate study team brainstorming. A broader "Facility and Stationary Source Siting Checklist," in question and answer format, can be found in Appendix F of the CCPS Concept Book *Revalidating Process Hazard Analyses*.[10] When using such checklists, the hazard evaluation team should provide specific justification to demonstrate that all combinations of major process hazards (energy sources) and people (e.g., in occupied buildings, in temporary structures, and off-site) have been identified and evaluated and adequate protection and controls are in place.

Table 9.11 Facility siting considerations related to personnel and property protection (see text)

Considerations related to preventing incidents	❑ Onsite traffic patterns, speeds, clearances, guards ❑ Process and piping relative to construction/maintenance activities ❑ Area electrical classification, ignition source control ❑ Prevention of flammable liquid/vapor entering sewer system ❑ Segregation of incompatible materials, dedicated loading/unloading ❑ Adjacent processes that could affect process
Considerations related to mitigating general incident impacts	❑ Location, minimization of large inventories ❑ Location of reaction vessels ❑ Location of passageways and pedestrian traffic ❑ Where to handle leaking/overpressurized transport containers ❑ Means and location of emergency shutdowns, isolation valves ❑ Large-release containment, pathway, destination, capacity ❑ Emergency relief discharge treatment capability (scrubber, flare, etc.) ❑ Vent, drain, and relief valve discharge locations ❑ Incident command center location, alternative location ❑ Effectiveness of emergency warning system(s) ❑ Dominant wind direction, downwind populations ❑ Evacuation routes, emergency exits, rally points ❑ Location, accessibility of emergency equipment, SCBA ❑ Distance, capability of offsite emergency support services
Considerations primarily related to mitigating fire/explosion impacts	❑ Plant layout, separation distances, avoidance of domino effects ❑ Flammable gas detector locations, actions ❑ Flammable liquid spill drainage, BLEVE avoidance ❑ Likely missile/fragment directionality; orientation of horizontal vessels ❑ Degree of congestion and confinement (vapor cloud explosion potential) ❑ Critical controls, communications, utilities functional after incident ❑ Access routes for fire fighting and emergency equipment ❑ Hydrant, monitor, hose, foam, extinguisher locations ❑ Deluge valve location, accessibility ❑ Construction of buildings/control rooms, blast and fire resistance ❑ Flammable vapor/smoke/fire detection in building or fresh air intake ❑ No heavy equipment/material storage above occupied rooms ❑ Exterior windows vs potential for injury to occupants
Considerations primarily related to mitigating toxic release impacts	❑ Toxic release detection, actions ❑ Spill neutralization, absorption, treatment ❑ Volatile spill covering, aerosol impingement, fume mitigation ❑ Building and control room air intake locations, sensors, isolation ❑ Wind socks visible to all affected personnel ❑ Onsite temporary safe havens, escape respirators ❑ Positive pressure in control room and temporary safe havens ❑ Monitoring, alarming loss of positive pressure or unsafe atmosphere

Source: Unwin Company, Columbus, Ohio

Section 9.6 References

1. Center for Chemical Process Safety, *Guidelines for Facility Siting and Layout*, ISBN 0-8169-0899-0, American Institute of Chemical Engineers, New York, 2003.
2. Center for Chemical Process Safety, *Guidelines for Chemical Process Quantitative Risk Analysis, 2nd Ed.*, ISBN 0-8169-0720-X, American Institute of Chemical Engineers, New York, 1999.
3. UK Health & Safety Executive, *Risk Criteria for Land-Use Planning in the Vicinity of Major Industrial Hazards,* ISBN 0118854917, HSE Books, 1989.
4. OSHA Instruction CPL 2-2.45A (Revised), "Process Safety Management of Highly Hazardous Chemicals--Compliance Guidelines and Enforcement Procedures," U.S. Occupational Safety and Health Administration, Washington DC, 1994.
5. IRI IM.2.5.2, *Plant Layout and Spacing for Oil and Chemical Plants,* HSB Industrial Risk Insurers, Hartford, Connecticut, 1991.
6. American Institute of Chemical Engineers, *Dow's Fire & Explosion Index Hazard Classification Guide, 7th Ed.,* ISBN 0-8169-0623-8, Wiley, New York, 1994.
7. Center for Chemical Process Safety, *Guidelines for Consequence Analysis of Chemical Releases*, ISBN 0-8169-0786-2, American Institute of Chemical Engineers, New York, 1999.
8. Center for Chemical Process Safety, *Guidelines for Evaluating Process Plant Buildings for External Explosions and Fires*, ISBN 0-8169-0646-7, American Institute of Chemical Engineers, New York, 1996.
9. API RP 752, *Management of Hazards Associated with Location of Process Plant Buildings, 2nd Ed.*, American Petroleum Institute, Washington DC, November 2003.
10. W. L. Frank and D. K. Whittle, *Revalidating Process Hazard Analyses*, ISBN 0-8169-0830-3, American Institute of Chemical Engineers, New York, 2001.
11. API RP 753, *Management of Hazards Associated with Location of Process Plant Portable Buildings*, American Petroleum Institute, Washington DC, June 2007.

Guidelines for Hazard Evaluation Procedures

Third Edition

Part II

Worked Examples

Part II
Worked Examples

Preface to the Worked Examples

Since 1985, organizations have dramatically increased their use of hazard evaluation techniques through implementation of process safety management (PSM and RMP) programs. Recognizing that this would result in a greater demand for qualified personnel, CCPS has produced these worked examples to help organizations train people to be skilled hazard analysts.

Part II of these *Guidelines for Hazard Evaluation Procedures*—the *Worked Examples*—illustrates the methods used to identify and assess the process safety hazards inherent in operations that use or produce hazardous chemicals. The *Worked Examples* also show how hazard evaluations can be performed throughout the life of a process. However, these approaches are not limited to the chemical manufacturing industry. They are also appropriate for use in any industry where there is the potential to harm workers or the public; damage equipment and facilities; or threaten the environment through chemical releases, fires, or explosions.

The *Worked Examples* contain a management overview, fourteen chapters, and a bibliography. The following list describes the organization of the *Worked Examples*.

Management Overview of the Worked Examples discusses the instructional strategy of the *Worked Examples* and lets managers know what they can realistically expect from engineers who use the *Worked Examples* and the *Hazard Evaluation Procedures* guidelines as training aids.

Chapter 10 — Introduction to the Worked Examples describes how hazard evaluation techniques fit into an overall process safety management program; shows how hazard evaluation techniques can be used throughout the lifetime of a process or operation; and describes how inexperienced analysts can use the Worked Examples to become more proficient in the use of hazard evaluation techniques.

Chapter 11 — Description of the Example Facility and Process provides basic information about the example process and facility used in the Worked Examples to illustrate hazard evaluation techniques. This chapter includes written descriptions, drawings, and technical data. It also outlines the evolution of the hazard evaluation process.

Chapter 12 — Hazard Identification for the Example Process describes how hazard identification techniques are used on the example process, and discusses the need for periodic re-examination of process hazards at key points during the process lifetime.

Chapters 13 through 22 — Hazard Evaluation Examples illustrates how hazard evaluation techniques can be used throughout the lifetime of a process. The examples cover the following aspects for each application: problem definition, analysis description, discussion of results, analysis follow-up, and conclusions and observations.

Preface to the Worked Examples

The *Worked Examples* provide the novice hazard analyst with realistic examples of the various hazard evaluation techniques that can be used throughout the life of a process. However, the *Worked Examples* also contain information useful to the accomplished analyst. Experienced hazard analysts selected to provide in-house training may find the *Worked Examples* extremely helpful as they develop training programs. Moreover, practitioners should also find the *Worked Examples* helpful in designing and executing the hazard analysis provisions of corporate process safety management programs.

Management Overview of the Worked Examples

Hazard evaluations should be performed throughout the life of a process as an integral part of an organization's process safety management program. Using hazard evaluations can help organizations make risk management decisions during each phase of the life cycle of a process: research and development (R&D), conceptual design, pilot plant operation, detailed engineering, construction/start-up, routine operation, plant expansion, incident investigation, and decommissioning phases. Using this "life cycle" approach along with other process safety activities can efficiently reveal deficiencies in design and operation before a facility is sited, built, operated, modified, investigated and mothballed, thus making the most effective use of resources spent toward ensuring the safe and productive life of a facility.

These *Worked Examples* have been developed for two groups of people: first, those who wish to become more proficient in the use of hazard evaluation techniques; and second, those responsible for training personnel to use these methods. The *Worked Examples* demonstrate how to perform a hazard evaluation using the following techniques:

- Safety Review
- Checklist Analysis
- Relative Ranking
- Preliminary Hazard Analysis
- What-If Analysis
- What-If/Checklist Analysis
- Hazard and Operability Study
- Failure Modes & Effects Analysis
- Fault Tree Analysis
- Event Tree Analysis
- Human Reliability Analysis

In the *Worked Examples*, most of these techniques are applied to a vinyl chloride monomer (VCM) manufacturing operation. As the VCM plant evolves, specific hazard evaluation techniques are used to evaluate process hazards during each of the following phases of the plant's lifetime:

- Research and development
- Conceptual design
- Pilot plant operation
- Detailed engineering
- Construction/start-up
- Routine operation
- Plant expansion
- Incident investigation
- Decommissioning

Using the complete *Guidelines* should help beginners understand the basics of hazard evaluation and enable them to perform hazard evaluations of simple processes using the less-complicated hazard evaluation techniques. Then, with guidance from the worked examples and with experience gained from actual studies, a hazard analyst should be able to scope, organize, lead, and document a hazard evaluation of most types of processes and operations with little outside assistance.

However, performing high quality hazard evaluations throughout the lifetime of a process cannot guarantee that incidents will not occur. When used as part of an effective process safety management program, hazard evaluation techniques can provide valuable input to managers who are deciding how to best manage the risk of chemical operations. With techniques such as these, organizations will be able to continually improve process safety.

10

Introduction to the Worked Examples

A hazard evaluation is an organized effort to identify and analyze the significance of hazards associated with a process or activity. Specifically, hazard evaluations are used to pinpoint weaknesses in the design and operation of facilities—weaknesses that could lead to chemical releases, fires, explosions, and other undesirable events. An effective hazard evaluation program is one of the cornerstones of an organization's process safety management efforts.

Each organization's process safety management program should require that hazard evaluations be performed throughout the life of a process. These studies can provide information that will help organizations make decisions about improving safety and managing the risk of their operations. Using this "life cycle" approach, along with other process safety management activities, can help organizations make the most effective use of resources spent to ensure the safe and productive life of a facility.

The dramatic increase in the number of companies who use hazard evaluation techniques has created a large demand for skilled hazard analysts. To become an expert hazard analyst, one needs the practical knowledge and insights that normally come only through years of industrial experience.

The Center for Chemical Process Safety (CCPS) recognized the increased need for analysts who are proficient in the use of hazard evaluation techniques. Thus, CCPS published the *Guidelines for Hazard Evaluation Procedures, Second Edition with Worked Examples,*[1] designed to help people understand the theory and application of the hazard evaluation techniques. The *Third Edition* updates the *Worked Examples* to better reflect current practice and advances in hazard evaluation understanding.

10.1 Purpose

The *Worked Examples* can help people become competent practitioners of hazard evaluation techniques, but first they must become familiar with standard hazard evaluation methods. Beginners can acquire basic knowledge about the techniques by using the *Hazard Evaluation Procedures Guidelines* (Part I). Then, they should use the *Worked Examples* to study the ways the techniques can be applied. To become seasoned analysts, however, they must hone their hazard analysis skills and intuition through participation in actual studies. Only then can organizations be confident that their hazard analysis leaders have acquired the necessary expertise for performing high quality hazard evaluations.

The purpose of the *Worked Examples* is to illustrate the use of commonly accepted hazard evaluation techniques on a realistic process. The *Worked Examples* demonstrate how each technique can be applied and how the results can be documented. Although these examples describe typical applications of the techniques, they should not be considered as the only way these methods could be used. The ultimate aim of the *Worked Examples* is to give the beginner a look at the kinds of problems that arise when these techniques are actually applied in industry.

10.2 Instructional Strategy

A fictional yet realistic process has been defined for use in the *Worked Examples*. A vinyl chloride monomer (VCM) plant will be constructed by ABC Chemicals, Inc. in Anywhere, U.S.A. Hazard evaluations are to be performed on the example process as the project evolves through the various phases of a typical process lifetime. During each phase, the reader is challenged with the following questions:

- Why is a hazard evaluation needed? What are its objectives?
- What information and resources are available at this stage of the project?
- What hazard evaluation technique(s) should be used? Are there alternatives?
- How should the results be documented? Who will be responsible for follow-up?

To help answer these questions, each chapter provides the following information to the reader:

Problem definition gives the background for the proposed hazard evaluation; provides information about the status of the VCM process, available resources, and selection of hazard evaluation techniques; and describes study preparation.

Analysis description discusses pertinent details of the analysis and presents portions of models (e.g., fault trees, event trees) along with narratives of typical hazard evaluation team meetings.

Discussion of results presents salient results of portions of the analysis and includes documentation examples.

Follow-up on recommended actions discusses the importance of developing a management response to the findings of each hazard evaluation and integrates safety improvements into the process for use during the next phase of the example.

Conclusions and observations presents the conclusions of the study, includes observations on how the study was performed, and describes successes and difficulties experienced by the hazard evaluation team.

Each chapter ties together all phases of the project to develop a realistic picture of the evolution of the process. Essential data for the various applications of hazard evaluation techniques are summarized in each chapter.

10.3 How to Use the Worked Examples

Chapters 11 and 12 give the necessary background for understanding the example process and its inherent hazards. Appropriate hazard evaluation techniques are selected and applied to a portion of the VCM plant during each of the nine phases of its life. Table 10.1 lists the various phases of this example project along with the hazard evaluation techniques chosen for each phase.

Readers who want to focus on a particular technique may skip to the appropriate chapter; however, some confusion may occur because essential information that was presented in an earlier chapter may not be repeated in the chapter of interest. To help avoid this problem, readers are cautioned to review Chapters 11 and 12 before proceeding to the example application of a particular technique. Chapter 11 describes the example VCM process and facility. Chapter 12 gives the results of the initial hazard identification for the VCM process. Chapters 13 through 22 illustrate the use of the hazard evaluation techniques at different points in the lifetime of the process.

Chapter 10 Reference

1. Center for Chemical Process Safety, *Guidelines for Hazard Evaluation Procedures, Second Edition with Worked Examples,* American Institute of Chemical Engineers, New York, 1992.

Table 10.1 Summary of example problems

Chapter	Project phase	Selected hazard evaluation techniques
13	Research and development	What-If Analysis
14	Conceptual design	Preliminary Hazard Analysis
15	Pilot plant operation	HAZOP Study
16	Detailed engineering	Fault Tree and Event Tree Analysis
17	Construction / start-up	Checklist Analysis and Safety Review
18	Routine operation	Safety Review for Management of Change
19	Routine operation	HAZOP Study for Cyclic Review
20	Plant expansion	Relative Ranking and HAZOP Study for a batch process
21	Incident investigation	FMEA and Human Reliability Analysis
22	Decommissioning	What-If/Checklist Analysis

11

Description of the Example Facility and Process

The *Worked Examples* illustrate how hazard evaluation techniques can be used throughout the life of a realistic chemical process. The *Worked Examples* simulate the evolution of a process from its inception through design, construction, start-up, and routine operation. It also shows a portion of this process reaching the end of its useful commercial life and being decommissioned after more than 30 years of operation. This chapter gives the reader general information about this hypothetical facility.

A vinyl chloride monomer (VCM) manufacturing process was chosen as the example process for several reasons. First, it is a common commercial process used by a number of chemical companies. Second, it possesses several inherent hazardous attributes that hazard evaluation techniques would customarily be used to address. For example, the materials used in the process are toxic, flammable, and potentially reactive. Finally, the example process contains both continuous and batch operations. All of these features combine to make the VCM process well suited for challenging the creative hazard evaluation abilities of prospective practitioners.

Although meant to be as realistic as possible, the reader is cautioned not to accept at face value the physical and chemical property data, process design information and plant operating characteristics used in the Worked Examples. Some safety problems were intentionally designed into the facility and the VCM process in order to more efficiently demonstrate the abilities of the various hazard evaluation techniques to discover process safety problems. Moreover, the company, facility, and process described in these examples are purely fictional, and any resemblance to actual companies, facilities or commercial process designs is purely coincidental. Each user of the Worked Examples remains solely and exclusively responsible for the use of this information and any conclusions they reach about the safety of the design features and operating philosophies of the example process.

The following sections give the general background of the hypothetical chemical company and facility, provide a brief overview of the VCM manufacturing process, and describe the operating lifetime of both the example facility and the process.

11.1 Company and Facility Background

ABC Chemicals, Inc. is a large U.S. corporation that produces a variety of commodity chemicals such as chlorine, caustic, sulfuric acid, and hydrochloric acid. ABC enjoys an excellent safety record earned over 50 years, and many of ABC's technical personnel are internationally recognized as experts in the manufacturing and handling of its chemical products. For a number of reasons, ABC has decided to expand into the VCM market. ABC wants to build a state-of-the-art, world-scale VCM manufacturing plant at their Anywhere, U.S.A. facility. A business team has been assembled to manage this three-year venture project. As a part of their recently formalized company policy on process safety management,

ABC will perform hazard evaluations at appropriate times throughout the development and operation of the plant.

11.2 Process Overview

The ABC business team performed an extensive literature review and patent search concerning VCM manufacturing technology. They have tentatively decided to use a VCM production process involving vapor phase dehydrochlorination of ethylene dichloride (EDC) at elevated temperatures. The intermediate EDC is produced through catalytic direct chlorination of ethylene. Later in the life of the facility, ABC decides to expand the plant to include polyvinyl chloride (PVC) production. Table 11.1 lists the primary feed, intermediate, and product materials for this process, along with an indication of their hazardous characteristics.

ABC has no experience with manufacturing VCM and limited experience with some of the process materials. Thus, they are justifiably cautious in proceeding with this venture, and plan to use state-of-the-art control and safety system technology in the VCM process. Figure 11.1 is a diagram of the proposed VCM manufacturing process. In the *Worked Examples*, only portions of the process are used to demonstrate the hazard evaluation techniques. Process safety information, drawings, and data are provided as needed to support the examples in the individual chapters. Appendix C gives definitions of the abbreviations and symbols used in the example drawings.

11.3 Description of the Process Lifetime

The best process safety management programs are ones in which process safety activities are performed throughout the life of a process. Managers and hazard analysts face the challenge of finding creative ways to perform the necessary hazard evaluations in an efficient and high-quality manner. Fortunately, the hazard evaluation techniques discussed in the *Hazard Evaluation Procedure Guidelines* can be used in many of the process safety activities that take place throughout the lifetime of a facility.

One important factor that influences the execution of hazard evaluations is the life-cycle phase a process is in when a hazard evaluation is needed. A primary objective of the *Worked Examples* is to illustrate how various hazard evaluation techniques can be used to perform hazard evaluations at any stage of a process lifetime.

Table 11.1 Primary VCM process materials and their primary hazards

Material	Hazardous properties
Chlorine	Toxicity, reactivity, oxidizer, liquefied gas
Ethylene	Flammability, reactivity
Ethylene dichloride	Flammability, toxicity, potential occupational carcinogenicity
Hydrogen chloride	Toxicity, reactivity, corrosivity
Vinyl chloride monomer	Flammability, toxicity, carcinogenicity, reactivity, liquefied gas

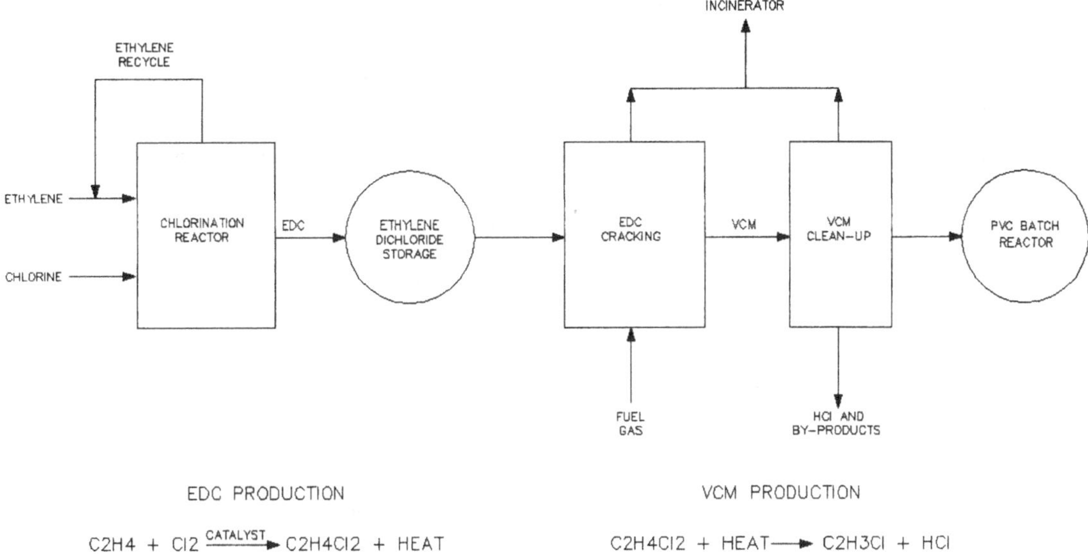

Figure 11.1 Schematic of the example VCM manufacturing process

The *Worked Examples* present a fictional history of ABC Chemicals, Inc.; its Anywhere, U.S.A. facility; and its VCM manufacturing process. Nine specific points in the life of the VCM process have been chosen as the times where the company's management thought a hazard evaluation should be performed. While the primary purposes of these hazard evaluations are to determine design or operating weaknesses and to find ways to improve the safety of the process, the motivation and focus of individual studies are different for each stage in its life.

Table 11.2 lists the nine lifetime stages used in the *Worked Examples*. Some of these are customary phases in the development and operation of any industrial facility, such as conceptual design, detailed engineering, and start-up. Others are unique event points that create a need for performing a hazard evaluation such as plant expansion, incident investigation and decommissioning. Also listed are the relative time points (start-up occurs at year 0) when these studies occur, as well as their purposes/motivations.

In reality, a specific company might not perform formal hazard evaluations at all of these points. In fact, organizations may have fewer or more specific phases leading to the construction and start-up of a new process. Thus, the example lifetime phases and hazard evaluations should not be viewed as a standard for dictating when hazard evaluations must be done. In any case, companies will likely need to perform many periodic reviews of the hazards of their processes during the operating life of a facility. These recurrent reviews are an important component of a facility's process safety management program. They provide necessary insights to organizations managing the risk of their chemical operations.

Chapter 12 outlines the hazards associated with the VCM process. Subsequent chapters illustrate the use of the hazard evaluation techniques covered in the *Hazard Evaluation Procedure Guidelines*.

Table 11.2 Summary of lifetime phases for the example VCM process

Lifetime phase	Time point	Purpose / motivation for the study
Research and development	Year – 3	Check safety feasibility; support preliminary process and material selection
Conceptual design	Year – 2.5	Provide input to site selection and preliminary process layout
Pilot plant operation	Year – 2	Satisfy potential safety concerns involving pilot plant operation; provide input to initial safety systems design for the world-scale process
Detailed engineering	Year –1.5	Finalize safe design before procuring equipment
Construction / start-up	Year 0	Verify process was constructed according to design intent and all previous hazard evaluation action items have been resolved
Routine operation	Year +2	Perform periodic hazard review required by company process safety management program; evaluate the potential safety impacts of a series of process changes that have occurred since start-up (other periodic reviews are performed on the plant during its life, but are not listed in this table)
Plant expansion	Year +5	Evaluate the safety aspects of a proposed installation of a polyvinyl chloride (PVC) batch reactor system
Incident investigation	Year +20	Determine the probable causes of a major release of toxic material
Decommissioning	Year +30	Identify any safety problems dealing with the dismantling of one VCM cracking furnace and the use of its tube bundle at another facility

12

Hazard Identification for the Example Process

Before performing a hazard evaluation, the analyst or hazard evaluation team must have a reasonably firm understanding of the inherent hazards of the process or activity being analyzed. Chapter 3, "Hazard Identification Methods," discusses various ways to identify the hazards associated with industrial processes. This chapter summarizes the results of the hazard identification study of the hypothetical vinyl chloride monomer (VCM) manufacturing process. These results are used in the various hazard evaluations of the VCM process illustrated in the *Worked Examples*.

12.1 Analysis of Material Properties

The VCM manufacturing process requires a number of different chemicals. Table 12.1 lists the chemicals ABC process engineering uses with the VCM process. To begin the hazard identification, ABC reviews the following literature to identify properties of these chemicals, and in particular, their hazardous characteristics.

- *Kirk-Othmer Encyclopedia of Chemical Technology* [1]
- *Perry's Chemical Engineer's Handbook* [2]
- *Sax's Dangerous Properties of Industrial Materials* [3]
- *CHRIS Hazardous Chemical Data* [4]
- NFPA *Fire Protection Guide on Hazardous Materials* [5]
- *Bretherick's Handbook of Reactive Chemical Hazards* [6]
- NFPA *Manual of Hazardous Chemical Reactions* [7]
- *NIOSH Pocket Guide to Chemical Hazards* [8]

In addition, the ABC engineers contact the American Institute of Chemical Engineers (AIChE) Design Institute for Physical Property Data (DIPPR), the Chlorine Institute, and suppliers of raw materials to be used in the plant. These professional groups and trade associations provide some additional hazardous material data. Table 12.2 summarizes the hazards for each chemical identified through the literature review.

Table 12.1 VCM process materials

▪ Catalyst	▪ Hydrogen chloride
▪ Caustic	▪ Natural gas
▪ Chlorine	▪ Propylene (refrigerant)
▪ Ethylene	▪ Oxygen
▪ Ethylene dichloride (EDC)	▪ Vinyl chloride monomer (VCM)
▪ Hydrogen	▪ Water

12.2 Review of Experience

The VCM plant uses complex technology and many chemicals that are new to ABC Chemicals, Inc. In particular, ABC personnel do not have much experience handling and processing hydrocarbons such as ethylene, and chemicals such as EDC and VCM. Thus, they rely heavily on literature reviews, chemical suppliers, outside consultants, and analysis of material properties to identify and evaluate hazards associated with these materials. They will also institute a research and development program and eventually build and operate a pilot plant to better understand the process and the hazards associated with these materials.

Table 12.2 Hazardous properties of VCM process materials

	Hazardous properties							
Process material	Asphyxiant	Acute toxic	Chronic toxic	Corrosive	Flammable/ explosive	Reactive	Skin irritant	Oxidizer
Catalyst			✓					
Caustic		✓		✓			✓	
Chlorine		✓		✓		✓	✓	✓
Ethylene	✓				✓	✓		
Ethylene dichloride		✓	✓		✓		✓	
Hydrogen	✓				✓			
Hydrogen chloride		✓		✓			✓	
Natural gas	✓				✓			
Propylene	✓				✓	✓		
Oxygen								✓
VCM	✓	✓	✓		✓	✓	✓	
Water	✓							

On the other hand, ABC has extensive experience with typical chlor-alkali plant chemicals such as chlorine, caustic, hydrogen chloride, and water, as well as with selected hydrocarbons. As part of the hazard identification process, ABC engineers review incident reports from several of their chlor-alkali plants. This review identifies chlorine as a highly reactive chemical—more reactive than some of the ABC engineers originally thought. In particular, they note that chlorine reacts with:

- Most hydrocarbons (thus, care must be taken to ensure process and instrument fluids and lubricants used are not reactive with chlorine)
- Small amounts of water
- Carbon steel at high temperatures (steel will burn in the presence of chlorine at elevated temperatures)
- Dry titanium at all temperatures.

12.3 Interaction Matrix

An interaction matrix provides the analyst with a structure to identify incompatibility problems between two different chemicals or a chemical under a certain process condition. Using the compatibility chart from Appendix E and input from an ABC chemist, the ABC engineering group develops an interaction matrix for the VCM plant. The matrix is constructed by listing all the plant chemicals on the horizontal and vertical axes. In addition, a few more parameters that concerned some of the engineers are listed on the vertical axis. After creating the matrix skeleton, the ABC engineers and chemist review the possible interactions illustrated by the matrix and flag the potential incompatible interactions. Figure 12.1 is the interaction matrix for the VCM plant chemicals. The X-marks indicate there is a potential incompatibility problem with the associated materials/process conditions. The team developing the matrix provides footnotes to give a further explanation for some of the incompatibility problems identified. The question marks indicate combinations of materials whose compatibility were unknown. These items, as well as other questions raised by the team (Table 12.3), are researched after the matrix is developed.

The team notes that some tertiary chemical interactions may be important. However, the interaction matrix only depicts binary incompatibility problems. ABC's chemist will investigate tertiary combinations outside of the meeting.

Table 12.3 Sample questions from the interaction matrix

- Will HCl react with the catalyst used in the process?
- Will VCM and caustic react?
- What are the reaction kinetics for chlorine-oxygen interactions?
- What is the reactivity of ethylene with itself? Of propylene with itself?
- Will the process operate at temperatures where carbon steel will burn in the presence of chlorine? Are there potential process upsets that will exceed the initiation temperature?

	Catalyst	Caustic	Chlorine	Ethylene	EDC	H_2	HCl	Nat.gas	Propylene	O_2	VCM	Water
Catalyst												
Caustic												
Chlorine		x										
Ethylene			x	?								
EDC		x										
Hydrogen			X									
HCl	?	x										
Natural gas			x									
Propylene			x						?			
Oxygen	x		?	x	x	x		x	x			
VCM		?								x	x	
Water	x	x[a]	x				x[a,e]					
High pressure				x[c]		x[d]					x	
High temperature			x[b]	x							x	
Contaminant (rust, oil)	x		x								x	
Environment (regulatory impact)		x	x		x		x				x	

[a]Heat of dilution energy release; [b]Carbon steel burns in the presence of chlorine at high temperatures; [c]Ethylene decomposes at high pressure; [d]Joule-Thompson effect; [e]Anhydrous HCl alone is not corrosive, but HCl is corrosive in the presence of water

Figure 12.1 Interaction matrix for VCM process materials

12.4 Hazard Evaluation Techniques Used for Hazard Identification

ABC analysts will use a variety of hazard evaluation techniques throughout the life of the VCM project to both identify and evaluate hazards. In the R&D and conceptual design phases, ABC will primarily use hazard evaluation techniques to identify hazards. The What-If Analysis and Preliminary Hazard Analysis techniques are applied to the VCM process "idea" during these phases. The application of these techniques and their results are described in Chapters 13 and 14. As shown in Chapter 14, some site-specific hazards (e.g., flooding) are identified that were not found by using the interaction matrix or by reviewing operating experience and material properties.

12.5 Summary

The results of the hazard identification efforts (Tables 12.2 and 12.3, Figure 12.1, and Chapters 13 and 14) helped ABC focus future hazard evaluation efforts. The hazard identification work showed ABC that (1) they must increase their understanding of ethylene and its interactions with other chemicals and (2) they must pay particular attention to chlorine/hydrocarbon interactions. This work also revealed that the potential for fires and explosions should be closely examined in future hazard evaluations.

Chapter 12 References

1. *Kirk-Othmer Encyclopedia of Chemical Technology*, Vol. 25, Wiley-Interscience, New York, 2007.
2. R. H. Perry and D. W. Green, *Perry's Chemical Engineer's Handbook, 8th Ed.*, McGraw-Hill, New York, 2007.
3. R. J. Lewis, *Sax's Dangerous Properties of Industrial Materials, 11th Ed.,* Wiley-Interscience, New York, 2005.
4. *Chemical Hazard Response Information System Manual (CHRIS),* (Commandant Instruction MI6465, 12A, available from U.S. Government Printing Office, Washington, DC, 20402), U.S. Coast Guard, Washington, DC, 1984, www.chrismanual.com.
5. *Fire Protection Guide to Hazardous Materials, 13th Ed.,* National Fire Protection Association, Quincy, Massachusetts, 2002.
6. P. G. Urben (ed.), *Bretherick's Handbook of Reactive Chemical Hazards, 6th Ed.,* 2 vols., ISBN 0-7506-3605-X, Butterworth Heinemann, Oxford, UK, 1999. Also available on CD-ROM as *Bretherick's Reactive Chemical Hazards Database - Version 3.0,* ISBN 0-7506-4342-0, Butterworth-Heinemann, Oxford, UK and online by subscription [http://www.chemweb.com].
7. NFPA 491, Manual of Hazardous Chemical Reactions, in *Fire Protection Guide to Hazardous Materials, 13th Ed.*, National Fire Protection Association, Quincy, Massachusetts, 2002.
8. National Institute for Occupational Safety and Health, *NIOSH Pocket Guide to Chemical Hazards and Other Databases,* U.S. Department of Health and Human Services, Centers for Disease Control and Prevention, DHHS (NIOSH) Publication No. 2001-145, August 2001.

13

Research and Development Phase
An Illustration of the What-If Analysis Method

13.1 Problem Definition

Background

ABC Chemicals has been producing chlorine, caustic, sulfuric acid, hydrochloric acid, and other chemicals for over 50 years. The company has an excellent safety record, and many of its personnel are internationally recognized experts on handling these chemicals safely.

Competition in the chlorine market (ABC's principal product) has increased significantly over the past five years as more companies have entered the market. And environmental regulations have increased chlorine's production costs and decreased its demand as users substitute other chemicals, consequently reducing ABC's profit margin. ABC has been looking to expand its product line into other areas where the market is stronger and the profit margin greater. Market research has identified vinyl chloride monomer (VCM) production as a rapidly growing and highly profitable market for ABC to consider.

Based on marketing's findings, ABC management is considering building a VCM plant at one of its existing chlorine plant sites. The final decision to actually build the plant is on hold, pending further investigation of the problems and costs associated with operating this kind of plant. However, the VCM project is now scheduled for start-up in three years.

Per ABC's company policy, a business team has been assembled to evaluate the available manufacturing technology for making vinyl chloride monomer. The team will direct all experimental research, production capability assessments, and safety reviews of the VCM production process. Some preliminary work has already been completed by this team, including a review of VCM incidents on record in the U.S., collection of literature on VCM production, basic chemistry work, and an initial identification of VCM process hazards. At this juncture, the business team leader has decided that a hazard evaluation of the VCM production process is needed to further identify and evaluate safety hazards. Results of this evaluation will be used by the team and ABC management to help determine if the perceived safety hazards associated with this venture are manageable.

Available Resources

The business team has just begun its investigation into the VCM venture project; thus, engineering design work has not yet been performed. ABC chemists define the basic process for manufacturing vinyl chloride monomer (depicted in Figure 13.1) as follows:

$$C_2H_4 + Cl_2 \rightarrow C_2H_4Cl_2 + \text{heat}$$

$$C_2H_4Cl_2 + \text{heat} \rightarrow C_2H_3Cl + HCl$$

The business team initiates an extensive literature review to gather information on ethylene (C_2H_4), chlorine (Cl_2), ethylene dichloride ($C_2H_4Cl_2$), hydrogen chloride (HCl), and vinyl chloride monomer (C_2H_3Cl). In addition, the team performs a patent search and gathers material safety data sheets (MSDSs) on these chemicals from other suppliers. Table 13.1 summarizes some key characteristics of these chemicals from the information gathered.

The team has also identified several experts within ABC Chemicals who are familiar with the safety hazards of some of the chemicals used in VCM production. In particular, ABC has Cl_2 and HCl experts, plant safety experts, and environmental specialists. However, they have little or no experience with ethylene, EDC, and VCM.

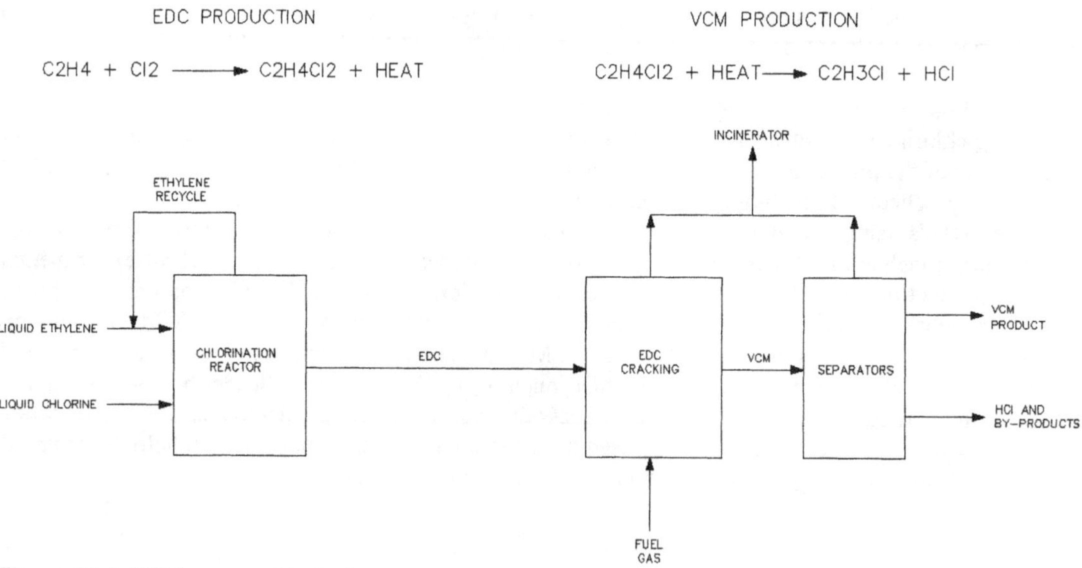

Figure 13.1 VCM process block diagram

Table 13.1 Summary of key characteristics of chemicals used in the VCM manufacturing process[a]

Properties	Chlorine	Ethylene	Ethylene dichloride	Hydrogen chloride	Vinyl chloride monomer
Boiling point (at one atm)	–29.4 °F (–34.1 °C)	–154.7 °F (–103.7 °C)	182.3 °F (83.5 °C)	–121 °F (– 85 °C)	7.2 °F (–13.8 °C)
Flash point	Not flammable	–213 °F C.C.[b]	60 °F O.C.[b]	Not flammable	–110 °F O.C.[b]
Threshold Limit Value	1 ppm	Simple asphyxiant	10 ppm	5 ppm	5 ppm
Short-term exposure limit	3 ppm for 5 min	Not pertinent	200 ppm for 5 min during any 3-hour period	5 ppm for 5 min	500 ppm for 5 min
DOT class	Nonflammable gas	Flammable gas	Flammable liquid	Nonflammable gas	Flammable gas
Flammable limits in air at standard conditions (% by vol.)	Not flammable	2.75% to 28.6%	6.2% to 15.6%	Not flammable	4% to 26%
Lethal or toxic concentration low[c]	873 ppm for 30 min (LC_{LO}-human)	950,000 ppm for 5 min (LC_{LO}-mammal)	Not available	1300 ppm for 30 min (LC_{LO}-human)	550 ppm for 4 hours (LC_{LO}-human)
Safety precautions	Avoid contact with liquid and vapor. Wear goggles, SCBA and rubber overclothing. Use water spray to control vapor. Moderately toxic if inhaled. Poisonous gases produced in fire.	Avoid contact with liquid and vapor. Wear SCBA. Use water spray to control vapor. Flashback may occur along vapor trail.	Avoid contact with liquid and vapor. Wear goggles, SCBA and rubber overclothing. Use water spray to control vapor. Use foam, dry chemicals or CO_2 to extinguish fire. Poisonous gases produced in fire.	Avoid contact with liquid and vapor. Wear chemical protection suit with SCBA. Use water spray to control vapor. Moderately toxic if inhaled or swallowed.	Avoid contact with liquid and vapor. Use water spray to control vapor. Flashback may occur along vapor line. Let fire burn itself out. Poisonous if inhaled.

[a]All information in this table, with the exception of the lethal or toxic concentration data, was obtained from the U.S. Coast Guard, *CHRIS Hazardous Chemical Data*, United States Government Printing Office, Washington, DC, 1984

[b]C.C. = closed cup, O.C. = open cup

[c]Lethal or toxic concentration data in this table were obtained from N. I. Sax and R. J. Lewis, Sr., *Dangerous Properties of Industrial Materials*, *7th Edition,* Van Nostrand Reinhold, New York, 1990

Selection of Hazard Evaluation Technique

The business team decides that a hazard evaluation is needed, but does not have the experience to determine what technique should be used to perform the evaluation. The business team asks ABC's process hazards analysis group to recommend a hazard evaluation technique. The VCM project is not yet a well-defined process; thus, the process hazards analysis group quickly rules out hazard evaluation techniques that generally require detailed drawings of a well-defined system as input data. Techniques eliminated from further consideration include HAZOP Study, Failure Modes and Effects Analysis, Event Tree Analysis, Fault Tree Analysis, Cause-Consequence Analysis, and Human Reliability Analysis. Techniques that are usually applied to an existing (already constructed) system, such as a Safety Review, are also eliminated from consideration.

ABC's process hazards analysis group narrows its choice of hazard evaluation techniques to Checklist Analysis, Relative Ranking, Preliminary Hazard Analysis, and What-If Analysis. The Relative Ranking and Preliminary Hazard Analysis methods are discarded because the team has little idea of the plant layout, equipment types and sizes, and chemical inventories at this stage in the design. The Checklist Analysis method appears acceptable; however, the team does not select it because they cannot identify a good checklist to follow for this project. The What-If Analysis technique is recommended for the hazard evaluation since it does not require detailed information of the VCM plant design and has broad flexibility for identifying and evaluating hazards.

Study Preparation

Mr. Al Chemist, a member of the business team, has been chosen to organize the What-If Analysis. Since he has never performed a What-If Analysis, Mr. Chemist asks Mr. Sam Safety, the safety and emergency coordinator for ABC's local chlorine plant, to lead the What-If Analysis. Mr. Safety has led or participated in several different hazard evaluations at this chlorine plant, including What-If analyses. Mr. Safety has selected the following skilled professionals for the What-If Analysis:

Chemist	—	A chemist, familiar with chlorine, hydrochloric acid, ethylene, EDC, VCM, etc., is needed to help identify unwanted and potentially dangerous chemical interactions. Mr. Al Chemist will fill this position. (*This is also a good informal way to communicate the What-If Analysis findings to the business team. However, ABC also requires that a formal report be sent to this team.*)
Chlorine and VCM expert	—	An expert on chlorine and VCM is needed to help identify hazards associated with handling large quantities of these materials. Ms. Kim Engineer, a 10-year veteran from the local ABC chlorine plant, will fill this position.
Ethylene expert	—	An expert is also needed to help identify hazards associated with handling and processing large quantities of ethylene. ABC Chemicals has no experience in this area. Using the CCPS *Directory of Chemical Process Safety Services*, Mr. Safety identifies several firms with expertise in handling ethylene. After gathering some information on these companies through phone calls, Mr. Safety hires Mr. Pete Petros of Hydrocarbons Consulting, Inc. to assist in the What-If Analysis.

Safety specialist	—	A safety specialist is needed to help understand and identify particular safety requirements associated with the new project. Hopefully, this specialist (as well as other team members) will also know about past incidents, near misses, and safety concerns pertinent to the VCM project. Mr. Sam Safety will fill this position.
Metallurgical engineer	—	An expert in the materials of construction is needed to help understand the types of materials that will be best suited for this new process. This could be done as a separate review; however, it is invaluable to have this person participate in the study so he/she can have a better understand of the critical process parameters. Mr. Matt Steele will fill this role.

Mr. Safety will lead the What-If Analysis. Based on the information provided to him by the research team, and the fact that the project is just beginning, Mr. Safety estimates that the What-If Analysis will take one or two days. He schedules the meeting to take place at a convenient location for the What-If Analysis team—in this case, the training room at the local plant. Two weeks before the scheduled meeting, Mr. Safety sends each member a half-page summary of the VCM production process and the information the research team has gathered for the project (i.e., simplified block diagram and normal operating ranges for critical parameters such as temperature, pressures, and flows).

One of Mr. Safety's last tasks in preparing for the review is to develop a preliminary list of What-If questions to ask in the meeting. While the What-If Analysis method is designed to encourage team members to verbalize their safety concerns, Mr. Safety has found that having a few questions ready to "get the ball rolling" is very helpful. Asking these questions early in the review also helps the group focus on safety issues. Referring to Figure 13.1 and Table 13.1, Mr. Safety spends less than an hour developing What-If questions for the review (Table 13.2).

Table 13.2 Sample What-If questions for the R&D phase example

What if...

- the ethylene feed is contaminated?
- the chlorine feed is contaminated?
- the chlorination reaction is too fast?
- a furnace explosion occurs?
- ethylene drops out with HCl by-product?
- the wrong material of construction is used?
- a line ruptures?
- large quantities of chlorine carry downstream with the EDC?
- EDC flow to the furnace is interrupted?
- VCM drops out into the by-products stream?
- by-products are sent to VCM storage?

13.2 Analysis Description

Two weeks later, on a Monday morning at 9:00 a.m., the hazard evaluation team assembles in the local plant training room. The What-If Analysis meeting is now ready to begin. First, Mr. Safety asks the other members of the review team to introduce themselves and describe their specialties. Mr. Chemist has agreed to record the minutes of the meeting. (Note: It is acceptable to have a person outside of the technical personnel act as the scribe, so that the hazard evaluation team can fully concentrate on the process hazards. This "scribe" should have some familiarity with process safety terms.) Mr. Safety then reviews the schedule for the day, indicating that refreshments and lunch will be brought in and that breaks will be taken hourly. Next, Mr. Safety briefly reviews the background of the VCM venture project, the chemistry involved, and why this What-If Analysis is being performed. Before beginning the review, Mr. Safety outlines the following ground rules: (1) all team members will have equal say; (2) any concern, no matter how inconsequential it appears, is fair to suggest; (3) all team members are to contribute; (4) spin-off questions and ideas derived from original What-If questions will be given priority before moving on; (5) detailed analysis and criticism of questions or ideas is not allowed; and (6) the focus of the review is to identify hazards, not to find solutions to these problems. He states that the findings of this review will be documented by Mr. Chemist and himself, reviewed by the group, and then transmitted as a formal report to the business team.

The following is an excerpt of the discussions from the What-If Analysis team meeting.

Mr. Safety — Well, let's begin by looking at the chemist's concept for producing VCM. Does anyone have a What-If question? [*Appropriate pause.*] Okay, well what if the ethylene feed into this process is contaminated?

Ms. Engineer — What are the contaminants?

Mr. Petros — The typical contaminant in ethylene is light oil. Water is also a remote possibility.

Mr. Chemist — From what we have seen in our initial experiments and in the literature, oil and chlorine will energetically react. That could be a problem. How dirty is the typical ethylene supply?

Mr. Petros — That obviously depends on your supplier. Assuming you get it by pipeline, however, there will probably be only trace amounts of contaminants in the ethylene supply. I can check on this if you like.

Mr. Safety — I think we should. We should also see how severe the reaction might be if contaminants are present. Mr. Chemist, are your meeting minutes up with us?

Mr. Chemist — Yes. For small amounts of oil, I think the large volumes of chlorine, ethylene, and EDC in the reactor would probably quench the oil/chlorine reaction. But I'll make a note to verify this.

Mr. Safety — Any new What-If's? [*Pause.*]

Mr. Chemist — What if the chlorine feed is contaminated?

Ms. Engineer — The major contaminant we have in chlorine is water. However, we would know if the chlorine has water in it at the chlorine plant well before it is fed to this process. Any significant amount of water would begin destroying equipment and piping from the hydrochloric acid it formed.

Mr. Safety — Yes, but didn't we say that water might come in with the ethylene feed? This might be a way to get hydrochloric acid into the VCM process. Guess we *had* better check on ethylene contaminants.

Mr. Chemist — I agree. I don't think small quantities of water will cause any problem, but I will check it out.

Mr. Steele — What if one of the supply pipelines breaks?

Mr. Petros — Well, if it's the ethylene line you will likely have a major fire or flammable vapor cloud on your hands. If it's the chlorine line, you have a toxic release problem.

Mr. Safety — I think we could handle the chlorine line problem since we have the equipment, people, and know-how to deal with it. The ethylene line is a different story. By the way, is the chlorine supply liquid or vapor?

Mr. Chemist — Right now we were thinking of liquid chlorine supply. But I see your thinking—a vapor feed would reduce the amount of chlorine released if a pipe ruptured. I'll have the research team check into this possibility. What about the ethylene line rupture?

Mr. Petros — I think that's something your company will have to decide whether they are comfortable handling. Obviously a number of firms safely handle ethylene daily. And there are plenty of things you can do to prevent or mitigate a release, like installing remotely operated isolation valves. However, your company definitely has some learning to do in this area and probably some additional fire-fighting equipment needs.

Mr. Safety — Mr. Chemist, be sure to note this. Management should definitely address this issue. Are you caught up with us?

Mr. Chemist — No, could you repeat that last recommendation?

Mr. Safety — Sure. ABC needs to determine what training and equipment is needed to safely handle ethylene, including fire-fighting equipment. [*Pause.*] Any more What-If's?

Ms. Engineer — What if the feed rates are out of balance? Will the reaction run away and blow up the reactor?

Mr. Chemist — I can't answer that question now. However, I'll make sure research and development addresses it.

Mr. Steele — What if the EDC cracking furnace is too hot?

Mr. Chemist — Again, we have not done our homework on that yet. From what I've read, we will make more VCM and more by-products.

Mr. Safety — How hot will the furnace be?

Mr. Chemist — Don't know yet.

Mr. Safety — You know, steel will burn in the presence of chlorine above some temperature. Will it exceed this temperature? [*No answer.*] Does anyone here know this temperature? [*Stark silence.*] We need to check this out.

Mr. Chemist — I'll note the materials of concern. At this stage in the project though, I think we will probably pass this information on to engineering.

Mr. Safety — That's fair enough. But my point is, can you consume all the chlorine in the direct chlorination reactor to avoid this problem?

Mr. Chemist — In theory, yes. Practically speaking, I doubt it. I'll bring it up with the research and development group.

The What-If discussions continue on in this fashion throughout the day. All team members supply questions or respond to them. The questioning continues until the team has no other What-If's or until the questions asked always appear to cover the same ground as before.

13.3 Discussion of Results

The What-If Analysis results are documented in tabular form. Included in the table are the What-If questions asked, the team's response to the questions, and the actions, if any, suggested by the team. ABC Chemicals has also included columns in their table for assigning responsibility for the action item follow-up and for the status of follow-up activities. While not in ABC's table, one other column that often appears is a resolution column where a description of how the action items are addressed appears. This column is filled in as the appropriate personnel resolve the action items they are assigned. Note that the hazard evaluation action items usually fall to the responsibility of one of the hazard evaluation team members. However, this does not mean they will complete the actions within the team. Many action items produced by hazard evaluations are resolved by other specialized experts that did not participate in the hazard evaluation.

Many of the questions raised by the team were quickly resolved in the meeting. Questions that were not quickly resolved were assigned to various team members to investigate outside of the questioning session. The assigned team members reported their findings back to the team leader for inclusion in the What-If Analysis report. The actions may be to follow up on concerns raised in the review or to obtain answers to unanswered questions.

During the What-If Analysis, Mr. Chemist kept notes on chart pads of the questions raised, items discussed, and actions suggested. He later transcribed these notes into a more formal report for the business team. He also included in this report the findings and recommendations made by team members who investigated questions left unanswered in the What-If meeting. Table 13.3 is an example of the formally documented What-If Analysis results. As shown by this example, the responses to What-If questions and suggested actions are condensed versions of the actual discussions. This table, recorded by Mr. Chemist during the meeting and edited by him shortly after the meeting, was reviewed and approved by all the team members before it was submitted to the VCM business team.

Table 13.3 Sample What-If Analysis results for the R&D phase

What-If Analysis Worksheet PROCESS / LOCATION: VCM plant concept TOPIC INVESTIGATED: Safety hazards EQUIPMENT/TASK INTENTIONS: Direct chlorination reactor feeds	DATE: ##/##/## TEAM: Mr. Safety (Leader—Anywhere Plant Safety Coordinator), Mr. Chemist (ABC Research & Development), Ms. Engineer (Anywhere Plant Engineer), Mr. Petros (Consultant), Mr. Steele (ABC Metallurgical Engineer)		PAGE: 1	
What if . . .?	Hazard / Consequence	Recommendation	Responsible Individual	Initial and date when resolved
1. Ethylene feed is contaminated	1. Typical contaminant in ethylene is oil. Oil will react energetically with chlorine. However, the amount of oil in ethylene is usually small, and the large quantity of ethylene dichloride (EDC) in the reactor should quench any oil/chlorine reaction. Water is also a trace contaminant.	1.a Verify availability of high-purity ethylene and reliability of supply 1.b Determine the reaction kinetics for oil/chlorine reactions. Examine the reaction kinetics for chlorine/water reactions	1.a Ethylene expert 1.b Chemist	
2. Cl_2 feed is contaminated	2. Typical contaminant in chlorine is water. Large quantities of water in chlorine would cause equipment damage in the chlorine plant and cause a shut-down well before it made it to the VCM plant. Small quantities of water should be no problem.	2.a Verify water content in chlorine supplies is very low	2.a Chlorine expert	
3. A feed pipeline breaks	3. Chlorine—A large release of liquid chlorine would occur, which would create a large chlorine vapor cloud Ethylene—A large liquid ethylene release would occur, resulting in a large ethylene vapor cloud, which is a potential fire and explosion hazard	3.a Consider supplying chlorine vapor to the VCM plant 3.b Evaluate ABC's abilities to handle highly flammable materials. Consider additional fire safety training and protection equipment 3.c Consider remote-controlled feed	3.a Chemist 3.b Plant fire chief, corporate training officer 3.c Engineer	
4. The feed rates are out of balance	4. A runaway reaction may be possible. An acceptable operating range is not yet known	4.a Examine the reaction rates at various ethylene/chlorine feed ratios	4.a Chemist (to communicate with research)	

Mr. Chemist also has written a brief two-page executive summary of the What-If Analysis. This summary lists where and when the What-If Analysis meeting was held, who attended the meeting, and what actions/concerns were expressed by the team. This summary was also reviewed by the What-If team prior to its release.

13.4 Follow-up

The What-If team assigned responsibility for the suggested action items both to members of the team and to other appropriate business personnel. Mr. Chemist, who is a member of the business team coordinating this VCM project, has agreed to contact appropriate business personnel and ensure that action items are addressed. Mr. Safety decides another meeting is not needed for follow-up. Results of follow-up actions will be incorporated into the What-If Analysis results and will be documented in the report sent to the business team for consideration.

13.5 Conclusions and Observations

The What-If Analysis proved highly successful. It succeeded because (1) an experienced leader kept the discussions on track and was *prepared* to ask What-If questions when the discussions lagged, (2) the appropriate experienced personnel attended and actively participated, and (3) the team did not try to immediately solve all the problems they identified. The What-If team did not determine whether to go forward with the VCM venture project. Nor did they decide whether the safety risks involved were acceptable. The What-If team did identify a number of safety and environmental issues that ABC management should consider in evaluating the viability of the VCM project. The time required to perform the What-If Analysis was relatively brief. Table 13.4 summarizes the effort required for this What-If Analysis.

Table 13.4 What-If Analysis staff requirements for the R&D phase

Personnel	Preparation (h)	Evaluation (h)	Documentation (h)
Leader	4	8	16
Team member[a]	1	12[b]	1

[a]Average per team member

[b]Team members spent some time outside the meeting obtaining answers to some questions raised

The findings of the What-If team varied, from specific recommendations to ensure that the quality needed in feed materials is available, to general concerns over whether the company could safely handle the new materials required for the process and respond to the potential emergencies that could be created by the use of these new materials. At this stage of a process, it is not unusual that such general concerns arise. The What-If simply highlighted areas where ABC Chemicals lacks expertise. Particular concerns raised by the team include the following:

- Does ABC Chemicals understand the environmental regulations concerning VCM, EDC, and other chemicals associated with this process? Will management commit to the additional staff needed to support safe VCM operations?

- ABC has never used ethylene before. How will ABC employees be trained to work safely with it? Also, extensive fire-fighting training and equipment is necessary if ABC begins to use ethylene.

- Monomers are also new to ABC; therefore, extensive training in this area will also be needed. ABC should consider hiring people with direct VCM experience to supplement the staff they already have.

- Any vent system for this new process must be capable of handling toxics and flammables.

In conclusion, management determined that the What-If Analysis team did not identify any unmanageable hazards that would warrant ABC abandoning the venture project. The actions suggested will help the business team direct future research, assess ABC's ability to manufacture VCM, and determine what future hazard evaluations are needed.

14

Conceptual Design Phase

An Illustration of the Preliminary Hazard Analysis Method

14.1 Problem Definition

Background

ABC Chemicals has decided to proceed further with the VCM project. The What-If Analysis performed during the R&D phase did not identify any unmanageable hazards. Follow-up of the actions suggested in this review indicated that ABC had or could hire people to adequately deal with the hazards identified in the VCM process. This information, combined with an encouraging marketing outlook, convinced ABC management that the VCM project was viable.

ABC Chemicals has chosen their Anywhere, U.S.A. chlorine plant as the site for the proposed VCM plant. This chlorine plant, while very similar to Mr. Safety's and Ms. Engineer's plant, is a different site from the one that was assumed to be the new plant location during the R&D What-If Analysis. The Anywhere, U.S.A. site was chosen instead of the other ABC plant sites for several reasons:

- The nearest population center is small (compared to other ABC plant sites).
- This site is relatively near an ethylene supply pipeline.
- This site has ample property to accommodate the new plant.
- This site is located in an area where other chemical facilities are located, so the pool of qualified new hires is better suited than the other ABC sites.

Additional information concerning the VCM plant is available in this project phase, including a tentative plant layout (Figure 14.1). ABC engineers have chosen to locate the VCM plant in what was "buffer" property for the chlorine plant, east of the existing unit. This site was chosen because (1) there is enough room to separate the plant and the administrative offices, (2) ABC already owns the property, (3) the site has the most advanced fire protection system in the company, (4) the site has the most advanced process safety management systems in the company due to the chlorine plant hazards, and (5) access to an ethylene supply is convenient.

With this new information in hand, the business team has decided another hazard evaluation is needed. While the business team wants to know about any additional hazards with the process, they also want the hazard evaluation to focus on the suggested VCM plant siting at this newly chosen location. The business team recommends that Mr. Safety again lead this effort since he is familiar with the project and did an excellent job in the last review. Mr. Chemist, however, convinces the business team that a new point of view is needed; therefore, Ms. Deal from the corporate process hazards analysis group is selected to lead this hazard evaluation.

Available Resources

The VCM plant has not yet been designed. However, engineering has provided Ms. Deal with a preliminary plant layout (Figure 14.1). Engineering and R&D have also provided the following:

- A preliminary list of raw materials, intermediate products, final products, and waste products (Table 14.1)
- A preliminary list of major vessels in the plant (Table 14.2)
- R&D reports on the VCM project
- Previously collected literature on EDC and VCM
- What-If Analysis results from the previous hazard evaluation
- A preliminary list of the systems that will interface with the Anywhere Chlorine plant.

Ms. Deal has also requested information on the Anywhere Chlorine plant. In particular, she has asked for emergency procedures, safety equipment such as fixed fire suppression systems and fixed chemical sensing systems, and a list of emergency shutdown for the plant. This information will be collected by an engineer at the Anywhere Plant.

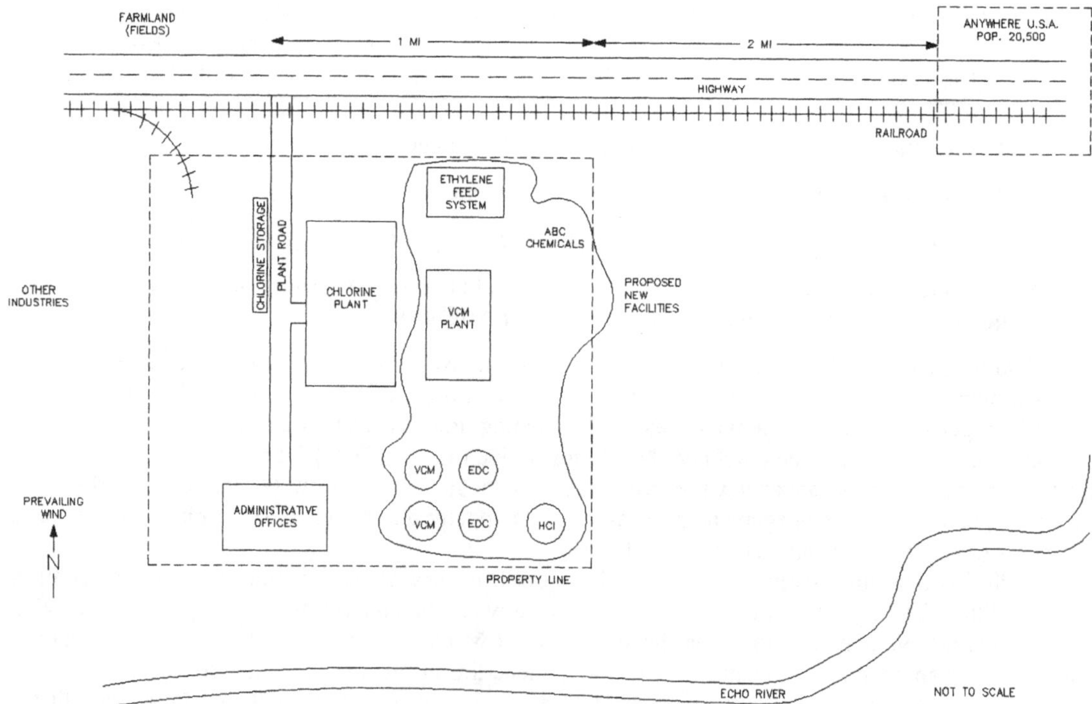

Figure 14.1 VCM plant layout

Table 14.1 Partial list of materials in the VCM plant

▪ Ethylene	▪ Vinyl chloride monomer	▪ Light hydrocarbons
▪ Chlorine	▪ Hydrogen chloride	▪ Heavy hydrocarbons
▪ Ethylene dichloride	▪ Water	▪ Natural gas

Selection of Hazard Evaluation Technique

The business team has left the selection of a hazard evaluation method up to Ms. Deal. They have, however, requested that the What-If Analysis technique not be chosen unless it is the only applicable method since it has been used once before.

Ms. Deal is familiar with most of the commonly used hazard evaluation methods. Because the process is still not well-defined, she quickly rules out HAZOP Study, FMEA, Event Tree Analysis, Fault Tree Analysis, Cause-Consequence Analysis, and Human Reliability Analysis. She also eliminates Checklist Analysis since she does not have a good checklist to use. What-If Analysis is a possibility, but Ms. Deal does not select it because it was previously used. Thus, the choice is narrowed to Relative Ranking and Preliminary Hazard Analysis. At this point, the selection is a difficult judgment call. Either method could be used; however, the Preliminary Hazard Analysis is better adapted to identifying hazards than is the Relative Ranking method, and Ms. Deal is more familiar with Preliminary Hazard Analysis, so she chooses it.

Table 14.2 Major equipment in the VCM plant

Equipment	Number	Chemical(s)	Volume
Direct chlorination reactor	1	Chlorine, ethylene	Large (liquid)
Compressor	1	Ethylene	Medium (gas)
Storage spheres	4	EDC, VCM	Large (liquid)
Storage tank	1	HCl acid	Large (liquid)
Distillation column	4	EDC, VCM, HCl, light hydrocarbons, heavy hydrocarbons	Large (liquid/gas)
Accumulators	*a*	EDC, VCM, HCl	Medium (liquid)
Furnace	1	EDC, VCM	Medium (gas)
Incinerator	1	Mixture	Medium (gas)

[a] Not yet determined

Study Preparation

Ms. Deal has selected two additional employees of ABC to assist with the Preliminary Hazard Analysis. Mr. Dennis is a process hazard analyst with ABC who works in the corporate engineering division. He is chosen by Ms. Deal because he has some VCM and ethylene experience with a previous employer. (Mr. Dennis was hired by ABC this past year, in part because of the VCM project.) Mr. Dennis and Ms. Deal are in the same working group within ABC Chemicals.

Mr. Scott, an experienced process engineer from the Anywhere chlorine plant, is the other team member. Mr. Scott has worked in most of the units in the chlorine plant and is very familiar with the personnel; the operating, safety, and emergency practices; and the layout of the Anywhere Plant facility.

A business team member will not participate in the Preliminary Hazard Analysis. To ensure that the business team's concerns are understood and that the Preliminary Hazard Analysis satisfies their objectives, Ms. Deal meets with Mr. Chemist to discuss these issues and to explain what the Preliminary Hazard Analysis will accomplish.

In preparation for the Preliminary Hazard Analysis, Ms. Deal sends an e-mail to Mr. Dennis and Mr. Scott announcing the Preliminary Hazard Analysis. Information in the e-mail includes: (1) the date, time (estimated to be one day), and location for the Preliminary Hazard Analysis; (2) a description of the Preliminary Hazard Analysis objectives; and (3) copies of previous hazard evaluations, R&D reports, and preliminary engineering studies for Preliminary Hazard Analysis team members to review. A one-day meeting is scheduled at the Anywhere site for the Preliminary Hazard Analysis. (The Anywhere site is chosen because the business team is particularly concerned about siting issues, and neither Ms. Deal nor Mr. Dennis has seen the Anywhere Plant.)

Before the Preliminary Hazard Analysis meeting, Ms. Deal will spend about a day reviewing the information on the VCM project, including the Anywhere site's capabilities to safely manage another highly hazardous process. Based on this review, she will formulate some questions (Table 14.3) to ask during the Preliminary Hazard Analysis. She will also prepare some blank Preliminary Hazard Analysis forms to use in the team meeting.

Table 14.3 Preliminary questions for the conceptual design Preliminary Hazard Analysis

- How close is the nearest public neighbor to the Anywhere Plant? The nearest industrial neighbor?
- What is due east of the plant? Due west? Due north? Due south?
- What emergency actions do you take in the event of a large chlorine release?
- Can the Echo River flooding threaten the plant site?
- What does the wind rose show for this location?
- What kinds of fire protection are available on-site? Off-site?

14.2 Analysis Description

The Preliminary Hazard Analysis study will involve the following tasks:

- A short tour of the Anywhere Chlorine plant and the proposed VCM site
- The Preliminary Hazard Analysis team meetings on the proposed VCM plant
- The review of the Preliminary Hazard Analysis results

Often, a tour of the area is not included in a Preliminary Hazard Analysis. However, since both Ms. Deal and Mr. Dennis are unfamiliar with the plant, a short tour (e.g., one-hour) is in order. During this tour, which is led by Mr. Scott, the team takes note of equipment and personnel locations at the existing chlorine plant and their proposed locations for the VCM plant. They also focus on what fixed emergency equipment is available (e.g., gas alarms and fire protection systems), where the equipment is located, and whether there is additional capacity available in the systems.

After the tour, the Preliminary Hazard Analysis team convenes in a plant conference room to begin the analysis. Ms. Deal tells the others that the Preliminary Hazard Analysis will take the rest of the day, breaks will be taken as needed, and refreshments and lunch will be brought in. Then she briefly describes the Preliminary Hazard Analysis technique to the others, shows them the Preliminary Hazard Analysis table to be filled in during the review, explains the hazard category definitions, and reminds the team to think about secondary equipment damage and system interactions that may result from causes of hazardous situations (e.g., a crane bumps into a line and ruptures it, releasing a chemical that eventually results in a vapor cloud explosion and ruptures a chemical storage tank). Ms. Deal also explains that to keep the Preliminary Hazard Analysis orderly she will suggest a hazard and then move through the process from the front end to the back end, looking for causes, effects, and suggestions for improvements. Ms. Deal then has the assembled team members introduce themselves and their backgrounds. Finally, Ms. Deal gives a brief overview of the VCM manufacturing process and then begins the analysis.

The following is a brief excerpt from the Preliminary Hazard Analysis review.

Ms. Deal — Let's start with a toxic release as a hazard. Beginning with the ethylene feed into the plant, what are some possible causes?

Mr. Dennis — Well, ethylene really isn't a toxic. It's much more an asphyxiant and a flammable hazard.

Ms. Deal — Okay. Let's consider some other toxics and come back to those hazards later. What about the chlorine line to the VCM plant?

Mr. Scott — As I understand it, liquid chlorine will be piped to the VCM plant. [*R&D determined that the chlorination reaction yield was higher using liquid chlorine.*] A flange or gasket leak would release chlorine, but probably only small amounts. A pipe rupture, say it's due to a crane incident, would release a lot of chlorine. Thermal expansion in a blocked-in line could also blow a gasket or valve packing, or cause the line to rupture.

Mr. Dennis	—	Perhaps we should suggest burying the line. That would remedy the crane problem.
Ms. Deal	—	Wait, guys. Let's get back on track. Before suggesting changes, let's get an idea of how bad this situation would be and what protective measures are provided.
Mr. Scott	—	Okay. Depending on wind direction, you could expose several operators or the administrative building personnel to high chlorine concentrations. However, the likely safeguards we would have in place include (1) temporary safe havens in the unit, (2) expansion chambers on the piping, (3) all-welded piping, (4) a very aggressive mechanical integrity inspection program on process piping and components, (5) crane lifts over active process lines requiring unit shutdown, (6) chlorine gas detectors, and (7) plant personnel trained to respond to a chlorine release. Should I go on?
Ms. Deal	—	What about public exposure?
Mr. Scott	—	The nearest residence is a mile and a half due east. Our emergency plans address this issue.
Ms. Deal	—	Do you see any special precautions we should take with the chlorine feed line?
Mr. Scott	—	We need to make sure the line is empty if the VCM plant is down for any length of time.
Mr. Dennis	—	What about a line break? How will you isolate it?
Mr. Scott	—	We need positive shutoff valves, perhaps with some kind of automatic isolation system to trip when a line rupture occurs. By the way, the VCM plant people will also have to learn our emergency procedures for chlorine releases and have personal protective equipment. Also, if the VCM plant suddenly shuts down, we will need to make sure the chlorine plant can handle this shock.
Mr. Dennis	—	What about burying the chlorine feed line?
Mr. Scott	—	No, that's not a good idea. Little chlorine leaks tend to become big ones in a short time. We would rather be able to see the pipe at all times and catch little leaks early.
Ms. Deal	—	I've noted these things in the Preliminary Hazard Analysis table. Using the categories I showed you at the start of this review, how would you categorize a line leak and line rupture?
Mr. Scott	—	Well, small chlorine leaks aren't uncommon and we have no trouble with them. I'd say a leak is a category one. A rupture is pretty serious. I'd call it a four.

Mr. Dennis — I think that sounds reasonable.

Ms. Deal — Okay, let's move on if there are no more ideas. [*Pause.*] The next major equipment item is the direct chlorination reactor.

Mr. Scott — Let me understand this thing. Liquid ethylene and chlorine are mixed together in this reactor to make ethylene dichloride. And the reaction is exothermic.

Ms. Deal — That's right.

Mr. Scott — Can it blow up?

Ms. Deal — I don't know. I'm sure engineering will design a pressure relief system for the worst credible case. Anyway, we should assume it could, but we should also consult with them concerning an appropriate design.

Mr. Scott — How much chlorine does the reactor have in it? And at what temperature?

Ms. Deal — The project is not far enough along to answer those questions. However, I believe it could be more than 10 tons. Don't know about the temperature at all.

Mr. Scott — If that reactor goes, you might expose Anywhere to a high concentration of chlorine. About 15 years ago, ABC moved our storage tanks and rail yard to the west side of Plant Road because they worried about a big chlorine release. This is definitely one of your category four hazards. I think some study that the environmental group did prompted this action. Anyway, it looks like this chlorination reactor may be in the same boat.

Mr. Scott — Our environmental group had to prepare a worst-case release scenario for our Risk Management Plan a few years ago. I will get that at our next break and that should give us some idea of what kind of worst-case consequences we are looking at with our current process and how far vapors could travel off-site. If we are talking about an exothermic reaction of chlorine with a highly flammable material, we may be looking at a significantly different scenario than the RMP worst case.

Ms. Deal — Good idea. What about safeguards on this reactor?

Mr. Dennis — We don't know what the reactor will have yet. We will need to put in appropriate preventive safeguards to avoid a runaway reaction. However, if a runaway isn't stopped, there will be a large chlorine and ethylene release. What about chlorine gas detectors?

Mr. Scott — The detectors are at select locations inside the chlorine plant. If the wind blows towards town, the detectors won't see the chlorine. If the rupture disk on the reactor bursts or a relief valve lifts, we would probably hear it.

Ms. Deal — So, do you suggest we move the VCM plant west of Plant Road?

Mr. Dennis — I think we should review the current RMP worst-case and alternative release scenarios, then have the environmental group model a large chlorine release from the proposed siting for the VCM plant under what we think could be the reasonable worst case conditions and see what the outcome is. If the public exposure is unacceptable, then I agree we should move the plant west to give us a larger buffer from any off-site impact.

Mr. Scott — I think we should just move the plant. The liquid chlorine piping is already there, so there is no new risk from a pipe rupture. The added distance helps protect the town. And our industrial neighbors to the west are better prepared to respond to a gas release.

Ms. Deal — It appears we may need to do a detailed siting study. If we move the plant west, we will probably have to buy more land. I will note both of your points and leave it to the business team to decide on a course of action. Let's move on to the EDC storage spheres. Can they contribute to a toxic release?

The discussions continue in this fashion until the team has reviewed all the major equipment items for causes of toxic releases. Ms. Deal then redirects the Preliminary Hazard Analysis team back to the beginning of the VCM process and focuses on causes of flammable gas and vapor releases. The process is reviewed from front to back. The Preliminary Hazard Analysis team cycles through the VCM plant several times, each time looking for causes of new hazards (e.g., decompositions, runaway reactions, cryogenic effects) and recommendations that address these hazards. (Note that other analysts might choose to identify all the hazards in one section of the plant and then move on to other plant sections to complete the Preliminary Hazard Analysis.)

At the end of the Preliminary Hazard Analysis review, Ms. Deal states that she will edit the Preliminary Hazard Analysis results (tables) and send them to the team members for review in two weeks. After receiving their comments she will send a report to the business team.

14.3 Discussion of Results

Table 14.4 lists an example of the Preliminary Hazard Analysis results. The table contains the hazards considered, causes of these hazards, and suggestions for corrective actions and preventive measures that ABC should consider. As with the earlier What-If Analysis, the Preliminary Hazard Analysis considered, but did not record, existing safeguards when evaluating hazards. Follow-up assignments for suggested corrective actions were not made either because the team did not have the authority or know the appropriate personnel to assign. Ms. Deal left this activity up to the business team.

In addition to this Preliminary Hazard Analysis table, Ms. Deal will prepare a brief executive summary and list everyone who participated in the Preliminary Hazard Analysis, as well as what was examined and the major findings. The results of the Preliminary Hazard Analysis proved very useful to the designers of the VCM plant, allowing them to correct inadequacies at an early phase of the project.

Table 14.4 Sample Preliminary Hazard Analysis results for the VCM plant conceptual design

AREA: VCM plant–conceptual design DRAWING: Figure 14.1		MEETING DATE: ##/##/## PAGE: 1 of 27 TEAM MEMBERS: Ms. Deal (Leader—ABC Process Hazards), Mr. Dennis (ABC Process Hazards), Mr. Scott (Anywhere Plant)		
Hazard	Possible initiating cause	Major effects	Hazard category*	Corrective / preventive measures suggested
Large inventory of high toxic hazard material	1. Chlorine line gasket/packing leak	1. Small chlorine release on-site	I	1. None
	2. Chlorine line rupture (i.e., vehicle collision, blocked-in line)	2. Large chlorine release, major on-site impact; potential off-site impacts	IV	2a. Verify chlorine line is evacuated whenever the VCM plant is down for extended time 2b. Provide safety instrumented systems to positively isolate the line in the event of a rupture 2c. Train VCM plant personnel to respond to chlorine releases 2d. Equip VCM plant personnel with PPE for chlorine 2e. Do not bury chlorine pipeline
	3. Direct chlorination reactor exotherm	3. Large chlorine/ EDC/ethylene release; depending on reactor size/ operating conditions, potential off-site impacts	IV	3a. Consider moving VCM plant west of Plant Road 3b. Perform dispersion studies to assess off-site impact of chlorine/EDC release due to exotherm 3c. Verify reactor pressure relief system can handle this release
	4. Direct chlorination reactor rupture	4.Large chlorine/ EDC/ethylene release; depending on reactor size/ operating conditions, potential off-site impacts	IV	4. Minimize inventory of chlorine/EDC in reactor
	5. Direct chlorination reactor relief valve lift	5. Potential large EDC/chlorine/ ethylene release	III	5. Verify the reactor pressure relief system incinerator and scrubber are sized to handle this release
	6. EDC storage sphere rupture	6. Large release of EDC, potential off-site impact; potential river contamination	IV	6. Consider moving EDC sphere away from river
	7. Flood damages EDC sphere	7. Large release of EDC, potential off-site impact; potential river contamination	IV	7a. Consider moving EDC sphere away from river 7b. Verify EDC (and other tanks) support structure designed to withstand flood conditions

* Hazard category: I - negligible, II - marginal, III - critical, IV - catastrophic

14.4 Follow-up

The Preliminary Hazard Analysis team's job is complete. The report submitted by this team has several suggestions for improvements and some unresolved issues, which will be addressed by the business team. The business team is responsible for assigning these issues to appropriate personnel and ensuring that each issue is resolved. The improvements, after review and approval by the business team, are passed on to the appropriate VCM project personnel within ABC Chemicals.

The suggestion of siting the VCM plant west of Plant Road requires immediate attention. If accepted, ABC must quickly begin purchasing land west of Plant Road. This action will increase the VCM project cost and may delay construction. However, siting the VCM plant to the east may increase the public risk. The business team requests an evaluation of the consequences of chlorine releases associated with the east site. However, ABC's process hazard group informs them that too little is known about chemical inventories and process conditions at this design stage to perform such an assessment. Based on Ms. Deal's review of the chlorine dispersion studies done for the Anywhere Plant's Risk Management Plan (which she received after the Preliminary Hazard Analysis meeting), she tells the business team that significant off-site effects are possible if the VCM plant is located at the east site. Therefore, the business team recommends the west site.

14.5 Conclusions and Observations

The Preliminary Hazard Analysis team developed several suggestions for addressing the hazards identified at the proposed VCM plant. Some of the more important suggestions were as follows:

- Consider moving the VCM plant to the west side of Plant Road, depending on plant-specific dispersion analysis results.
- Locate the EDC, VCM, and HCl storage tanks away from Echo River (the team was concerned about flooding and releases into the waterway).
- Minimize the inventory of EDC, VCM, and HCl stored on-site. This will require some equipment changes to closely couple the unit operations.
- Modify plant emergency plans to address flammable releases (plant traffic and existing equipment could ignite such releases).
- Position the chlorine rail car spur so that a VCM polymerization incident will not precipitate a chlorine rail car rupture.
- Verify that the Anywhere Plant firewater system has sufficient capacity for the proposed VCM plant and is protected against explosions.

Several observations can be made concerning this Preliminary Hazard Analysis. They include the following:

- No additional hazards (toxic materials, flammable materials, etc.) were identified. However, the proposed VCM plant might have created new hazards if it had been sited at another ABC facility, due to interactions between the proposed and existing plants.
- Many of the causes or consequences of hazardous situations are related to equipment location. Consequently, many of the suggestions made involved moving equipment.
- Several new, specific hazardous situations (e.g., flooding, rail car incidents) were identified that the R&D phase What-If Analysis did not find. The previous What-If Analysis could not identify these situations since information needed to find them (such as siting/plant layout) did not exist at the time the What-If Analysis was performed.
- Experienced Preliminary Hazard Analysis team members were key to the success of the review. In particular, Mr. Scott's recollection that large quantities of toxic materials were stored west of Plant Road (i.e., chlorine storage) may save ABC from doing unnecessary siting work.

The time required to perform the Preliminary Hazard Analysis was not substantial, principally because the project is still in an early design phase. Table 14.5 summarizes the effort required for this Preliminary Hazard Analysis.

In summary, the Preliminary Hazard Analysis team identified many causes of hazardous situations that the VCM plant designers should consider when they lay out the new facility. Addressing these hazards now will result in a safer plant and, they hope, will help ABC avoid costly modifications in the future.

Table 14.5 Preliminary Hazard Analysis staff requirements for the conceptual design phase

Personnel	Preparation (h)	Evaluation (h)	Documentation (h)
Leader	8	8	12
Team member[a]	1	8	2

[a]Average per team member

15

Pilot Plant Operation Phase
An Illustration of the HAZOP Study Method

15.1 Problem Definition

Background

The VCM venture project is now one year into its development. Engineering has designed a pilot plant for the VCM process. This small plant (Figure 15.1; see Appendix C for the abbreviations and symbols used) will manufacture EDC and VCM. ABC plans to run the pilot plant for only a few days at a time. During each campaign, plant engineers will vary process parameters (e.g., material feed rates and furnace temperatures) under controlled conditions and sample the EDC and VCM produced to optimize process performance and to detect possible production problems. The VCM, by-products and waste streams will be immediately destroyed in a state-permitted incinerator.

ABC will build the pilot plant in Anywhere, U.S.A. This site was chosen because the world-scale VCM production plant will be built there and ABC wants to begin acclimating the community and the plant staff to this new project. Also, even though the pilot plant will be run mainly by engineers and chemists from ABC's engineering development center, Anywhere Plant operators will assist in pilot plant operations. Thus, the pilot plant will provide some training for potential operators of the world-scale VCM plant.

ABC Chemicals and the VCM business team have required that a hazard evaluation of the pilot plant be performed before start-up. Past hazard evaluations have yielded very useful information for assuring and improving the safety of the VCM process. The business team believes another such evaluation will be just as fruitful. Also, the business team wants to be sure that all prudent measures are taken to minimize the likelihood of an incident in the pilot plant phases. A significant loss event at this phase could derail the project timeline due to negative feedback from the community, so it is imperative that ABC demonstrate to the community and the site employees and contractors that the process can be run safely.

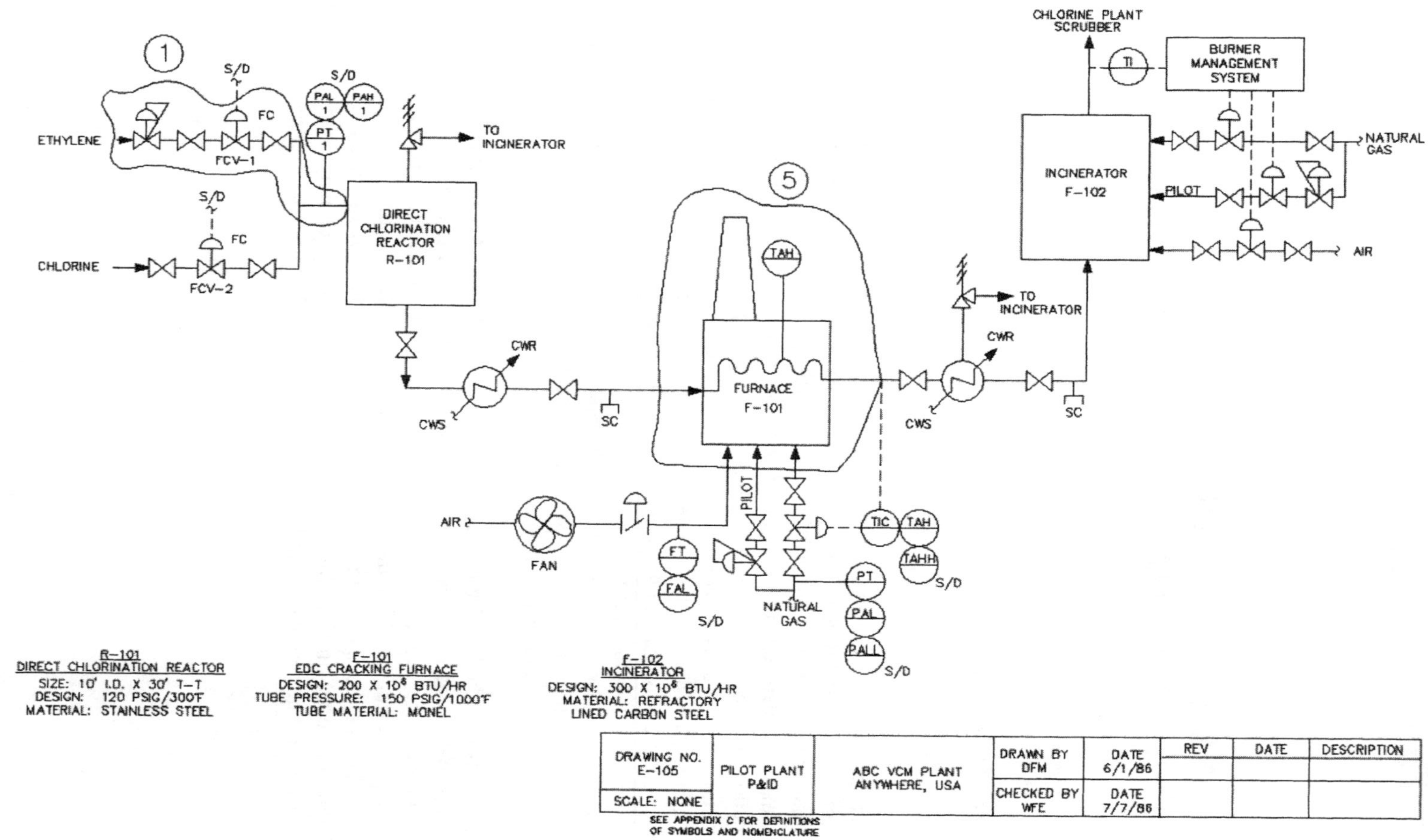

Figure 15.1 VCM pilot plant P&ID

Available Resources

Over the past year, ABC has gained a lot of knowledge and laboratory experience with VCM. Several ABC engineers and chemists have worked on the project and have become very familiar with many of the hazards associated with VCM production. Moreover, ABC has accumulated numerous resources describing the VCM process. The following information is available for the pilot plant hazard evaluation:

- Pilot plant process flow diagrams (PFDs) including operating parameters and limits
- Pilot plant piping and instrumentation diagrams (P&IDs)
- Pilot plant mechanical integrity data such as materials of construction and relief system design
- Laboratory development reports and experience
- Plot plan and equipment layout drawings
- Previous What-If Analysis and Preliminary Hazard Analysis reports (and information used in these studies—MSDSs, lab reports, literature on VCM, etc.)
- Available operating procedures (start-up procedures for the furnace and incinerator)
- Major component design specifications.

All of this information will be useful for the hazard evaluation. Of these items, however, the operating parameters and limits and a set of accurate P&IDs will be most important to the review effort.

Selection of Hazard Evaluation Technique

Mr. Chemist of the business team is assigned responsibility for seeing that a hazard evaluation of the pilot plant is completed. He asks Ms. Deal of ABC's process hazards analysis group to lead this effort, and then asks her to select an appropriate hazard evaluation method.

The pilot plant is a well-defined system with sufficient detail available to allow the option of using any of several hazard evaluation methods, since it is intended to represent the future VCM plant on a much smaller scale but using similar pressures, temperatures, flows, etc. that are now known as part of the pilot plant design.

The entire pilot plant is to be examined to identify process safety hazards. For this reason, Ms. Deal eliminates Fault Tree Analysis, Event Tree Analysis, Cause-Consequence Analysis, and Human Reliability Analysis from further consideration. While these methods could be used in this situation, Ms. Deal has found them more effective in examining one specific type of incident situation. The Relative Ranking method is not selected because the pilot plant is small and all the major equipment will be examined. The Preliminary Hazard Analysis method is not chosen because it was used in the last review. Ms. Deal also feels that Preliminary Hazard Analysis is better suited for examining broad issues such as plant siting. The Checklist Analysis method is not selected because ABC does not have a comprehensive checklist for such a plant or materials (although a checklist could have been developed for part of the facility using applicable codes and standards).

Thus, Ms. Deal narrows her choice to the What-If Analysis, FMEA, and HAZOP Study techniques. Any of these hazard evaluation methods may be adequate for this particular study. Ms. Deal has the most experience with the HAZOP Study method, so she chooses to do a HAZOP Study.

Study Preparation

As a first step, Ms. Deal selects a team of people to participate in the HAZOP Study. She determines that the following skilled people are needed for the HAZOP Study to succeed:

Leader	—	A person experienced in leading a HAZOP Study. Ms. Deal will be the leader.
Scribe	—	A technically astute person who can record information quickly and accurately. Joe Associate, who has recorded two other Anywhere chlorine HAZOP Studies, will be the scribe. (*Ms. Deal insisted that a person with HAZOP Study experience fill this job.*)
Process designer	—	A person knowledgeable of the pilot plant design (and the basis for design specifications) and how it might respond to process transients. Mr. Jim Design, the chief design engineer of this pilot plant, will fill this position.
Chemist	—	A person who knows what chemical interactions might occur between feed chemicals, intermediate products, waste streams, and materials of construction in the VCM process. Mr. Chemist, who will be responsible for operating the pilot plant, will fill this position, and he will also serve as the liaison to the VCM business team.
Operator	—	An experienced person who can explain how operators will detect process upsets and how they will respond to them. ABC has no operators experienced with the VCM process; therefore, Ms. Opera, a 10-year veteran operator at the Anywhere chlorine plant, will fill this position. Note: The Anywhere Plant is a U.S. OSHA VPP STAR facility and Mr. Safety has many contacts at other VPP sites in the chemical industry around the country. Through this network, he may be able to locate a VCM facility that is also in OSHA's VPP program and request some talent from this facility to assist in this HAZOP. This would be done under the VPP Mentoring Program.
Instrumentation and controls engineer	—	A person familiar with the control and shutdown strategies for the plant is needed to help the HAZOP Study team understand how the pilot plant will respond to process deviations. Ms. Control from the corporate engineering office will fill this position.

Safety specialist	—	A person familiar with the Anywhere Plant safety and emergency procedures. Mr. Scott, who also participated in the Preliminary Hazard Analysis, will fill this position.
Environmental specialist	—	A person who understand the current environmental constraints under which the Anywhere site operates, as well as the potential impact the pilot plant operations may have on the environmental systems and permits. Mr. E. P. Avery will be available to fill this role on an as-needed basis.

Before the HAZOP meeting, Ms. Deal and Mr. Design divide the pilot plant into sections, also called HAZOP Study "nodes". They identify about a dozen sections on the P&IDs, highlight these sections with different colored markers, and provide each team member with a set of marked P&IDs. In defining sections, Ms. Deal and Mr. Design look for significant changes in process conditions, in stream composition, and in equipment function. They define the following study nodes for the pilot plant:

- Ethylene feed line
- Chlorine feed line
- Direct chlorination reactor
- EDC supply line to the furnace
- Furnace
- Air supply line to the furnace
- Pilot gas supply line to the furnace
- Main gas supply line to the furnace
- VCM supply line to the incinerator
- Incinerator
- Air supply line to the incinerator
- Pilot gas supply to the incinerator
- Main gas supply to the incinerator

Based on her experience, Ms. Deal estimates the HAZOP Study will require about four hours per section, or about six days for the review of the pilot plant.

Next, Ms. Deal sends an e-mail to all the HAZOP team members. The e-mail describes where and when the HAZOP Study of the pilot plant will occur and a brief explanation as to the purpose and objectives of this hazard evaluation. The transmittal also contains a PFD of the pilot plant. Since some of the team members have never participated in a HAZOP Study, Ms. Deal includes a brief description of the HAZOP Study method. Also listed are the plant sections to be analyzed and the preliminary design intention of each section. Finally, Ms. Deal reminds all team members to bring any pertinent information they have on the pilot plant to the meeting.

To further prepare for the meeting, Ms. Deal and Mr. Associate develop a blank HAZOP table containing the names of the process sections. Since she plans to use ABC's HAZOP Study software to document the review, this table is saved as a computer file. Such preparation makes the HAZOP meeting go more smoothly. However, as Ms. Deal knows, this sectioning should not automatically be viewed as complete. The leader should always be prepared to restructure process sections or add new ones as the HAZOP Study progresses based upon input from the team. Ms. Deal also reserves one of the conference rooms at the site, as well as a projector so that the study software can be displayed on a large screen so all participants can view the recording as the study progresses.

15.2 Analysis Description

The HAZOP Study begins Monday morning at 8:00 a.m. To open the meeting, Ms. Deal has everyone introduce themselves and mention their areas of expertise. She then outlines the schedule for the next three days, including breaks and lunch plans. She also informs the group that they will probably be very tired at the end of each day and encourages them to press on so that the review can be completed on schedule. (Normally, Ms. Deal would have only four- to five-hour meetings over an extended number of days; however, several team members are from out of town and cannot attend for more than three days. Also, Ms. Deal would usually have the team tour the plant as part of the review; however, the pilot plant has not yet been built.) Finally, Ms. Deal reiterates the objectives of the HAZOP Study—to identify any safety or operability issues with the pilot plant. She then goes over the ground rules for the review: (1) all team members will have equal say; (2) everyone should focus on potential problems, not solutions; and (3) any process deviation is fair game to explore.

To begin the HAZOP Study, Ms. Deal spends about 30 minutes reviewing the HAZOP Study technique with the group. She informs the team that a deviation-by-deviation analysis is the manner in which she will facilitate this hazard evaluation, and describes that particular approach. She also points out that she will record team recommendations on a flip chart pad (in addition to the scribe's entering the recommendations into the HAZOP Study software) to help ensure all recommendations are accurately reported. She then asks Mr. Design to review the layout, process flow, and operation of the pilot plant equipment. After the team asks some brief questions, they begin to analyze the first section: the ethylene feed line to the direct chlorination reactor. Ms. Deal and Mr. Design sectioned the pilot plant in the order that process materials flow through the plant. The following is an excerpt of the discussions:

Ms. Deal — The ethylene feed line is designed to supply ethylene vapor to the DC (*direct chlorination*) reactor at 100 psig and ambient temperature. (*R&D determined that the liquid ethylene/chlorine reaction was too hard to control and that the yields with larger-scale experiments were much lower than anticipated. Also, previous hazard review teams suggested using chlorine and ethylene gases because of the added safety benefit from lower material masses. Thus, they chose to use a gas phase reaction.)* The supply rate is regulated by a flow control valve (*FCV-1*) and is 1100 scfm. Let's use the HAZOP Study guide words and this intention to generate a set of deviations to examine. The guide words are No, Less, More, Part Of, As Well As, Reverse, and Other Than. What are the resulting deviations?

Mr. Design — Applying these words to your design intent gives No Flow, Low (*less*) Flow, High (*more*) Flow, Low (*less*) and High (*more*) Pressure, and Low (*less*) and High (*more*) Temperature.

Ms. Deal — Okay. How about the guide words Reverse, Other Than, As Well As, and Part Of?

Ms. Opera — Well, there's Reverse Flow. Is Other Than Ethylene something we should consider?

Ms. Deal — Sure, good deviation! Remember our ground rule—anything's fair game right now. Any more deviations? [*Long pause.*] How about Reverse Pressure; that is, vacuum?

Mr. Design — As long as you have ethylene present that won't happen. Ethylene's vapor pressure is too high to allow a vacuum at ambient conditions. However, I guess it is a deviation.

Ms. Deal — Any As Well As or Part Of deviations? [*Long pause.*]

Ms. Control — Ethylene As Well As Contaminants would be a deviation. Although I don't know what kind of contaminants you find in ethylene.

Ms. Deal — We'll want to define the contaminants when we get to this deviation. Any other deviations anyone wants to suggest? [*None suggested.*] Okay, let's start with the first deviation. What's the consequence if the plant is running normally and something happens so there is now No Ethylene Flow?

Mr. Chemist — The process is designed to consume all the chlorine in the direct chlorination reactor. If the ethylene is cut off, then pure chlorine will flow through the furnace and incinerator to the plant's scrubber.

Ms. Deal — So what's the potential effect?

Mr. Chemist — Well, the scrubber will handle this flow for a while. Eventually you'll deplete the caustic and have a chlorine breakthrough. That large of a chlorine flow may also damage the furnace tubes. I really don't know the metallurgy well enough to say.

Ms. Deal — Okay, so our consequences are a potential chlorine release and possible furnace damage. If a furnace tube ruptures, then we have a chlorine release in the area. Correct? [*Nods of approval.*] What are the causes of No Ethylene Flow?

Ms. Control — Looking at the P&ID, I'd say the pressure control valve failing closed or the flow control valve (*FCV-1*) failing closed would do the trick.

Mr. Design — A loss of ethylene supply would cause the problem also.

Ms. Deal — Any others? [*None suggested.*] Safeguards?

Ms. Opera — The P&ID shows a pressure sensor on the feed line to the reactor, with a low pressure alarm (*PAL-1*) and a protective system to shut off chlorine to the reactor on low ethylene supply pressure. Also, it's not shown here, but we have an alarm on the chlorine plant scrubber to detect a breakthrough.

Ms. Deal — Joe, will you make a note in the actions for someone to verify the alarm exists and is functional?

Mr. Associate — Got it. Wasn't there also a recommendation to check on the furnace tube metallurgy with pure chlorine present?

Ms. Deal — Yes. That's right. Good catch. Are there other safeguards?

Mr. Design — We plan to take samples every 30 minutes during a run. We'd see the chlorine in these samples.

Mr. Chemist — Would the operator collecting a sample be at risk if it were pure chlorine?

Mr. Design — Shouldn't be. They'll be required to wear appropriate safety gear, including a respirator, whenever samples are collected.

Mr. Scott — Wait a minute. The respirators we use around the plant probably aren't much good for such a high chlorine concentration. We may want to recommend using fresh air when samples are taken. We should also train the operators to detect and respond to high chlorine content in a sample.

Ms. Deal — Any more safeguards or action items?

Mr. Design — I've been staring at this low pressure alarm (*PAL-1*). I'm not so sure it would trip on No Ethylene Flow. Looks like the chlorine pressure would maintain an adequate pressure to avoid an alarm. Anyway, I suggest someone look at this and verify that the alarm setpoints are adequate.

Ms. Deal — Ms. Control, what do you think? [*Ms. Control has been very quiet up until now.*]

Ms. Control — I agree, we should examine the setpoints. Jim may be right.

Ms. Deal — Any more recommendations?

Mr. Avery — Maybe we should also consider a chlorine detector on the furnace stack or a gas chromatograph on the process line to detect high chlorine.

Ms. Deal — These are good ideas. Instead of designing a specific solution, let's just note that we should examine ways to detect high chlorine in the lines downstream of the incinerator. Other recommendations? [*None suggested.*] Okay, that takes care of the No Ethylene Flow deviation. What is another deviation for this feed line?

Mr. Design — Wait, I've got another recommendation. We should examine if this high chlorine flow will damage the scrubber.

Mr. Associate — Wait for me to get that down. Your recommendation is to examine the impact of high chlorine flow on the scrubber. Are you concerned about a chlorine breakthrough and scrubber damage due to the high chlorine/caustic reaction?

Mr. Design — Both, but principally the latter problem.

Ms. Deal — Okay, what's another deviation?

Ms. Opera — Well, according to the list of deviations we initially generated for this line, Low Flow of ethylene.

Ms. Deal — What's the consequence of low ethylene flow?

Mr. Design — Pretty much the same as No Ethylene Flow — a potential chlorine release through the scrubber and possible damage to the furnace tubes.

Ms. Deal — Causes?

Ms. Opera — The ethylene pressure regulator failing partially closed, the flow control valve failing partially closed, and low supplier pressure. By the way, a loss of plant power should also cause no ethylene flow because the valve fails closed if power is lost.

Ms. Deal — Safeguards?

Mr. Design — The same as before. [*Mr. Design repeats the safeguards. Others agree.*]

Ms. Deal — Any new recommendations, besides the ones we already have? [*None suggested.*] Next deviation is High Flow. Consequences?

Mr. Chemist — The direct chlorination reactor runs ethylene-rich. A high ethylene flow means we waste more ethylene, but results in no real damage that I can see.

Ms. Deal — Can the incinerator handle this high flow?

Mr. Design — I checked the design specifications for the incinerator. It should be okay. The incinerator can handle the maximum possible ethylene flow rate.

Ms. Deal — Let's move on to Reverse Flow. Any consequences?

Mr. Design — Sure, we contaminate the supplier pipeline. That's going to be big bucks. Plus who knows where the chlorine and EDC backed into this line might go. There certainly could be a safety issue. For example, another user downstream might receive our chlorine and suffer an explosion. Or, the chlorine might mix with water in someone else's process, corrode their piping, and cause a release. We'll have to look into this matter in more detail outside of this meeting.

Ms. Deal — Causes?

Ms. Opera — Low ethylene supply pressure. Also high pressure in the reactor.

Ms. Deal — Others? [*None suggested.*] Safeguards?

Mr. Design — Again we have our questionable pressure alarm and shutoff (*PAL/H-1*) on the feed line. We also have a very high supply pressure of ethylene, which should prevent reverse flow. In fact, the reactor relief valve should lift well before the backpressure in the reactor would back materials into the ethylene feed line. In addition, low ethylene supply pressure, and subsequently low ethylene flow, would upset the unit. Operators should easily detect this and shut off the ethylene supply before a flow reversal occurs.

Ms. Control — The pressure alarm does not indicate reverse flow and shouldn't be a safeguard. However, the operator intervention sounds okay.

Ms. Deal — I agree. Any other safeguards? [*Pause.*] Any action items? [*Pause.*] How about it, Ms. Control? Any suggestions?

Ms. Control — We might want to put a check valve on the ethylene supply line.

Ms. Deal — Okay.

Mr. Associate — Is that a recommendation?

Ms. Deal — Yes. Any more recommendations?

This discussion continues on through the day. Based on this discussion, Ms. Deal has decided to examine all High deviations first, then all the Low deviations, and so on. She found that the team did a more thorough and consistent job of finding causes when they stayed on one deviation type for an extended time. The following is a brief excerpt from the discussions on the second day.

Ms. Deal — Let's move on to section five — the furnace. Mr. Design, would you give us a quick overview of its operation?

Mr. Design — The furnace is designed to heat the EDC to about 900 °F and crack it to form VCM. The hotter the furnace, the more complete the cracking, and the more by-products produced. We'll vary the temperature some to see what works best. Anyway, the furnace is fired with natural gas regulated by a TIC valve on the product discharge. A high-high discharge temperature will trip the TIC valve closed, as well as the feed valves to the direct chlorination reactor. A low-low natural gas pressure or a low air flow will also cause the plant to trip in this way. We also put a thermocouple on the furnace tube with a high skin temperature alarm, but no shutdown.

Ms. Deal — The first deviation we will examine is High EDC Flow. What are some causes?

Ms. Opera — Well, as we said earlier, high ethylene or chlorine feeds will increase flow through the system. The chlorine or ethylene flow control valve failing open is a cause.

Ms. Deal — Let's just note this as high pressure in the DC reactor since that's what we discussed before (*i.e., high pressure in the last upstream component causes high EDC flow to the furnace*). Do you have that down Joe? [*Nods yes.*]

Any more causes? [*Silence.*] Okay, what are the consequences?

Mr. Chemist — Like we said yesterday, we would have high EDC flow through the furnace, low cracking if the TIC does not keep up, and high carryover of EDC to the incinerator — potentially an EDC release. Also, the process material will cool down if the TIC can't keep up.

Ms. Deal — Safeguards? [*No comment.*] Actions?

Mr. Safety — I really think we had better see if the incinerator can handle a high EDC flow. Maybe we should also have an incinerator stack monitor for EDC.

Mr. Avery — A stack monitor is not environmentally required in this state since we are running for such short time periods.

Ms. Control — The TIC alarms will tell us of a problem. Besides, the incinerator may handle high EDC flows anyway.

Ms. Deal — Time out, folks. We are designing solutions. Let's leave it to engineering to see if this is a problem and let them solve it. Other actions? Next deviation is High Natural Gas Flow.

Ms. Opera — The natural gas fuel valve tripping open will cause high gas flow. So will the TIC/thermocouple outputting a false signal.

Mr. Scott — High natural gas supply pressure will also raise the process temperature.

Mr. Design — That can't happen. The TIC will reduce the gas flow as the process gets hotter. Also, the tube TAH will alarm if the gas flow control valve is open too far.

Ms. Deal — Wait a minute. Let's stick to causes. Those are safeguards that, for the moment, we'll assume don't work. Now, any more causes? [*none*] What's the consequence of the fuel valve failing open?

Mr. Design — If we don't catch it soon enough, the tubes melt and the furnace catches on fire. Even if we don't melt a tube, we may plug a tube with by-products and blow a relief valve somewhere. I guess it's also possible that the flames lift off the burners and extinguish. That could lead to an explosion.

Ms. Deal — If the TIC/thermocouple failed, wouldn't this failure increase the fuel flow while disabling the TAH and the shutdown?

Ms. Control — Yes, the TIC drifting low would do that. We should consider installing an independent high-high temperature alarm and shutdown on the discharge from the furnace.

Ms. Deal — Got that recommendation Joe? [*Nods yes.*] What are the existing safeguards?

Mr. Design — Right now we have a TAH and a TAHH on the furnace discharge as well as a tube skin TAH. Also, the tubes are designed to withstand very high temperatures. As long as there is flow, it will take a long time to melt a tube.

Ms. Deal — Joe, let's record these safeguards under high temperature since they directly apply to that deviation and cross-reference back to high gas flow.

Mr. Scott — We also have a fire monitor in the area and a well-trained emergency response team. Plus, you can shut off the chlorine and ethylene feeds from the control room.

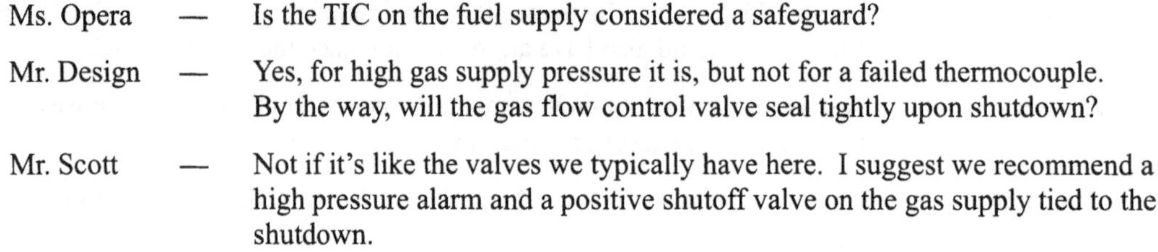

Ms. Opera — Is the TIC on the fuel supply considered a safeguard?

Mr. Design — Yes, for high gas supply pressure it is, but not for a failed thermocouple. By the way, will the gas flow control valve seal tightly upon shutdown?

Mr. Scott — Not if it's like the valves we typically have here. I suggest we recommend a high pressure alarm and a positive shutoff valve on the gas supply tied to the shutdown.

The HAZOP meeting continues in this fashion until the team has examined all the deviations for each process parameter that they can imagine for the furnace. Ms. Deal then instructs the team to reexamine the furnace for the start-up mode of operation. Before beginning, she asks Mr. Design to review the start-up procedure for the furnace (Table 15.1). The team then begins postulating deviations during start-up and asks the usual HAZOP questions: what are the causes, consequences, and safeguards for this deviation, and are the safeguards considered adequate? The following is a brief excerpt from this portion of the HAZOP Study.

Ms. Deal — Let's begin with deviations from Step 1. Suppose there is More (*too long*) Purge. What are the consequences?

Mr. Scott — I don't see a problem with too long a purge, other than wasting time. [*Others agree.*]

Ms. Deal — What about Less (*too short*) or No Purge?

Mr. Scott — That could be a problem. If gases have accumulated in the furnace, then you could have an explosion. Again, I think we need a positive shutoff valve to help prevent fuel leaking into the furnace.

Ms. Deal — Our scribe has that as a recommendation noted and we can reiterate it here. Let's get back to causes.

Mr. Design — The fan could be dead. Or the air damper could fail closed. Or power to the fan could be lost.

Table 15.1 Furnace start-up procedure

Step	Action to take
1	Start furnace fan and purge firebox for 10 minutes.
2	Verify furnace pilot is lit (visual check).
3	Verify incinerator is operating.
4	Begin ethylene flow through the furnace to the incinerator.
5	Start natural gas flow to the furnace burners (slowly move TIC to setpoint).
6	Verify main burners are lit (process temperature is rising).

Mr. Scott — The operator could just inadvertently skip this step.

Ms. Deal — Any more causes? [*Silence.*] What are the safeguards?

Mr. Scott — We have a low flow alarm on the air line and mechanical stops to allow minimum air flow. Also, a loss of power will trip all valves to their safe positions.

Mr. Design — If it were my unit, I'd want a lot more safeguards like tying gas flow to air flow, having a positive shutoff valve on the gas line, and continuously sweeping the furnace with air.

Ms. Deal — Those are good ideas. Why don't we suggest to engineering that they design a verifiable purge system for the furnace and give them your ideas to consider. Any other actions? [*Quiet.*] What about Other Than Step 1?

Ms. Control — Looks like the same problem as No Purge. [*Others agree.*]

Ms. Deal — Is the deviation Part Of Step 1 the same as less purge? [*Team nods yes.*] We'll cover the Reverse deviation when we get to Step 2. How about the deviation Step 1 As Well As Another Step? Any consequences?

Ms. Control — Doing Step 1 As Well As Step 2, 3, or 4 looks okay. Doing Step 1 As Well As Step 5 looks like it would have the same consequences as the no purge deviation.

Ms. Deal — What are the causes?

Ms. Control — Operator error.

Ms. Deal — Others? [*None suggested.*] Safeguards? [*None suggested.*] Actions?

Ms. Control — Maybe we should interlock the natural gas flow so that it cannot be started until a 10-minute purge is complete.

Ms. Deal — Okay. Got that Joe? [*Nods yes.*] The next step is to verify that the pilot is lit. What are consequences of More Verification or the pilot being lit too long?

Ms. Control — We plan to always have the pilot lit, so this won't be a problem. And extra verification of the pilot being lit is good.

Ms. Deal — Okay, what's another deviation? What about No or Less Verification?

Mr. Scott — If the pilot's lit anyway, then no problem. If it's out, then you may have an explosion once an ignition source is found.

Ms. Deal — Causes?

Ms. Opera — Low gas pressure. Pilot PCV fails closed. Operator inadvertently blocks in line. High gas pressure blows out the pilot.

Ms. Deal	—	Any of those factors along with the operator forgetting to check the pilot causes a major problem—a potential explosion. Safeguards?
Mr. Design	—	Just a high and low pressure alarm and shutdown on the gas supply line.
Mr. Scott	—	I suggest we recommend a flame scanner with shutdowns.
Mr. Associate	—	I've got that down.
Ms. Deal	—	Any other actions?
Ms. Control	—	We might want to consider a high and low pressure alarm on the pilot line.
Ms. Deal	—	Good recommendation. Others? [*none*] Let's examine the deviation Reversing Step 2. That is, doing it out of order.
Ms. Opera	—	Depends on how much out of order. If the pilot is checked first, then the air purge may blow it out and we have the same gas buildup problem as before. If the operator checks too late in the sequence and tries to relight the pilot, assuming it is out, then he or she may get seriously injured, if not killed.
Ms. Deal	—	Safeguards?
Mr. Scott	—	I guess we should stress the importance of this procedure in our operator training. We also need to train operators to always purge before even lighting a pilot. In fact, we probably should develop a start-up checklist for the furnace.

The HAZOP Study continues until all deviations for the pilot plant are examined. Both routine and emergency shutdown procedures for the incinerator are also examined. At the end of each day, Ms. Deal reviews the recommendations she recorded on the chart pad to make sure they are accurately recorded and a recommendation is not missed. At the end of the meeting, she thanks the team members for their active participation and asks them to review the HAZOP report she will send them in the next few weeks.

15.3 Discussion of Results

Tables 15.2 and 15.3 list examples of the HAZOP Study results. Table 15.2 lists the deviations examined for the furnace during normal operation, along with the causes, consequences, and safeguards identified by the team. The table is printed in the deviation-by-deviation format as described in Section 5.3. That is, all the causes, consequences, safeguards, and actions for a particular item deviation are compressed; there is not necessarily a one-to-one correspondence between a particular cause, consequence, safeguard, and action. Actions suggested by the team are referred to in Table 15.2 and explicitly defined in Table 15.3. As with the previous hazard evaluation methods used, these tables give a brief, direct synopsis of the discussions in the hazard evaluation.

Table 15.2 Sample HAZOP Study results for the VCM pilot plant (deviation-by-deviation approach)

DRAWING: VCM Pilot Plant, Revision 0 MEETING DATE: ##/##/#### PAGE: 56 of 124			TEAM: Ms. Deal (Leader – ABC Process Hazard Analysis), Mr. Associate (ABC Anywhere Plant), Mr. Design (ABC Engineering), Mr. Chemist (ABC Research and Development), Ms. Opera (ABC Anywhere Plant), Ms. Control (ABC Engineering), Mr. Scott (ABC Anywhere Plant), Mr. Avery (ABC Environmental)		
Item	**Deviation**	**Causes**[a]	**Consequences**[b]	**Safeguards**	**Actions**
5 Furnace – VCM Furnace (normal operation; raise EDC to 900 °F, 160 psig to make VCM; flow 1200 lb/h)					
5.1	High flow – EDC	High pressure in DC reactor (Item 3.5)	Low conversion of EDC to VCM. High carry-over of EDC to incinerator with potential EDC release to the environment Low process temperature in the VCM furnace (Item 5.8)		1
5.2	High flow – natural gas	Gas FCV fails open Natural gas supply pressure high TIC fails low signal	Possible loss of flame and potential explosion High process temperature in the VCM furnace (Item 5.7) Potential tube damage and possible fire if tube ruptures. Excess by-products in product stream (Item 5.12)	Natural gas supplier has been very reliable over past 15 years TIC controls gas supply	2, 3, 4
5.3	High flow – air	Air damper fails full open Furnace skin leaks	Poor combustion in furnace. Low conversion of EDC to VCM. High carry-over of EDC to incinerator with potential EDC release to the environment Low process temperature in the VCM furnace (Item 5.8)	TIC controls gas supply Fixed-speed fan	1
5.4	Low flow – EDC	Low pressure in the DC reactor (Item 3.6) EDC sample connection (upstream) left open Fouling of EDC cooler	High process temperature in the VCM furnace (Item 5.7) High production of by-products during EDC cracking. Potential furnace tube damage and possible fire if tube ruptures. EDC release to the environment (Item 5.12)	Operator present when EDC sample connection used	5
5.5	Low flow – natural gas	Gas FCV fails closed Natural gas supply pressure low TIC fails high signal	Low conversion of EDC to VCM. High carry-over of EDC to incinerator with potential to release EDC to the environment Low process temperature in the VCM furnace (Item 5.8)	Natural gas supplier has been very reliable over past 15 years Natural gas supply line PAL TIC controls gas supply	1
5.6	Low flow – air	Air damper fails closed Air filter plugged Air blower fails off	Low process temperature in the VCM furnace (Item 5.8) Poor combustion in furnace. Low conversion of EDC to VCM. Possible to extinguish furnace flame with potential for a fire or explosion (buildup of gas in furnace). High carry-over of EDC to incinerator with potential to release EDC to the environment	Minimum stop on air damper TIC controls gas supply Air FAL with hardwired shutdown	1

Table 15.2 Sample HAZOP Study results for the VCM pilot plant (deviation-by-deviation approach)

DRAWING: VCM Pilot Plant, Revision 0 MEETING DATE: ##/##/#### PAGE: 56 of 124			TEAM: Ms. Deal (Leader – ABC Process Hazard Analysis), Mr. Associate (ABC Anywhere Plant), Mr. Design (ABC Engineering), Mr. Chemist (ABC Research and Development), Ms. Opera (ABC Anywhere Plant), Ms. Control (ABC Engineering), Mr. Scott (ABC Anywhere Plant), Mr. Avery (ABC Environmental)		
Item	**Deviation**	**Causes**[a]	**Consequences**[b]	**Safeguards**	**Actions**
5.7	High temperature	High natural gas flow (Item 5.2) Low EDC flow (Item 5.4)	High pressure in the VCM furnace tubes (Item 5.9) High production of by-products during EDC cracking. Potential furnace tube damage and possible fire if tube ruptures. Furnace damage (Item 5.12)	Furnace tube skin TAH Product discharge TAH and TAHH with hardwired shutdown of plant Furnace tubes designed to withstand very high temperatures	3
5.8	Low temperature	High EDC flow (Item 5.1) High air flow (Item 5.3) Low natural gas flow (Item 5.5) Low air flow (Item 5.6)	Low conversion of EDC to VCM. High carry-over of EDC to incinerator with potential to release EDC to the environment		1
5.9	High pressure		No significant safety concerns identified. See high temperature (Item 5.7)		
5.10	Low pressure		No significant safety concerns identified		
5.11	Contaminant		No significant safety concerns identified		
5.12	Tube leak/rupture	Fouling Corrosion Bad weld High flow – natural gas (Item 5.2) Low flow – EDC (Item 5.4) High temperature (Item 5.7)	Furnace fire with subsequent release of EDC to the environment Potential major equipment damage	Fire monitors in furnace area Emergency response team trained for fire fighting Ethylene and chlorine supplies can be shut off remotely Tubes inspected and welds x-rayed prior to service Furnace will only be run for short periods (few days) over the next year. Fouling should be minor Tube material of construction adequate for EDC, chlorine, and ethylene service	6
5.13	Furnace wall leak/rupture		No significant safety concerns identified		

[a]All causes do not necessarily cause all consequences to occur.
[b]All causes or consequences are not necessarily prevented or mitigated by all safeguards listed.

Table 15.3 Sample action items from the VCM pilot plant HAZOP Study

List number	Action to be considered	Responsibility	Status
1	Determine if incinerator has the capacity to destroy EDC when EDC/VCM conversion is low. Consider installing an EDC stack monitor. (Items 5.1, 5.3, 5.5, 5.6, 5.8)		
2	Consider a PAH and high pressure shutdown for natural gas using a positive isolation valve (Item 5.2)		
3	Consider installing an independent TAHH and high furnace discharge temperature shutdown (Items 5.2, 5.7)		
4	Consider installing a flame scanner and loss of flame shutdown (Item 5.2)		
5	Consider installing a PAL on the direct chlorination reactor (Item 5.4)		
6	Verify adequate quality assurance program for all furnace tubes (Item 5.12)		

The following are some of the more important findings of the HAZOP Study:

- A hazard evaluation should be performed on the start-up procedures for all the units in the pilot plant (the team found a potentially fatal incident situation with the furnace start-up procedures).
- The capability of the incinerator to destroy EDC at high feed rates should be verified.
- The furnace controls and shutdowns should be more highly automated.
- A separate scrubber should be considered for the pilot plant (pilot plant upsets may trip the chlorine plant scrubber and consequently the chlorine plant).
- Procedures for disposing of pilot plant samples need to be developed (an environmental concern).
- Operators taking pilot plant samples should wear personal protective equipment appropriate to the hazards of the material being sampled and possible deviations from standard procedure.

The report describing the HAZOP Study is prepared by Ms. Deal. This report lists the team members, their job titles, and which meetings they attended; the drawings, procedures, etc., used in the meeting; a summary of the team's findings and recommendations; and the detailed HAZOP table. This report is sent to the team members for review and comment before it is transmitted to the business team.

15.4 Follow-up

During the HAZOP Study, several questions were raised that could not be immediately answered by the team. These questions often revolved around the structural strength of a component (e.g., could the vessel withstand full vacuum), the design basis for relief valve sizing, and instrumentation sensing ranges. Whenever possible, a HAZOP team member contacted the appropriate individual during a meeting break to resolve the question. Ms. Deal attempted to obtain answers to questions that were left unanswered during the meeting before issuing the HAZOP report to the business team. The unanswered questions that remained became findings for follow-up, listed in Ms. Deal's report.

The VCM business team reviewed all the findings and recommendations made by the HAZOP team, as well as the basis for each. Most were accepted and became the basis for follow-up action items. The business team prioritized these actions into two categories: those that must be implemented before pilot plant start-up, and those that should be implemented as quickly as is convenient. The business team documented their reasons for rejecting certain suggestions made by the HAZOP team. These reasons, along with the HAZOP report, were placed in the VCM project file.

Mr. Chemist accepted responsibility for following up on the accepted HAZOP actions. He has assigned appropriate ABC personnel the task of resolving these actions and reporting the resolutions. ABC has a computerized tracking system that Mr. Chemist uses to help him follow the status of implementation. He uses this system to prompt him to check monthly on the status of the action items. Mr. Chemist places a description of each resolution in the file with the HAZOP report as he receives them.

15.5 Conclusions and Observations

The HAZOP Study went very well, principally because Ms. Deal is an excellent leader and she had knowledgeable personnel on the team. She kept the team from talking about non-pertinent subjects and designing solutions. She also used compliments to tactfully cut "rambling" short while still encouraging participation.

The HAZOP team identified safety, operability, and environmental concerns. Had the scope of the HAZOP Study been more narrowly focused (e.g., had they examined safety issues only), the review could have been completed in less time (and at lower cost). However, ABC Chemicals thinks the operability and environmental issues identified at this early stage will save them more than the study cost. In addition, many of the operability issues identified had safety implications also. Thus, the HAZOP Study should lead to both a safer and smoother-running plant.

HAZOP Study software was used to document the review. This was not required, but Ms. Deal chose to use the software to expedite documenting the HAZOP meeting minutes. Ms. Deal summarized the comments of the HAZOP team on each deviation and Mr. Associate (the scribe) typed them into the computer. As the meeting progressed, Mr. Associate became proficient at summarizing the team's commentary and recording it on the computer, which helped Ms. Deal prepare the HAZOP report quickly and efficiently. Ms. Deal chose to review and edit these tables before the team's review. However, she would have asked the team to review each day's tables had the meetings been shorter and the time between meetings longer (e.g., a HAZOP meeting every other day). The time required to perform the HAZOP Study is summarized in Table 15.4.

Table 15.4 HAZOP Study staff requirements for the VCM pilot plant

Personnel	Preparation (h)	Evaluation (h)	Documentation (h)
Leader	32	40	20
Scribe	8	40	16
Team member[a]	4	40	2

[a]Average per team member

15.6 Cause-by-Cause Alternative Approach

The cause-by-cause HAZOP approach discussed in Section 5.3 is an alternative means of analyzing and documenting HAZOP deviations that can be explicitly scenario-based, thus making it more amenable to risk-based scenario analysis as described in Chapter 7. The analysis discussion of Section 15.2 would proceed somewhat differently for a cause-by-cause analysis, with the divergence indicated by the dotted line below when the causes begin to be discussed. At this point, the consequences are examined one cause at a time, and the safeguards are then associated with each unique initiating cause/loss event pair. A sampling of findings and recommendations is listed in Table 15.5 and a portion of the HAZOP Study documentation that might result using a cause-by-cause approach is presented in Table 15.6. Note that for the first initiating cause there are three different loss events, each with a different severity of consequences and with different safeguards for each initiating cause / loss event combination. An unscheduled emergency shutdown could be considered a fourth loss event, depending on whether it would have significant business or equipment damage impact and whether it was within the study scope.

Table 15.5 Sample action items from the VCM pilot plant HAZOP Study (cause-by-cause approach)

List number	Action to be considered	Responsibility	Status
1	Determine whether low ethylene pressure safeguards would be effective to protect against loss of ethylene flow to reactor, given possibility that chlorine pressure would prevent alarm and shutdown from being actuated. (Items 1.1, 1.2, 1.3)		
2	Check furnace tube metallurgy in the presence of pure chlorine vapor. If affected, determine whether effect is gradual enough that 2/h reactor sampling would warn against loss of ethylene flow before furnace tube damage would be realized. (Items 1.1, 1.2, 1.4, 1.5)		
3	Verify presence of alarm on the chlorine plant scrubber to detect a breakthrough; update P&ID to reflect. (Items 1.3, 1.6)		
4	Examine ways to detect high chlorine in the lines downstream of the incinerator. (Items 1.3, 1.6)		
5	Consider installing redundant pressure or flow sensor on ethylene feed, separate from BPCS. (Items 1.4, 1.5, 1.6)		

Table 15.6 Sample HAZOP Study results for the VCM pilot plant (cause-by-cause approach)

DRAWING: VCM Pilot Plant, Revision 0 MEETING DATE: ##/##/#### PAGE: 1 of 124			TEAM: Ms. Deal (Leader – ABC Process Hazard Analysis), Mr. Associate (ABC Anywhere Plant), Mr. Design (ABC Engineering), Mr. Chemist (ABC Research and Development), Ms. Opera (ABC Anywhere Plant), Ms. Control (ABC Engineering), Mr. Scott (ABC Anywhere Plant), Mr. Avery (ABC Environmental)		
Item	**Deviation**	**Initiating Cause**	**Consequences**	**Safeguards**	**Actions**
1 Ethylene Feed Line (supply 1100 scfm ethylene vapor to DC reactor at 100 psig, ambient temperature)					
1.1.1	No Flow – ethylene	FCV-1 fails closed or is commanded to close	Unreacted chlorine to furnace; possible damage to furnace tubes from chlorine contact	[] Alarm, shutdown on low pressure at PT-1 [] Detection of loss of ethylene flow by 2/h reactor sampling before furnace tubes damaged	1, 2
			Unreacted chlorine to furnace; possible failure of furnace tubes from chlorine contact damage; hot chlorine vapor release from furnace	[] Alarm, shutdown on low pressure at PT-1 [] Detection of loss of ethylene flow by 2/h reactor sampling before furnace tube(s) fail	1, 2
			Unreacted chlorine through furnace and incinerator to plant scrubber; eventual chlorine breakthrough; chlorine release from scrubber stack	[] Alarm, shutdown on low pressure at PT-1 [] 2/h reactor sampling [] Scrubber breakthrough alarm	1, 3, 4
1.1.2		PCV-1 fails closed or is commanded to close	Unreacted chlorine to furnace; possible damage to furnace tubes from chlorine contact	[] Detection of loss of ethylene flow by 2/h reactor sampling before furnace tubes damaged	2, 5
			Unreacted chlorine to furnace; possible failure of furnace tubes from chlorine contact damage; hot chlorine vapor release from furnace	[] Detection of loss of ethylene flow by 2/h reactor sampling before furnace tube(s) fail	2, 5
			Unreacted chlorine through furnace and incinerator to plant scrubber; eventual chlorine breakthrough; chlorine release from scrubber stack	[] 2/h reactor sampling [] Scrubber breakthrough alarm	3, 4, 5
1.1.3		Loss of ethylene supply			
1.1.x		Ethylene feed line or connection failure			
Etc.					

Ms. Deal — The ethylene feed line is designed to supply ethylene vapor to the DC (*direct chlorination*) reactor at 100 psig and ambient temperature.

(R&D determined that the liquid ethylene/chlorine reaction was too hard to control and that the yields with larger-scale experiments were much lower than anticipated. Also, previous hazard review teams suggested using chlorine and ethylene gases because of the added safety benefit from lower material masses. Thus, they chose to use a gas-phase reaction.)

The supply rate is regulated by a flow control valve (*FCV-1*) and is 1100 scfm. Let's use the HAZOP Study guide words and this intention to generate a set of deviations to examine. The guide words are No, Less, More, Part Of, As Well As, Reverse, and Other Than. What are the resulting deviations?

Mr. Design — Applying these words to your design intent gives No Flow, Low (*less*) Flow, High (*more*) Flow, Low (*less*) and High (*more*) Pressure, and Low (*less*) and High (*more*) Temperature.

Ms. Deal — Okay. How about the guide words Reverse, Other Than, As Well As, and Part Of?

Ms. Opera — Well, there's Reverse Flow. Is Other Than Ethylene something we should consider?

Ms. Deal — Sure, good deviation! Remember our ground rule—anything's fair game right now. Any more deviations? [*Long pause.*] How about Reverse Pressure, that is, vacuum?

Mr. Design — As long as you have ethylene present that won't happen. Ethylene's vapor pressure is too high to allow a vacuum at ambient conditions. However, I guess it is a deviation.

Ms. Deal — Any As Well As or Part Of deviations? [*Long pause.*]

Ms. Control — Ethylene As Well As Contaminants would be a deviation. Although I don't know what kind of contaminants you find in ethylene.

Ms. Deal — We'll want to define the contaminants when we get to this deviation. Any other deviations anyone wants to suggest? [*None suggested.*] Okay, let's start with the first deviation. What's the consequence if the plant is running normally and something happens so there is now No Ethylene Flow?

Mr. Chemist — The process is designed to consume all the chlorine in the direct chlorination reactor. If the ethylene is cut off, then pure chlorine will flow through the furnace and incinerator to the plant's scrubber.

Ms. Deal — So what are the potential consequences of this happening?

Mr. Chemist — Well, the scrubber will handle the flow of chlorine for a while. Eventually you'll deplete the caustic and have a chlorine breakthrough. That large of a chlorine flow may also damage the furnace tubes. I really don't know the metallurgy well enough to say.

Ms. Deal — Okay, so our consequences are a potential chlorine release and possible furnace damage. If a furnace tube ruptures, then we have a chlorine release in the area. Correct? [*Nods of approval.*] Ok, now what would be a cause of "No Flow" of Ethylene?

Ms. Control — Looking at the P&ID, I'd say the pressure control valve being closed or the flow control valve (*FCV-1*) being closed would do the trick.

Ms. Deal — Great job, but let's take them one at a time and discuss what kind of safeguards we have or may need to adequately protect us against each of these causes.

Ms. Control — Well, the most obvious cause of "no flow" would be that FCV-1 is closed not allowing ethylene to flow.

Ms. Deal — Good! What safeguards do we have to prevent FCV-1 from being closed when it should be open, or to protect against loss of flow?

Mr. Design — Right now, my engineering spec sheets are showing this is a fail-closed valve. It is also an automated valve so operators would not normally be manipulating this valve. That reduces the likelihood of it being inadvertently closed.

Ms. Opera — The P&ID shows a pressure sensor on the feed line to the reactor, with a low pressure alarm (*PAL-1*) and a shutdown of chlorine to the reactor on low ethylene supply pressure. Also, it's not shown here, but we have an alarm on the chlorine plant scrubber to detect a breakthrough.

Ms. Deal — Joe, will you make a note in the actions for someone to verify the alarm exists and is functional?

Mr. Associate — Got it. Wasn't there also a recommendation to check on the furnace tube metallurgy with pure chlorine present?

Ms. Deal — Yes. That's right. Good catch. Are there other safeguards?

Mr. Design — We plan to take samples every 30 minutes during a run. We'd see the chlorine in these samples.

Ms. Deal — Do we feel these safeguards are enough, considering the potential consequences? [*Nods of approval.*] Ok, what's next?

Mr. Chemist — The pressure control valve (PCV) could fail closed, causing us to have no flow of ethylene.

Mr. Design — Yes, but that valve is also automated, so we have the same preventive safeguard in that scenario.

Ms. Control — But the safeguards won't be the same, since the pressure control loop uses the same sensor as the low pressure alarm and shutdown. Good practice would be to have an independent pressure or flow sensor for the alarm and shutdown system, located such that it would be sure to detect loss of ethylene supply.

Ms. Deal — This is very good discussion. Let's make sure we capture that item, and that we associate the preventive safeguards with only the combinations of causes and consequences to which they pertain. [*Discussion continues, and Mr. Associate is given time to complete the No Flow scenarios with everyone looking on.*]

Ms. Deal — Does everyone see how a cause-by-cause analysis is done? [*Nods of approval.*] Anyone think of other causes of no flow? [*pause*]

Mr. Design — A loss of ethylene supply would cause the problem also.

Ms. Deal — OK, what safeguards do we have in place or need in place to prevent a loss of ethylene supply?

Ms. Opera — Operators monitor the supply pressure via the DCS, as well as check local gauges on their rounds three times each shift. We would generally get advance warning from the pipeline company, also. So losing flow because we lost the entire supply without warning seems pretty remote.

Ms. Deal — And preventive safeguards would be the same as if the pressure control valve closed. Everyone agree? [*Nods of approval.*] Anyone think of any other reasonable cause of no ethylene flow?

Ms. Control — How about if the line was to fail or be left open after maintenance work?

Ms. Deal — OK, what would be the consequences if we had a line opening, whether it is caused by a failure of the line or the line is left open after maintenance work?

Mr. Design — With that segment of the line we would have a release of either chlorine or ethylene, depending on what was happening in the process at that time. This would cause either a significant chemical hazard from chlorine or could cause an explosion from the ethylene.

Ms. Deal — OK, what safeguards do we have in place should we be faced with this situation or to prevent the line from being open?

Mr. Design — Our piping is designed to proper engineering specs to prevent any type of overpressuring of the pipe, piping is protected from vehicle traffic, and the piping gets inspected on a regular basis. As for the line being left open after maintenance, our line break SOP includes a fit-for-duty return to service checklist. Should we have an incident, the process areas are protected by sprinkler systems and fixed monitor nozzles, as well as the areas being equipped with LFL and chlorine sensors to detect flammable vapors or chlorine.

Ms. Deal — Sounds like we have a lot of preventive and mitigative safeguards. Do we think they are adequate for the potential consequences? [*Nods of approval.*] OK, anyone think of anything else that would cause no flow? [*Silence.*]

Ok then, moving along to the next deviation – Low Flow. Since the control point is 1100 scfm, "Low Flow" means the ethylene flow rate going below 1100 scfm.

The HAZOP meeting again continues in this fashion until the team has examined all the deviations for each process parameter that they can imagine for the furnace. The team would then similarly reexamine the furnace for the start-up mode of operation.

15.7 Extension of Cause-by-Cause Approach Using Scenario Risk Estimates

Table 15.7 illustrates how the cause-by-cause scenarios developed in the previous section might be analyzed by a hazard review team using the scenario risk analysis approach outlined in Chapter 7.

- Cause frequencies are given as annual frequency exponents, such that a **Freq** value of **-1** for each of the two initiating causes in Table 15.7 represents a cause frequency estimate of 10^{-1} events/year, or about once every ten years.

- Each of the three loss events (furnace tube damage requiring early replacement or repair; failure of furnace tubes with hot chlorine vapor release from furnace; chlorine release from scrubber stack) has a different **Impact** (severity of consequences). The first loss event is considered outside the scope of the study, and therefore is not further evaluated. The second loss event was assessed to have a more severe impact than the third, with impact magnitudes of **4** and **3**, respectively, using a scale such as in Table 7.2. Note that separate impacts could have been determined and documented for on-site (worker) health effects, off-site (public) effects, and environmental impacts by assigning an impact magnitude to each of these three categories for each scenario.

- An order-of-magnitude measure of the risk reduction afforded by each safeguard in preventing the loss event from ensuing, given the occurrence of the initiating event, is documented in square brackets. A **[1]** therefore represents a risk reduction factor of one order of magnitude, or ten-fold. If a safeguard does not provide protection independent of the initiating event or other safeguards already credited, then it is assigned zero effectiveness. The combined effectiveness of the safeguards is obtained by adding the numbers in square brackets, which is equivalent to multiplying the risk reduction factors. Thus, for the second scenario in this example, the scenario risk is reduced by three orders of magnitude by virtue of the two independent safeguards of the alarm and shutdown on PT-1 low pressure (one order of magnitude, or 90% effective) and the detection of loss of ethylene flow by twice-an-hour reactor sampling before furnace tube(s) fail (two orders of magnitude, or 99% effective).

- The overall scenario frequency magnitude **SFreq** is obtained by reducing the cause frequency magnitude **Freq** by the total orders of magnitude afforded by the safeguards taken together. Thus, for the second scenario in Table 15.7, **SFreq** = **-1** – (**[1]**+**[2]**) = **-4**. This represents a scenario frequency that is on the order of 10^{-4} per year, or one chance in 10,000 per year of operation that the particular loss event will occur as a result of the specific initiating cause of FCV-1 failing closed or being commanded to close in error.

- For each scenario, the scenario risk magnitude **SRisk** is the sum of **Impact** and **SFreq**. The scenario risk magnitude and its significance by comparison with company-established action levels can also be determined by use of a risk matrix such as Figure 7.3. The examples in Table 15.7 show three scenarios in an intermediate risk range (**SRisk** = -1 or 0) where risk reduction is generally warranted unless the risks can be shown to be as low as reasonably practicable, and one scenario in a high risk range (**SRisk** > 0) where prompt risk-reduction action would typically be required.

- In addition to pointing out which scenarios warrant further risk reduction, these scenario risk calculations can also be used as one input to prioritizing risk-reduction actions. Thus, actions required to reduce the scenario found to be in the high risk range would likely warrant higher priority than those in lower risk ranges.

The numerical factors in Table 15.7 are provided for illustrative purposes only, and are not intended to be considered normative for the particular initiating causes, consequences and safeguards used in the example.

Table 15.7 Sample HAZOP Study results for the VCM pilot plant, with scenario risk estimates using frequency and impact magnitudes

DRAWING: VCM Pilot Plant, Rev. 0
MEETING DATE: ##/##/####
PAGE: 1 of ###

TEAM: Ms. Deal (Leader – ABC Process Hazard Analysis), Mr. Associate (ABC Anywhere Plant), Mr. Design (ABC Engineering), Mr. Chemist (ABC Research and Development), Ms. Opera (ABC Anywhere Plant), Ms. Control (ABC Engineering), Mr. Scott (ABC Anywhere Plant), Mr. Avery (ABC Environmental)

Item	Deviation	Initiating Cause	Freq	Consequences	Impact	Safeguards	SFreq	SRisk	Actions
1 Ethylene Feed Line (supply 1100 scfm ethylene vapor to DC reactor at 100 psig, ambient temperature)									
1.1.1	No Flow – ethylene	FCV-1 fails closed or commanded to close	-1	Unreacted chlorine to furnace; possible damage to furnace tubes from chlorine contact; replace/repair	Not in scope	[1] Alarm, shutdown on PT-1 low pressure [1] Detection of loss of ethylene flow by 2/h reactor sampling before furnace tubes damaged	---	---	1, 2
				Unreacted chlorine to furnace; possible failure of furnace tubes from chlorine contact damage; hot chlorine vapor release from furnace	4	[1] Alarm, shutdown on PT-1 low pressure [2] Detection of loss of ethylene flow by 2/h reactor sampling before furnace tube(s) fail	-4	**0** (Med)	1, 2
				Unreacted chlorine through furnace and incinerator to plant scrubber; eventual chlorine breakthrough; chlorine release from scrubber stack	3	[1] Alarm, shutdown on PT-1 low pressure [1] Detection of loss of ethylene flow by 2/h reactor sampling before chlorine release [0] Scrubber breakthrough alarm	-4	**0** (Med)	1, 3, 4
1.1.2		PCV-1 fails closed or commanded to close	-1	Unreacted chlorine to furnace; possible damage to furnace tubes from chlorine contact; replace/repair	Not in scope	[1] Detection of loss of ethylene flow by 2/h reactor sampling before furnace tubes damaged	---	---	2, 5
				Unreacted chlorine to furnace; possible failure of furnace tubes from chlorine contact damage; hot chlorine vapor release from furnace	4	[2] Detection of loss of ethylene flow by 2/h reactor sampling before furnace tube(s) fail	-3	**+1** (High)	2, 5
				Unreacted chlorine through furnace and incinerator to plant scrubber; eventual chlorine breakthrough; chlorine release from scrubber stack	3	[1] Detection of loss of ethylene flow by 2/h reactor sampling before chlorine release [1] Scrubber breakthrough alarm	-3	**0** (Med)	3, 4, 5

16

Detailed Engineering Phase
An Illustration of the Fault Tree and Event Tree Analysis Methods

16.1 Problem Definition

Background

The pilot plant test runs have been completed, and ABC is now ready to build their first world-scale VCM plant. ABC's design engineering company has completed the first drawings (Revision 0) of the plant. Before making further drawing changes, ABC performed a HAZOP Study on the plant design to help identify safety and operability concerns that could be remedied during this phase of the project.

One concern that repeatedly arose in the HAZOP Study was the potential for incinerator explosions—several parts of the plant were targeted as having the potential for creating upsets that could lead to an incinerator explosion. While the HAZOP team noted that the incinerator had several shutdowns to safeguard against such an event, they were not sure that these safeguards would provide an adequate level of protection. Therefore, they recommended that Fault Tree and Event Tree Analyses be performed to investigate potential incinerator incidents.

The business team accepted the HAZOP team's recommendation. But the business team worried about making changes so late in the design process; the incinerator is a packaged system, and ABC must place their order for such a system about two years before they need it. The business team asked that ABC's process hazards analysis group complete the analysis quickly in order to avoid project delays, but (1) the process hazards analysis group does not have personnel available to perform this "rush" job and (2) their analysts have limited experience in performing Fault Tree and Event Tree Analyses. The group recommended hiring a contractor that has hazard evaluation and incinerator experience to perform the analysis.

Available Resources

The volume of information on the VCM plant is now quite extensive. However, the consultant has indicated that only certain information will be needed for the analysis. In particular, the following available information will be used:

- Piping and instrumentation diagram of the incinerator (Figure 16.1)
- Documentation from previous hazard evaluations (including the most recent HAZOP Study)

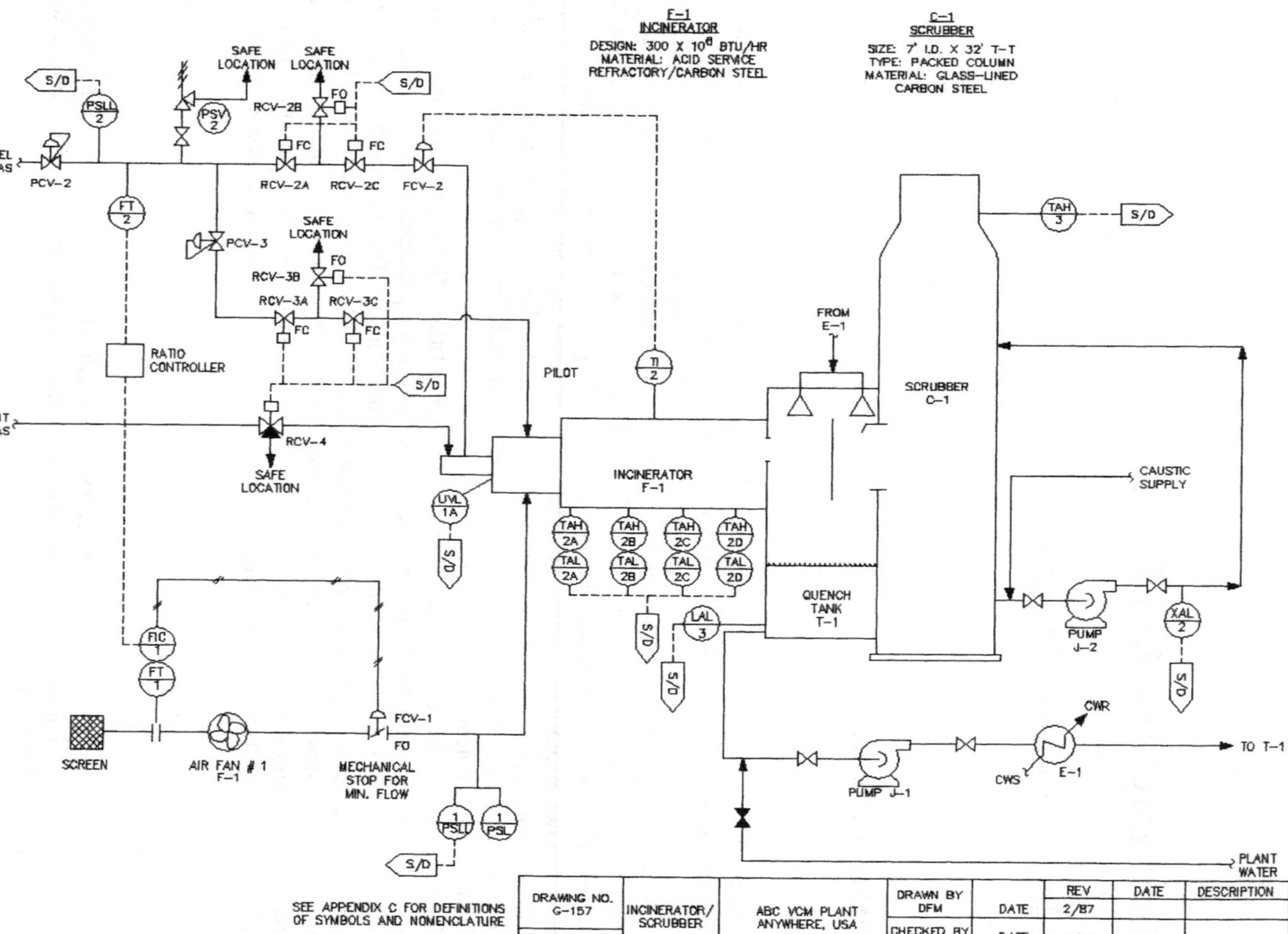

Figure 16.1 VCM plant incinerator P&ID

- A description of the incinerator and its operating procedures (provided by the vendor)
- A description of the incinerator shutdowns (provided by the vendor, Table 16.1)
- Design specifications of the incinerator, quench tank, and scrubber

Selection of Hazard Evaluation Technique

For this particular phase of the VCM project, two hazard evaluations were to be performed: one on the Revision 0 plant design and another on the potential incinerator explosions. The HAZOP Study technique was selected for the first study because a HAZOP Study can accurately identify potentially significant safety and operability issues. The technique is also broad in scope and is applicable to a large number of processes. ABC personnel performed the HAZOP Study on the Revision 0 design of the plant; an engineer from the design engineering firm participated in this review.

The HAZOP team suggested that the Fault Tree Analysis technique be used for evaluating potential incinerator explosions. While ABC's process hazards analysis group thought that other hazard evaluation methods might be just as effective, they concurred that a Fault Tree Analysis would best suit their needs. Specifically, the hazards analysis group stated that they would consider only hazard evaluation methods designed to analyze a particular problem in a highly complex and redundant system; the Fault Tree Analysis approach is the most commonly selected method in this case. However, the process hazards analysis group also noted that they had not performed many Fault Tree Analyses and suggested a consultant be used for this particular effort. ABC hired Mr. Joe Consultant of Fault Tree, Inc. to perform a Fault Tree Analysis of potential incinerator explosions.

Table 16.1 VCM plant incinerator shutdowns

Shut down incinerator on...	Instrument number
Low-low air fan discharge pressure	PSLL-1
No flame detected	UVL-1
Low-low fuel gas pressure	PSLL-2
High incinerator temperature (3-out-of-4 sensors high)	TAH-2A/B/C/D
Low incinerator temperature (3-out-of-4 sensors low)	TAL-2A/B/C/D
Low quench tank level	LAL-3
Low scrubber pH	XA-2
High scrubber stack temperature	TAH-3

Study Preparation

Mr. Consultant has performed numerous Fault Tree Analyses. Preparation for each analysis involves the following steps: (1) understanding the system design *and* operation, (2) defining the problem to be analyzed, and (3) defining the scope of the analysis. To prepare for the analysis, Mr. Consultant reviews both the available documentation on the incinerator system (P&IDs, system description, shutdown list, vendor-supplied operating manual, etc.) and the previous hazard evaluations completed on the VCM plant. Since the plant is still being designed, Mr. Consultant reviews only the design model and drawings to understand the VCM unit layout. He also visits the plant site for a day to review the plant layout and to discuss with operators how they would operate the incinerator under both normal and emergency conditions. The information gathered during this visit helps him understand how operating actions (or lack thereof) may prevent or contribute to potential incinerator incidents. Mr. Consultant also interviews some ABC operators who are likely to transfer to the new plant. Finally, he contacts the design engineering firm to discuss the design bases for the incinerator package.

At the end of the site visit, Mr. Consultant meets with ABC's process hazard analyst, who is serving as ABC's main technical contact, to further define the problem and its scope. Initially, ABC requested that the team model only incinerator explosions caused by plant upsets. However, Mr. Consultant points out that incinerator fires, toxic releases, and inadvertent shutdowns may also be of concern. He also notes that the actions taken by preventive safeguards that help avoid an explosion (or other incident) depend on the cause of the incinerator upset. In similar studies, Mr. Consultant found the Event Tree Analysis method to be an efficient way to identify the combinations of protection system failures that can lead to adverse consequences.

After further discussion, Mr. Consultant and the process hazards analysis group agree that fault tree and event tree models should be developed for all major safety incidents involving the incinerator system. However, operational problems (e.g., an inadvertent shutdown) and external events (e.g., floods, aircraft crashes, earthquakes) will not be considered, and utility system failures will not be modeled in detail. Also, the analysis will assume the plant and incinerator are initially operating normally. Mr. Consultant and ABC's process hazard analyst agree that start-up and shutdown operations are also very important; however, too little information exists to analyze these operations at this time.

16.2 Analysis Description

The analysis procedure Mr. Consultant uses for this study involves the steps listed in Table 16.2. The first step, defining the problem, is completed during the preparation phase. But before Mr. Consultant develops the preliminary event trees (step two), he reviews the design and operation of the incinerator system. He finds that the incinerator consists of three distinct components: the incinerator firebox, the quench tank, and the scrubber. The firebox is designed to burn all the flammable materials released into the vent header from the VCM plant. After combustion, the hot gases are cooled by a water spray in the quench tank and then scrubbed to remove toxic materials.

Table 16.2 Steps in a combined Fault Tree and Event Tree Analysis

- Define the problem(s) of concern
- Develop event tree model(s)
- Review and revise event tree model(s)
- Develop fault tree model(s)
- Review and revise fault tree model(s)
- Identify incident scenarios leading to consequences of concern
- Determine minimal cut sets for incident sequences
- Evaluate results and make recommendations

The HAZOP Study of the VCM plant design identified some VCM unit process upsets that would lead to large material releases to the incinerator. The team hypothesized that some of these releases might overwhelm the incinerator, extinguishing the flame and possibly causing an explosion. This factor must also be incorporated into the event trees.

To develop the event trees, Mr. Consultant first composes a list of initiating causes (initiating events) that would challenge the incinerator safety system; he composes this list by reviewing the HAZOP Study results and by using his own experience. For each of these initiating causes, he then identifies the incinerator system functions that would safely mitigate the upset. With this information, he draws event trees that portray the conditions, system successes, and system failures that lead to significant safety consequences. Note that if only one consequence is of concern, and all incident initiating causes challenge the same safeguards, then Fault Tree Analysis alone may be sufficient to model the incinerator's risk.

Figure 16.2 is one of the event trees Mr. Consultant developed. This event tree for an undefined process upset initiating cause includes the following safety system functions and process circumstances:

1. A significant unit upset occurs.
2. The upset does (or does not) extinguish the firebox flame.
3. The incinerator shutdown system trips and shuts down the incinerator. (Note: The HAZOP team did not identify any process upsets that would overheat the incinerator.)
4. The quench system cools the hot gases from the incinerator.
5. The scrubber effectively removes the toxic materials from the waste gas.

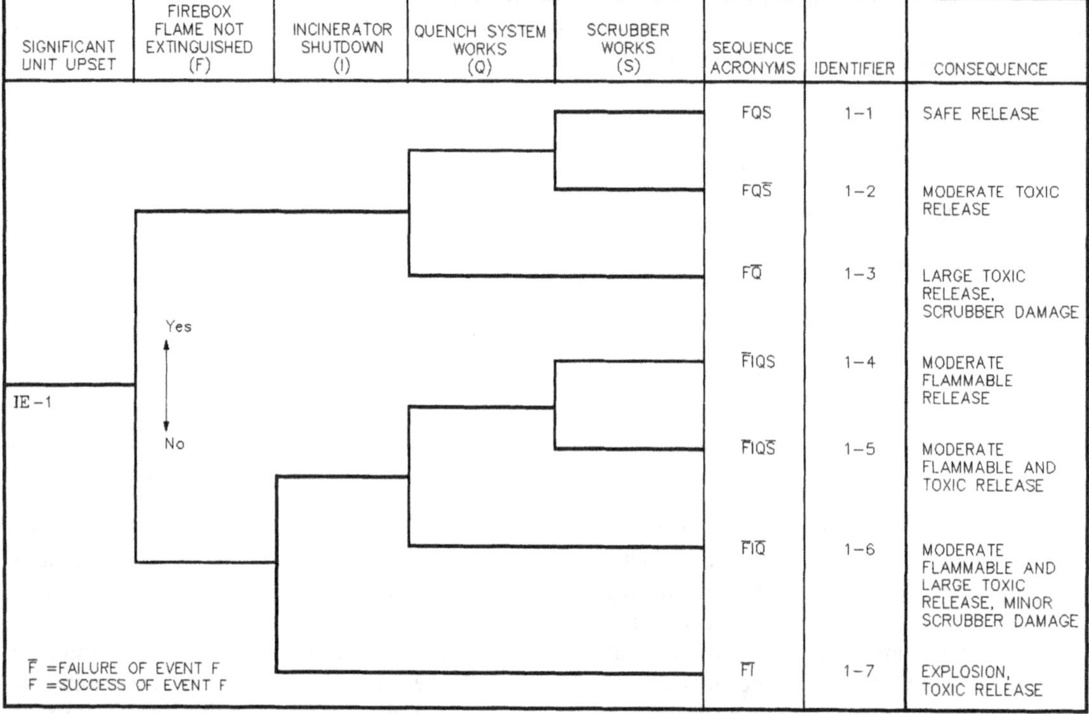

Figure 16.2 Example event tree for the VCM plant—generic process upset initiating cause

Not all of these actions/conditions apply to every incident scenario. For example, incident scenario 1-3 (i.e., an upset occurs that does not extinguish the firebox flame) shows that if the quench system fails, then the scrubber cannot effectively remove toxic materials from the waste gas. This same scenario shows that if the firebox flame is not extinguished, then the incinerator shutdown system is not challenged (but the quench system is).

Figure 16.3 shows another event tree for the system. This time, the initiating cause is low fuel gas pressure. Initially, Mr. Consultant thought this event tree would be identical to Figure 16.2, with the exception of the initiating cause. However, during consultation with ABC design engineers and the incinerator vendor, he learned that low fuel gas pressure would probably force the flame in the firebox to go out. Thus, he deletes the "flame not extinguished" branch from event trees for all initiating causes that challenge the incinerator safety system.

After developing event trees, Mr. Consultant reviews and revises these event trees with the ABC process hazard analyst and process design team before developing the fault trees. During this review, Mr. Consultant and ABC's engineers discuss the basis for the event tree logic. In particular, he traces through each incident sequence depicted on each event tree, defining which upset initiated the problem, which safety functions are working or have failed, and which incinerator flame conditions exist. With this information, the review team determines a qualitative description of the consequences of each incident scenario depicted in the event trees.

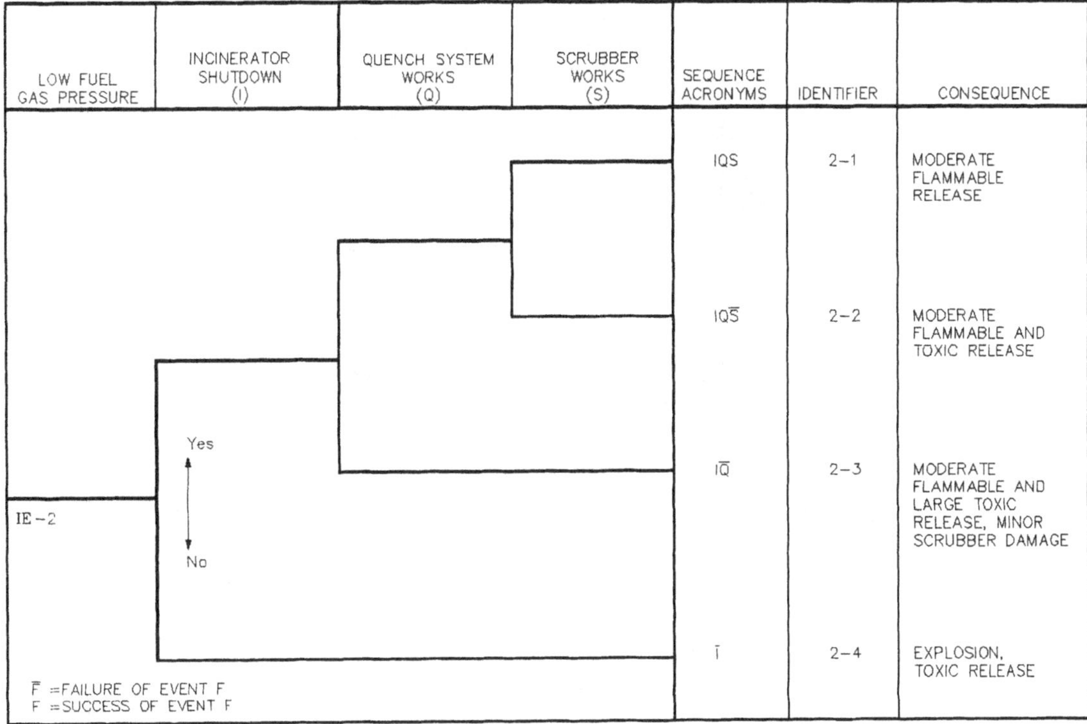

Figure 16.3 Example event tree for the VCM plant—low fuel gas pressure initiating cause

Next, Mr. Consultant develops fault trees for each of the system failures represented by branch points in the event trees. The initiating causes are well understood; thus, fault tree models are not developed for these events. However, the following system failures are modeled:

- The incinerator safety system fails to shut down the incinerator, given a flameout condition exists.
- The quench system fails to adequately cool the waste gas.
- The scrubber system fails to adequately remove toxic materials from the waste gas.

In developing the fault trees, Mr. Consultant assumes each system is capable of performing its design intention if it operates reliably. Thus, the fault trees graphically illustrate combinations of equipment failures and human errors that incapacitate the systems. Figure 16.4 is the preliminary fault tree Mr. Consultant develops for the incinerator shutdown system. The fault tree logic proceeds from the top down, expanding fault events into the logical combinations of component failures, human errors, and subsystem failures that will not be developed further because they contain adequate detail for identifying system weaknesses or because they are outside the boundary conditions of the study. Table 16.3 lists the procedure he follows in developing this fault tree.

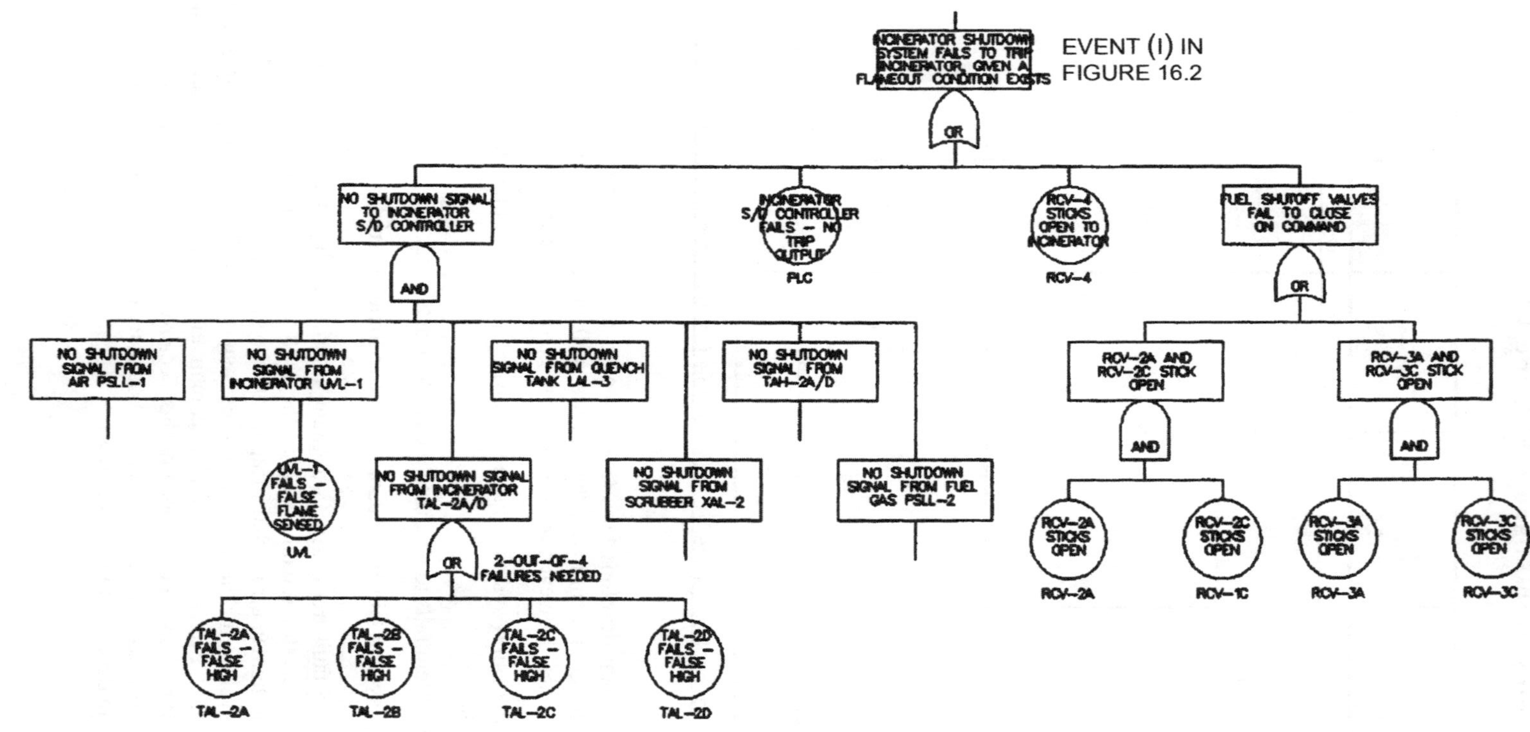

Figure 16.4 Preliminary fault tree developed for the incinerator shutdown system

Table 16.3 Fault Tree Analysis steps

- Define the Top event (system failure of concern)
- Lay out the tree top structure
- Revise the tree top structure to account for initiating causes and safety system failures
- Develop each fault event to its basic causes

Mr. Consultant has already defined the Top events based on his Event Tree Analysis. To develop the tree top structure for the incinerator shutdown system, he first follows the flow of shutdown signals through this system. For the shutdown system to work, it must (1) sense the upset condition, (2) process the trip signal(s) properly, and (3) close the fuel gas and vent gas shutoff valves. Failure to perform any of these functions defeats the shutdown system. Thus, Mr. Consultant's tree top consists of an OR gate with the failure of these functions as inputs (Figure 16.4). Developing the fault events in this tree one level further, Mr. Consultant notes that all the appropriate sensors (with the exception of TAH-2A/D) must fail in order to lose the needed input shutdown signal (as indicated by AND logic). However, the shutoff valves remaining open on the pilot gas line, the supply gas line, or the vent line will defeat the valve closure function (as indicated by OR logic).

At this stage in the fault tree development, Mr. Consultant reviews and revises the model to account for initiating causes. In this particular case, the initiating cause is a process upset that extinguishes the incinerator flame. Reviewing the sensors tied to the shutdown system, Mr. Consultant determines that the high furnace temperature switch, the fuel supply low pressure switch, the quench tank low level switch, and the scrubber pH sensor could not be expected to detect the incinerator problem, even if working properly. Thus, he eliminates these fault events from the fault tree. He develops the remaining fault events to their basic causes.

Subsequently, Mr. Consultant visits ABC's Anywhere Plant to review the fault tree with the ABC engineers. In this review, the ABC engineers verify that the fault tree reflects the incinerator system as they understand it will be built and operated. They also review the reasonableness of assumptions made by Mr. Consultant in constructing the fault tree. The ABC engineers identify an error in the preliminary fault tree—the low-low air fan pressure switch (PSLL-1) should not have been counted as a safeguard when a process upset extinguishes the incinerator flame. Mr. Consultant revises the fault tree model as appropriate. The final fault tree is shown in Figure 16.5 as it is presented as output from a commercial FTA program.

Separate fault trees for the quench tank and scrubber portions of the system are also developed. These models are also reviewed by ABC for logic errors, inconsistencies, and inappropriate assumptions, and are revised as needed.

The fault tree and event tree models enabled Mr. Consultant to identify combinations of equipment failures and human errors (minimal cut sets) that result in an undesirable consequence (Section 5.5 of the *HEP Guidelines* describes methods for finding minimal cut sets). For example, the fault tree in Figure 16.6 describes combinations of conditions and failures involving scenario 1-7 from Figure 16.2 that result in an incinerator explosion. Solving this fault tree yields explosion incident sequence cut sets. These cut sets are reviewed by Mr. Consultant and ABC to ensure that they accurately reflect explosion incidents. This information, along with insights gained during the construction of the fault trees and event trees, will be used to develop recommendations for system improvements.

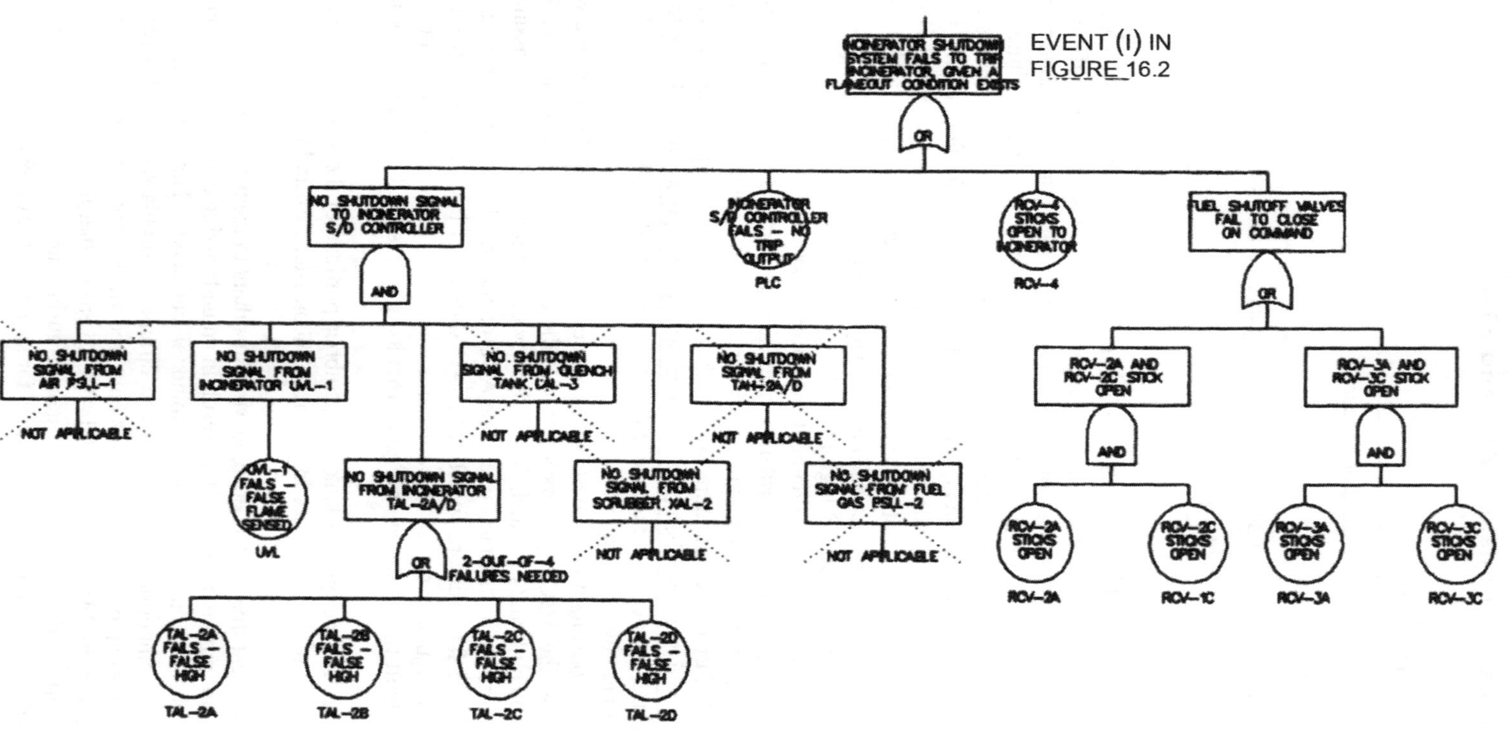

Figure 16.5 Final fault tree for the incinerator shutdown system

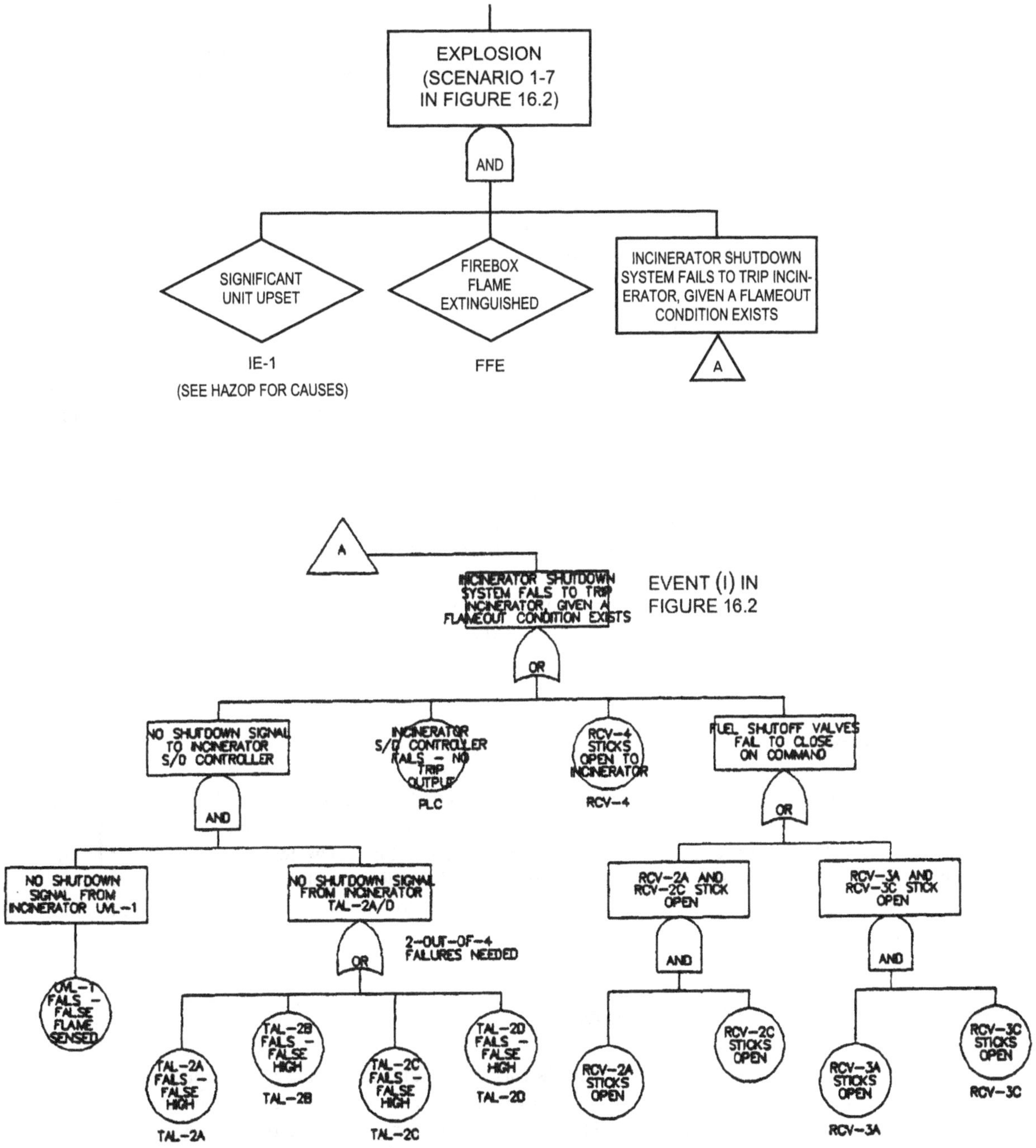

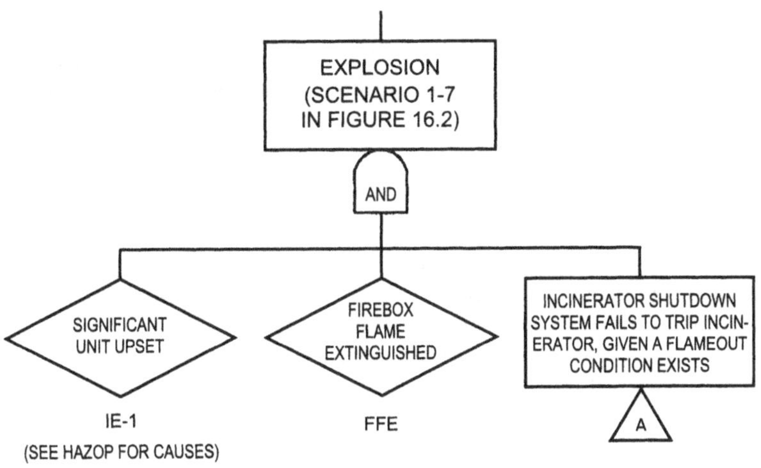

A
INCINERATOR SHUTDOWN SYSTEM FAILS TO TRIP INCINERATOR, GIVEN A FLAMEOUT CONDITION EXISTS
EVENT (I) IN FIGURE 16.2
OR
NO SHUTDOWN SIGNAL TO INCINERATOR S/D CONTROLLER
INCINERATOR S/D CONTROLLER FAILS – NO TRIP OUTPUT
PLC
RCV-4 STICKS OPEN TO INCINERATOR
RCV-4
FUEL SHUTOFF VALVES FAIL TO CLOSE ON COMMAND
AND
OR
NO SHUTDOWN SIGNAL FROM INCINERATOR UVL-1
NO SHUTDOWN SIGNAL FROM INCINERATOR TAL-2A/D
OR
2-OUT-OF-4 FAILURES NEEDED
RCV-2A AND RCV-2C STICK OPEN
AND
RCV-3A AND RCV-3C STICK OPEN
AND
UVL-1 FAILS – FALSE FLAME SENSED
TAL-2A FAILS – FALSE HIGH
TAL-2A
TAL-2B FAILS – FALSE HIGH
TAL-2B
TAL-2C FAILS – FALSE HIGH
TAL-2C
TAL-2D FAILS – FALSE HIGH
TAL-2D
RCV-2A STICKS OPEN
RCV-2C STICKS OPEN
RCV-3A STICKS OPEN
RCV-3A
RCV-3C STICKS OPEN
RCV-3C

Figure 16.6 Fault tree for incident scenario 1-7— explosion

While not shown here, Mr. Consultant also used the event trees and fault trees to find incident sequence cut sets that lead to other significant safety impacts (e.g., flammable releases, toxic releases). Again, this information provides insight into system weaknesses and places where improvements might be most effective.

16.3 Results

The Fault Tree and Event Tree Analyses resulted in a set of failure logic models, lists of incident sequence cut sets for each significant safety impact of concern, and recommendations for improving the system. Example fault tree and event tree models are shown in the previous section. Table 16.4 lists some of the incident sequence cut sets (using the basic event names listed on the fault trees) identified for explosions. This list shows that three or more failures are needed to create an incinerator explosion. However, closer examination of this list reveals two things: (1) if the PLC or vent isolation valve (RCV-4) alone is faulty, and a severe unit upset (IE-1) is severe enough to extinguish the flame (FFE), this may lead to an explosion; and (2) if the thermocouples (TAH/L-2A/B/C/D) cannot quickly detect a flameout (via low temperature), then the flame scanner (UVL-1) is the only effective sensor for protecting the incinerator.

Table 16.5 lists some of the recommendations Mr. Consultant made. Most of these recommendations are based on evaluation of the incident sequence cut sets.

Mr. Consultant documents the results of the Fault Tree and Event Tree Analyses and sends them to ABC's process hazard analyst. The report that includes the results describes the scope of the problem analyzed, the analysis methods used, and Mr. Consultant's results and recommendations. It also includes the fault tree and event tree models. ABC's analyst (Ms. Deal) prepares an executive summary of this report for the VCM business team.

16.4 Follow-up

The VCM business team reviewed Mr. Consultant's recommendations and acted on them. In particular, they asked the design engineering firm to incorporate all the suggested design changes, with the exception of the double block and bleed valve on the vent gas line. The team decided to delay any action on this recommendation, pending input from ABC's environmental group, who must decide whether they will allow vent gases to be released directly to the atmosphere in the event of an emergency.

16.5 Conclusions and Observations

The combined Fault Tree and Event Tree Analysis was very useful in helping ABC understand some of the weaknesses of the incinerator design. However, Fault Tree Analysis and Event Tree Analysis methods are narrowly focused and require skilled personnel to apply them; therefore, ABC does not use these methods often, and prefers to use consultants when one of these methods is chosen. There is another reason ABC doesn't often use these methods: the more broad-brush methods, such as HAZOP Study, FMEA, Checklist Analysis, etc., have been effective hazard evaluation methods for evaluating most of ABC's systems.

Table 16.4 Sample incident sequence minimal cut sets — incinerator explosion

Minimal cut set no. 1
- Significant unit upset
- Firebox flame extinguished
- Incinerator shutdown controller fails – no trip output

Minimal cut set no. 2
- Significant unit upset
- Firebox flame extinguished
- RCV-2A sticks open
- RCV-2C sticks open

Minimal cut set no. 3
- Significant unit upset
- Firebox flame extinguished
- RCV-3A sticks open
- RCV-3C sticks open

Minimal cut set no. 4
- Significant unit upset
- Firebox flame extinguished
- UVL-1 fails – false high
- TAH/L-2A fails – false high
- TAH/L-2B fails – false high

Minimal cut set no. 5
- Significant unit upset
- Firebox flame extinguished
- UVL-1 fails – false high
- TAH/L-2A fails – false high
- TAH/L-2C fails – false high

Minimal cut set no. 6
- Significant unit upset
- Firebox flame extinguished
- UVL-1 fails – false high
- TAH/L-2A fails – false high
- TAH/L-2D fails – false high

Minimal cut set no. 7
- Significant unit upset
- Firebox flame extinguished
- UVL-1 fails – false high
- TAH/L-2B fails – false high
- TAH/L-2C fails – false high

Minimal cut set no. 8
- Significant unit upset
- Firebox flame extinguished
- UVL-1 fails – false high
- TAH/L-2B fails – false high
- TAH/L-2D fails – false high

Minimal cut set no. 9
- Significant unit upset
- Firebox flame extinguished
- UVL-1 fails – false high
- TAH/L-2C fails – false high
- TAH/L-2D fails – false high

Minimal cut set no. 10
- Significant unit upset
- Firebox flame extinguished
- RCV-4 sticks open to incinerator

Table 16.5 Incinerator safety improvement alternatives

- Consider using a PLC that has self-checking capability in the shutdown system
- Consider installing a redundant flame scanner
- Perform a Common Cause Failure Analysis (CCFA) of the redundant valves and redundant instrumentation (see *Guidelines for Chemical Process Quantitative Risk Assessment* for CCFA details)
- Determine if the incinerator thermocouples can detect a flameout (due to low temperature) quickly enough to initiate a shutdown before explosive volumes of gas collect in the incinerator
- Consider tripping the fuel gas flow control valve closed when a shutdown signal occurs
- Consider installing double block and bleed valves on the vent gas line

The time required to perform the combined Fault Tree and Event Tree Analysis is summarized in Table 16.6. Note that many fault trees and event trees were developed in this analysis, only a few of which are shown in this chapter. The staff requirements shown in Table 16.6 reflect the time it took to develop, analyze, and document all of these models.

It is worth noting that, in this example, the use of one hazard evaluation resulted in a recommendation to use another hazard evaluation method. The HAZOP team who reviewed the detailed design of the VCM plant recognized that the incinerator system was a complex, highly redundant design. While the HAZOP Study method can identify single failures with potentially significant consequences, it is not as thorough and precise of a method for finding multiple failures that lead to serious incidents. The HAZOP team quickly recognized that the incinerator could fail in ways that had very serious consequences. However, because of the high level of redundancy in the incinerator's safeguards, the team recommended that this system be examined in detail with another, more appropriate hazard evaluation tool—Fault Tree Analysis, in this case.

Finally, the incinerator system was evaluated in a detailed manner, in part because it was not designed by ABC. It is commendable that ABC wants to thoroughly understand the vendor's incinerator package and ensure the highest level of safety. However, ABC should also be sure to apply the same rigorous reviews to their own work.

Table 16.6 Combined Fault Tree/Event Tree Analysis staff requirements for the detailed engineering phase

Personnel	Preparation (h)	Evaluation (h)	Documentation (h)
Consultant	24	150	100
ABC reviewers[a]	8	40	16

[a]Average per ABC reviewer

17

Construction/Start-up Phase

An Illustration of the Checklist Analysis and Safety Review Methods

17.1 Problem Definition

Background

It is the third year of ABC's project and they are finishing their world-scale plant in Anywhere, U.S.A. Most of the plant's major equipment has been installed. As the work progresses, ABC's general contractor inspects the site and equipment installations to ensure the work is completed according to the contract. The plant is only a few months away from its targeted start-up date.

ABC wants the plant to start up safely and on time. To help ensure this, ABC's business team asks their process hazards analysis group to perform a HAZOP Study of the plant's construction. Mr. Dennis, of ABC's process hazards analysis group, believes that a HAZOP Study of the construction is inappropriate, but he does think it is a good idea to review or audit the facility (1) to ensure it is built according to design and (2) to identify any additional safety concerns that may have been overlooked in previous hazard evaluations. Therefore, he suggests a person from the engineering department perform this hazard evaluation using a combination of the Checklist Analysis and Safety Review techniques. This person must be experienced with plant construction and start-up to effectively perform the assessment.

The business team contacts the ABC engineering department with their request for an experienced start-up engineer to perform a Checklist Analysis/Safety Review of the VCM plant. Mr. Buildit, a 15-year veteran of ABC, is selected. Mr. Buildit has participated in three plant start-ups (two with ABC) and is thoroughly familiar with ABC's design codes and standards. He will use a checklist to help him identify potential safety concerns, not just to verify that as-built equipment complies with ABC's standards. He will also tour the entire site as part of a Safety Review to identify safety concerns.

Available Resources

A great deal of information exists on the plant. This information includes:

- PFDs and P&IDs
- Construction drawings
- Operating procedures
- R&D reports
- Plot plans
- Equipment specifications
- Wiring diagrams
- Previous hazard evaluations

Mr. Buildit briefly scans through this information to see what will be most useful. He reviews the PFDs, plot plans, and process descriptions to gain a general understanding of the plant and VCM production technology. He also reviews MSDSs and previous hazard evaluations to gain an understanding of the material hazards, the process hazards the team identified, the action items, and the resolution of these items. However, the most useful information is the P&IDs, the construction drawings, and the equipment specifications. Mr. Buildit plans to use this information extensively while actually performing the hazard evaluation.

During his preliminary review, Mr. Buildit notes that ethylene is used in the VCM process. While he is familiar with the types of equipment that will handle ethylene (pumps, piping, instruments, etc.), he is unfamiliar with the physical and chemical properties of ethylene. He suggests that ABC have a different person review the plant areas that involve the handling of the ethylene feed material.

Selection of Hazard Evaluation Technique

The purpose of performing a hazard evaluation at this phase of the project is to verify that the plant is built as designed, that it is built according to accepted construction industry practices, and that safety hazards have not been overlooked. The VCM plant design has already been reviewed several times using a variety of hazard evaluation techniques and, according to Mr. Dennis, should be safe. However, ABC wants to ensure that the VCM safeguards the hazard evaluation teams reviewed and recommended are actually built according to specification and that additional safety hazards have not been created or overlooked.

In compliance with Mr. Dennis's recommendation, Mr. Buildit will use the Checklist Analysis method to perform this hazard evaluation. ABC has built many plants and through the course of these construction projects has developed many checklists to audit the facility's construction. These checklists are based on company design standards and experience gained from reviewing errors, incident reports, and insights from past hazard evaluations.

In addition, Mr. Buildit will perform a Safety Review of the construction site. In this assessment, he will rely on his construction and start-up experience to help him identify unsafe construction practices and determine whether the process is ready to begin start-up testing.

Study Preparation

The Checklist Analysis/Safety Review will be performed primarily by Mr. Buildit. In preparation for this work, Mr. Buildit gathers together the checklists ABC has for construction and start-up. These checklists include generic field inspection items for fired heaters, pumps, fans, piping, vessels, columns, valves, and so forth. All of these checklists are available from ABC's Engineering Division.

Next, Mr. Buildit spends several days reviewing the construction drawings of the VCM plant. During this review he uses the checklists to help him prepare specific questions concerning the materials of construction, installation, and testing of various equipment. He notes these questions on the construction drawings he will use during the on-site review.

Finally, Mr. Buildit schedules a visit to the Anywhere site with ABC's construction manager. He provides the manager with a list of areas he will review and a schedule for examining each area. He asks that both ABC and contractor personnel from each area be available to answer questions during the visit. In particular, Mr. Buildit will need to talk with the:

Construction Engineer — The contractor's engineer responsible for the installation of the equipment being examined

Electrician — The contractor's supervisor responsible for electrical equipment and wiring in the areas being examined

Site Inspector — The contractor and/or ABC employee responsible for verifying that the equipment being examined is properly installed.

He estimates that he will not spend more than an hour per equipment item with these people, and states that he prefers to meet with them as a group.

17.2 Analysis Description

Mr. Buildit arrives at the Anywhere Plant at 7:30 a.m. Monday morning to begin his review. He meets with Mr. Star, ABC's manager for this project, to discuss the purpose and activities that will take place during his one-week visit. Mr. Buildit points out that the VCM project is receiving a lot of attention in the company, particularly from a safety perspective. As Mr. Star knows, the VCM plant has been the subject of numerous hazard evaluations, which identified several ways to improve the safety of the plant design. The objective of Mr. Buildit's visit is to ensure that this well-designed plant is also built as designed. To accomplish this, he will meet with the people responsible for installing each major equipment item and ask them a number of questions from his checklist. He also plans to visually inspect each major piece of equipment and, in fact, prefers to review his checklist with the personnel at the equipment site.

Following this discussion, Mr. Buildit takes a tour of the site. During this tour he takes note of any unsafe practices he observes and immediately brings them to the attention of the appropriate personnel.

Mr. Buildit is now ready to begin his Checklist Analysis. The first item to examine is the HCl storage tank (shown in Figure 17.1). Participating in the review are Mr. McNamara (the contractor's engineer for the tank farm) and Mr. Quay (the contractor's inspector). Using his storage tank checklist as a guide, Mr. Buildit begins the review.

Mr. Buildit — To start with, has an NDE (*nondestructive examination*) of the HCl tank been performed?

Mr. Quay — Yes, two weeks ago.

Mr. Buildit — How was this done?

Mr. Quay — The NDE was performed by Leaktest, Inc. They have done most of the NDE work for us on the site. For this tank, my records show that they radiographed all the welds. I personally checked the radiographs and also visually inspected the welds.

Mr. Buildit — Are the Leaktest examiners certified by ASME? And is the NDE method they used ASME-approved?

Mr. Quay — Yes and yes. We checked their certification before hiring them.

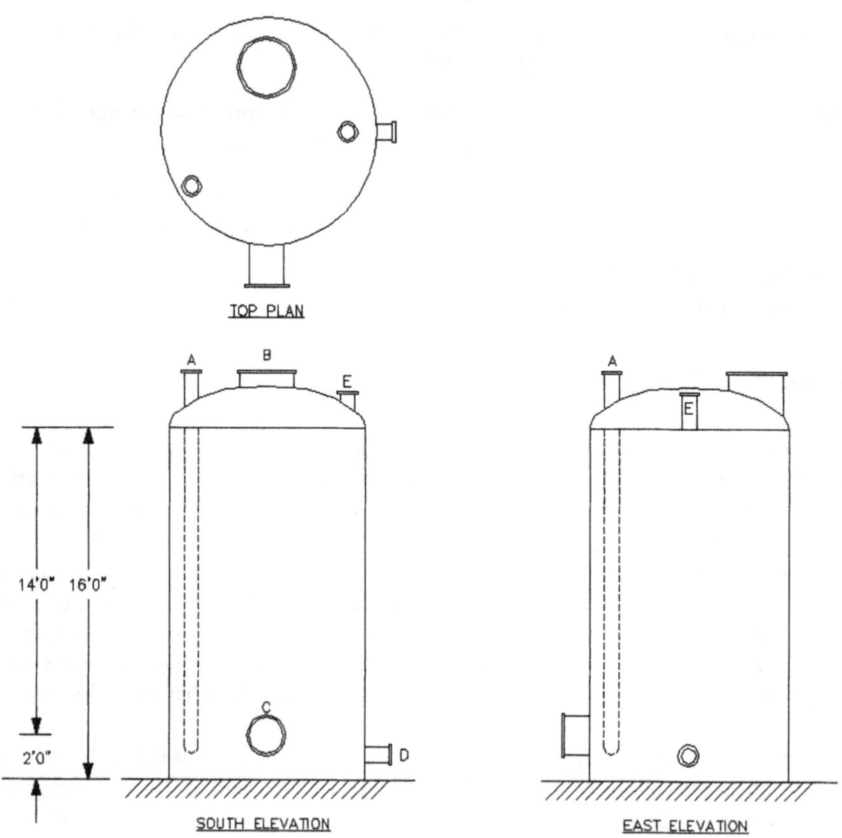

NOZZLE SCHEDULE			
MARK	SIZE	TYPE	DESCRIPTION
A	2"	150#	FILL
B	20"	150#	MANWAY
C	16"	150#	MANWAY
D	2"	150#	DISCHARGE
E	2"	150#	VENT

DESIGN DATA

DIMEN:
8' DIA. X 16' STR. SIDE
CAPACITY: 13,500 GAL
DES. PRESSURE
SHELL: 50 PSIG
OPERATING PRESSURE
SHELL: ATMOS
DESIGN TEMPERATURE
SHELL: 250°F
OPERATING TEMPERATURE
SHELL: 120°F
TEST PRESSURE
SHELL: 100 PSIG
SHELL THICKNESS 5/16"
MATERIAL OF CONSTR.
CARBON STEEL
REMARKS:
FIBERGLASS LINED

DRAWING NO. H-107	HCl STORAGE TANK	ABC VCM PLANT ANYWHERE, USA	DRAWN BY DFM	DATE 9/8/88	REV	DATE	DESCRIPTION
SCALE: NONE			CHECKED BY WGB	DATE 10/20/88			

Figure 17.1 Schematic of the HCl storage tank

Mr. Buildit — Do you have the results of the NDE?

Mr. Quay — Sure, they're with the engineering files for the HCl tank. Would you like me to get them from the trailer?

Mr. Buildit — Later. Let's move on. Did you heat treat the welds?

Mr. Quay — Yes, according to code. And we performed Brinell hardness testing on samples of the steel used to construct the tank.

Mr. Buildit — And the results are on file?

Mr. Quay — Same file as the NDE results. I'll get that later also.

Mr. Buildit — Okay. Mr. McNamara, did you check the foundation installation and tank support anchors?

Mr. McNamara — Yes, we did that both before and after tank installation.

Mr. Buildit — What about the materials of construction? Did the tank material, welding material, and gaskets comply with the equipment specs?

Mr. Quay — Yes, as you can see, the equipment list and construction drawing specifies a fiberglass-lined, carbon steel tank. Note the nameplate on the tank [*points to nameplate*]; it indicates carbon steel.

Mr. Buildit — By the way, it looks like the ground inside the diked area slopes towards the tank. Shouldn't it be the other way?

Mr. McNamara — You're right. I'll make a note to correct this.

Mr. Buildit — I don't have the grading drawings with me. How large is the dike supposed to be?

Mr. McNamara — One-and-a-half tank volumes.

Mr. Buildit — Okay, hang on a minute. [*Mr. Buildit walks off the dimensions of the dike and then estimates its volume. Satisfied that it is about right, he moves on to other items.*] Let's see, you already said all the welds were tested and inspected. How about tack welds on vertical joints? Were they removed?

Mr. McNamara — We chipped them all off and smoothed the surface. I think Mr. Quay checked this also. [*Mr. Quay nods yes.*]

Mr. Buildit — Okay, I see the tank plate looks fine and accurate. Did you do a hydrotest?

Mr. Quay — Yes, that was done and I have the test results on file.

Mr. Buildit — Did you see any settling due to the hydrotest?

Mr. McNamara — No. We checked the tank foundation elevation after the hydrotest and it was within specs.

Mr. Buildit — Have you checked the physical dimensions of the tank? Proper height, diameter, and roundness?

Mr. McNamara — Yes, that was all done.

Mr. Buildit — How about the elevation and orientation of the nozzles? Is the tank plumb and on centerline?

Mr. McNamara — Yes. In fact, I did that myself.

Mr. Buildit — I see you've got your tape measure there. Let's do a rough check on the distance between the top and bottom nozzles. [*McNamara measures the distance and it comes out okay.*] Looks like the grouting is okay. Are the foundation bolts tightened?

Mr. McNamara — Yes. They should be torqued as recommended on the construction drawing.

Mr. Buildit — Think that pipefitter over there [*points to left*] will loan us his torque wrench?

Mr. McNamara — Not a chance. But he'll check it out for us. Hey Joe, can you check the torque on a bolt for me real quick? [*Joe comes over and verifies that the foundation bolts are properly torqued.*]

Mr. Buildit — Let's see now. Item 11 on my list is ladders and platforms. They look right. Item 12, trays level and correct orientation. That doesn't apply. Internal pipes installed with correct bolts and gaskets? [*McNamara nods yes.*] Let's go up top and look in the manway.

Mr. Quay — Okay, but there's only one ladder. Only one person at a time should be on the ladder. [*Mr. Buildit climbs to the top and looks in the manway.*]

Mr. Buildit — From what I can see, the dip leg looks okay. Hey McNamara, isn't this tank supposed to be lined?

Mr. McNamara — Sure, it's an HCl storage tank.

Mr. Buildit — I think you'd better take a look. [*Mr. McNamara exchanges places with Mr. Buildit.*]

Mr. McNamara — I can't believe it! They forgot the #*?x$! lining!

Mr. Quay — That's a major item. I'll have to go back now and check our paperwork to see what happened.

Mr. Buildit — I agree. However, I've only got a few more items. Let's go ahead with these items and finish the tank. What kind of gaskets are on the tank?

Mr. McNamara — The equipment specs say paper, and that's what we have. This is an atmospheric tank.

Mr. Buildit — Internal tray manways closed. That doesn't apply. Packing installed—doesn't apply. Is this tank to be insulated?

Mr. McNamara — Not according to the specs.

Mr. Buildit — Let's verify the instrument taps and atmospheric vent and I'll be done. [*The team visually checks all the storage tank taps and the vent.*]

Upon completing his review of the HCl storage tank, Mr. Buildit moves on to the HCl transfer pump, which is in the same area. Using a checklist for pumps, he begins the review process over again. This examination of equipment continues throughout the week, with Mr. Buildit and appropriate teams of construction inspection personnel. During times when he is not in the field, Mr. Buildit verifies some of the QA paperwork for equipment items that had an unusual number of deficiencies during the Checklist Analysis. At the end of each day he documents the deficiencies found for Mr. Star's followup.

17.3 Discussion of Results

For the most part, Mr. Buildit documents the Checklist Analysis and Safety Review results by initialing his checklist form as he performs the review (Table 17.1 is an example of a partially completed checklist). Checklist items requiring further attention are noted on his form and described in a separate table attached to the checklist (Table 17.2).

Important findings from the Checklist Analysis were (1) the tank lining was missing and (2) the ground inside the diked area needed to be sloped away from the tank. In addition, Mr. Buildit made several observations during his visual audit of the construction site. These include the following:

- Some hot work areas were not free of flammable materials.
- Heavy equipment work using cranes was not always adequately monitored.
- Completed, inspected work was not always properly tagged (leaving it vulnerable to further, perhaps uninspected alterations).

Mr. Buildit provides his checklist results to Mr. Star at the conclusion of the site visit.

17.4 Follow-up

The Checklist Analysis/Safety Review was used to audit the construction of the VCM plant, and the deficiencies identified were addressed. Usually, the appropriate contractor personnel were present whenever the Checklist Analysis uncovered a deficiency, and they initiated actions to correct the problem. To ensure this, Mr. Star assigned ABC personnel at the site the responsibility of verifying that every problem identified by Mr. Buildit was corrected. Mr. Star tracked the corrective actions weekly, meeting with appropriate personnel to determine the status of these corrections.

Table 17.1 Checklist Analysis results for the HCl storage tank inspection

Item to check	O.K. (sign and date)
1. Nondestructive examination (NDE) performed	
a. NDE examiners are ASME-certified	I.M.B. ##/##/##
b. Approved ASME NDE method used	I.M.B. ##/##/##
c. NDE results in engineering file	I.M.B. ##/##/##
2. Postweld heat treatment and hardness testing performed	
a. Postweld heat treatment in accordance with ASME code	I.M.B. ##/##/##
b. Brinell hardness testing performed	I.M.B. ##/##/##
c. Test results in engineering files	I.M.B. ##/##/##
3. Vessel foundation elevation and slope checked	Action required
4. Vessel material and construction materials in compliance with specifications and job requirements	I.M.B. ##/##/##
5. All welds inspected and tested	I.M.B. ##/##/##
6. All tack welds in vertical joints properly removed	I.M.B. ##/##/##
7. Vessel wall plate in good condition (or properly repaired if damaged) and contains all pertinent information	I.M.B. ##/##/##
8. Vessel hydrotested	I.M.B. ##/##/##
9. Dimensional check of vessel performed	I.M.B. ##/##/##
10. Elevation and orientation of nozzles checked. Vessel is on centerlines, is level, and is properly grouted. Foundation bolts tightened	I.M.B. ##/##/##
11. Ladders and platforms installed as per drawings	I.M.B. ##/##/##
12. Trays level and correct orientation. Downcomer clearance, weir height, drain holes, gaskets, bolts, etc., installed per specification	Not applicable
13. Internal pipes installed with correct bolts and gaskets	I.M.B. ##/##/##
14. Internal lining intact	Action required
15. Internal tray manways closed	Not applicable
16. Packing installed	Not applicable

Table 17.2 Action items from the HCl storage tank Checklist Analysis

No.	Action	Responsibility	Date
3	Verify HCl storage tank basin is graded away from the tank		
14	Verify lining is installed in HCl storage tank		

17.5 Conclusions and Observations

The Checklist Analysis/Safety Review proved quite successful in uncovering construction deficiencies, primarily because Mr. Buildit had good checklists and extensive plant construction experience. In addition, the review was successful because Mr. Buildit personally checked the equipment to ensure it met specifications rather than relying solely on the experience of plant personnel.

Several other important observations can be made about the Checklist Analysis:

- The Checklist Analysis can be used both as an investigative tool to identify hazards and an audit tool to verify designs and installations. In this example, the design had been thoroughly reviewed with other hazard evaluation methods, thus the checklist was used primarily as an audit.
- The Checklist Analysis/Safety Review is not all-inclusive and should prompt additional questions from the reviewer. Mr. Buildit did not limit himself to the checklist or the Safety Review, but asked additional questions as he thought of them.
- Mr. Buildit's verification of the responses to checklist questions varied. As he gained confidence in the team's answers (by checking some of the records on the equipment item), he did less physical checking of the equipment.
- An experienced person is needed to perform a Checklist Analysis, especially if the review is being done as an audit.

For the amount of equipment examined, the Checklist Analysis/Safety Review went very quickly. Table 17.3 summarizes the time required to perform this review.

Table 17.3 Checklist Analysis and Safety Review staff requirements for the construction/start-up phase

Personnel	Preparation (h)	Evaluation (h)	Documentation (h)
Leader [a]	1	0.5	0.25
Team member [a]	—	0.5	—

[a]Average per equipment item

18

Routine Operation Phase

An Illustration of the Safety Review Method for Management of Change

18.1 Problem Definition

Background

The Anywhere VCM plant has been operating for over six months without an incident besides subtle control issues and several abnormal situations that were successfully dealt with. The start-up of the process has gone smoothly, but there has been a need for changes to increase the reliability and productivity of the process. As required by regulation and corporate process safety standards, the site has developed and implemented a management of change (MOC) procedure to ensure that the impact of any change on safety and health is fully evaluated and that all changes are properly authorized, documented and communicated.

The site MOC procedure requires the initiator of the change to assemble an appropriate team to review the proposed change. All process and equipment changes except like-for-like replacements require a technical review. The team may be one person in addition to the initiator for a small change, or a multidisciplinary team for a more substantial change. At this facility, a "small change" is defined as one that costs less than $10,000 and does not involve any regulated material or process. All changes, large or small, go through the rigorous MOC procedure. In this case, the process supervisor must sign off on the MOC form, indicating that a review has been done.

Alex, an engineer who worked on the start-up, has proposed the addition of a caustic supply connection and a low pH alarm to be added to the quench tank to allow for pH control. The original design called for drums of caustic to be manually added to the quench tank. Samples were taken manually and analyzed in the laboratory for pH monitoring. An engineering analysis indicated that maintaining a constant pH in the quench tank would allow for tighter control of the overall process.

Available Resources

From the start-up preparation, the following materials are available to the MOC team to review:

- Piping and instrumentation diagrams
- Process flow diagrams

- Documentation of previous hazard evaluations of the VCM plant (including the HAZOP Study of the original incinerator design)
- Operating and emergency procedures
- Vendor design specifications, including design bases for relief valves
- Incident reports for the incinerator.

In addition, the team decided it also needed:

- Piping specifications for the caustic supply line
- Valve specifications for the connecting valve
- Instrument specifications on the pH monitor
- Pump specifications for the pump on the caustic supply line.

Selection of Hazard Evaluation Technique

The site MOC procedure allows for any appropriate hazard evaluation technique to be used for the MOC review. For most MOC reviews, the Safety Review technique is used with the help of a checklist. Alex has selected the Safety Review technique because of the simplicity of the system. Also, the operator and instrument technician are familiar with this technique. For more complex process changes, other hazard evaluation methods including What-If Analysis, What-If/Checklist, HAZOP Study, or FMEA may be employed. Specialized or supplemental studies might occasionally involve Fault Tree Analysis, Event Tree Analysis, Cause-Consequence Analysis or Human Reliability Analysis. Preliminary Hazard Analysis and Relative Ranking would not normally be employed for MOC reviews, unless in conjunction with other methods for special situations such as comparing F&EI exposure radii of alternatives.

Study Preparation

Alex, as the engineer initiating the change, has chosen Bob, the furnace operator, and Joe, the instrument technician, to be the review team members. Bob was selected as a team member because he has operated the furnace since start-up and was very familiar with the furnace operation including the quench tank. Joe was selected as a team member because of his experience with pH control systems. Alex is a chemical engineer with five years of experience in the plant and worked as an engineer on the VCM start-up team. The total project cost of $15,000 makes it a significant change requiring a multidisciplinary team.

To prepare for the review, Alex provides the team with an updated P&ID showing the process change, a process flow diagram, piping specifications for the caustic line, proposed changes to the operating procedures, and the instrument specification for the pH monitoring system. She also provides the team with an extensive checklist that had been developed for Safety Reviews (Appendix B). Alex tells the team members that the sections in the checklist on piping and valves, pumps, and instrumentation would be especially pertinent to this change. The team takes a tour of the plant where the proposed change would be installed just prior to beginning the review.

18.2 Analysis Description

The safety review begins at 8:30 a.m. after the area tour. Alex begins the review by asking questions from the checklist.

Alex — I will take notes on any action items that we come up with. I'll ask the questions, and we will all discuss each one until we are satisfied that it has been answered correctly. Is that okay with everyone?

Joe — Who will be responsible to follow up on action items? I hope it's not me.

Alex — Don't worry; I will take care of that. I will also be sure that the Operations manager signs off on the MOC form. I'll assign each action item to a team member for follow-up, and together we can make sure the list is completed before start-up.

Alex — For the first item: is the piping specification suitable for the process conditions?

Bob — It says here in the pipe spec that Schedule 40 carbon steel pipe is okay. I know we use it in other processes with 50% caustic.

Joe — That has been my experience also.

Alex — I'm sure that the pipe spec is right. I got the information from the Technology Center manager.

Bob — It gets down to 20 below in the winter. What will we do to prevent the caustic from freezing and plugging the system?

Alex — Good point. We need to be sure that we put electric tracing on the caustic line. I will check with Operations to see what they do.

Alex — Are there any special considerations, for either normal or abnormal conditions, that could promote pipe failure?

Joe — We already considered the tracing idea. I think also we should avoid low points in the line that could accumulate caustic and freeze.

Alex — I'll make a note to avoid low points when routing the piping. We should also slope the line back to the transfer pump where we can drain it for maintenance.

Alex — Are the contents of the line identified?

Bob — You can give that one to me. I am checking to see that all the lines in the furnace area are properly labeled.

Alex — I will put it on the list. Are block valves or double block and bleed valves required?

Bob — If we need to do maintenance on the caustic line, we'll need a drain valve and isolation valve at the pump. I believe that we have them but I will check in the field.

Alex — I will put that down as another action item.

Alex continues asking questions from checklist on piping and valves. She then moves on to the pump checklist questions.

Alex — Unless there are any more issues on piping and valves, let's move on to pump questions. Can the pump discharge pressure exceed the design pressure of the casing?

Joe — The caustic pump is a centrifugal pump that is designed for the deadhead pressure. I don't believe we have a problem here.

Alex — OK. Can the pump discharge pressure exceed the design pressure of the downstream piping or equipment?

Bob — I don't think so. The piping is Schedule 40 and is good for the pump deadhead pressure of 60 psig, and the quench tank is vented to the atmosphere.

Alex — Can the design temperature of the pump be exceeded?

Bob — Shouldn't we have deadhead protection on the pump in case someone leaves the pump running with the valve closed?

Alex — The P&IDs show a thermocouple in the casing with a temperature switch to shut off the pump if it is deadheaded. I think we are covered on this, but we should check in the field. I will put in on the action item list.

Alex — Let's move on to instrumentation. What would be the effect of a faulty pH probe? How would it be detected? This one is for you, Joe.

Joe — If the pH reading was faulty, we may have too much caustic in the quench tank. The real problem is we may not have enough caustic to neutralize the vent gases and we may have an environmental incident. We need to calibrate the probe weekly. It needs to be on the schedule for instrument calibration.

Alex — Good point. Another action item.

The Safety Review continues through the morning. When the team is satisfied that all pertinent issues have been discussed, Alex brings the MOC form to the Operations manager for his signature. After reviewing the action items, the Operations manager signs off.

18.3 Discussion of Results

Alex goes back to her office and types up the action items on her computer. The action items are entered in the plant-wide data base for tracking. Each action item is assigned to a team member. The MOC form is filed in the plant office. Table 18.1 lists some action items from the MOC review.

Table 18.1 Sample MOC review action items

List number	Action item	Responsibility	Status
1	Check on electric tracing for the caustic line.	Alex	To be completed before restart
2	Avoid low point in caustic line; slope back to the pump to drain.	Alex	To be completed before restart
3	Label the caustic line.	Bob	To be completed before restart
4	Check for block and bleed at the pump.	Bob	To be completed before restart
5	Check for deadhead protection on the caustic pump.	Alex	To be completed before restart
6	Put the pH monitoring system on the instrument calibration schedule.	Joe	To be completed within 30 days after restart

18.4 Follow-up

Alex assigns each action item to a team member and personally checks each one before restart, except for the one item authorized to be completed after start-up (putting the pH monitoring system on the instrument calibration schedule). She reports to the Operations manager when all pre-start items are completed. The Operations manager is pleased with the smooth restart and the improved control. However, the pH monitoring system somehow did not get onto the instrument calibration schedule until the still-open action item was flagged by a later audit. To avoid the same thing happening again with another change, the Anywhere plant manager has since instituted a practice to delay restart until all MOC action items are complete except updating P&IDs when a redlined copy is available.

18.5 Conclusions and Observations

The Safety Review approach using the checklist as an aid was effective and efficient. Alex had assembled the right team members for the nature of this review. The staff requirements for the MOC review are given in Table 18.2. For a more complex process change, the team makeup would need to be expanded and a more structured and comprehensive technique is likely to be appropriate. Additional forms and checklists for management of change safety reviews are given in Appendices A1 and A2.

Table 18.2 Safety Review staff requirements for the MOC review

Personnel	Preparation (h)	Evaluation (h)	Documentation (h)
Leader	2	2	2
Team member [a]	1	2	0

[a]Average per team member

19

Routine Operation Phase

An Illustration of the HAZOP Study Method for Periodic Review

19.1 Problem Definition

Background

The Anywhere VCM plant has been operating for the past two years without an incident. ABC corporate management believes that the hazard evaluations performed on the plant during its design and after its construction have contributed to this good operating history. To help ensure that this operating record is maintained, ABC management is requiring the Anywhere Plant to perform periodic hazard evaluations of all operating units (in fact, ABC has required this of all its facilities). Also, in the U.S., two regulatory agencies require periodic updates of hazard evaluations on processes covered under the OSHA Process Safety Management Standard and the Environmental Protection Agency's Risk Management Plan Rule. These updates are due at least every five years, and more frequently as needed. As illustrated here, many organizations revalidate their hazard evaluations on a more frequent basis, not relying on regulatory requirements to drive their process safety efforts.

The Anywhere plant manager will phase in the hazard evaluations over the next three years. The first unit to be reexamined is the plant's incinerator. This unit was selected because it has undergone several minor design changes over the past two years to correct deficiencies discovered during early operations. Specifically, the following changes have been made:

Process Change A — A redundant air supply fan, designed to automatically start upon low air flow (FIC-1), was added.

Process Change B — A redundant flame scanner (UVL-1B) was added to the shutdown system.

Process Change C — A second incinerator temperature indicator (TI-3) was added, as well as a temperature controller that averages TI-2 and TI-3 to regulate fuel gas flow.

Process Change D — A caustic supply connection and a low pH alarm were added to the quench tank to allow pH control.

All of these design changes were made to solve the operability problems discovered during the first two years. Plant engineering and safety personnel reviewed and approved each individual change prior to installation, in accordance with the Anywhere Plant's management of change program.

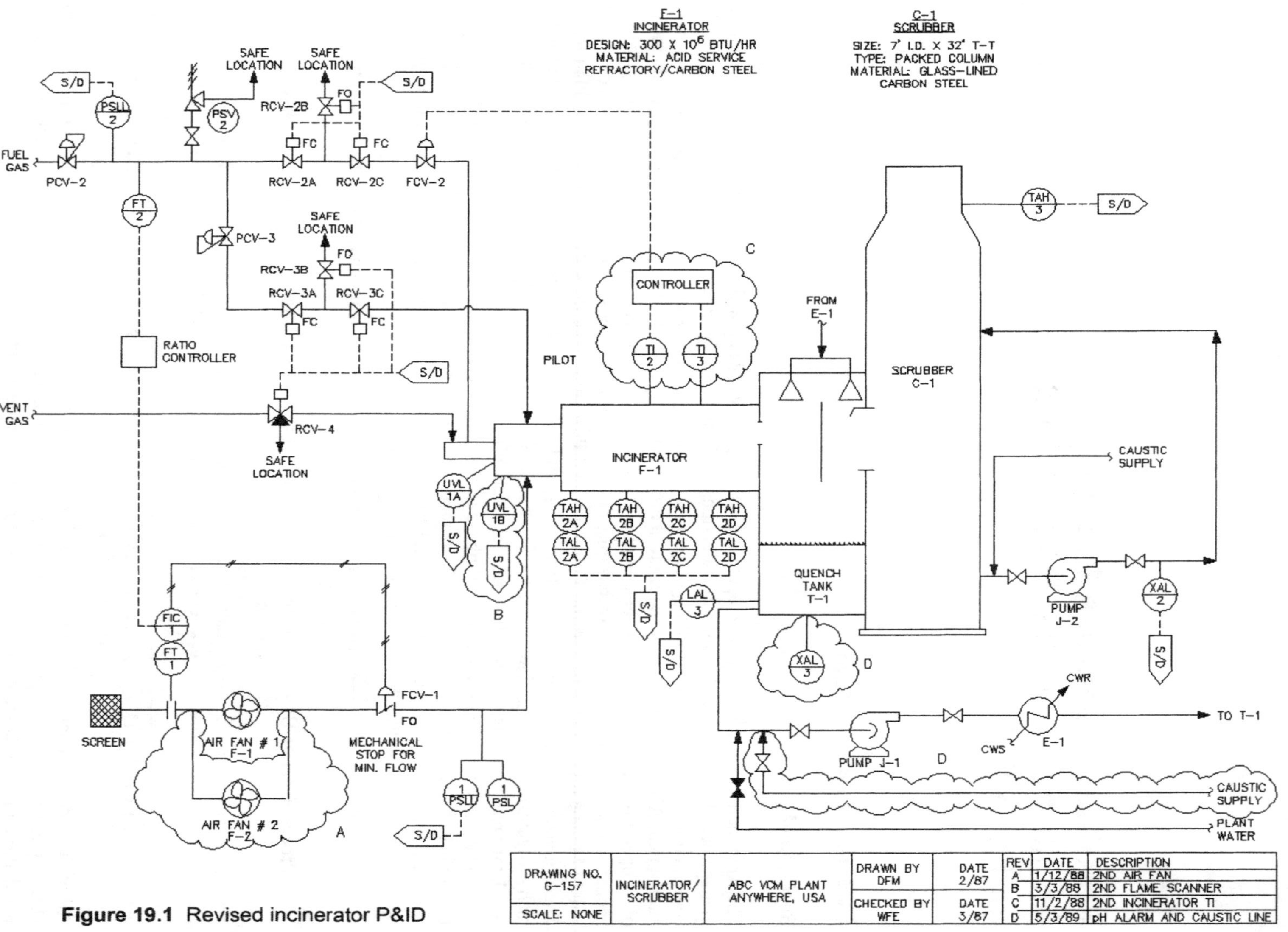

DRAWING NO. G-157	INCINERATOR/ SCRUBBER	ABC VCM PLANT ANYWHERE, USA	DRAWN BY DFM	DATE 2/87
SCALE: NONE			CHECKED BY WFE	DATE 3/87

REV	DATE	DESCRIPTION
A	1/12/88	2ND AIR FAN
B	3/3/88	2ND FLAME SCANNER
C	11/2/88	2ND INCINERATOR TI
D	5/3/89	pH ALARM AND CAUSTIC LINE

Figure 19.1 Revised incinerator P&ID

Available Resources

ABC now has a great deal of material describing the VCM incinerator. They also have two years of operating experience with this unit. The following material is available for the hazard evaluation team to review:

- Piping and instrumentation diagrams (Figure 19.1)
- Process flow diagrams
- Documentation of previous hazard evaluations of the VCM plant (including the HAZOP Study of the original incinerator design)
- Operating and emergency procedures
- Maintenance procedures
- Vendor design specifications, including design basis for relief valves
- Incident reports for the incinerator.

Selection of Hazard Evaluation Technique

The Anywhere VCM plant manager has designated Mr. Smart to perform the hazard evaluation of the incinerator. Mr. Smart is the process engineer responsible for EDC production. However, he has also worked in the incinerator area during construction, start-up, and the first six months of incinerator operation. In addition, Mr. Smart has participated in several hazard evaluations (HAZOP Study and What-If Analysis) of the Anywhere chlorine units during his five years at the plant.

The incinerator is a well-defined system with detailed documentation. With the information available, almost any technique could be used for the hazard evaluation. Mr. Smart's objective, though, is to identify any new hazards that may have been created because of changes in the incinerator design. With such a broad objective, Mr. Smart quickly rules out the selection of hazard evaluation techniques that are better suited to focusing on a specific problem (e.g., Fault Tree Analysis, Event Tree Analysis, Cause-Consequence Analysis, and Human Reliability Analysis). He also does not choose experience-based methods such as the Checklist Analysis and Safety Review techniques, because audits were performed when each change was made. Thus, he narrows his choice of hazard evaluation methods to What-If Analysis, What-If/Checklist Analysis, FMEA, HAZOP Study, Preliminary Hazard Analysis, and Relative Ranking. The Preliminary Hazard Analysis and Relative Ranking methods are discarded because they are too general to use at this stage of the process lifetime. Of the remaining methods, Mr. Smart is most experienced with the HAZOP Study technique and therefore chooses it.

Study Preparation

Mr. Smart must select appropriately skilled personnel to assist in the HAZOP Study. Since all the design changes to the incinerator were engineered at the plant, Mr. Smart selects all the HAZOP team members from plant personnel. He selects the following people for this review:

Leader	—	A person experienced in leading HAZOP Studies. Mr. Smart will be the leader.
Scribe	—	A person who understands technical terms and can record information quickly and accurately. Joe Associate has served as the scribe for ABC HAZOP Studies before and will fill this position.
Process Engineer	—	A person knowledgeable of the incinerator design and how it responds to process transients. Brenda Piper, a chemical engineer responsible for this unit the last one-and-a-half years, will fill this position.
Operator	—	An experienced person from Operations who knows how operators detect and respond to incinerator upsets. David Stedman, an operator for the incinerator since start-up, will fill this position.
Instrumentation and Controls Expert	—	A person familiar with the control and shutdown strategies for the incinerator. Mr. Volt, who designed the instrument changes for the incinerator, will fill this position.

In preparation for the HAZOP meeting, Mr. Smart divides the incinerator into process sections. The incinerator was completely analyzed using the HAZOP Study technique prior to construction. To expedite this review, Mr. Smart chooses to analyze only the design changes; therefore, he sections only those portions of the incinerator that had design changes. In particular, the sections Mr. Smart decides to review in this HAZOP Study are:

- The air supply line to the incinerator (since a second fan was added to this line)
- The fuel gas supply line to the incinerator (since the FCV-2 control strategy was changed)
- The quench tank water recirculation line (since a caustic supply was connected to this line).

The UVL-1B and XAL-3 instruments provide additional safeguards for the incinerator and quench tank, respectively. Thus, Mr. Smart decides he will review the previous HAZOP Study of the incinerator system and insert these new safeguards where appropriate.

Next, Mr. Smart sends an e-mail to all the HAZOP team members informing them of the time and place of the HAZOP meetings. Mr. Smart has scheduled a one-day HAZOP Study of the incinerator changes to be held in the plant's training building. Attached to the e-mail are an updated drawing of the incinerator, a list of sections to be examined, the design intention of each section, and a brief description of the HAZOP Study method. Mr. Smart asks the team members to bring pertinent information on the incinerator (e.g., operating procedures, incident reports, shutdown instrumentation loop sheets) for reference.

As a last preparation step, Mr. Smart prepares a blank HAZOP worksheet to use in the review. He also prepares a preliminary list of deviations to use. This list will be supplemented with other deviations that the HAZOP team identifies during the hazard evaluation.

19.2 Analysis Description

The HAZOP Study begins at 8:00 a.m. Wednesday morning when the team tours the unit. Afterwards, they gather in the plant's training room and Mr. Smart starts the meeting by reviewing ABC's corporate policy regarding periodic hazard evaluations of operating units. He also notes that the incinerator was the first unit chosen for reevaluation at the VCM plant, not because of any specific safety concerns, but because it had been modified several times over the past two years. Mr. Smart reviews the schedule for the day and the ground rules for the HAZOP Study. (He dispenses with introductions since everyone knows each other.) He then briefly reviews how the technique will work and how Mr. Associate will record the results of the review. Mr. Smart also describes the sections of equipment (or nodes) the team will analyze, the design intent of each section, and the order in which they will review these sections.

To start the review, Mr. Smart asks Mr. Stedman to describe (1) the operation of the air supply line to the incinerator and (2) the reason the second air fan was added. (Note: Mr. Smart purposely wanted to involve the operator, who is normally a quiet person, in the discussions; thus, he asks Mr. Stedman to speak first on a familiar subject.) Mr. Smart then begins the HAZOP Study meeting, starting with the deviation "No Flow of air."

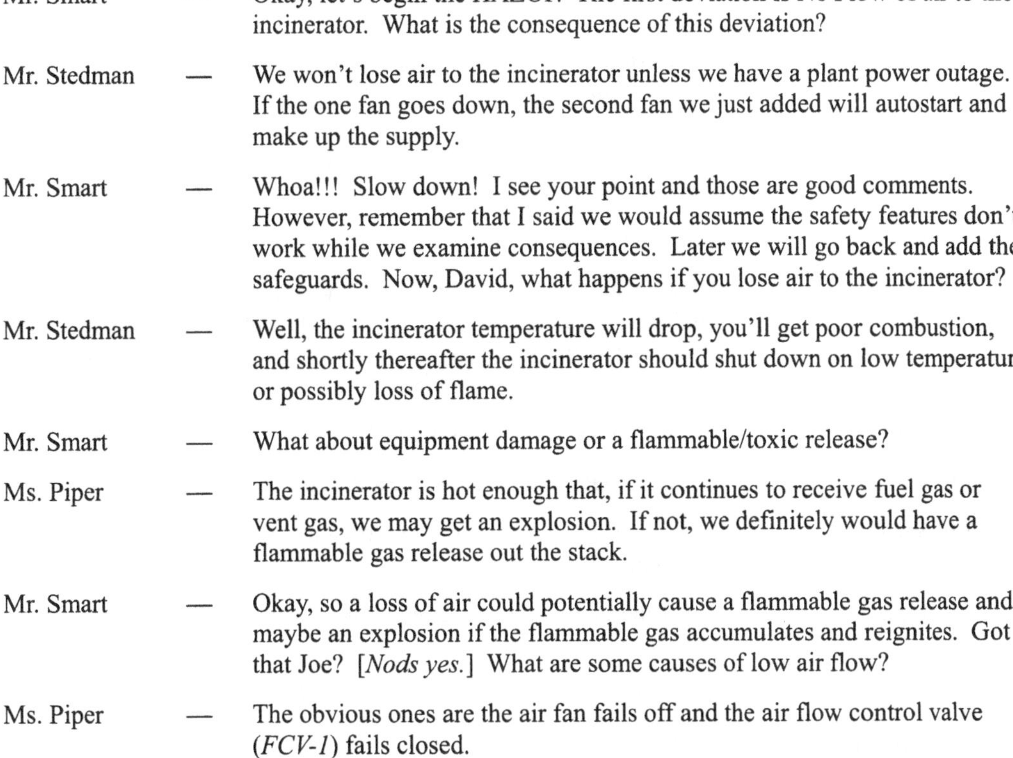

Mr. Smart — Okay, let's begin the HAZOP. The first deviation is No Flow of air to the incinerator. What is the consequence of this deviation?

Mr. Stedman — We won't lose air to the incinerator unless we have a plant power outage. If the one fan goes down, the second fan we just added will autostart and make up the supply.

Mr. Smart — Whoa!!! Slow down! I see your point and those are good comments. However, remember that I said we would assume the safety features don't work while we examine consequences. Later we will go back and add the safeguards. Now, David, what happens if you lose air to the incinerator?

Mr. Stedman — Well, the incinerator temperature will drop, you'll get poor combustion, and shortly thereafter the incinerator should shut down on low temperature or possibly loss of flame.

Mr. Smart — What about equipment damage or a flammable/toxic release?

Ms. Piper — The incinerator is hot enough that, if it continues to receive fuel gas or vent gas, we may get an explosion. If not, we definitely would have a flammable gas release out the stack.

Mr. Smart — Okay, so a loss of air could potentially cause a flammable gas release and maybe an explosion if the flammable gas accumulates and reignites. Got that Joe? [*Nods yes.*] What are some causes of low air flow?

Ms. Piper — The obvious ones are the air fan fails off and the air flow control valve (*FCV-1*) fails closed.

Mr. Volt — A low signal from the fuel gas flow indicator (*FT-2*) or a high signal from the air flow transmitter (*FT-1*) will also close the air flow control valve (*FCV-1*).

Mr. Stedman — A plugged air screen will choke the air flow.

Mr. Smart — Joe, are you keeping up? [*Joe shakes head no.*] Okay, the causes so far are: a plugged air screen, an air fan failure, a false high signal from FT-1, a false low signal from FT-2, and air flow control valve FCV-1 failing closed. Any more? [*No answer. Joe breathes a sigh of relief.*] What about a loss of power?

Mr. Stedman — That will stop the fan, but it should also cause the incinerator to shut down because the RCVs on the fuel lines (*RCV-2A/B and RCV-3A/B*) should close and the incinerator is instrumented to trip on a power loss. If it's only a local power loss, then the spare fan should make up the air flow.

Mr. Smart — Okay, that's a safeguard for partial loss of air. Are there any other causes? [*Quiet.*] Okay, what other safeguards are in place to protect against no air flow?

Ms. Piper — The air flow control valve (*FCV-1*) has a mechanical stop to prevent it from fully closing, we have redundant fans with one on autostart, the air screen is cleaned weekly, there is a low-low air pressure (*PSLL-1*) shutdown, there are numerous other safety shutdowns on the incinerator, and a loss of power will automatically shut down the incinerator.

Mr. Smart — How often is the autostart fan tested?

Ms. Piper — Well, we put the spare fan in because we occasionally had vibration problems with fan #1 and we can't run the plant without the incinerator up. We've actually used fan #2 three times in the last 18 months. I don't know if it is on a test schedule, though.

Mr. Smart — Do I hear a recommendation here?

Mr. Volt — Yes, we should routinely test the autostart for the fan.

Mr. Smart — What about the fuel/air ratio controller and the fuel/air flow transmitter arrangement.

Ms. Piper — I think the safeguards we have are more than adequate. [*Others nod agreement.*]

Mr. Smart — The next deviation is Low (*Less*) Flow of air. Consequences?

Mr. Stedman — What is low?

Mr. Smart — Outside the normal operating limits. Say the air flow control valve (*FCV-1*) is closed down to the mechanical stop.

Mr. Stedman — You probably have the same consequences as before. Perhaps less severe.

Ms. Piper — I agree. The safeguards are still the same and the causes are the same.

Mr. Smart — Joe, read back the causes and safeguards you just wrote down. [*Joe reads list.*] Are these okay?

Mr. Stedman — Everything but loss of power and the fan failing off are okay for causes. The safeguards are fine.

Mr. Smart — Everyone agree? [*Nods yes.*] Any recommendations? [*None suggested.*] The next deviation is High Flow. Consequences?

Ms. Piper — If the fuel/air ratio is too lean, you again have poor combustion. Since the air flow is high, you probably will sweep unburned flammable gas out the stack. If the flow is high enough you may blow out the flame and cause an incinerator shutdown.

Mr. Smart — Any other damage?

Mr. Stedman — Hopefully, the air will sweep any fuel gas or vent gas out the stack. However, we always worry about a potential explosion whenever we lose the flame in the incinerator.

Mr. Smart — What are some causes for high air flow?

Ms. Piper — Well for starters, the opposite causes of low flow. Air flow control valve FCV-1 failing open, fuel flow transmitter FT-2 outputting a false high signal, or air flow transmitter FT-1 outputting a low signal will do the trick.

Mr. Stedman — How about both fans running at full speed. Can that cause the problem?

Ms. Piper — No, the autostart will not turn on fan #2 until fan #1's power is interrupted. Also, air flow transmitter FT-1 would close the damper.

Mr. Volt — I don't think that's correct. According to the instrument loop sheets, the autostart for fan #2 is triggered by a low flow signal from FT-1. And besides, an operator could put fan #2 on manual and start it with fan #1 running.

Mr. Smart — So air flow transmitter FT-1 outputting a false low signal will start fan #2 and open air flow control valve FCV-1?

Mr. Volt — Yes, I believe that's right. I recommend we trigger the autostart from low pressure switch PSL-1. We will need to make sure the second fan gets up to speed before we get a low-low trip.

Mr. Smart — Joe, are you keeping up? [*Joe asks the group to wait a minute.*] We jumped ahead to recommendations. Let's back up and discuss the safeguards.

Mr. Stedman — The fans are fixed-speed. Also, you have all the shutdowns on the incinerator as before.

Mr. Smart — Any others?

Mr. Stedman — We monitor the air flow as part of our board checks. A good operator would note a faulty air flow reading and correct the problem.

Mr. Smart — Would it be caught before the incinerator shuts down?

Mr. Stedman — Maybe. Depends on how much time it takes until an incinerator shutdown is reached.

Ms. Piper — If you blow the flames off the burners, just seconds.

Mr. Smart — Any recommendations?

Ms. Piper — I agree with Mr. Volt that we should trigger the autostart off pressure switch PSL-1. We should also consider installing a high flow alarm to alert the operator to possible FT-1/FCV-1 failures. By the way, we should use a separate flow transmitter for this alarm.

Mr. Smart — Okay, but let's not design the solution. I've got your ideas noted and we'll leave the design work to engineering and to instrumentation and controls. Other recommendations? [*Pause.*] The next deviation is Low Air Temperature. Consequences?

Ms. Piper — We use ambient air. Even on the coldest day, I don't believe the incinerator would see any impact other than using slightly more fuel.

Mr. Smart — Okay, so there is no significant consequence. Let's move on to the deviation High Air Temperature?

Ms. Piper — Again, I don't think it would do anything.

The HAZOP questioning continues throughout the day until all the deviations postulated by the team are examined for the three process sections (air line, fuel line, and quench tank water recirculation line). At this point, Mr. Smart is ready to close the meeting by reviewing the recommendations made. However, Ms. Piper raises an additional issue.

Ms. Piper — Before we stop, I think we should examine the added flame detector. We've discussed all the equipment and instrument changes except this one.

(Note: The team discussed the new pH monitor [XAL-3] when they examined the new caustic line.)

Mr. Smart — I didn't include the UVL instrument change as a separate item in the HAZOP Study because it is solely an added safety feature. I didn't see any safety implications with this change.

Ms. Piper — I see your point. But I would have said the same thing about the second air fan before we examined it.

Mr. Smart — You're right. However, rather than HAZOP the entire incinerator or shutdown system, why don't we just perform a failure modes and effects analysis on this change? [*The others agree after Mr. Smart explains what an FMEA is.*] Okay, how can the UVL detector fail?

Mr. Volt — It can fail to detect loss of flame and it can inadvertently initiate a shutdown signal when no problem exists.

Mr. Smart — Any other failure modes? [*No answer.*] What is the effect of the UVL failing to detect loss of flame?

Ms. Piper — If the incinerator is operating normally, none. If there is a flameout, then you only have one detector to protect you. You may get an explosive mixture in the incinerator.

Mr. Smart — Safeguards?

Ms. Piper — We have a second flame detector. Also, we have the temperature shutdowns on the incinerator.

Mr. Smart — Recommendations? [*None suggested.*] What is the effect of a false flameout signal?

Mr. Volt — A false signal from either UVL will cause an incinerator shutdown. As long as the fuel is shut off, there's no danger.

Mr. Smart — And if fuel isn't shut off, do you have an explosion potential?

Ms. Piper — Yes! The incinerator will become fuel-rich for a period of time. But operators will eventually shut off the fuel supply. We may then reach a point as air leaks into the incinerator where the fuel/air mix is flammable.

Mr. Smart — What safeguards exist?

Mr. Stedman — The UVLs haven't given us any problem thus far. I think the second UVL was put in just to give us a better scan of the fire.

Mr. Smart — Okay, we'll note that the UVLs have been reliable. Any other safeguards? [*None suggested.*] Any recommendations?

Mr. Volt — We might want to consider a voting system for the UVLs like we have for the incinerator thermocouples.

Mr. Smart — Why don't we suggest that the reliability of the UVLs be examined to determine if a voting system is needed? Does that sound okay? [*Others agree.*]

Mr. Smart then closes the meeting by reviewing the recommendations made during the course of the day. He also asks that the team members review the report he will prepare for management when it is ready. Finally, he thanks the team for their participation and compliments them on the excellent review.

19.3 Discussion of Results

The results of the HAZOP Study and the FMEA were entered into hazard evaluation software by Mr. Associate. This tabulation describes, on a cause-by-cause basis, the results of the HAZOP Study. Tables 19.1, 19.2, and 19.3 list a portion of the team's findings. (Note: These examples illustrate qualitative hazard evaluations; see Table 15.6 for a worked example that includes scenario risk calculations.)

Some of the important findings from this review are the following:

- Ensure the redundant air fan autostart will not be triggered by FT-1 malfunctions.
- Verify the UVL reliability, and consider a UVL voting system if the reliability is not acceptable.
- Program the controller associated with TI-2 and TI-3 to ignore an out-of-bounds temperature signal.
- Verify that the design and construction materials of circulation pump J-1 and heat exchanger E-1 can withstand high pH service.

Next, Mr. Smart prepares a report of the HAZOP Study and the FMEA. This brief report contains a list of the team members, a list of information used, a summary of the team's recommendations, and detailed HAZOP and FMEA tables. Also included in this report is a description of the review's scope and a copy of the updated P&ID used in the study. After the team members review this report, Mr. Smart sends it to plant management.

19.4 Follow-up

All of the questions raised during the HAZOP Study were eventually resolved. Also, the team found no problems that needed immediate attention by management. After completing and transmitting the report to the plant managers, the HAZOP team was finished. Plant management reviewed the recommendations made, accepted them, and assigned the incinerator area supervisor the responsibility of resolving each recommendation. The supervisor assigned the recommendations to appropriate personnel and entered them into ABC's computer tracking system. He checked on the status of implementation monthly until all recommendations were resolved. The HAZOP report and resolutions of each recommendation (documented by the supervisor) were placed in the engineering files for the incinerator.

Table 19.1 Sample HAZOP Study results for the routine operation phase

P&ID No.: E-250, Revision D Meeting Date: ##/##/##				Team: Mr. Smart, Mr. Associate, Ms. Piper, Mr. Stedman, Mr. Volt (all from the ABC Anywhere Plant)	
1.0 Line — Air supply line to incinerator (Intention: Supply 15,000 scfm air to incinerator at ambient temperature and 3 in. WC)					
Item	**Deviation**	**Cause**	**Consequences**	**Safeguards**	**Actions**
1.1.1	No Air Flow to Incin-erator	Air fan #1 fails off or fan #1 power off	Loss of combustion. Release out the stack	[] Redundant fan on standby with autostart [] Incinerator shutdowns on low-low air pressure PSLL-1, low temperature, or flameout	1, 2
			Incinerator explosion	[] Redundant fan on standby with autostart [] Incinerator shutdowns on low-low air pressure PSLL-1, low temperature, or flameout [] Accumulation, then reignition of gas+air required for damaging pressure to develop	1, 2
1.1.2		FCV-1 fails closed, FT-1 fails high signal, or FT-2 fails low signal	Loss of combustion. Release out the stack	[] Mechanical stop on FCV-1 [] Incinerator shutdown on low-low air pressure PSLL-1, low temperature, or flameout	2
			Incinerator explosion	[] Mechanical stop on FCV-1 [] Incinerator shutdown on low-low air pressure PSLL-1, low temperature, or flameout [] Accumulation, then reignition of gas+air required for damaging pressure to develop	2
1.1.3		Loss of electric power	Loss of combustion. Release out the stack	[] Incinerator automatic shutdown on loss of electric power	
			Incinerator explosion	[] Incinerator automatic shutdown on loss of electric power [] Accumulation, then reignition of gas+air required for damaging pressure to develop	
1.1.4		Plugged air screen	Loss of combustion. Release out the stack	[] Air screen cleaned weekly [] Incinerator shutdown on low-low air pressure PSLL-1, low temperature, or flameout	2
			Incinerator explosion	[] Air screen cleaned weekly [] Incinerator shutdown on low-low air pressure PSLL-1, low temperature, or flameout [] Accumulation, then reignition of gas+air required for damaging pressure to develop	2

Table 19.1 Sample HAZOP Study results for the routine operation phase

P&ID No.: E-250, Revision D Meeting Date: ##/##/##				Team: Mr. Smart, Mr. Associate, Ms. Piper, Mr. Stedman, Mr. Volt (all from the ABC Anywhere Plant)	
1.0 Line — Air supply line to incinerator (Intention: Supply 15,000 scfm air to incinerator at ambient temperature and 3 in. WC)					
Item	Deviation	Cause	Consequences	Safeguards	Actions
1.2.1	Low Air Flow to Incin-erator	FCV-1 fails partially closed, FT-1 fails high signal, or FT-2 fails low signal	Loss of combustion. Release out the stack	[] Mechanical stop on FCV-1 [] Incinerator shutdown on low-low air pressure PSLL-1, low temperature, or flameout	2
			Incinerator explosion	[] Mechanical stop on FCV-1 [] Incinerator shutdown on low-low air pressure PSLL-1, low temperature, or flameout [] Accumulation, then reignition of gas+air required for damaging pressure to develop	2
1.2.2		Plugged air screen	Loss of combustion. Release out the stack	[] Air screen cleaned weekly [] Incinerator shutdown on low-low air pressure PSLL-1, low temperature, or flameout	2
			Incinerator explosion	[] Air screen cleaned weekly [] Incinerator shutdown on low-low air pressure PSLL-1, low temperature, or flameout [] Accumulation, then reignition of gas+air required for damaging pressure to develop	2
1.3.1	High Air Flow to Incin-erator	FCV-1 fails open, FT-2 fails high signal, or FT-1 fails low signal	Poor combustion. Release of flammable gas out the stack	[] Incinerator shutdown on low temperature or flameout	3
			Incinerator explosion	[] Incinerator shutdown on low temperature or flameout [] Accumulation, then reignition of gas+air required for damaging pressure to develop	3
1.3.2		Operator inadver-tently starts fan #2 (fan #1 still running)	Poor combustion. Release of flammable gas out the stack	[] FT-1 closing to reduce air flow [] Incinerator shutdown on low temperature or flameout	2, 3
			Incinerator explosion	[] FT-1 closing to reduce air flow [] Incinerator shutdown on low temperature or flameout [] Accumulation, then reignition of gas+air required for damaging pressure to develop	2, 3
1.4	Low Temperature		No consequence of concern		
1.5	High Temperature		No consequence of concern		

Table 19.2 Sample action items from the routine operation phase

List number	Action to be considered	Responsibility	Status
1	Establish a periodic program for testing the autostart fan (Item 1.1)	Ms. Piper	To be completed within 50 days
2	Consider using PSL-1 to autostart the standby fan (Item 1.1 and 1.2)	Mr. Volt	To be completed before restart
3	Consider installing a high flow alarm on the air supply line independent of FT-1 (Item 1.3)	Mr. Volt	To be completed within 30 days

Table 19.3 Sample FMEA results for the routine operation phase

DRAWING: E-250, Revision D TEAM: Mr. Smart, Mr. Associate, Ms. Piper, Mr. Stedman, Mr. Volt (Anywhere VCM Plant)					PAGE: 1 of # MEETING DATE: ##/##/##
Item	Component	Failure Mode	Effects	Safeguards	Actions
1	Flame scanner UVL-1B	No signal change	Loss of capability to initiate an incinerator shutdown upon loss of flame. Potential incinerator fire or explosion if flame extinguished	Redundant UVL Multiple incinerator shutdowns (temperature, fuel, air)	Safeguards considered to be adequate
		False flameout signal	Inadvertent incinerator shutdown. Potential incinerator explosion if incinerator fuel not shut off	Shutdown is alarmed. Operators verify shutdown actions Double block and bleed valves in fuel lines Three-way shutoff valve in vent line	Verify reliability of the UVLs

19.5 Conclusions and Observations

The HAZOP Study (and FMEA) went well because Mr. Smart was prepared, the right team of skilled personnel was assembled, and Mr. Smart involved everyone in the review. The HAZOP Study took only six hours. To expedite the review, Mr. Smart examined consequences of deviations before inquiring about safeguards. If the consequences were of no concern, he moved on to other deviations. Mr. Smart also kept the team focused, quickly curtailing unproductive side discussions and avoiding excessive time spent on designing solutions to problems. Table 19.4 summarizes the time required to perform the HAZOP Study.

Mr. Smart also expedited the review by not arbitrarily sticking with one hazard evaluation method. Upon realizing that the flame scanner must also be reviewed, he chose the FMEA technique to examine this one piece of hardware. The FMEA method is well-suited to examining the impacts of hardware failures.

An alternative approach that Mr. Smart avoided was using the previous incinerator HAZOP Study as the basis for the current HAZOP Study. The previous HAZOP Study results were shared with the current HAZOP Study team members before they began the second review, but these results were not used as a "checklist" for the current HAZOP Study. Although using the results of an earlier study as a checklist appears to be an effective way to review a system, Mr. Smart attended some HAZOP Studies in the chlorine plant where this approach proved to be a mistake. More often than not, the team tended to analyze only whether the changes could cause process upsets rather than thinking of other ways such upsets could occur.

One final observation is that the team did develop a few recommendations with respect to the changes made. ABC had modified the incinerator slightly over the past two years to remedy some operability problems. The modifications improved the availability of the incinerator and apparently enhanced its safety. However, the HAZOP Study identified some failures related to these changes that could create safety problems—even after the modifications had been through the usual plant engineering reviews.

Table 19.4 HAZOP Study staff requirements for the routine operation phase

Personnel	Preparation (h)	Evaluation (h)	Documentation (h)
Leader	4	6	12
Scribe	2	6	16
Team member [a]	1	6	1

[a]Average per team member

20

Plant Expansion Phase

An Illustration of the Relative Ranking and HAZOP Study Methods for a Batch Process

20.1 Problem Definition

Background

The VCM plant has proven to be a successful enterprise. Five years into its operation, ABC's marketing group has recommended expanding the product line by making polyvinyl chloride (PVC) from vinyl chloride monomer (VCM). After ABC's business team reviews this recommendation, they decide to move forward with this venture. To expedite this expansion, ABC has chosen to buy the PVC technology from American Industrial Chemical Enterprises, Inc. (AICE), a major developer and licensor of PVC manufacturing technology. They are also recognized as a safety leader in the industry, and, after preliminary review of the PVC reactor design, ABC believes AICE's design and safety standards are as good as or better than ABC's.

As with all the other units at the VCM plant, the business team and company management require that a hazard evaluation be performed on this new unit. ABC is particularly concerned because PVC production is a batch process, rather than a continuous one like most of ABC's other processes, and evidence has shown that batch processes are more prone to safety incidents. The business team asks ABC's process hazards analysis group to perform the hazard evaluation.

ABC is considering two locations for the PVC reactor (Figure 20.1). Location #1 is near the VCM storage area. It is the preferred location because its proximity to the VCM supply would reduce piping and pumping costs. A negative aspect of site #1 is its proximity to the local highway: catastrophic incidents in the PVC reactor may threaten passing motorists. Also, the unit would be outdoors, which may pose a problem with freezing and plugging, since the PVC process uses water as a carrier.

The second site being considered is near the VCM purification area. This location is an ample distance from the public. Also, the reactor could be located in an existing heated building, thus reducing the likelihood of freezing and plugging problems. However, PVC reactor incidents may threaten personnel in the plant's administrative offices.

In addition to the two locations, ABC is also considering using a decommissioned reactor vessel it has in storage. This vessel, while only designed to a 150-psig pressure, is adequate for use as the PVC reactor (the operating pressure of the PVC reactor is 75 psig). However, AICE has recommended a reactor with a 250-psig design pressure.

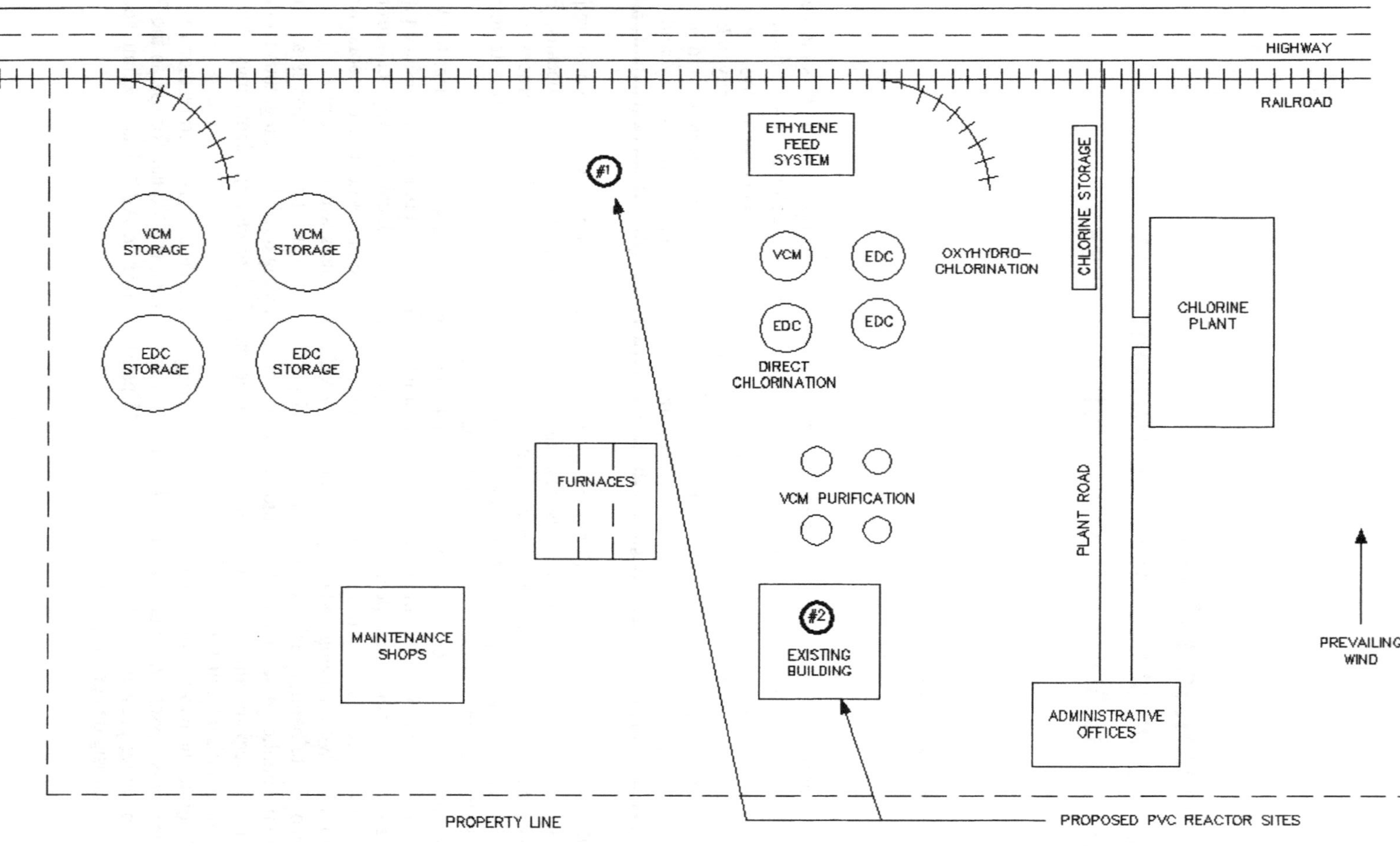

Figure 20.1 VCM plant layout—PVC siting alternatives

Available Resources

While ABC has some experience producing VCM, they have no experience producing PVC. AICE, however, has a wealth of experience in this area. In particular, they can provide the following information for ABC's hazard evaluation:

- Typical P&IDs (Figure 20.2)
- Design specifications
- Process flow diagrams
- Operating experience
- Hazard evaluations on PVC technology
- Operating procedures
- Alarm setpoints and safety shutdown specifications
- MSDSs of all materials in the PVC process
- Emergency response plan

In addition, AICE offers to supply ABC with one of its PVC process designers for the hazard evaluation.

Selection of Hazard Evaluation Techniques

Mr. Dennis, of ABC's process hazards analysis group, is assigned to supervise the hazard evaluation of the new PVC production process. After an initial review of the PVC reactor design, Mr. Dennis determines that two studies are needed: one hazard evaluation to help determine where the PVC reactor should be sited, and another to examine the process safety hazards associated with the unit. Only the PVC reactor poses a significant siting concern because of its large VCM inventory; after the VCM is converted to PVC, the material presents a minimal hazard.

For the siting issue, Mr. Dennis quickly narrows his choices of hazard evaluation methods to Preliminary Hazard Analysis and Relative Ranking. These techniques are selected because they focus on both the equipment at hand and on other equipment and buildings in the area. To select from among the siting alternatives, Mr. Dennis decides he needs a more definitive result than the qualitative ranking that a Preliminary Hazard Analysis normally provides. With this in mind, he chooses the Relative Ranking method, which gives a "hazard score" for each siting location.

The detailed hazard evaluation of the PVC unit should cover a broad range of hazards. Thus, Mr. Dennis decides that What-If Analysis, FMEA, HAZOP Study, and Checklist Analysis methods may be applicable. Not having a good checklist for this new technology, Mr. Dennis rules out this method. (Mr. Dennis has some generalized process hazard checklists, but he is not comfortable using them on a technology unknown to him.) Of the remaining choices, Mr. Dennis believes the HAZOP Study method will work best because the guide word technique can easily be applied to batch procedures. Thus, he chooses the HAZOP Study method.

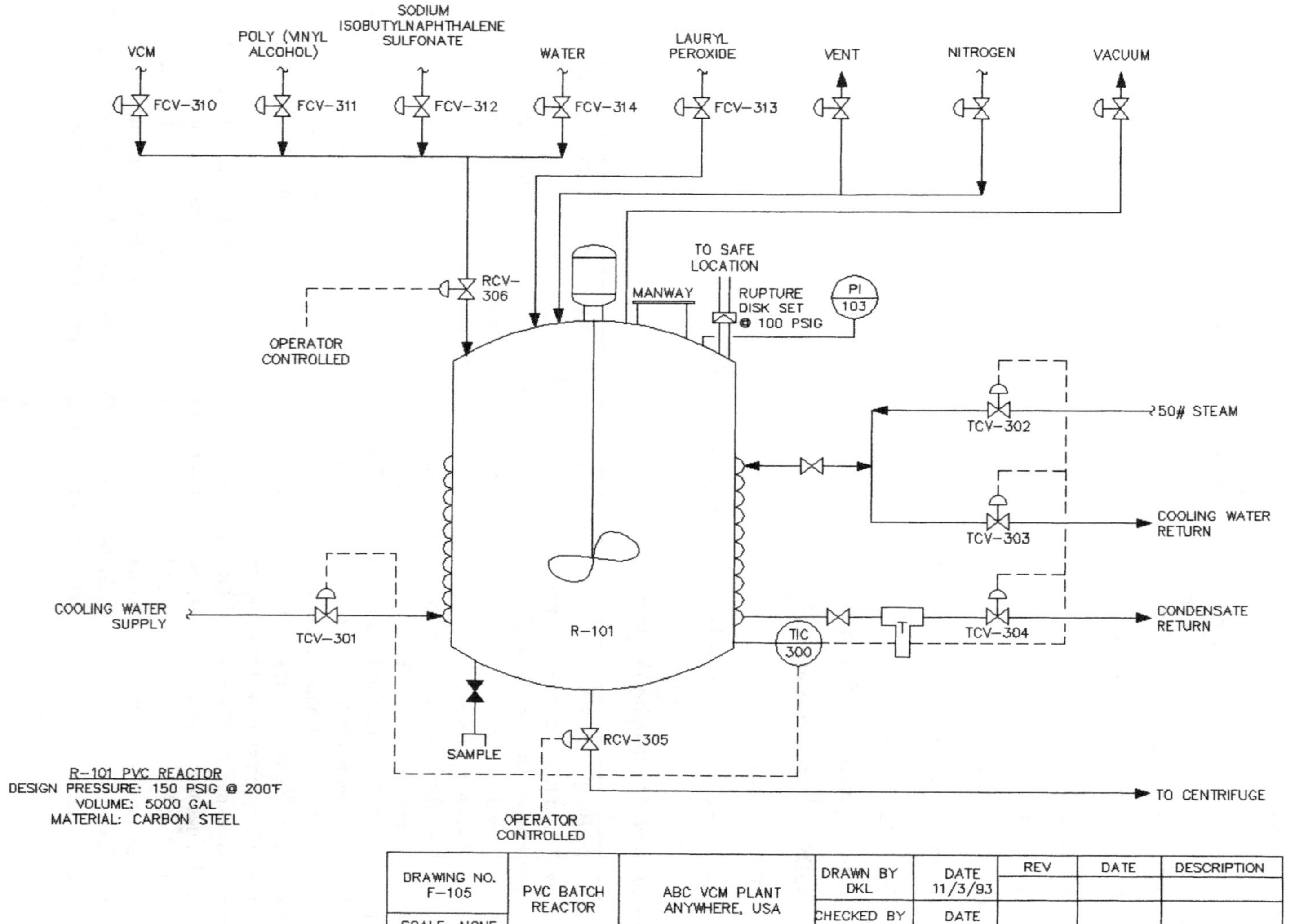

Figure 20.2 PVC batch reactor P&ID

Study Preparation

Relative Ranking. Mr. Dennis chooses to use the Dow Fire and Explosion Index (F&EI) method in his relative ranking hazard evaluation. To perform the ranking, he needs a plot plan of the site, a PFD of the new unit, and *Dow's Fire & Explosion Index Hazard Classification Guide.*[1] He also needs an inventory of the flammable and/or toxic materials in the new unit. Mr. Dennis obtains a plot plan with the proposed new unit locations from the Anywhere VCM plant. The remaining information is obtained from AICE. Mr. Dennis reviews this information and the F&EI *Guide* to see if any additional data are needed. He notes that the credit factors require information on process control features and fire protection, information that he does not have. He will obtain this information through telephone interviews with AICE and Anywhere Plant personnel.

HAZOP Study. As with other HAZOP Studies he has led, Mr. Dennis selects appropriately, skilled personnel to participate in the review. Since most of the ABC personnel who will participate have little or no PVC production experience, Mr. Dennis focuses on obtaining people with related experience (polymer production, batch operations, etc.). He also decides that it is critical that an experienced AICE engineer participate in the study. The following skilled personnel are selected for the review:

Leader	—	A person experienced in leading a HAZOP Study. Mr. Dennis will be the leader.
Scribe	—	A person with a good understanding of technical terminology and the ability to quickly and accurately record information. Ms. Selda Slough will fill this position.
Process Engineer	—	A person who is thoroughly familiar with the design of the process and how it responds to process transients. Mr. Thurmand Plastic of AICE, who helped design the PVC reactor, will fill this position.
Operator	—	An experienced person who knows how operators detect and respond to reactor upsets. ABC has no experienced operators for this reactor system. Mr. Jim Smith, who has worked as a VCM plant operator for the past four years, will fill this position. Jim formerly worked for AICE on some of their batch reactors.
Safety Expert	—	A person familiar with the safety features at the VCM plant. Mr. Fred Scott, a chemical engineer with over 15 years of experience at the Anywhere site (both at the chlorine and at the VCM plant), will fill this position.

In preparation for the HAZOP meeting, Mr. Dennis, working with Mr. Plastic, divides the unit into sections. For each section, he defines a design intention—the function performed by that section and its normal operating parameters. Since it is a batch operation, the design intention of the sections may change throughout the batch. To address this, Mr. Dennis leads the HAZOP Study of the batch procedure and of the process equipment.

Next, Mr. Dennis notifies the team of the schedule and place for the HAZOP Study. Mr. Dennis chooses the Anywhere plant site for the meeting, thus allowing him to review the site for the new unit. (Mr. Dennis plans to complete the Relative Ranking before the HAZOP Study; the HAZOP Study visit will allow him to double-check the siting and protection information he used in the Relative Ranking.) Included in the notification is a set of information provided by AICE (P&IDs, operating procedures, etc.) and the sections and intentions developed by Mr. Dennis and Mr. Plastic.

As a last step, Mr. Dennis asks Ms. Slough to prepare a blank HAZOP table to use in the review. Ms. Slough records the HAZOP Study minutes using commercially available HAZOP software.

20.2 Analysis Description

Relative Ranking

Mr. Dennis performs the Relative Ranking alone, and then has his work reviewed by another member of the process hazards analysis group. Since the PVC reactor is the only major unit with a significant inventory of flammable material, he selects this vessel for the hazard evaluation. The batch reaction used to make PVC requires large volumes of VCM and water. For simplicity (and to be conservative), he only considers the VCM in the batch reactor's mixture when performing the Relative Ranking (i.e., he does not take into account the lower ranking that may occur if water is in the mixture).

Having reviewed the F&EI *Guide*, Mr. Dennis summarizes much of the information needed for the ranking (Table 20.1). He also calls Mr. Scott (to discuss the safety features in place at the two proposed sites) and Mr. Plastic (to discuss safety shutdown systems on the reactor). Based on these discussions, Mr. Dennis determines that the loss control credit factors that are part of the F&EI *Guide* will be about the same for both locations. He chooses to estimate these factors because they affect the damage estimate for an incident, and thus are a factor in the site selection.

The F&EI *Guide* describes a step-by-step procedure for estimating a fire and explosion index and, subsequently, a radius of exposure for the different reactors. In calculating the factors, Mr. Dennis:

Table 20.1 PVC reactor/site information

Volume: 5,000 gal
Normal fill: 80%
Mixture: ~43%VCM, 56% water, 1% other
Operating pressure: 75 psig
Rupture disk setting: 200 psig (new vessel) 100 psig (used vessel)
Outdoor unit (site #1) Indoor unit (site #2)

Distance to:	Site #1 (ft)	Site #2 (ft)
Administrative offices	1500	800
Maintenance shops	1200	850
Highway	800	1500
VCM storage	180	800
Furnaces	500	200
Ethylene feed area	150	900
OHC reactors	300	600
Direct chlorination reactors	200	450
VCM purification columns	350	100

1. Estimates the material factor (MF) for VCM. (VCM is listed in the *Guide* and Mr. Dennis locates its factor in the appropriate table.)

2. Estimates the general process hazard factor (F_1). Following the instructions in the *Guide*, Mr. Dennis determines penalties for exothermic chemical reactions, material handling, enclosed process unit, access, and drainage. Because site #2 is indoors and has fair-to-poor drainage, Mr. Dennis penalizes this PVC reactor more heavily than the reactor at site #1.

3. Estimates the special process hazard factor (F_2). Again, Mr. Dennis follows the *Guide*'s instructions. He notes that the pressure component of this factor changes slightly, depending on which reactor vessel is used. He also penalizes site #2 for its proximity to the cracking furnaces.

4. Estimates the unit hazard factor (F_3), fire and explosion index, and radius of exposure. These calculations involve simple multiplication and addition of MF, F_1, and F_2. Note that Mr. Dennis calculates the F&EI twice for the site #2 cases because the unit hazard factor exceeds the recommended cutoff of 8.0. Since Mr. Dennis only wants to compare causes rather than use the specific values, he calculates the F&EI using his estimated unit hazard factor and the cutoff value of 8.0.

5. Estimates the loss control credit factor. Mr. Dennis gives additional credit to site #1 for drainage and to site #2 for the building sprinkler system. Otherwise, both sites have the same credits. Credit values are selected in accordance with the Guide instructions, including a credit for other hazard evaluations (HAZOP Study).

Figures 20.3 through 20.10 at the end of this chapter summarize Mr. Dennis's calculations for the two proposed PVC sites and the different reactors. Table 20.2 summarizes his findings. Using the estimated radius of exposure and the Anywhere VCM plant plot plan, Mr. Dennis then identifies the equipment and personnel most likely to be affected should a fire or explosion occur in the PVC unit.

Table 20.2 Relative Ranking results for the plant expansion phase

Option	F&E Index	Loss Control Credit Factor	Radius of Exposure (ft)
Site #1, 150 psig	187	0.70	157
Site #1, 250 psig	181	0.70	152
Site #2, 150 psig	192 (250)*	0.65	161 (210)*
Site #2, 250 psig	192 (242)*	0.65	161 (203)*

*Value assuming the unit hazard factor is not truncated at 8.0. The actual calculated value is used.

HAZOP Study

As the HAZOP Study begins, Mr. Dennis introduces the team members and their areas of expertise. He then reviews the analysis schedule. Normally, Mr. Dennis would also review the HAZOP Study method with the team; however, all of the team members, including Mr. Plastic, have previously participated in HAZOP Studies. After discussing the objectives of the HAZOP Study, Mr. Dennis asks Mr. Plastic to explain the PVC production process and the batch reactor operating procedure (Table 20.3).

Mr. Plastic — The PVC production process is relatively straightforward (*see Figure 20.11, at the end of this chapter*). To begin, we prepare the PVC reactor for feeds according to procedure. Once the reactor is ready, we add process water, the dispersants, the monomer, and the initiator. The reactor is heated to 50 °C to initiate the reaction. The pressure at the beginning of PVC production is about 75 psig, but it falls to around 7 psig when the conversion is complete. We then heat the reactor to drive off excess monomer, and follow this step by dropping the reactor contents into the centrifuge. In the centrifuge, we remove excess water from the PVC. The PVC pellets are then conveyed by air to a hot air dryer, sieved to remove oversized particles, and finally stored.

Mr. Dennis — Thanks. To start the review, we have agreed to HAZOP the batch procedure. Mr. Plastic, could you walk us through the details of the batch?

Table 20.3 PVC batch reactor operating procedure

Step	Description
1	Verify that all feed line valves to the PVC reactor are closed and reactor is empty.
2	Open the vacuum line valve and reduce reactor pressure to 10 psi.
3	Open the water line valve and add 2250 gal of water to the reactor (set water flow totalizer).
4	Start the agitator.
5	Close the vacuum line valve.
6	Open the poly (vinyl alcohol) line valve and add 7 gal of poly (vinyl alcohol). Open the sodium isobutylnaphthalene sulfonate line valve and add 1 gal.
7	Open the lauryl peroxide line valve and add 7 gal.
8	Open the VCM line valve and add 1700 gal of monomer.
9	Set the reactor temperature controller to 50 °C and heat up the reactor for about 8 hours. Maintain reactor temperature at 50 °C for 8 hours.
10	When the reactor pressure drops to 7.5 psig, open the vacuum line valve for 15 minutes.
11	Cool the reactor to 16 °C (about 6 hours).
12	Open the nitrogen supply line valve and vent to atmospheric pressure.
13	Open the discharge valve and drop the reactor contents into the centrifuge.

Mr. Plastic — Sure. Each of you has a copy of the batch procedure (*see Table 20.3*), so I will just explain the reasoning behind each step. First, we verify that all feed valves are closed to help ensure that we start with an empty reactor. Then we pull a vacuum on the reactor to minimize the oxygen concentration in it, thus reducing the potential for creating a flammable concentration in the gas phase. Next we fill the reactor approximately half full with water, which acts as the suspension phase for the polymerization. After that the agitator is started, the reactor is bottled up, and dispersants are added. Finally, we add lauryl peroxide, an initiator, to the batch and then add the monomer. With all the ingredients present, we raise the reactor temperature to 50 °C (the reactor pressure will be about 75 psig initially) with steam heating on the jacket, and maintain the temperature until about 90% conversion is reached. It takes about eight hours. At that point, the pressure is about 7.5 psig. The vacuum line is opened and the excess monomer is vented off and recovered. Usually 5 minutes is all that is needed for this step, but we recommend 15 minutes for a margin of safety. After recovering excess monomer, the reactor is cooled; then the reactor is padded with nitrogen, vented, and the PVC/water mixture is dropped into the centrifuge to begin the drying process.

Mr. Dennis — Okay. The first guide word is No. Applying this to the first batch step gives the deviation No Verification. What is the consequence of no verification of closed feed valves?

Mr. Smith — If you ran the previous batch correctly, I wouldn't think anything would happen. But, if you didn't dump the previous batch, you'll eventually overfill the reactor with water. The reactor will be solid with PVC and water, which you back up into one of the chemical feed lines, the vacuum line, or the nitrogen line. That would be a terrible clean-up problem, but I don't see any safety consequences.

Mr. Plastic — I agree. However, that error would probably only force you to take longer in centrifuging and restarting the batch. On the other hand, if one of the chemical feed valves is open, then you may add too much dispersant, which lowers product quality, or too much monomer, which could be a fire hazard. This problem exists even if the last batch went through okay.

Ms. Slough — I don't understand. Where is the fire hazard?

Mr. Plastic — If you have too much monomer, then you could have monomer left in the batch after the recovery step. The problem arises when this monomer gets to the hot air dryer.

Mr. Dennis — What are causes for skipping step 1?

Mr. Smith — Well, I'll point the finger at myself. An operator could forget this step.

Mr. Dennis — Others? [*None suggested.*] Safeguards?

Mr. Scott — I'm sure we'll have extensive operator training before starting this process. Also, we'll require written procedures for the batch.

Mr. Dennis — Any more safeguards? [*None suggested.*] Actions?

Mr. Scott — We should develop a checklist for the operator—a simple one-page check-off like this list you gave us. I also don't see any load cells or a level indicator on the reactor P&ID. Perhaps we should add (1) a level indicator so the operator can catch a valve being open at the wrong time and (2) a high level alarm.

Mr. Dennis — Ms. Slough, do you have those recommendations? [*Nods yes.*]

Mr. Plastic — The water and monomer lines are the only big lines to the reactor. A level indicator and alarm may catch a problem with these lines, but I doubt it will help with the other feeds.

Mr. Dennis — Wait a minute. We are already designing solutions. Let's just note that the operators need a way to positively verify and acknowledge that the valves are closed and the reactor is empty. If there are no more ideas, let's move on.

Mr. Scott — The flow meters on the feed lines would be a better way to catch an open valve. That's what you should alarm.

Mr. Dennis — I get the idea. Again, let's not attempt to design the solution. Another possible No deviation is No Step; that is, a written step prior to verification is needed. Is this the case here? [*Heads shake no.*] Okay, do you think the High and Low guide words apply to step 1? [*Team says no.*] How about Reversing the step sequence, doing something Other Than this step, or doing only Part Of this step?

Mr. Plastic — Reverse this Step would be checking the reactor first and the valves second. That should be no problem. Doing only Part Of the step or something Other Than this step is the same as not doing step 1. You may have too much dispersant, initiator, or monomer that could lead to poor product quality or a fire hazard in the dryer.

Mr. Dennis — What are the causes?

Mr. Plastic — I think the causes, safeguards, and actions we just suggested apply again. [*Others agree.*]

Ms. Slough — Let me read them back to you to be sure. [*She reads them and team agrees they are okay.*]

Mr. Dennis — I plan to cover the "As Well As" deviation as we cover the later steps in the process. The next step is to reduce the reactor pressure to about 10 psi. Is there a step missing between steps 1 and 2? [*Team agrees no.*] What happens if this step is skipped?

Mr. Plastic — Then you will have air in the head space during the batch. That is a potential fire and explosion hazard.

Mr. Dennis — Causes?

Mr. Smith — Operator error.

Mr. Dennis — What else? [*Period of silence.*] What about a failed vacuum system or a stuck valve on the vacuum line?

Mr. Scott — Those make sense, although it seems as if that would be alarmed.

Mr. Plastic — The failure of the vacuum system would trigger an alarm, and the vacuum valve position is indicated in the control room. I guess you count those as safeguards. Another possible cause is a faulty reactor pressure gauge.

Mr. Dennis — Safeguards?

Mr. Plastic — You've got the alarms and indicators I've just mentioned, and you've got the operator training mentioned earlier.

Mr. Dennis — Others? [*None suggested.*] Actions? [*None suggested.*] Okay, let's look at the High deviation. What's the consequence of pulling too much vacuum?

Mr. Plastic — I don't think the vacuum system can pull the reactor down much lower than 4 psi.

Mr. Dennis — So, the potential exists to collapse the reactor. Ms. Slough, make a note in the recommendations to determine the maximum vacuum capability of the vacuum system.

Mr. Plastic — Wait a minute. Although not rated for full vacuum, the 250 psig reactor is large enough and has sufficient wall thickness to withstand full vacuum. I'm not sure about the 150 psig reactor.

Mr. Scott — It probably can take full vacuum, but we should check it out. By the way, that 10 psi, is it gauge or absolute?

Mr. Plastic — It's absolute. But I see your point. An operator could get it wrong. I guess an action is to update the procedure to fix this.

Mr. Dennis — Come on now—let's slow down or we'll get lost. Before making recommendations, what are the causes of too much vacuum?

Mr. Smith — Operator error.

Mr. Dennis — Okay, but it's not always operator error. Jim, what else?

Mr. Smith — Well, a bad pressure gauge on the reactor would possibly mislead an operator. Also, plugging is a common problem.

Mr. Dennis — Other causes? [*None suggested.*] Safeguards?

Mr. Scott — Operator training. Once this reactor is up and running, I'm sure operators will know the amount of time it takes to get to 10 psia.

Mr. Dennis — Recommendations?

Mr. Smith — Besides the things you have already said, it seems like another pressure gauge would help since pressure seems pretty important in this process. Plus, the second gauge may reveal a plugging problem.

Ms. Slough — Just a minute so I can catch up on these recommendations.

Mr. Dennis — Okay. Other recommendations? [*Quiet.*] What about Too Little Vacuum?

Mr. Scott — Well, I think you have the same things as with skipping this step, only maybe not as bad.

Mr. Dennis — Do all of you agree? [*Others nod approval.*] What if you Reverse steps 1 and 2?

Mr. Plastic — If the last batch went through okay, no problem. Otherwise, if you have material left in the reactor, a leaky or open feed valve, or a leaky agitator seal, then you won't reach your 10 psia and you'll end up throwing away some materials through the vacuum line. Since we pull excess monomer out this way anyway, I don't see much more than an economic penalty.

Mr. Scott — What about removing the air from the reactor? Won't we have the same consequence as the No Vacuum?

Mr. Plastic — Maybe. I don't know if we remove enough air in this case to be okay or not. Maybe we should take a closer look at this!

Mr. Dennis — Any other causes besides operational error for the deviation? [*None suggested.*] Some safeguards are, again, operator training and the pressure gauge on the reactor. Others? [*Nothing said.*] Recommendations?

Mr. Scott — Why doesn't this batch reactor have a PLC (*programmable logic controller*) to sequence through the steps?

Mr. Plastic — That can certainly be done. I can only guess that ABC didn't want the cost.

Mr. Scott — I think we should recommend a PLC. We are barely into this and it looks like there is plenty of room for operator error.

Mr. Dennis — Any more recommendations? [*None.*] Ms. Slough, are you keeping up? [*Nods yes.*] Let's see. Part Of would be the same as Low Vacuum. Is there anything an operator might do Other Than open the vacuum line?

Ms. Slough — There's no incentive for an operator to take any other action here, and there are no parallel reactors that might be evacuated by mistake.

Mr. Dennis — Okay. Let's look at As Well As. What if we do step 2 As Well As one of the next steps?

Mr. Plastic	—	Pulling a vacuum As Well As feeding materials will result in not evacuating the air as planned, and thus we have the fire hazard. If the vacuum line is open during the batch, you'll lose a lot of monomer and will probably lift the relief valve on the condenser for monomer recovery. Also, the reactor pressure will drop.
Mr. Dennis	—	Causes besides operator error? [*None suggested.*] Safeguards?
Mr. Plastic	—	You've got the same operator training factors as before, as well as the pressure gauge on the reactor. There's no reason for the operator to take a shortcut like that.
Mr. Dennis	—	Recommendations?
Mr. Scott	—	I'll reiterate my PLC recommendation.

This review continues until all the steps of the batch are examined. The team then evaluates each section of equipment in the new unit during normal operation to identify and evaluate any additional hazards. At the conclusion of the HAZOP Study, Mr. Dennis thanks all the participants and reminds them that he will send them a HAZOP report for their review and comment in a few weeks.

20.3 Discussion of Results

Relative Ranking

The Relative Ranking method was used to estimate a radius of exposure for each of the two proposed PVC unit sites. The F&EI calculations Mr. Dennis performed showed an exposure radius of about 155 ft for site #1 and 161 ft (or 203-210 using Mr. Dennis's unit hazard factor instead of the cutoff value) for site #2, regardless of which reactor vessel (the low pressure or the high pressure) was used. According to the plant's plot plan, the site #2 exposure zone encompasses other major equipment—the VCM purification columns. More importantly, the exposure zone for site #1 does ***not*** involve the VCM storage tanks, although it does reach the ethylene feed area. Large populations of people are not affected at either site. The safety factors at either location are about equal. Thus, site #1 appears to be the better choice.

HAZOP Study

Table 20.4 illustrates some of the HAZOP Study results. Ms. Slough generates the HAZOP table, transcribing her notes into the HAZOP software. Because the PVC reactor design had not previously been examined and because it was not designed to ABC's standards, the HAZOP team suggests a number of improvements. Table 20.5 lists some of the more important recommendations.

Note that the HAZOP discussion would have needed to be more risk-oriented and the documentation expanded if the HAZOP Study was performed using the scenario risk analysis approach given in Chapter 7. In particular, the HAZOP discussion would need to:

Table 20.4 Sample HAZOP Study results for the PVC reactor

Item	Deviation	Causes	Consequences	Safeguards	Actions
P&ID No.: E-101, Revision 5 **Procedure Dated:** ##/##/## **Meeting Date:** ##/##/##			**Team:** Mr. Dennis (Leader — Process Hazard Analysis), Ms. Slough (Anywhere VCM Plant), Mr. Smith (Anywhere VCM Plant), Mr. Scott (Anywhere VCM Plant), Mr. Plastic (AICE)		
Item	**Deviation**	**Causes**	**Consequences**	**Safeguards**	**Actions**
1.0 PVC reactor batch — Step 1: Verify all feed line valves are closed and reactor is empty.					
1.1	No Verification	Operator error — skips step	Increased potential for material to leak into reactor when air is present. Potential fire hazard (monomer and air)	Operator training Written procedures	Develop a batch checklist Provide instruments to verify valve positions and reactor status Add a reactor level indicator and high level alarm
1.2	Part of Verification	Operator error — skips step	Increased potential for material to leak into reactor when air is present. Potential fire hazard (monomer and air)	Operator training Written procedures	Same as 1.1 above
2.0 PVC reactor batch — Step 2: Open the vacuum line valve and reduce reactor pressure to 10 psi.					
2.1	No Evacuation	Operator error — skips step Vacuum system failure Vacuum valve sticks closed Reactor pressure gauge fails — false low	Potential flammable/ explosive mixture in reactor head space	Operator training Written procedures Vacuum system alarm Pressure indicated in control room Vacuum valve position indicated in control room	Safeguards considered to be adequate
2.2	High Evacuation (high vacuum)	Operator error Reactor pressure gauge fails — false high	Potential reactor damage (assuming lower pressure reactor is used)	Operator training Reactor pressure indicated in control room	Evaluate the capability of the vacuum system to damage the reactor Revise the batch procedure to make the units clear (psia vs psig) Install a second pressure gauge on the reactor
2.3	Low Evacuation	Operator error — skips step Vacuum system failure Vacuum valve sticks closed Reactor pressure gauge fails — false low	Potential flammable/ explosive mixture in reactor head space	Operator training Written procedures Vacuum system alarm Pressure indicated in control room Vacuum valve position indicated in control room	Consider using a PLC to sequence the batch
2.4	Reverse Steps 1 and 2	Operator error	Potential flammable/ explosive mixture in reactor head space	Operator training Written procedures	Consider using a PLC to sequence the batch
2.5	Step 2 As Well As another step	Operator error	Potential large release of hot monomer to the atmosphere through the overhead condenser relief valve Potential fire hazard	Operator training Written procedures	Consider using a PLC to sequence the batch

- Break out the grouped causes and safeguards in Items 2.1 to 2.3 to reflect specific scenarios using the cause-by-cause approach instead of the deviation-by-deviation approach as shown.
- Recognize that "Operator training" and "Written procedures" are not safeguards but control measures that would only affect the likelihood of the operator error causes.
- Take the Consequences all the way to specific loss events (e.g., whether the "potential flammable/explosive mixture in reactor head space" would lead to a contained internal deflagration or a vessel rupture explosion if ignited) and evaluate the loss event impacts.
- Evaluate the probability of ignition of the flammable vapor/air mixtures as part of the overall scenario frequency.
- Using a risk matrix or other method, combine qualitative or order-of-magnitude quantitative estimates of the cause frequency, preventive safeguards probability of failure on demand, and loss event impacts to evaluate where the scenario risks fall with respect to predetermined action levels.

The brief worked example in Section 15.7 illustrates the type of risk-based approach that could have been employed to provide a framework for addressing these issues.

Table 20.5 Sample recommendations from the HAZOP Study of the PVC batch reactor

- Run the batch reactor with a programmable logic controller; program verification checks into the PLC to verify proper reactor conditions before proceeding to the next batch step
- Install a redundant pressure gauge and redundant temperature indicator on the reactor
- Install an oxygen analyzer and alarm on the reactor
- Rewrite the batch procedure to clearly state valve positions and process conditions at each batch step
- Develop a checklist for operators to use when making a batch
- Define procedures for safe entry into the reactor to clear a plug in the discharge line
- Install a redundant trip on the centrifuge to shut down the centrifuge on high vibration
- Locate the centrifuge in an unoccupied room with reinforced walls (to protect against shrapnel in the event of centrifuge mechanical failure during operation)
- Examine the materials onsite to see if VCM reactive materials could inadvertently be added to the reactor
- Evaluate capability of vacuum system to damage reactor
- Add reactor level indicator and alarm
- Provide instruments to verify valve position and reactor status

20.4 Follow-up

Relative Ranking

The Relative Ranking portion of the hazard evaluation was completed entirely by Mr. Dennis and reviewed by Ms. Deal (also with ABC's process hazards analysis group) and Mr. Plastic with AICE. The results, along with Mr. Dennis' recommendations, were transmitted to the VCM business team for evaluation. The business team used these results to help them select site #1 for the new PVC reactor.

HAZOP Study

Some unresolved issues remained following the HAZOP Study. These issues were assigned to various team members to resolve and were entered into ABC's tracking system. Mr. Dennis checked with these members weekly until every issue was resolved. The recommendation for the PLC was accepted and the team was later reassembled to perform an FMEA on the PLC.

Mr. Dennis, along with Ms. Slough, prepared a report of the HAZOP Study. The report described the scope of the review, who participated, what information (drawings, procedures, etc.) was used, and the findings of the team. This report was circulated among team members for comment, updated by Mr. Dennis, and then sent to the VCM business team for evaluation. The business team was made responsible for following up on each recommendation. Using ABC's computerized tracking system, Mr. Chemist (of the business team) checked monthly on the implementation of the recommendations until all recommendations were completed.

20.5 Conclusions and Observations

Relative Ranking

The Relative Ranking hazard evaluation was performed to determine which site should be used for the new PVC reactor. While the fire and explosion indexes for the two sites produced nearly the same results, the analysis still proved to be quite useful. Site #1 was the business team's preference of the two sites; however, they were concerned with putting the high pressure reactor near the VCM storage, the ethylene supply system, and the highway. Site #2, while more distant from large fuel supplies and the highway, had the drawback of being close to the furnaces and the administration building. The Relative Ranking analysis showed, however, that a major fire and explosion in the PVC reactor at site #1 should not significantly threaten the equipment, buildings, and roadways near the site. This was not true for site #2. Thus, the business team recommends site #1. The business team also notes that, provided it is adequately inspected and tested, the lower-pressure reactor vessel should be adequate for this service. The time required to perform the Relative Ranking is summarized in Table 20.6.

Table 20.6 Relative Ranking staff requirements for the plant expansion phase

Personnel	Preparation (h)	Evaluation (h)	Documentation (h)
Analyst [a]	4	2	4

[a]Analyst has used the F&E Index

HAZOP Study

The HAZOP Study proved to be a very important part of the hazard evaluation because the new technology, not designed or previously evaluated by ABC, was found to have a number of safety deficiencies. The review team identified a number of recommendations that the business team should carefully consider with the acquisition of the new technology.

The HAZOP Study meetings went quite smoothly, both because Mr. Dennis is a highly experienced leader and because all the team members were experienced and cooperative HAZOP participants. It is important to note that Mr. Plastic was key to the success of the review. Without his depth of experience with the PVC process ABC was acquiring, the team would have struggled to understand the design. The time required to perform the HAZOP Study is summarized in Table 20.7.

Table 20.7 PVC batch reactor HAZOP Study staff requirements for the plant expansion phase

Personnel	Preparation (h)	Evaluation (h)	Documentation (h)
Leader	8	40	24
Scribe	4	40	20
Team member [a]	2	40	2

[a]Average per team member

Chapter 20 Reference

1. American Institute of Chemical Engineers, *Dow's Fire & Explosion Index Hazard Classification Guide, 7th Edition,* ISBN 0-8169-0623-8, Wiley, New York, June 1994.

FIRE & EXPLOSION INDEX

AREA / COUNTRY	DIVISION	LOCATION	DATE
USA		ANYWHERE	
SITE	MANUFACTURING UNIT	PROCESS UNIT	
ANYWHERE PLANT	VCM-SITE #1	150 PSIG PVC	

PREPARED BY:	APPROVED BY: (Superintendent)	BUILDING
DD		
REVIEWED BY: (Management)	REVIEWED BY: (Technology Center)	REVIEWED BY: (Safety & Loss Prevention)
AD		FS

MATERIALS IN PROCESS UNIT: VCM, WATER, DISPERSANTS, INITIATOR

STATE OF OPERATION: __ DESIGN __ START UP ✓ NORMAL OPERATION __ SHUTDOWN

BASIC MATERIAL(S) FOR MATERIAL FACTOR: VCM

	Penalty Factor Range	Penalty Factor Used(1)
MATERIAL FACTOR (See Table 1 or Appendices (A) or B) Note requirements when unit temperature over 140 °F (60 °C)		24
1. General Process Hazards		
Base Factor	1.00	1.00
A. Exothermic Chemical Reactions	0.30 to 1.25	0.50
B. Endothermic Processes	0.20 to 0.40	0.00
C. Material Handling and Transfer	0.25 to 1.05	0.50
D. Enclosed or Indoor Process Units	0.25 to 0.90	0.00
E. Access	0.20 to 0.35	0.20
F. Drainage and Spill Control ____ gal or cu.m.	0.25 to 0.50	0.25
General Process Hazards Factor (F_1)		2.45
2. Special Process Hazards		
Base Factor	1.00	1.00
A. Toxic Material(s) N(H) = 2	0.20 to 0.80	0.40
B. Sub-Atmospheric Pressure (< 500 mm Hg)	0.50	0.00
C. Operation In or Near Flammable Range __ Inerted __ Not Inerted		
1. Tank Farms Storage Flammable Liquids	0.50	0.00
2. Process Upset or Purge Failure	0.30	0.30
3. Always in Flammable Range	0.80	0.00
D. Dust Explosion (See Table 3)	0.25 to 2.00	0.00
E. Pressure (See Figure 2) Operating Pressure 75 (psig) or kPa gauge; Relief Setting 100 (psig) or kPa gauge		0.32
F. Low Temperature	0.20 to 0.30	0.00
G. Quantity of Flammable/Unstable Material: Quantity 13000 (lb) or kg; H_C = 8000 (BTU/lb) or kcal/kg		
1. Liquids or Gases in Process (See Figure 3)		0.16
2. Liquids or Gases in Storage (See Figure 4)		0.00
3. Combustible Solids in Storage, Dust in Process (See Figure 5)		0.00
H. Corrosion and Erosion	0.10 to 0.75	0.10
I. Leakage – Joints and Packing	0.10 to 1.50	0.30
J. Use of Fired Equipment (See Figure 6)		0.10
K. Hot Oil Heat Exchange System (See Table 5)	0.15 to 1.15	0.00
L. Rotating Equipment	0.50	0.50
Special Process Hazards Factor (F_2)		3.18
Process Unit Hazards Factor ($F_1 \times F_2$) = F_3		7.79
Fire and Explosion Index ($F_3 \times MF$ = F&EI)		187

Figure 20.3 Fire and explosion index calculations for low-pressure PVC reactor site #1

LOSS CONTROL CREDIT FACTORS

1. Process Control Credit Factor (C_1)

Feature	Credit Factor Range	Credit Factor Used(2)	Feature	Credit Factor Range	Credit Factor Used(2)
a. Emergency Power	0.98	0.98	f. Inert Gas	0.94 to 0.96	1.00
b. Cooling	0.97 to 0.99	1.00	g. Operating Instructions/Procedures	0.91 to 0.99	1.00
c. Explosion Control	0.84 to 0.98	1.00	h. Reactive Chemical Review	0.91 to 0.98	0.91
d. Emergency Shutdown	0.96 to 0.99	0.98	i. Other Process Hazard Analysis	0.91 to 0.98	0.94
e. Computer Control	0.93 to 0.99	1.00			

C_1 Value(3) 0.82

2. Material Isolation Credit Factor (C_2)

Feature	Credit Factor Range	Credit Factor Used(2)	Feature	Credit Factor Range	Credit Factor Used(2)
a. Remote Control Valves	0.96 to 0.98	0.96	c. Drainage	0.91 to 0.97	0.95
b. Dump/Blowdown	0.96 to 0.98	1.00	d. Interlock	0.98	1.00

C_2 Value(3) 0.91

3. Fire Protection Credit Factor (C_3)

Feature	Credit Factor Range	Credit Factor Used(2)	Feature	Credit Factor Range	Credit Factor Used(2)
a. Leak Detection	0.94 to 0.98	1.00	f. Water Curtains	0.97 to 0.98	1.00
b. Structural Steel	0.95 to 0.98	1.00	g. Foam	0.92 to 0.97	1.00
c. Fire Water Supply	0.94 to 0.97	0.97	h. Hand Extinguishers/Monitors	0.93 to 0.98	0.97
d. Special Systems	0.91	1.00	i. Cable Protection	0.94 to 0.98	1.00
e. Sprinkler Systems	0.74 to 0.97	1.00			

C_3 Value(3) 0.94

Loss Control Credit Factor = C_1 X C_2 X C_3(3) = 0.70 (Enter on line 7 below)

PROCESS UNIT RISK ANALYSIS SUMMARY

Item		
1. Fire & Explosion Index (F&EI)........................ (See Front)	187	
2. Radius of Exposure ....(Figure 7)	157 ft or m	
3. Area of Exposure..	77,500 ft² or m²	
4. Value of Area of Exposure		$MM
5. Damage Factor ....(Figure 8)	0.89	
6. Base Maximum Probable Property Damage – (Base MPPD) [4 x 5]		$MM
7. Loss Control Credit Factor..........................(See Above)	0.70	
8. Actual Maximum Probable Property Damage – (Actual MPPD) [6 x 7]		$MM
9. Maximum Probable Days Outage – (MPDO)......(Figure 9)	days	
10. Business Interruption – (BI)		$MM

(2) For no credit factor enter 1.00.

(3) Product of all factors used.

Figure 20.4 Radius of exposure calculations for low-pressure PVC reactor site #1

FIRE & EXPLOSION INDEX

AREA / COUNTRY	DIVISION	LOCATION	DATE
USA		ANYWHERE	
SITE	**MANUFACTURING UNIT**	**PROCESS UNIT**	
ANYWHERE PLANT	VCM-SITE #1	250 PSIG PVC	
PREPARED BY:	**APPROVED BY: (Superintendent)**	**BUILDING**	
DD			
REVIEWED BY: (Management)	**REVIEWED BY: (Technology Center)**	**REVIEWED BY: (Safety & Loss Prevention)**	
AD		FS	

MATERIALS IN PROCESS UNIT
VCM, WATER, DISPERSANTS, INITIATOR

STATE OF OPERATION
___ DESIGN ___ START UP ✓ NORMAL OPERATION ___ SHUTDOWN

BASIC MATERIAL(S) FOR MATERIAL FACTOR
VCM

MATERIAL FACTOR (See Table 1 or Appendices (A) or B) Note requirements when unit temperature over 140 °F (60 °C): **24**

1. General Process Hazards	Penalty Factor Range	Penalty Factor Used(1)
Base Factor	1.00	1.00
A. Exothermic Chemical Reactions	0.30 to 1.25	0.50
B. Endothermic Processes	0.20 to 0.40	0.00
C. Material Handling and Transfer	0.25 to 1.05	0.50
D. Enclosed or Indoor Process Units	0.25 to 0.90	0.00
E. Access	0.20 to 0.35	0.20
F. Drainage and Spill Control ______ gal or cu.m.	0.25 to 0.50	0.25
General Process Hazards Factor (F_1)		**2.45**
2. Special Process Hazards		
Base Factor	1.00	1.00
A. Toxic Material(s) N(H)=2	0.20 to 0.80	0.40
B. Sub-Atmospheric Pressure (< 500 mm Hg)	0.50	0.00
C. Operation In or Near Flammable Range ___ Inerted ___ Not Inerted		
1. Tank Farms Storage Flammable Liquids	0.50	0.00
2. Process Upset or Purge Failure	0.30	0.30
3. Always in Flammable Range	0.80	0.00
D. Dust Explosion (See Table 3)	0.25 to 2.00	0.00
E. Pressure (See Figure 2) Operating Pressure 75 (psig) or kPa gauge; Relief Setting 200 (psig) or kPa gauge		0.22
F. Low Temperature	0.20 to 0.30	0.00
G. Quantity of Flammable/Unstable Material: Quantity 13000 (lb) or kg; H_C = 8000 (BTU/lb) or kcal/kg		
1. Liquids or Gases in Process (See Figure 3)		0.16
2. Liquids or Gases in Storage (See Figure 4)		0.00
3. Combustible Solids in Storage, Dust in Process (See Figure 5)		0.00
H. Corrosion and Erosion	0.10 to 0.75	0.10
I. Leakage – Joints and Packing	0.10 to 1.50	0.30
J. Use of Fired Equipment (See Figure 6)		0.10
K. Hot Oil Heat Exchange System (See Table 5)	0.15 to 1.15	0.00
L. Rotating Equipment	0.50	0.50
Special Process Hazards Factor (F_2)		**3.08**
Process Unit Hazards Factor ($F_1 \times F_2$) = F_3		**7.55**
Fire and Explosion Index ($F_3 \times MF$ = F&EI)		**181**

Figure 20.5 Fire and explosion index calculations for high-pressure PVC reactor site #1

LOSS CONTROL CREDIT FACTORS

1. Process Control Credit Factor (C_1)

Feature	Credit Factor Range	Credit Factor Used(2)	Feature	Credit Factor Range	Credit Factor Used(2)
a. Emergency Power	0.98	0.98	f. Inert Gas	0.94 to 0.96	1.00
b. Cooling	0.97 to 0.99	1.00	g. Operating Instructions/Procedures	0.91 to 0.99	1.00
c. Explosion Control	0.84 to 0.98	1.00	h. Reactive Chemical Review	0.91 to 0.98	0.91
d. Emergency Shutdown	0.96 to 0.99	0.98	i. Other Process Hazard Analysis	0.91 to 0.98	0.94
e. Computer Control	0.93 to 0.99	1.00			

C_1 Value(3) 0.82

2. Material Isolation Credit Factor (C_2)

Feature	Credit Factor Range	Credit Factor Used(2)	Feature	Credit Factor Range	Credit Factor Used(2)
a. Remote Control Valves	0.96 to 0.98	0.96	c. Drainage	0.91 to 0.97	0.95
b. Dump/Blowdown	0.96 to 0.98	1.00	d. Interlock	0.98	1.00

C_2 Value(3) 0.91

3. Fire Protection Credit Factor (C_3)

Feature	Credit Factor Range	Credit Factor Used(2)	Feature	Credit Factor Range	Credit Factor Used(2)
a. Leak Detection	0.94 to 0.98	1.00	f. Water Curtains	0.97 to 0.98	1.00
b. Structural Steel	0.95 to 0.98	1.00	g. Foam	0.92 to 0.97	1.00
c. Fire Water Supply	0.94 to 0.97	0.97	h. Hand Extinguishers/Monitors	0.93 to 0.98	0.97
d. Special Systems	0.91	1.00	i. Cable Protection	0.94 to 0.98	1.00
e. Sprinkler Systems	0.74 to 0.97	1.00			

C_3 Value(3) 0.94

Loss Control Credit Factor = C_1 X C_2 X C_3(3) = 0.70 (Enter on line 7 below)

PROCESS UNIT RISK ANALYSIS SUMMARY

1. Fire & Explosion Index (F&EI) .... (See Front)	181	
2. Radius of Exposure .... (Figure 7)	152 ft or m	
3. Area of Exposure	72,600 ft² or m²	
4. Value of Area of Exposure		$MM
5. Damage Factor .... (Figure 8)	0.88	
6. Base Maximum Probable Property Damage – (Base MPPD) [4 x 5]		$MM
7. Loss Control Credit Factor .... (See Above)	0.70	
8. Actual Maximum Probable Property Damage – (Actual MPPD) [6 x 7]		$MM
9. Maximum Probable Days Outage – (MPDO) ...... (Figure 9)	days	
10. Business Interruption – (BI)		$MM

(2) For no credit factor enter 1.00. (3) Product of all factors used.

Figure 20.6 Radius of exposure calculations for high-pressure PVC reactor site #1

FIRE & EXPLOSION INDEX

AREA / COUNTRY	DIVISION	LOCATION	DATE
USA		ANYWHERE	
SITE	**MANUFACTURING UNIT**	**PROCESS UNIT**	
ANYWHERE PLANT	VCM - SITE #2	150 PSIG PVC	

PREPARED BY:	APPROVED BY: (Superintendent)	BUILDING
DD		
REVIEWED BY: (Management)	**REVIEWED BY: (Technology Center)**	**REVIEWED BY: (Safety & Loss Prevention)**
AD		FS

MATERIALS IN PROCESS UNIT
VCM, WATER, DISPERSANTS, INITIATOR

STATE OF OPERATION	BASIC MATERIAL(S) FOR MATERIAL FACTOR
__ DESIGN __ START UP ✓ NORMAL OPERATION __ SHUTDOWN	VCM

MATERIAL FACTOR (See Table 1 or Appendices (A) or B) Note requirements when unit temperature over 140 °F (60 °C)		24
1. General Process Hazards	**Penalty Factor Range**	**Penalty Factor Used(1)**
Base Factor	1.00	1.00
A. Exothermic Chemical Reactions	0.30 to 1.25	0.50
B. Endothermic Processes	0.20 to 0.40	0.00
C. Material Handling and Transfer	0.25 to 1.05	0.50
D. Enclosed or Indoor Process Units	0.25 to 0.90	0.45
E. Access	0.20 to 0.35	0.30
F. Drainage and Spill Control ______ gal or cu.m.	0.25 to 0.50	0.50
General Process Hazards Factor (F_1)		3.25
2. Special Process Hazards		
Base Factor	1.00	1.00
A. Toxic Material(s) N(H) = 2	0.20 to 0.80	0.40
B. Sub-Atmospheric Pressure (< 500 mm Hg)	0.50	0.00
C. Operation In or Near Flammable Range __ Inerted __ Not Inerted		
1. Tank Farms Storage Flammable Liquids	0.50	0.00
2. Process Upset or Purge Failure	0.30	0.30
3. Always in Flammable Range	0.80	0.00
D. Dust Explosion (See Table 3)	0.25 to 2.00	0.00
E. Pressure (See Figure 2) Operating Pressure 75 (psig) or kPa gauge; Relief Setting 100 (psig) or kPa gauge		0.32
F. Low Temperature	0.20 to 0.30	0.00
G. Quantity of Flammable/Unstable Material: Quantity 130000 (lb) or kg; H_C = 8000 (BTU/lb) or kcal/kg		
1. Liquids or Gases in Process (See Figure 3)		0.16
2. Liquids or Gases in Storage (See Figure 4)		0.00
3. Combustible Solids in Storage, Dust in Process (See Figure 5)		0.00
H. Corrosion and Erosion	0.10 to 0.75	0.10
I. Leakage – Joints and Packing	0.10 to 1.50	0.30
J. Use of Fired Equipment (See Figure 6)		0.12
K. Hot Oil Heat Exchange System (See Table 5)	0.15 to 1.15	0.00
L. Rotating Equipment	0.50	0.50
Special Process Hazards Factor (F_2)		3.20
Process Unit Hazards Factor (F_1 x F_2) = F_3		10.4
Fire and Explosion Index (F_3 x MF = F&EI)		192*

*The F&EI *Guide* recommends a maximum Process Unit Hazard Factor F_3 of 8. The F&EI is calculated twice, using the truncated F_3 = 8.0 resulting in an F&EI of 192, and the actual F_3 = 10.4 resulting in an F&EI of 250

Figure 20.7 Fire and explosion index calculations for low-pressure PVC reactor site #2

LOSS CONTROL CREDIT FACTORS

1. Process Control Credit Factor (C_1)

Feature	Credit Factor Range	Credit Factor Used(2)	Feature	Credit Factor Range	Credit Factor Used(2)
a. Emergency Power	0.98	0.98	f. Inert Gas	0.94 to 0.96	1.00
b. Cooling	0.97 to 0.99	1.00	g. Operating Instructions/Procedures	0.91 to 0.99	1.00
c. Explosion Control	0.84 to 0.98	1.00	h. Reactive Chemical Review	0.91 to 0.98	0.91
d. Emergency Shutdown	0.96 to 0.99	0.98	i. Other Process Hazard Analysis	0.91 to 0.98	0.94
e. Computer Control	0.93 to 0.99	1.00			

C_1 Value(3) 0.82

2. Material Isolation Credit Factor (C_2)

Feature	Credit Factor Range	Credit Factor Used(2)	Feature	Credit Factor Range	Credit Factor Used(2)
a. Remote Control Valves	0.96 to 0.98	0.96	c. Drainage	0.91 to 0.97	1.00
b. Dump/Blowdown	0.96 to 0.98	1.00	d. Interlock	0.98	1.00

C_2 Value(3) 0.96

3. Fire Protection Credit Factor (C_3)

Feature	Credit Factor Range	Credit Factor Used(2)	Feature	Credit Factor Range	Credit Factor Used(2)
a. Leak Detection	0.94 to 0.98	1.00	f. Water Curtains	0.97 to 0.98	1.00
b. Structural Steel	0.95 to 0.98	1.00	g. Foam	0.92 to 0.97	1.00
c. Fire Water Supply	0.94 to 0.97	0.97	h. Hand Extinguishers/Monitors	0.93 to 0.98	0.97
d. Special Systems	0.91	1.00	i. Cable Protection	0.94 to 0.98	1.00
e. Sprinkler Systems	0.74 to 0.97	0.88			

C_3 Value(3) 0.83

Loss Control Credit Factor = C_1 X C_2 X C_3(3) = 0.65 (Enter on line 7 below)

PROCESS UNIT RISK ANALYSIS SUMMARY

1. Fire & Explosion Index (F&EI) .... (See Front)	192 (250)*	
2. Radius of Exposure .... (Figure 7)	161 (210)* ft or m	
3. Area of Exposure	81,700 ft² or m²	
4. Value of Area of Exposure		$MM
5. Damage Factor .... (Figure 8)	0.89	
6. Base Maximum Probable Property Damage – (Base MPPD) [4 x 5]		$MM
7. Loss Control Credit Factor .... (See Above)	0.65	
8. Actual Maximum Probable Property Damage – (Actual MPPD) [6 x 7]		$MM
9. Maximum Probable Days Outage – (MPDO) ...... (Figure 9)	days	
10. Business Interruption – (BI)		$MM

(2) For no credit factor enter 1.00. (3) Product of all factors used.

*Depending on F&EI used

Figure 20.8 Radius of exposure calculations for low-pressure PVC reactor site #2

FIRE & EXPLOSION INDEX

AREA / COUNTRY	DIVISION	LOCATION	DATE
USA		ANYWHERE	
SITE	**MANUFACTURING UNIT**	**PROCESS UNIT**	
ANYWHERE PLANT	VCM-SITE #2	250 PSIG PVC	
PREPARED BY:	**APPROVED BY: (Superintendent)**	**BUILDING**	
DD			
REVIEWED BY: (Management)	**REVIEWED BY: (Technology Center)**	**REVIEWED BY: (Safety & Loss Prevention)**	
AD		FS	

MATERIALS IN PROCESS UNIT
VCM, WATER, DISPERSANTS, INITIATOR

STATE OF OPERATION
__ DESIGN __ START UP ✓ NORMAL OPERATION __ SHUTDOWN

BASIC MATERIAL(S) FOR MATERIAL FACTOR
VCM

MATERIAL FACTOR (See Table 1 or Appendices (A) or B) Note requirements when unit temperature over 140 °F (60 °C): **24**

1. General Process Hazards	Penalty Factor Range	Penalty Factor Used(1)
Base Factor	1.00	1.00
A. Exothermic Chemical Reactions	0.30 to 1.25	0.50
B. Endothermic Processes	0.20 to 0.40	0.00
C. Material Handling and Transfer	0.25 to 1.05	0.50
D. Enclosed or Indoor Process Units	0.25 to 0.90	0.45
E. Access	0.20 to 0.35	0.30
F. Drainage and Spill Control ______ gal or cu.m.	0.25 to 0.50	0.50
General Process Hazards Factor (F_1)		3.25
2. Special Process Hazards		
Base Factor	1.00	1.00
A. Toxic Material(s) N(H)=2	0.20 to 0.80	0.40
B. Sub-Atmospheric Pressure (< 500 mm Hg)	0.50	0.00
C. Operation In or Near Flammable Range __ Inerted __ Not Inerted		
1. Tank Farms Storage Flammable Liquids	0.50	0.00
2. Process Upset or Purge Failure	0.30	0.30
3. Always in Flammable Range	0.80	0.00
D. Dust Explosion (See Table 3)	0.25 to 2.00	0.00
E. Pressure (See Figure 2) Operating Pressure 75 (psig) or kPa gauge; Relief Setting 200 (psig) or kPa gauge		0.22
F. Low Temperature	0.20 to 0.30	0.00
G. Quantity of Flammable/Unstable Material: Quantity 130000 (lb) or kg; H_C = 8000 (BTU/lb) or kcal/kg		
1. Liquids or Gases in Process (See Figure 3)		0.16
2. Liquids or Gases in Storage (See Figure 4)		0.00
3. Combustible Solids in Storage, Dust in Process (See Figure 5)		0.00
H. Corrosion and Erosion	0.10 to 0.75	0.10
I. Leakage – Joints and Packing	0.10 to 1.50	0.30
J. Use of Fired Equipment (See Figure 6)		0.12
K. Hot Oil Heat Exchange System (See Table 5)	0.15 to 1.15	0.00
L. Rotating Equipment	0.50	0.50
Special Process Hazards Factor (F_2)		3.10
Process Unit Hazards Factor ($F_1 \times F_2$) = F_3		10.1
Fire and Explosion Index (F_3 x MF = F&EI)		192* (242)

(1) For no penalty use 0.00.

*The F&EI *Guide* recommends a maximum Process Unit Hazard Factor F_3 of 8. The F&EI is calculated twice, using the truncated F_3 = 8.0 resulting in an F&EI of 192, and the actual F_3 = 10.1 resulting in an F&EI of 242

Figure 20.9 Fire and explosion index calculations for high-pressure PVC reactor site #2

LOSS CONTROL CREDIT FACTORS

1. Process Control Credit Factor (C_1)

Feature	Credit Factor Range	Credit Factor Used(2)	Feature	Credit Factor Range	Credit Factor Used(2)
a. Emergency Power	0.98	0.98	f. Inert Gas	0.94 to 0.96	1.00
b. Cooling	0.97 to 0.99	1.00	g. Operating Instructions/Procedures	0.91 to 0.99	1.00
c. Explosion Control	0.84 to 0.98	1.00	h. Reactive Chemical Review	0.91 to 0.98	0.91
d. Emergency Shutdown	0.96 to 0.99	0.98	i. Other Process Hazard Analysis	0.91 to 0.98	0.94
e. Computer Control	0.93 to 0.99	1.00			

C_1 Value(3) 0.82

2. Material Isolation Credit Factor (C_2)

Feature	Credit Factor Range	Credit Factor Used(2)	Feature	Credit Factor Range	Credit Factor Used(2)
a. Remote Control Valves	0.96 to 0.98	0.96	c. Drainage	0.91 to 0.97	1.00
b. Dump/Blowdown	0.96 to 0.98	1.00	d. Interlock	0.98	1.00

C_2 Value(3) 0.96

3. Fire Protection Credit Factor (C_3)

Feature	Credit Factor Range	Credit Factor Used(2)	Feature	Credit Factor Range	Credit Factor Used(2)
a. Leak Detection	0.94 to 0.98	1.00	f. Water Curtains	0.97 to 0.98	1.00
b. Structural Steel	0.95 to 0.98	1.00	g. Foam	0.92 to 0.97	1.00
c. Fire Water Supply	0.94 to 0.97	0.97	h. Hand Extinguishers/Monitors	0.93 to 0.98	0.97
d. Special Systems	0.91	1.00	i. Cable Protection	0.94 to 0.98	1.00
e. Sprinkler Systems	0.74 to 0.97	0.88			

C_3 Value(3) 0.83

Loss Control Credit Factor = C_1 X C_2 X C_3(3) = 0.65 (Enter on line 7 below)

PROCESS UNIT RISK ANALYSIS SUMMARY

1. Fire & Explosion Index (F&EI)......................... (See Front)	192 (242)*	
2. Radius of Exposure ....(Figure 7)	161 (203)* ft or m	
3. Area of Exposure..	81,700 ft² or m²	
4. Value of Area of Exposure		$MM
5. Damage Factor ....(Figure 8)	0.89	
6. Base Maximum Probable Property Damage – (Base MPPD) [4 x 5]................................		$MM
7. Loss Control Credit Factor............................(See Above)	0.65	
8. Actual Maximum Probable Property Damage – (Actual MPPD) [6 x 7].............................		$MM
9. Maximum Probable Days Outage – (MPDO)......(Figure 9)	days	
10. Business Interruption – (BI)..		$MM

(2) For no credit factor enter 1.00. (3) Product of all factors used.

*Depending on F&EI used

Figure 20.10 Radius of exposure calculations for high-pressure PVC reactor site #2

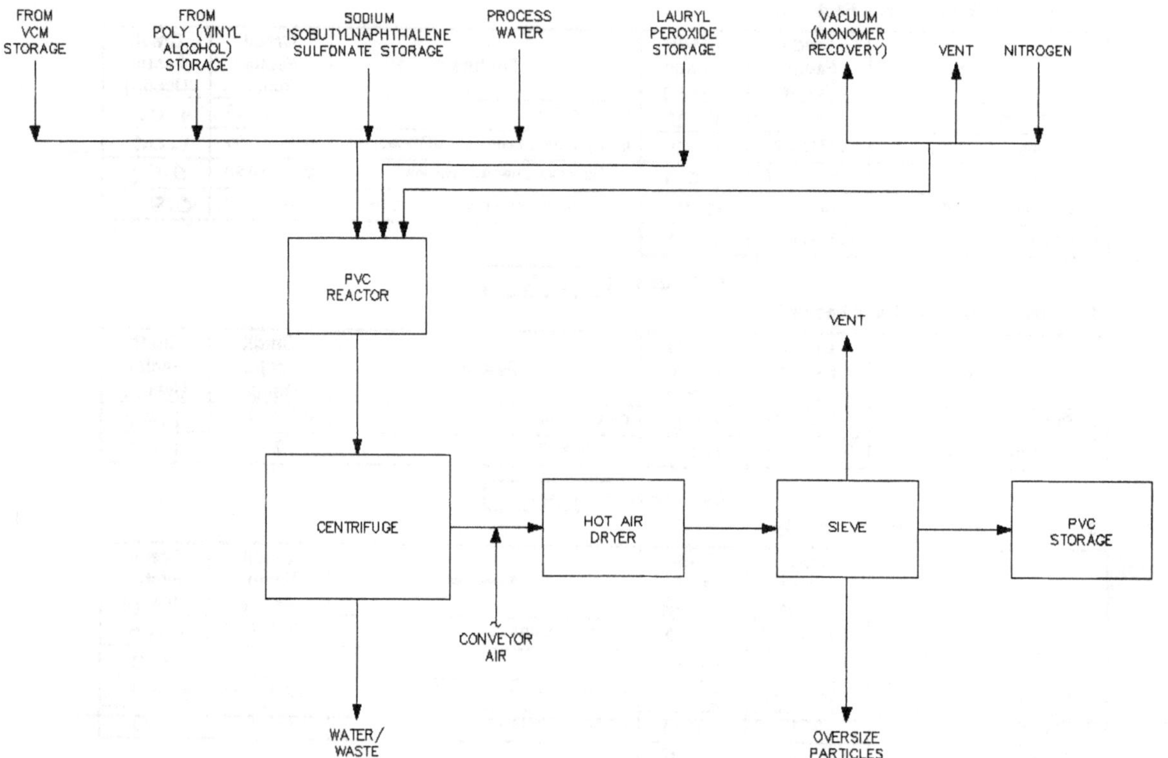

Figure 20.11 PVC unit block diagram

21

Incident Investigation Phase

An Illustration of the FMEA and HRA Methods

21.1 Problem Definition

Background

The VCM plant is now 20 years old. Overall, the plant has had a good operating record. A few minor incidents have occurred that have caused temporary plant shutdowns. However, until now no incident has caused major equipment damage or severe employee injury.

Unfortunately, the VCM plant recently experienced an unintentional release of HCl from its HCl distillation column relief valves. As a result, the column was shut down and has not been restarted. (HCl is one of the by-products that must be removed from the VCM product stream.) ABC reported the incident to the State Environmental Control Board (SECB), Local Emergency Planning Committee (LEPC), and the National Response Center and stated it was taking corrective measures to prevent another such incident. ABC's corporate health and safety officer is also pressuring the plant to quickly resolve this problem since he wants to maintain good relations with the regulatory agencies. (In fact, he has promised the agencies that the specific cause of this problem will be corrected before the VCM plant will operate again.)

The VCM plant manager has assigned Ms. Piper the responsibility of investigating this incident. Her specific objective is to identify and fix the cause of this recent HCl release. However, she has also been asked to identify other reasons HCl might be released from this column and to suggest corrective actions.

Available Resources

With 20 years of operating experience behind them, ABC has a wealth of information on the HCl column. This information includes process flow diagrams, P&IDs, operating procedures, operator logs, maintenance logs, incident histories, and previous hazard evaluations of the plant. It is crucial that the plant be restored as quickly as possible, but all of this information cannot be reviewed quickly. To expedite the investigation, Ms. Piper collects the following information:

- HCl column P&ID (Figure 21.1)
- HCl column operating procedures
- Operator logs for the day of the incident
- Column DCS recordings on the day of the incident (column temperature, reboiler steam flow)

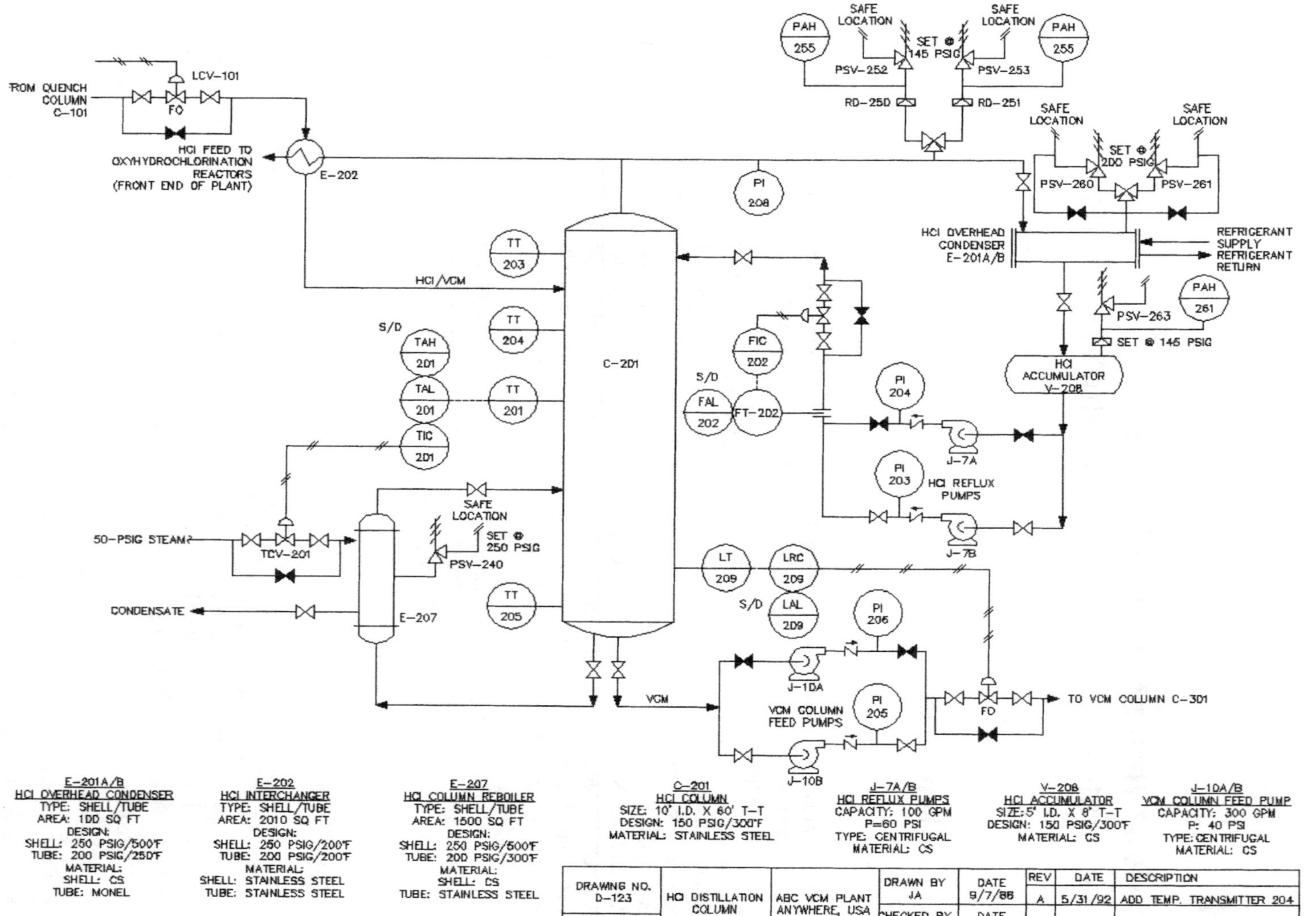

Figure 21.1 HCl column P&ID

Ms. Piper will supplement this information with the knowledge of experienced personnel from the distillation area.

Selection of Hazard Evaluation Technique

Ms. Piper is not a professional process hazard analyst, but she has worked as a chemical engineer at the VCM plant for the past 20 years. Since she began working for ABC, she has participated in and led several hazard evaluations. (These studies were usually HAZOP Studies, What-If Analyses, and FMEAs.)

In selecting a technique to use for the incident investigation, Ms. Piper first calls ABC's process hazards analysis group for advice. Since Ms. Piper only wants to examine HCl releases from the HCl distillation column, the hazard analysis group quickly eliminates the broader and more general review methods such as Preliminary Hazard Analysis, Relative Ranking, and Safety Review. They also eliminate the Fault Tree Analysis, Event Tree Analysis, Cause-Consequence Analysis, and HRA methods because Ms. Piper has little experience with these methods. Nor is the Checklist Analysis method selected; while it works well to help ensure safety is built into a system, it has not proved effective for investigating incidents (because the checklist questions tend to be too general). Thus, they suggest Ms. Piper use the What-If Analysis, HAZOP Study, or FMEA method for her investigation. Since she is concerned with only one effect, the FMEA technique would probably be her best choice. Ms. Piper has previously performed FMEAs and is comfortable with this choice.

Study Preparation

Investigation of the HCl release incident must be executed quickly. Ms. Piper has selected a team of individuals, all from the Anywhere Plant, to participate in the investigation. These skilled personnel are the following:

Leader	—	A person experienced in leading hazard evaluations. Ms. Piper will be the leader.
Scribe	—	A technically astute person who can quickly and accurately record information. Ms. Sally Quick, an engineer from the VCM storage area, will fill this position. (Sally did an excellent job on the last HAZOP Study she scribed for Ms. Piper.)
Process Engineer	—	A person knowledgeable of the HCl distillation column design and operation. Mr. Trey Process, an engineer assigned to the distillation area for the last two years, will fill this position.
Operator #1	—	The control room operator on duty at the time the HCl release occurred. Mr. Vic Martin is this operator.
Operator #2	—	The outside operator on duty at the time the HCl release occurred. Mr. Q. T. Action is this person.
Safety	—	A person familiar with the Anywhere VCM plant safety and emergency procedures. Ms. Opera is the safety manager for the VCM plant and will

fill this position. (Ms. Opera has been at the Anywhere site for over 30 years.)

Outsider — A person familiar with the hazards associated with the HCl distillation process in a VCM plant. Mr. Clean, from ABC's health, safety, and environment group, will fill this position.

A meeting to investigate the HCl release using the FMEA method will occur in two days. There is too little time for Ms. Piper to prepare and distribute an information package to the team members. Instead, she sends everyone an e-mail describing where and when the one-day FMEA of the HCl column will take place. She attaches to this e-mail a list of information she will have at the meeting (see ***Available Resources*** earlier in this section) and requests that team members bring any additional information they feel is appropriate.

In preparation for the meeting, Ms. Piper numbers all of the equipment items in the HCl distillation column. She prepares several blank FMEA tables stored in computer files to complete during the review, and she prepares lists of potential failure modes to examine for each component type in the HCl column. To help develop these lists, Ms. Piper reviews some old FMEAs.

Finally, Ms. Piper visits the distillation area to gain a mental picture of the equipment to be reviewed. During the visit, Ms. Piper also walks through the control room to gain an understanding of its layout. Ms. Piper will use the information collected during this visit to help her formulate questions regarding operator interfaces with equipment.

21.2 Analysis Description

The Failure Modes & Effects Analysis of the HCl column begins promptly at 8:00 a.m. Ms. Piper opens the meeting by asking Mr. Clean to discuss ABC's concerns regarding the recent HCl release. (All the team members know each other, so introductions are skipped.) Mr. Clean notes that ABC has had an excellent operating record in this state until recently. Within the past year, however, three of its plants have had releases. While no single release was particularly significant, the regulatory agencies are beginning to closely scrutinize ABC's plants. ABC's vice president of manufacturing has promised that every reasonable step will be taken to prevent any more incidents. Thus, the Anywhere plant staff must do its utmost to prevent any other unintentional releases.

Ms. Piper next reviews the FMEA plan for the day. The FMEA will focus only on the HCl column area, and within this area, only on failures that lead to an HCl release. To perform the FMEA, she will postulate failures for each component in the system. For each failure, the team will determine the effect. If it is an effect of interest, the team will identify existing safeguards and possible improvement actions. Otherwise, Ms. Piper will move on to another failure mode. Also, Ms. Piper will ask the team to suggest component failure modes in addition to the ones she has postulated.

Ms. Piper instructs the team to clearly identify the causes of the most recent release, as well as any other potential causes of HCl releases. She also tells them to identify safeguards that would mitigate a release or prevent it from occurring. Ms. Piper also strongly emphasizes that the purpose of this investigation is ***not*** to blame anyone, but to learn from this incident so measures can be taken to reduce the chance of it happening again. To open the discussions, Ms. Piper asks Mr. Vic Martin to recount the events that occurred just before and after the release.

Mr. Martin — Well, the unit had been running fairly smoothly for the last few weeks. But at about 3:45 p.m., I noticed a slight rise in the column bottoms temperature (*TI-205*). Since the other column temperatures were okay, I thought I should just check it more frequently. A few minutes later an alarm sounded for the incinerator—a low pH alarm in the quench tower. Since we had been having trouble with false alarms from this sensor, I silenced the alarm and called Q. T. (the outside operator) to go check the pH. When I next scanned the HCl column temps, the bottoms temperature had risen several degrees, and the other column's TIs had risen, except for TIC-201, which read low.

At that point I closed TCV-201 to stop the reboiler steam and LCV-101 to stop feed to the column. I radioed Q. T. to start the other HCl overhead condenser. However, the column pressure had already risen above 145 psig and the relief valve lifted. The relief valve reseated in about 15 seconds.

(*Mr. Process follows this discussion with a review of the strip chart recordings and the Engineering Department's analysis of these recordings.*)

Ms. Piper — I want to first focus on the recent HCl release and its cause. After we have resolved that issue, we will review the column for other causes of HCl releases. From what Mr. Martin said, it sounds like too much heat was put into the column. Let's start the FMEA with the valve controlling steam to the reboiler. What are some failure modes for this valve?

Mr. Process — The control valve (*TCV-201*) could fail in position, fail open, fail closed, or leak externally.

Mr. Action — The control valve (*TCV-201*) could leak through also, allowing steam to the reboiler when you don't want it.

Ms. Piper — Okay, let's take these failure modes one at a time. What is the effect of TCV-201 failing open?

Mr. Martin — That would increase the reboiler duty, raise the column temperature and pressure, and eventually lift a relief valve (*PSV-252 or PSV-253*) if the shutdown system and operator did not intervene.

Ms. Piper — So, we might have an HCl release. What safeguards exist?

Mr. Martin — From the P&ID, you can see that we have multiple temperature indicators on the column that we monitor continuously. There is a temperature sensor *(TT-201)* that will send a signal to alarm (*TAH-201*) and shut down the column on high temperature, as well as a low reflux flow alarm (*FAL-202*) and a low column level alarm (*LAL-209*). Also, a loss of the refrigerant compressor will shut down the column. And of course we have redundant rupture disks and relief valves on the column and extra overhead cooling capacity with redundant HCl overhead condensers and the interchanger.

Ms. Piper — That sounds like a lot of safeguards! But which of those would actually help you prevent a release if TCV-201 failed open?

Mr. Martin — Well, the TIs and TAH-201 definitely help us catch the problem. Valving in the spare overhead condenser would probably provide enough excess cooling to avoid lifting a relief valve if done in time. The other alarms probably wouldn't matter if flows around the column were normal.

Ms. Piper — Does the team believe that TCV-201 failing open caused this incident?

Mr. Martin — No, because I closed this valve when the column temperature got too high and the valve worked.

Ms. Piper — How do you know the valve worked?

Mr. Martin — The limit switch on the valve was made, indicating a fully closed valve. Also, shortly thereafter the column temperature started dropping.

Ms. Piper — Okay. Before we leave this equipment failure mode, are there any recommendations?

Mr. Process — I think we need a high pressure alarm on the column. Right now, operators have only the temperature alarm to alert them to this problem. Also, we should have an emergency checklist that instructs operators to put the spare overhead condenser in service whenever the column temperature is too high.

Ms. Opera — That checklist should also specify keeping LCV-101 open, since this stream is a coolant to the column.

Ms. Piper — Another component failure is TCV-201 closes. What's the effect of this failure?

Mr. Martin — That would cut off steam to the reboiler and cool down the column. HCl would drop out with the VCM, but it wouldn't cause an HCl release.

Ms. Piper — Okay, let's move on. What is the effect of an external valve leak?

Mr. Martin — As long as it's not a major leak, the column should run normally. We just waste steam. If it's a big leak, the column would cool down and you have the same impact as a closed valve. But not an HCl release.

Ms. Piper — What about the valve leaking through?

Mr. Process — That failure really only makes sense if the column is shut down. If we were on hot standby, this leakage could cause the column temperature to rise slowly and eventually cause an HCl release.

Ms. Piper — Safeguards?

Mr. Process — The same as with TCV-201 failing open. Plus, operators should have a lot of time to diagnose and correct this problem.

Ms. Piper — Recommendations? [*None suggested.*] Let's move on to the instrumentation that controls TCV-201, the temperature transmitter (*TT-201*), and controller (*TIC-201*). The failure modes for this equipment are (1) a false high temperature signal output, (2) a false low temperature signal output, and (3) no change in signal even though the column temperature has changed. Any other failure modes? [*No answer.*] Okay, what is the effect if TT-201 or TIC-201 outputs a false high signal?

Mr. Martin — The HCl column will get cold as steam to the reboiler is cut off. This would cause HCl to drop out the bottom with the VCM and lower the quality of our product.

Ms. Piper — Would it cause an HCl release?

Mr. Martin — No.

Ms. Piper — Then let's move on. What's the effect of a false low signal?

Mr. Martin — I think that's the cause of our incident. The false low would cause TCV-201 to open, raising the temperature of the column without sounding the alarm. If we don't catch it in time, the column relief valves will lift. Also, the failed transmitter or TIC would disable the high temperature shutdown for the column.

Ms. Piper — Well that's certainly not good design practice, using the same transmitter to control, alarm, and shut down the column. We should put the alarms and safety shutdown on another transmitter. Have you tested TT-201 and TIC-201?

Mr. Process — It's being done as we speak.

Ms. Piper — Okay, so the temperature transmitter/controller failure may be the cause of our HCl release. What are the safeguards? And did any of them fail?

Mr. Process — The safeguards are the same as Mr. Martin stated earlier—the column TIs and TAH-201. Well, I guess we can't count the temperature alarm.

Ms. Piper — Any other safeguards?

Ms. Opera — The excess cooling capacity of the overhead condensers and the interchanger might be considered safeguards since they can help prevent an overpressure event. Like I said earlier, stopping HCl/VCM feed through the interchanger reduced our overhead cooling capacity and probably contributed to the release.

Ms. Piper — Does the team agree with this scenario? [*Nods of agreement.*] Okay, let's assume that this is the cause of the release pending the outcome of the instrument testing. What recommendations are there?

Mr. Process — Well, I reiterate my previous recommendations. We should put a high pressure alarm on the column and we should develop an emergency checklist for operators to follow when the column temperature is too high. I also agree with your suggestion to use a different transmitter for the column temperature alarms and safety shutdown.

Mr. Martin — Another item we should have on this checklist is to increase the reflux flow. That should buy us some additional time to get the spare condenser on line.

Ms. Piper — Those are good ideas, but let's not design the solution in detail. That can be done outside this group. Sally, did you get all of this? [*Nods yes.*] Any other recommendations? [*No response.*] The last failure mode for this instrumentation is no change in signal. What's the effect in this case?

Mr. Martin — If the flows around the column are steady, nothing would happen. If the flows changed some, the column will get a little hot or cold, depending on which way the flows drifted. I guess there is a slight chance the column pressure would rise too high, but it would be a slow rise.

Ms. Piper — Safeguards?

Ms. Opera — The same safeguards as with the false low signal. However, we should note that operators will have added time to diagnose and correct the problem. Also, operators can manually regulate the steam flow.

Ms. Piper — Recommendations?

Mr. Martin — The same ones we mentioned before, a high pressure alarm on the column, an emergency checklist, and a different temperature transmitter for the column alarms and safety shutdowns.

Ms. Piper — The next item I want to examine is the overhead condenser. The failure modes that I identified are a shell leak, a tube leak, and loss of cooling capability. Any others? [*None suggested.*] What's the effect of a shell leak?

Mr. Action — Process material is on the shell side. Thus, a shell leak would release HCl.

Ms. Piper — Safeguards?

Mr. Action — The design pressure of the shell is 250 psig, well above the point where the HCl column relief valves would lift. Gaskets on this exchanger are rated to 600 psig and the shell is tested annually at turnaround. We've never had any shell leaks with this condenser.

Ms. Piper — Any recommendations? [*None suggested.*] What is the effect of a tube leak?

Mr. Process — The refrigerant would leak into the process. Since it is a low boiler, it would act as a noncondensable in the column and cause the pressure to rise. If the leak were severe, we would lift the column relief valves. We have occasionally experienced a tube leak that has caused a pressure buildup. We just switch over to a spare condenser and repair the leaky tube. We vent the noncondensables to the relief header.

Ms. Piper — Safeguards?

Mr. Action — The tubes are Monel®, well-designed for the temperature and pressure service they see. They are rated to 200 psig and inspected annually. Also, the tube sheets are seal welded. We've never had anything but small tube leaks, which we always caught early by finding refrigerant in the product samples we take each shift or by seeing a slight pressure rise on the overhead PI.

Ms. Piper — Any recommendations? [*None suggested.*] What is the effect of the refrigerant system shutting down?

Mr. Martin — The column will quickly overpressurize and the relief valve should open. However, the refrigerant compressor shutting down is interlocked to shut down the distillation area, including the HCl column.

Ms. Piper — Okay, we have the effect of an HCl release and one of the safeguards. Any other safeguards?

Ms. Opera — You still have the other column safeguards—the high temperature safety shutdown, the pressure alarm, and the emergency checklist.

Ms. Piper — Wait a minute. The pressure alarm and emergency checklist were recommendations. We can't count on them being there yet. But we can count the high temperature shutdown. Any other safeguards? [*None suggested.*] Recommendations? [*None suggested.*] Sounds like the HCl column pressure is highly sensitive to the overhead condenser working. Before we leave this item, are there any other ways to lose refrigeration other than the compressor shutdown?

Mr. Action — Sure, we could run our refrigerant tank dry. But that would starve the compressor and automatically trip the distillation columns. The refrigerant tank also has a low level alarm. Another possibility is to rupture a refrigerant supply pipe, but that would be rare. These pipes are above ground, well labeled, and inspected annually.

Ms. Piper — How about unintentionally blocking in the active condenser?

Mr. Action — I guess that's a possibility, but it doesn't seem too likely. Outside operators for this unit should be the only ones to isolate a condenser,

		and we check the local column pressure indicator whenever we isolate a condenser when switching to the standby.
Ms. Piper	—	As far as safeguards, besides the procedural controls you already have, do any alarms warn you that a condenser has been inadvertently isolated?
Mr. Action	—	Sure, the column relief valve! [*Laugh.*]
Ms. Piper	—	I recommend we put a low flow alarm on the refrigerant supply to the condenser.
Mr. Process	—	Hold on. This isn't the only overhead condenser we have out there. To be consistent in the plant, we would have to install dozens of instruments and alarms, which I think would overload the control room operators. I think we should examine how likely this error is before we start adding instrumentation.
Ms. Piper	—	I see your point. Why don't we suggest that our computer process hazards analysis group perform a Human Reliability Analysis of switchovers on the overhead condensers to see if additional instrumentation and alarms are needed.
Mr. Process	—	Sounds okay to me. [*Others agree.*]
Ms. Piper	—	Let's stick with the overhead equipment. The next component is the HCl accumulator. What are some failure modes for it?
Mr. Action	—	The accumulator could leak, which causes an HCl release.
Ms. Piper	—	Okay. Others?
Mr. Process	—	The accumulator could be flooded.
Ms. Piper	—	Actually those are conditions or effects in the accumulator caused by the reflux pumps failing off, the reflux flow control valve failing closed, etc. We will examine these component failures and their effects in a few minutes.

The FMEA review continues throughout the day until all the components around the HCl column are examined. To expedite her review, Ms. Piper examines all the components in one process line, starting at the column and ending at the column (e.g., tracing through the reflux line equipment) or at another unit (e.g., tracing through the HCl column bottoms discharge line). Upon completing the FMEA, Ms. Piper briefly reviews the recommendations suggested by the team and gives them a printout of the FMEA table generated during the meeting. She will circulate a report of the team's findings for their comment and approval in a few days.

Then Ms. Piper contacts Mr. Dennis of ABC's process hazards analysis group about the Human Reliability Analysis requested by the team. Mr. Dennis states that he can do a small, qualitative HRA of the condenser maintenance procedure in just a few hours. Ms. Piper faxes him a drawing of the overhead condensers and the procedures for switching condensers.

The switchover procedures are as follows:

- The field operator (or lab) detects refrigerant in the product sample and notifies the control room operator.
- The control room operator radios the field operator to isolate the failed condenser to minimize product contamination. (It is implied that the operator should valve in the spare condenser.)
- The outside operator must manually open the process valves on the spare condenser and close the process and refrigerant valves on the leaking condenser. (Refrigerant flow is normally maintained through both condensers.)
- During rounds, the operator is to check a local column pressure gauge to verify proper column pressure.

Experience has shown that the switchover must be completed within a few hours of detecting the leak. Otherwise, the column must be shut down to avoid an overpressurization caused by refrigerant leaking into the column.

Reviewing these procedures, Mr. Dennis postulates the following failures can lead to a loss of condensing and subsequently to an HCl release:

- The outside operator makes a condenser switchover on the wrong column (thus, noncondensables continue to build up in the HCl column, eventually lifting a relief valve).
- The outside operator blocks in the active condenser before putting the standby condenser into service.
- The outside operator fails to block in the leaking condenser.
- Isolation valves on the leaking condenser fail to close.

Mr. Dennis then constructs an HRA event tree (Figure 21.2) to depict the potential human errors (represented by capital letters) and the potential equipment failures (represented by capital Greek letters) that lead to a loss of condensing. The failure pathways in this HRA event tree end with an "F" and are combinations of human and component failures that lead to a loss of condensing.

For example, Mr. Dennis' event tree shows that if the outside operator made a condenser switchover on the wrong column (event A) and failed to detect this error on rounds (event B), then the leaking condenser would continue to release refrigerant into the HCl column and overpressurize it. This is failure pathway F_1 in Figure 21.2. Another failure pathway, F_4, shows that the outside operator chooses the right column overhead condensers (event a) and tries to isolate the leaking condenser after putting the spare condenser into service (events c and d). However, the process inlet valve on the leaking condenser fails to fully close (event Σ) and the operator fails to detect the stuck valve (event E) and the continuing pressure rise in the column (event B).

The HRA event tree and its minimal cut sets, along with a brief report by Mr. Dennis, are transmitted back to Ms. Piper for evaluation by the FMEA team.

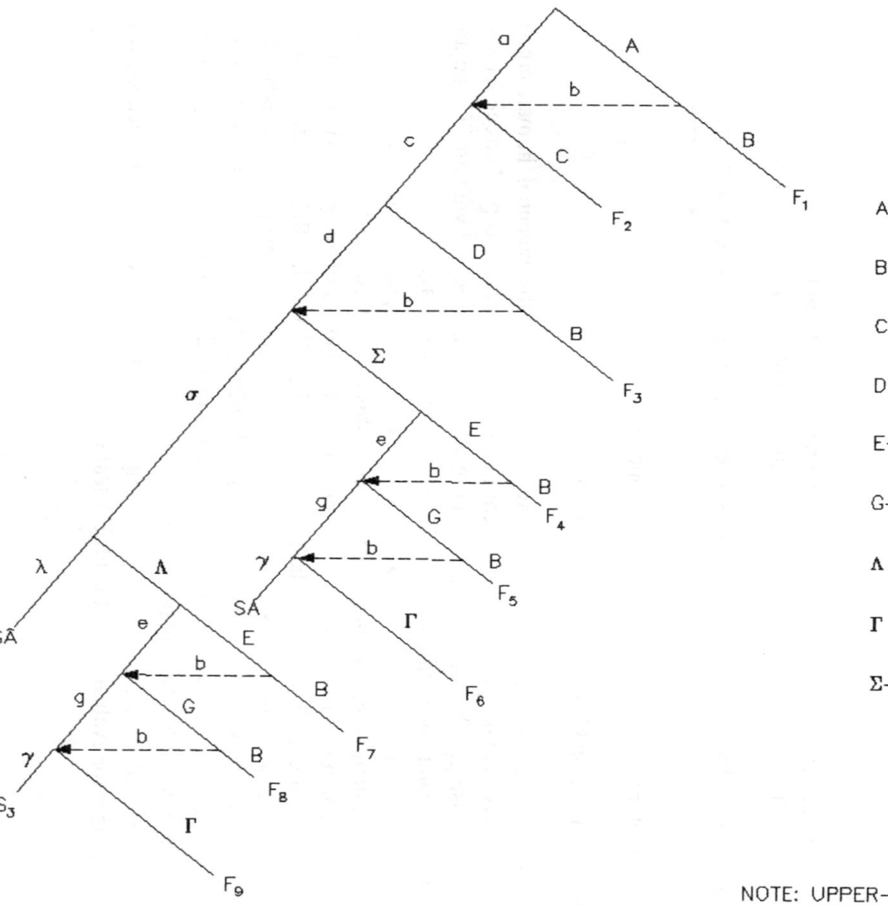

A – OPERATOR MAKES SWITCHOVER ON WRONG CONDENSER

B – OPERATOR ON ROUNDS FAILS TO DETECT CONTINUED HIGH COLUMN PRESSURE

C – ACTIVE CONDENSER ISOLATED FIRST (WRONG ORDER)

D – OPERATOR FAILS TO ATTEMPT TO ISOLATE LEAKY CONDENSER

E – OPERATOR FAILS TO DETECT STUCK VALVE

G – OPERATOR FAILS TO ATTEMPT TO ISOLATE REFRIGERANT SUPPLY/RETURN VALVES

Λ – CONDENSER PROCESS OUTLET VALVE FAILS TO CLOSE

Γ – REFRIGERANT SUPPLY/RETURN VALVE STICKS OPEN

Σ – CONDENSER PROCESS INLET VALVE FAILS TO CLOSE

NOTE: UPPER–CASE ENGLISH LETTERS REPRESENT OPERATOR ERRORS
UPPER–CASE GREEK LETTERS REPRESENT EQUIPMENT FAILURES
LOWER–CASE ENGLISH LETTERS REPRESENT OPERATOR SUCCESSES
LOWER–CASE GREEK LETTERS REPRESENT EQUIPMENT SUCCESSES
F_n INDICATES A FAILURE PATHWAY
S_n INDICATES A SUCCESS PATHWAY

Figure 21.2 HRA event tree for loss of overhead condensing

21.3 Discussion of Results

The results of the FMEA were recorded by Ms. Quick using FMEA software. A printout of the FMEA results was distributed to the team for their review at the end of the meeting. Table 21.1 lists a portion of the team's findings. Table 21.2 lists the minimal cut sets found using the HRA event tree.

Recommendations that the hazard evaluation team made based on the FMEA results include:

- Consider installing a high pressure alarm on the column.
- Consider developing a checklist of emergency actions to follow in the event of a high column temperature or pressure.
- Consider installing a separate flow transmitter and low flow alarm on the reflux line.
- Consider installing an autostart circuit for the spare HCl reflux pump.
- Consider using a different temperature transmitter on the column to alarm high and low column temperature and initiate an automatic column shutdown.

Based on their review of the HRA results, the hazard evaluation team also recommended:

- Clearly labeling the overhead condensers and their valves
- Requiring operators to monitor column pressure hourly for the first few hours following a condenser switchover.

A few days after the meeting, Ms. Piper prepares a report describing the team's findings. The report lists the team members and the information used in the review, describes the probable cause of the recent HCl release (the instrument technician's testing verified the team's consensus that a TT-201 failure initiated the incident), and gives recommendations the team made and the basis for each. The FMEA tables and the HRA event tree generated as part of the investigation are also included as attachments, as well as a copy of the HRA study performed by Mr. Dennis. This short report describes the HRA, the analyst's conclusions, and the HRA event tree. The incident report, after review by the hazard evaluation team, is sent to the plant manager and ABC's corporate health and safety officer.

21.4 Follow-up

The hazard evaluation team was able to address all the pertinent issues during the meeting. The HRA information, used by the team to judge whether additional instrumentation on the overhead condenser was needed, was made available by Mr. Dennis before the report was issued. Thus, the team was able to review these results and determine that only better labeling of the condensers and monitoring of pressure after switchovers were needed.

Issues concerning the HCl release incident have high priority within ABC. Thus, Anywhere's VCM plant manager personally took responsibility for ensuring that every recommendation was addressed. Using a computerized tracking system, the plant manager checked the status of each recommendation weekly until every recommendation was resolved. The FMEA, HRA report, and resolution of the recommendations were placed in the engineering files for the HCl distillation column.

Table 21.1 Sample results from the incident investigation FMEA

AREA: HCl Distillation Column TEAM: Ms. Piper (Leader C Anywhere VCM Plant), Ms. Quick, Mr. Process, Mr. Martin, Ms. Opera, and Mr. Clean (ABC Health, Safety, and Environment)					PAGE: 1 of # DRAWING: E-708, Rev. F MEETING DATE: ##/##/##
Item	Component	Failure Mode	Effects	Safeguards	Actions
1	Temperature control valve (TCV-201)	a. Fails open	Increased heating of the HCl column. Potential column overpressure and release of HCl	Multiple TIs on column High temperature alarm and shutdown on column Excess overhead condensing capacity (spare condenser)	Add a high pressure alarm Develop an emergency checklist for operators to follow on high temperatures
		b. Fails closed	HCl column cools down. No effect of interest	---	
		c. Leaks externally	Loss of steam. No effect of interest	---	
		d. Leaks internally	Increased heating of the HCl column; however, at a slow rate Potential column overpressure and release of HCl	Multiple TIs on column High temperature alarm and shutdown on column Excess overhead condensing capacity (spare condenser) Operators have ample time to diagnose problem and manually isolate TCV-201	
2	Temperature transmitter (TT-201)/ temperature controller (TIC-201)	a. False high output	HCl column cools down. No effects of interest	---	
		b. False low output	Increased heating of the HCl column. Potential column overpressure and release of HCl	Multiple TIs on column Excess overhead condensing capacity (spare condenser) Note: The temperature alarm and shutdown will not function in this scenario	Add a high pressure alarm Develop an emergency checklist for operators to follow on high temperature Move the high and low temperature alarms to another column transmitter
		c. No signal change	Possible increase in column temperature due to minor process upsets. Slight chance of an overpressurization and release of HCl	Multiple TIs on column Excess overhead condensing capacity (spare condenser) Ample time for operators to diagnose problem and manually control column temperature Note: The temperature alarm and shutdown will not function in this scenario	Add a high pressure alarm Develop an emergency checklist for operators to follow on high temperature Move the high and low temperature alarms to another column transmitter

Table 21.2 Minimal cut sets for the incident investigation HRA event tree

Minimal cut set no. 1
- Operator makes switchover on wrong condenser
- Operator fails to detect high column pressure on rounds

Minimal cut set no. 2
- Operator isolates active condenser first (wrong order)

Minimal cut set no. 3
- Operator fails to isolate leaky condenser
- Operator fails to detect high column pressure on rounds

Minimal cut set no. 4
- Condenser process inlet valve fails to close
- Operator fails to detect stuck valve
- Operator fails to detect high column pressure on rounds

Minimal cut set no. 5
- Condenser process inlet valve fails to close
- Operator fails to attempt to isolate refrigerant supply/return valves
- Operator fails to detect high column pressure on rounds

Minimal cut set no. 6
- Condenser process inlet valve fails to close
- Refrigerant supply/return valve sticks open

Minimal cut set no. 7
- Condenser process outlet valve fails to close
- Operator fails to detect stuck valve
- Operator fails to detect high column pressure on rounds

Minimal cut set no. 8
- Condenser process outlet valve fails to close
- Operator fails to attempt to isolate refrigerant supply/return valves
- Operator fails to detect high column pressure on rounds

Minimal cut set no. 9
- Condenser process outlet valve fails to close
- Refrigerant supply/return valve sticks open

21.5 Conclusions and Observations

The FMEA method proved to be effective for quickly examining the HCl column. In addition, Ms. Piper kept the review very narrowly focused on potential HCl releases. Failures that did not result in this effect were not explored.

The FMEA also succeeded because Ms. Piper is a skilled investigator. She did not depend on the FMEA tool alone to find the problems. Rather, the FMEA tool merely provided a systematic way to raise potential issues of concern. For example, Ms. Piper quickly received answers from the team on effects and safeguards related to a loss of refrigerant. However, suspecting there may be additional problems relating to the condensers, she probed for additional causes of cooling loss (rather than treating it as a "black box"). This resulted in additional recommendations that could help protect against losing a condenser.

Ms. Piper also recognized that FMEA was not always the right tool to use during the investigation. In particular, she recommended an HRA for the overhead condensers when she recognized that operator errors might play a significant role in losing a condenser. The time required to perform this investigation is summarized in Table 21.3.

One of Ms. Piper's shortcomings in this analysis was to not involve Mr. Clean in the discussions. Other than the few words he said on behalf of ABC's corporate safety and health group, he did not contribute to the meeting. Ms. Piper should have asked some simple, direct questions during the review to get Mr. Clean involved.

Table 21.3 FMEA staff requirements for the incident investigation

Personnel	Preparation (h)	Evaluation (h)	Documentation (h)
Leader	8	8	24[a]
Team Member[b]	1	8	1
Mr. Dennis	2	4	4

[a] Includes time working with Mr. Dennis on the HRA
[b]Average per team member

22

Decommissioning Phase

An Illustration of the What-If/Checklist Analysis Method

22.1 Problem Definition

Background

ABC Chemicals, Inc. has been producing VCM for over 30 years. Their expansion into the VCM market proved quite profitable; in addition to the Anywhere Plant, ABC has built VCM manufacturing facilities at three other locations. However, the Anywhere VCM plant has become less efficient as it has aged, and ABC has decided to reduce production at the plant in an effort to cut costs; the newer and more efficient plants will increase their VCM production.

As part of the reduction in capacity, ABC will decommission one of the plant's four furnaces used to crack EDC and make VCM. (This furnace was scheduled to have its tube bundle tested at the next turnaround.) Prior to decommissioning, the Anywhere plant manager has asked that a hazard evaluation of the decommissioning plan be performed.

No personnel at the Anywhere Plant have performed or participated in a hazard evaluation involving the decommissioning of a process. Thus, the plant manager asks ABC's process hazards analysis group for assistance. Mr. Dennis, the director of the group, assigns Mr. O. L. Timer to this project. Mr. Timer has performed numerous hazard reviews, including hazard evaluations of equipment being decommissioned. Through his experience, and the experience of others in ABC's process hazards analysis group, he has developed a checklist (Table 22.1) for reviewing decommissioning activities. Mr. Timer tentatively plans to use this checklist for the hazard evaluation of the Anywhere VCM furnace decommissioning activity.

Available Resources

ABC is considered by the industry to be a leader in VCM production. As a result of their vast experience, ABC has substantial information on the design and operation of VCM plants, including P&IDs and design specifications for the cracking furnace that is being decommissioned at the Anywhere Plant. In particular, the plant has the following information available for the hazard evaluation:

- Furnace P&IDs
- Furnace design specifications
- Process flow diagrams (Figure 22.1)
- Operating procedures and logs
- Maintenance procedures and logs
- Incident histories
- Previous hazard evaluations
- MSDSs for all VCM plant process materials
- Plot plans
- Environmental regulations

Table 22.1 Sample decommissioning checklist

Shutdown and isolation

1. Do procedures exist for shutting down the unit? Are personnel familiar with these procedures? Has the unit been shut down before? Is Operations aware of the decommissioning?
2. Do decommissioning procedures exist? Have they been reviewed for technical content?
3. Have equipment changes or modifications occurred that are not reflected in the system documentation? Have the potential effects of these changes on maintenance actions been addressed?
4. Will utility systems be disconnected from the unit? Do appropriate lockout/tagout procedures exist? Are the disconnections permanent? Will making these disconnections affect other units?
5. Will any safety or control feature be temporarily or permanently disabled? How will this affect other operating equipment? Will it possibly initiate a shutdown?
6. Will someone familiar with the decommissioning plans always be available for emergencies? Do emergency plans exist?
7. Is any special medical surveillance required during decommissioning?
8. Are any fire protection systems being disabled as part of decommissioning?
9. Will equipment always be electrically grounded?
10. How will the unit process lines be isolated from other plant systems? Will someone verify these isolations?
11. Do all isolated vessels have adequate relief protection? Are relief paths clear and operable during decommissioning?
12. Do any vessels require vacuum protection during decommissioning? Will any isolated vessels be cooled during decommissioning?

Draining

1. Do procedures exist for draining process material from the unit?
2. Is any special protective gear needed during draining operations?
3. Will lines have to be "broken" to drain the unit? Do adequate measures exist to ensure that very hot, very cold, or high pressure material is not in the line? Do adequate measures exist to protect against toxic or flammable releases? Are hot work permits or line break permits required?

Draining (cont'd)

4. How will the drained material be disposed of? Will this vessel contain incompatible materials?
5. Does the area have proper ventilation should a spill occur? Does it have proper drainage? Does it have adequate fire protection?
6. Will access to the area be limited during draining activities?
7. Is the area free of ignition sources? Is the area free of combustible materials?
8. Is the equipment used to drain the unit compatible with this process material?
9. Will confined space entry be required to drain the unit? Have permits been obtained?
10. Are reverse flows possible in the drain line?

Cleaning

1. Will the equipment in the unit be cleaned after draining?
2. Are the cleaning materials potentially reactive with any process materials? Can less hazardous materials be used?
3. Do the cleaning materials require special handling? Is personal protective equipment required?
4. Could the wrong cleaning material inadvertently be used?
5. How will the cleaning solution be disposed of?
6. If the cleaning material is combustible, have appropriate fire protection measures been taken?
7. Are there any concerns with residues left after cleaning?

Dismantling

1. Will heavy equipment be used to dismantle the unit? Are adequate measures in place to monitor heavy equipment movement? Are rigging checks required? Is the lighting adequate?
2. Are there hazardous or flammable materials in the area that could be released in an incident? Are adequate safety precautions in place?
3. Will equipment in the unit be reused? Is it adequately designed for the new service?
4. Is the equipment properly labeled?
5. Where will the equipment be stored? Are there any special storage requirements?
6. Are any special procedures needed to comply with environmental regulations?
7. Are there other units, lines, etc., in areas that may be hit by heavy equipment used for the dismantling?

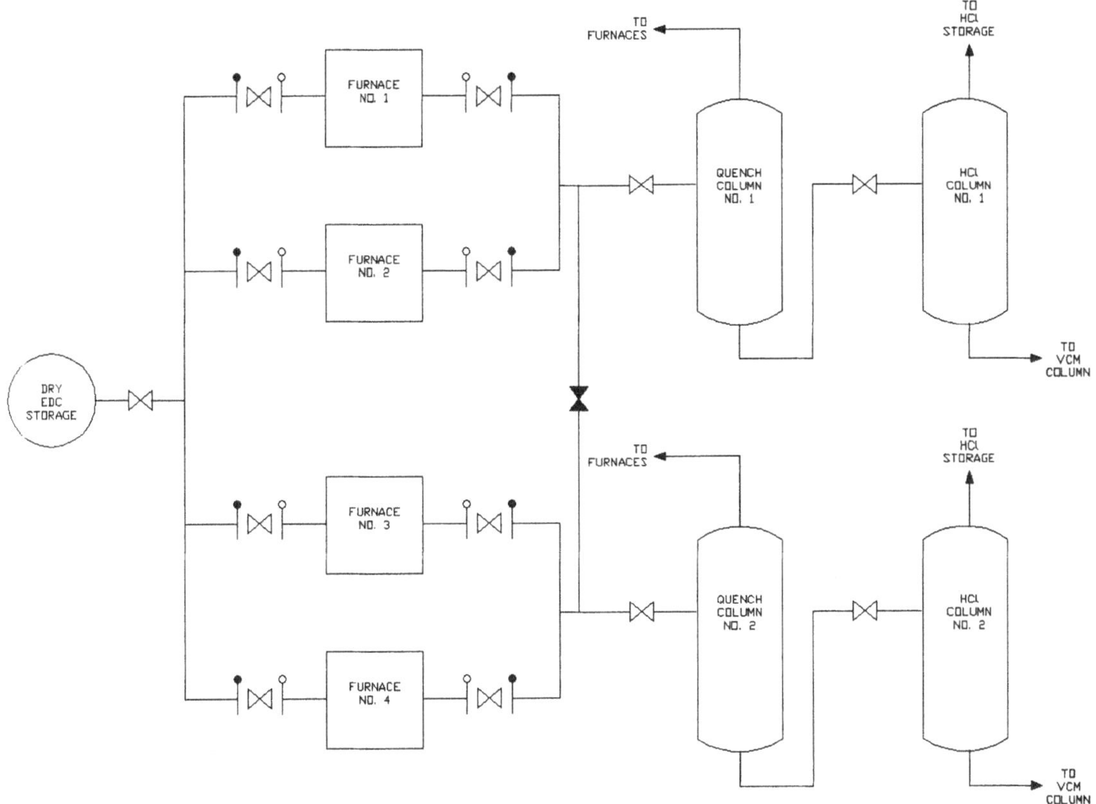

Figure 22.1 Process flow diagram for VCM furnace area

Anywhere Plant personnel have developed a preliminary plan for decommissioning the furnace. This information, supplemented with the knowledge of highly experienced Anywhere Plant personnel, will be used in the hazard evaluation of the furnace decommissioning.

Selection of Hazard Evaluation Technique

Decommissioning the furnace could expose personnel to many different hazards. For this reason, Mr. Timer decides to use a hazard evaluation technique that examines a broad range of hazards. Thus, he quickly narrows his choice to Preliminary Hazard Analysis, What-If Analysis, Checklist Analysis, HAZOP Study, and FMEA. Since the furnace is not operating during the decommissioning, the HAZOP Study and FMEA techniques are not strong candidates. Of the remaining three methods, Mr. Timer selects the Checklist Analysis method because (1) the Checklist Analysis method covers a broad range of hazards, (2) he has a decommissioning checklist that he has successfully used in the past, and (3) the Checklist Analysis method can be used quickly and easily.

However, Mr. Timer has never performed a Checklist Analysis of VCM equipment being decommissioned. For this reason, he decides to supplement the Checklist Analysis with a What-If

Analysis. To do this, Mr. Timer encourages the review team to ask any What-If questions as they move through his checklist. Mr. Timer has performed a What-If/Checklist Analysis in the past both by (1) examining the items on a checklist and then asking What-If questions and by (2) asking What-If questions as he examines items on a checklist. He prefers the latter method, so he uses it in this study. Mr. Timer hopes the checklist items will prompt team members to ask questions that reveal hazardous situations that may be present.

Study Preparation

To perform the What-If/Checklist Analysis, Mr. Timer needs skilled personnel familiar with the planned decommissioning activities. Because the government has placed increasingly complex and stringent environmental regulations on EDC and VCM over the years, Mr. Timer decides that an environmentalist should also participate in the review. The following skilled personnel are selected for the review:

Leader	—	A person skilled in leading What-If/Checklist Analysis. Mr. O. L. Timer will be the leader.
Process Engineer	—	A person thoroughly familiar with the operation of the VCM plant, and, in particular, the operation of the furnace area. Mr. C. Chem, the process engineer for the furnace area and an author of the decommissioning plan, will fill this position.
Maintenance Engineer	—	A person who knows plant maintenance practices for the furnace area. Mr. P. I. Fitter, the maintenance engineer who helped develop the decommissioning plan, will serve in this role.
Environmental Specialist	—	A person who understands environmental regulations regarding EDC and VCM. Ms. Kelly Green, the environmental engineer for the Anywhere Plant, will fill this position.

To prepare for the hazard evaluation, Mr. Timer reviews the information on the furnace. In particular, he focuses on the decommissioning plan, the plot plan (since heavy equipment will be needed to dismantle the furnace), and the P&ID (which shows how the furnace interfaces with other plant equipment). He then prepares an information package to send to each team member. This package contains appropriate drawings of the furnace area, the decommissioning plan, and Mr. Timer's checklist. Mr. Timer also includes in this package a description of the Checklist and What-If Analysis methods. He reserves a conference room and a projector at the plant for the one-day study.

A one-day What-If/Checklist Analysis is planned. Mr. Timer schedules the review at the Anywhere site so that the team can inspect the furnace area before the review. Choosing this location also enables the team to readily obtain additional information and personnel if needed for the hazard evaluation.

As a final note to the team members, Mr. Timer asks them to prepare some What-If questions before the review. He knows that the team discussions will prompt other ideas and concerns; however, he has found that if the team members do some homework before the meeting, the results of the hazard evaluation are generally better and the members are more prepared to answer questions as the review progresses.

22.2 Analysis Description

The What-If/Checklist Analysis begins at 9:00 a.m. with a brief tour of the furnace area at the Anywhere Plant. During this tour Mr. Chem and Mr. Fitter describe how the furnace will be isolated from the rest of the plant, emptied, flushed/purged, and dismantled. Following this tour the team reconvenes in the plant's training room to begin the study. Before starting through the checklist, Mr. Timer reviews the ground rules for the discussions. They are: (1) all team members have a right and responsibility to raise any issue that concerns them, (2) no issue is unimportant, (3) the aim is to identify safety concerns—not solve them, (4) criticism of team members is not allowed, and (5) all team members are equals. With the ground rules established, Mr. Timer opens the discussions.

Mr. Timer — As I stated in the memo to each of you, we will perform a What-If/Checklist Analysis of the decommissioning for the No. 4 furnace. To do this, we will discuss each of the items on the checklist I sent you (*Table 22.1, provided prior to this narrative*). However, the discussions are not limited to just the items on the checklist. I hope the checklist will prompt other What-If questions. I will make note on the chart pad of any important issues that arise so that you can keep me honest. Mr. Chem has also agreed to take notes as a backup. Any questions before we begin?

Ms. Green — I didn't have a chance to review the written decommissioning plans. Think we could have a brief overview before starting?

Mr. Chem — I can do that, Kelly. As you know, we are taking the No. 4 furnace out of service as part of the plant's reduction in capacity. To do so, we'll first shut down the furnace. This will be done in three steps: first, the inside operator will shut off fuel gas flow to the furnace; then, the outside operator will block in the gas supply to the furnace; and then the outside operator will block in the furnace process supply line. We've shut down the furnace this way many times. Then we'll blow the tube bundle clear with nitrogen and block in the process discharge line from the furnace. Once the furnace is cool, we'll open it up and remove the tube bundle. For this, of course, we've got to use a crane. The tube bundle will be taken to the maintenance shop for cleaning, and then will be shipped to the Somewhere plant for reuse. The rest of the furnace will be mothballed for now.

Ms. Green — [*Nods.*] Thanks for the overview.

Mr. Timer — Let's move on to the checklist. Do you have shutdown procedures for the furnace, and are the operators familiar with them?

Mr. Fitter — Yes and yes. As Mr. Chem just said, Operations has shut down this furnace many times in the past.

Mr. Timer — I see my next checklist item is already answered, then. How about the decommissioning plans. Have the operators discussed these plans?

Mr. Chem — Both P. I. and I have talked with the operators about decommissioning and considered their input in drafting the plan; however, they haven't yet reviewed it.

Ms. Green — Has anyone reviewed the plan?

Mr. Fitter — Just this group.

Mr. Timer — I recommend that you have this plan reviewed by the unit supervisor, the area maintenance manager, and the plant safety coordinator. [*Mr. Timer writes this recommendation on the flip chart for the group.*]

Mr. Fitter — Okay.

Mr. Timer — The next item on the list is equipment changes. Have you modified the furnace so that it might change your shutdown procedures or decommissioning plans?

Mr. Chem — The furnace hasn't been modified in over 10 years. And we have gone through several shutdowns without a hitch during that time.

Mr. Timer — Okay. Let's move over to the utilities for the furnace. Will they be disconnected permanently? And will these disconnections affect other units?

Mr. Fitter — For the electricals, we will lock out and tag out breakers supplying this furnace. As far as instrument air goes, there is only one pneumatic valve. We'll probably leave it connected. The natural gas supply to the furnace will be blocked in and blinded at the supply header.

Mr. Timer — Will these disconnections affect other units?

Mr. Fitter — I don't think there's any problem from looking at the P&IDs, and we've shut down furnaces in the past and it hasn't bothered the other units. Isn't that right, Mr. Chem? [*Nods yes.*]

Ms. Green — O. L., you said we can ask What-If questions. Mind if I propose some now?

Mr. Timer — No, go ahead. I'm glad you have other questions.

Ms. Green — What if the electricians lock out the wrong breakers or remove the locks at a later time?

Mr. Fitter — As far as the wrong breakers, Operations checks the lockout and verifies zero energy before Maintenance touches anything. And Operations wouldn't remove the control lock on the system until it is ready for "return to service." Seeing how this is not expected to be the case, this system will remain permanently locked out. In fact, we are working on an "Out of Service" program for this type of equipment, rather than leaving it under a "lockout".

Mr. Timer — But you don't plan to reactivate this furnace, at least not as far as we can see. Seems to me that you should just de-energize the breakers by having an electrician pull and cap the leads. No reason to even have the possibility of an electrical hazard.

Mr. Chem — I guess I agree. That way we won't have to worry about energizing a dead line in error. [*Mr. Timer writes the recommendation on his flip chart.*]

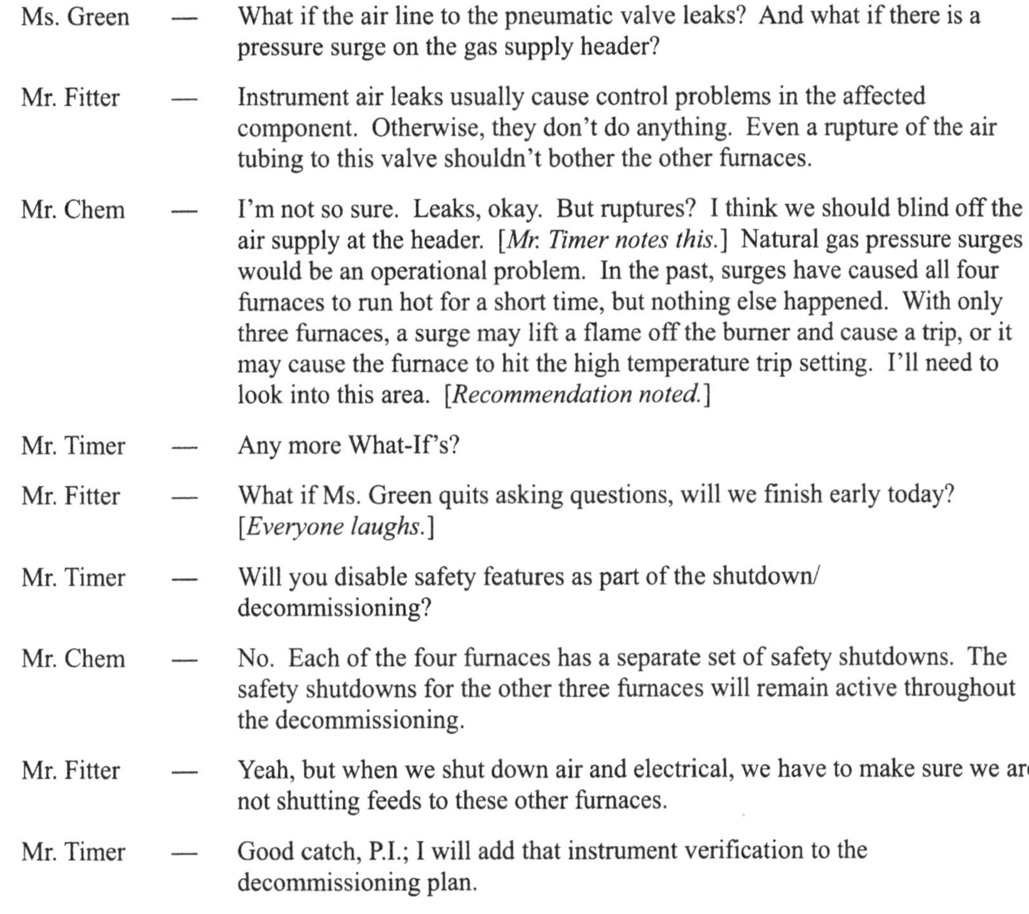

Ms. Green — What if the air line to the pneumatic valve leaks? And what if there is a pressure surge on the gas supply header?

Mr. Fitter — Instrument air leaks usually cause control problems in the affected component. Otherwise, they don't do anything. Even a rupture of the air tubing to this valve shouldn't bother the other furnaces.

Mr. Chem — I'm not so sure. Leaks, okay. But ruptures? I think we should blind off the air supply at the header. [*Mr. Timer notes this.*] Natural gas pressure surges would be an operational problem. In the past, surges have caused all four furnaces to run hot for a short time, but nothing else happened. With only three furnaces, a surge may lift a flame off the burner and cause a trip, or it may cause the furnace to hit the high temperature trip setting. I'll need to look into this area. [*Recommendation noted.*]

Mr. Timer — Any more What-If's?

Mr. Fitter — What if Ms. Green quits asking questions, will we finish early today? [*Everyone laughs.*]

Mr. Timer — Will you disable safety features as part of the shutdown/ decommissioning?

Mr. Chem — No. Each of the four furnaces has a separate set of safety shutdowns. The safety shutdowns for the other three furnaces will remain active throughout the decommissioning.

Mr. Fitter — Yeah, but when we shut down air and electrical, we have to make sure we are not shutting feeds to these other furnaces.

Mr. Timer — Good catch, P.I.; I will add that instrument verification to the decommissioning plan.

The What-If/Checklist Analysis continues throughout the day. The following is the last part of the review and the dialogue from the close of the meeting.

Mr. Timer — Okay, we've now got the tube bundle out and at the maintenance shop. The last major activity is to clean the tube bundle and ship it out. Let's see, are the cleaning materials reactive with any of the process materials?

Mr. Fitter — No, we've cleaned the tube bundles before with no problem.

Mr. Timer — How about switching cleaning solutions. Are less hazardous cleaners available?

Mr. Fitter — I think ABC looked into this a few years back when the environmental group was pushing all plants to review hazardous material inventories. We decided then that the cleaner we are now using is fairly nonhazardous.

Ms. Green — I don't have any problems with their cleaner from an environmental perspective.

Mr. Timer — Do you require protective clothing or special handling of the cleaner?

Mr. Fitter — Yes, we do, but because of the EDC and VCM in the tubes, not because of the cleaner. We will conform to the appropriate regulations when cleaning the tube bundle.

Mr. Timer — Any possibility you'd use the wrong cleaner?

Mr. Fitter — No, we've cleaned furnace tube bundles several times without a hitch.

Ms. Green — What if the wrong cleaner is used? Could a worker get hurt?

Mr. Fitter — I don't know. Depends on the cleaning solution.

Ms. Green — But there are other cleaners in the shop? [*Nods yes.*] I recommend someone review what cleaners are present to see if serious incidents could happen. [*Mr. Timer notes this recommendation on his chart.*]

Mr. Timer — How about disposal of the cleaning solution?

Mr. Fitter — The dirty solution is collected in 55 gallon drums, labeled, and shipped to a waste facility for disposal.

Mr. Timer — You said earlier the cleaning solution is not combustible, right? [*Nods yes.*] Any more questions?

Ms. Green — What if the tubes are not properly cleaned?

Mr. Fitter — The cleaning is done mainly to reduce any chance of employee exposure to EDC/VCM. We will fill the tube bundles with desiccant and seal them before shipment. If there is residue, it shouldn't matter since this bundle will serve as a spare for the VCM furnaces at the Somewhere plant.

Ms. Green — I think we should test for proper cleaning for two reasons: first, the tube bundle may be used in some other service and the residue may present a safety hazard; and second, we'll have to ship the bundle as a hazardous material at great cost if we don't prove it's clean. Besides, if a shipping incident occurs, we don't want to be publicized as having "spilled" a hazardous material. [*Mr. Timer records the recommendation.*]

Mr. Timer — Any more questions? [*Pause.*] Okay, let's quickly review the recommendations to be sure I got them all. [*Mr. Timer reviews the recommendations recorded on the chart pad.*] Well, I appreciate everyone's input. Mr. Fitter, as we discussed at lunch, I'll document the results of the review and, after the team has okayed the report, you'll follow up on the recommendations.

Mr. Fitter — Agreed.

22.3 Discussion of Results

The What-If/Checklist Analysis results consist of the table of checklist items covered in the review, as well as a table of recommendations (Table 22.2). The table of recommendations was generated by Mr. Timer from the notes he recorded. Mr. Timer recorded only answers to Checklist or What-If questions for which the review team recommended changes. He did note in the hazard evaluation report that all the items on the Checklist, as well as many What-If questions, were examined in the review.

22.4 Follow-up

The review team identified several issues and made several recommendations that ABC should examine before beginning the decommissioning. These issues and recommendations were reviewed by all the team members before they were transmitted to Anywhere's VCM plant manager. The plant manager accepted all of the recommendations and assigned Mr. Fitter the job of following up on these items. The decommissioning is scheduled to occur in three months. Mr. Fitter assigned the recommendations to various personnel at the Anywhere Plant and, using a computerized tracking system, checked on the resolution of each recommendation weekly until all the recommendations were addressed.

22.5 Conclusions and Observations

Mr. Timer ran an excellent What-If/Checklist Analysis of the decommissioning plan. All the team members participated throughout the day. It is interesting to note that the two team members least familiar with the furnace identified the most recommendations. Their knowledge of safety and environmental hazards, along with their inquiring natures, helped uncover a number of issues that might have otherwise been overlooked. The time required to perform this What-If/Checklist Analysis is summarized in Table 22.3.

Table 22.2 Sample recommendations from the furnace decommissioning What-If/Checklist Analysis

- Request a unit supervisor, area maintenance manager, and plant safety coordinator review the decommissioning plan.
- Disconnect power from the breakers supplying electricity to Furnace No. 4.
- Install blinds on the instrument air line to Furnace No. 4.
- Examine the impact of natural gas supply pressure surges on the three operating furnaces.
- Review the cleaning solutions used in the maintenance shops to determine if any are incompatible with EDC/VCM.
- Develop a plan to ensure the tube bundle is adequately cleaned for shipment.

Table 22.3 Decommissioning What-If/Checklist Analysis staff requirements

Personnel	Preparation (h)	Evaluation (h)	Documentation (h)
Leader	4	8	16
Team member[a]	1	14[b]	1

[a]Average per team member

[b]This includes time required by two analysts to research the answers to several questions that arose during the review

Appendices

A – **Additional Checklists and Forms**

B – **Supplemental Questions for Hazard Identification**

C – **Symbols and Abbreviations for Example Problem Drawings**

D – **Software Aids**

E – **Chemical Compatibility Chart**

F – **Organizations Offering Process Safety Enhancement Resources**

Appendix A
Additional Checklists and Forms

Included in this Appendix are the following supplemental materials that may prove useful in hazard evaluations performed for particular purposes:

A1 Example Checklist for Evaluating Changes

A2 Example Management of Change Hazard Review Form

A3 Example Reactivity Checklist

A4 An Inherently Safer Process Checklist

A1

Example Checklist for Evaluating Changes

(Courtesy Dow Chemical Company)

If a more thorough review needs to be done to evaluate a change, one or more of the techniques in Chapter 4 and 5 can be used for an in-depth safety analysis of the change, which may be in addition to a previous hazard evaluation completed for the operating facility. Table A1.1 contains an example of What-If checklist items that have been used in location-led hazard evaluations of facility or operational changes. Note that these particular checklist items focus on mechanical failures, so they may need to be expanded to fully cover operational issues. The more extensive listing of safety questions in Appendix B can also be used.

Typical steps in using this What-If checklist to evaluate changes are as follows.

- Define the scope of the process to study
- Assemble information (as necessary):
  - — Process Flow Diagrams and energy information
  - — P&IDs for the existing process and modification
  - — Plot Plan
  - — Equipment data and safety system logic
  - — Operations and maintenance procedures
  - — Operational Safety Standards and controls
  - — Previous hazard evaluation
  - — Completed checklist.
- Identify the review team - Number of persons and skills depend on modification.
- Conduct analysis using What-if/Checklist and record scenarios and analysis on a separate What-if worksheet (list: scenario, hazard/consequence, existing controls and recommendations, follow-up responsibility, and indicate when complete)
- Include this checklist and any What-if worksheets with the review package.

As for all hazard evaluations, proper consideration needs to be given to the appropriate skills and number of persons needed for the review, and sufficient unconstrained time allowed for creative, thoughtful review.

Table A1.1 Example What-If checklist used in evaluating hazards of facility/operational changes

A. Releases via mechanical failure causing emissions, fires, explosions:

Releases of materials to the surroundings that could result in pool fires, flash fires, vapor cloud explosions or toxic vapor clouds, dust clouds, or mist clouds, with significant acute exposure to personnel.

1. Vessel Failure

❑ a. Installation
   1) vibration
   2) fatigue
   3) embrittlement (e.g., cast iron/ steel, hydrogen)

❑ b. Impingement
   1) crane drop
   2) heavy equipment impact
   3) vehicle impact
   4) railcar/barge/tank truck collision

❑ c. Overpressure from
   1) process upsets
   2) common vent header
   3) pump/compressor
   4) nitrogen supply
   5) blowing lines into vessel
   6) steaming to clean
   7) ruptured tube
   8) homogeneous nucleation/ low boilers with high boilers
   9) overfill
   10) liquid filled/valved-in
   11) hydraulic hammer
   12) water freezing

❑ d. Natural forces
   1) lightning
   2) earthquake
   3) hurricane
   4) ice/snow load

❑ e. Corrosion/erosion
   1) wrong MOC
   2) stress corrosion cracking
   3) crevice corrosion/pitting
   4) internal wall
   5) external sweat zone
   6) lining/jacket failure
   7) erosion
   8) high temp. corrosion

❑ f. Vacuum collapse:
   1) sudden cooling and vapor cooling/condensation
   2) plugged filters/lines
   3) vacuum systems
   4) pumping contents from closed-in vessel

❑ g. High temperature:
   1) fire exposure/BLEVE
   2) heater malfunction
   3) operation above maximum allowable working temperature

❑ h. Low temperature fracture:
   1) embrittlement - operating below minimum design metal temp.
   2) low pressure flashing
   3) low ambient temperature

2. Piping System Failure

❑ a. Installation
   1) improper material of construction
   2) improper installation
   3) vibration
   4) fatigue

❑ b. Impingement
   1) crane drop
   2) heavy equipment impact
   3) vehicle impact
   4) third party intervention (e.g., backhoe)

❑ c. Natural forces
   1) earthquake
   2) high winds

❑ d. Corrosion/erosion
   1) chemical – improper material of construction
   2) stress cracking
   3) internal wall
   4) external wall (e.g., under insulation)
   5) lining failure
   6) erosion
   7) high temperature corrosion

❑ e. Overpressure from
   1) common vent header
   2) high pressure pump/ compressor
   3) reaction in line
   4) nitrogen supply
   5) blowing lines
   6) steaming to clean
   7) hydraulic expansion
   8) hydraulic hammer
   9) water freezing in
   10) solids plugging line/filter

❑ f. Temperature too high from
   1) fire exposure
   2) loss of coolant
   3) loss of fluid flow (e.g., in furnace tube)

❑ g. Temperature too low from
   1) liquid flashing
   2) low temperature fluid impingement

❑ h. Open vents and drains
   1) left open
   2) inadvertently opened

❑ i. Valve failure
   1) bonnet gasket/bolts failure
   2) packing blowout
   3) line gasket failure
   4) reactive/cryogenic material trapped in ball valve

3. Other Releases

❑ a. Sight glass b. Expansion joints
c. Swivel joints d. Hoses
e. Flare outage
f. Scrubber breakthrough
g. Incinerator failure

❑ h. Heat exchanger failure
   1) tube rupture – release through heating or cooling system
   2) tube rupture followed by jacket hydraulic failure

❑ i. Compressed gas cylinder failures
   1) valve broken off
   2) propelled if unsecured
   3) fusible plug melted/dislodged
   4) inappropriate heating
   5) wrong regulator/tubing used

❑ j. Pump failures
   1) packing blowout
   2) single mechanical seal rupture
   3) rupture of both double/tandem mechanical seals
   4) deadheaded
   5) positive displacement blocked in

❑ k. Compressor wreck
   1) liquid in suction
   2) lubrication failure
   3) sudden loss of load
   4) vibration
   5) turbine overspeed
   6) discharge blocked in (reciprocating)

❑ l. Drums
   1) improper (splash) filling
   2) loss of inerting
   3) overpressure with N2/Air
   4) puncture with forklift
   5) stacking too high
   6) improper thawing (e.g., glacial acrylic acid)

❑ m. Loss of utilities
   1) cooling
   2) heating
   3) electricity
   4) instrument air
   5) nitrogen
   6) steam

❑ n. Relief device, spurious opening

❑ o. Interfaces with
   1) raw materials
   2) waste streams
   3) distribution
   4) blowback
   5) energy systems
   6) laboratory
   7) product storage/handling

Table A1.1 Example What-If checklist (continued)

B. Releases via internal undesirable reactions and explosions:	C. Releases via internal flammable mixtures:	D. Personnel life-threatening events:
Undesirable chemical process reactions that could cause events such as overpressure, overtemperature or generation of toxic vapor clouds, or which could self-accelerate if not controlled; unstable compounds that could explode or detonate; highly reactive compounds that could undergo uncontrolled reaction upon contamination or catalysis.	Flammable or explosive mixtures of process materials inside vessels, enclosures, or confined spaces and buildings that could lead to fire or explosion, with ignition from sources such as static electricity, lightning, sparking from mechanical impact, friction or other energy sources.	The Life-Threatening Injury category includes special cases where life-threatening personnel injury could occur, but does not duplicate the injury potential implicit with fires, vapor cloud explosions, undesirable chemical reactions, equipment explosions, or any of the other categories listed above, where injury potential is recognized. Includes acute process health hazards, but not chronic process health hazards or personnel safety hazards.
❑ 1. Contamination: a. Via normal transfers b. From/to other units c. Back flow d. Common headers e. Utilities f. With air/water	❑ 1. Air into Hydrocarbons a. Improper venting b. Improper crossties c. Contamination of inerting medium d. Entrainment-with additives etc. e. Blower/compressors low suction pressure. f. Improper air addition (e.g., air hose) g. Breaking vacuum with air h. Failure to purge prior to maintenance	❑ 1. Insufficient oxygen 2. Toxic/corrosive chemical exposure 3. Thermal exposure a. Hot water b. Hot process chemicals 4. Mechanical energy a. Turbine overspeed b. Reciprocating compressor deadhead 5. Electrical energy
❑ 2. Mix Reactive Chemicals: a. Mistaken identity b. Feed too fast c. Wrong addition order d. Wrong feed ratio e. Improper mixing	❑ 2. Hydrocarbons to Air: a. Accumulation in buildings or other confined space b. Above tank internal floating roofs c. Sewers	Hazard evaluation team members for this review:
❑ 3. Exothermic Reactions: a. Decomposition b. Polymerization c. Isomerization d. Oxidation/reduction e. Heat of mixing/solution	❑ 3. Ignition Sources: a. Pyrophoric chemicals b. Unstable materials (e.g., acetylides) c. Static generation (Non- or semi-conductive fluids) d. Strong oxidizers e. Hot surfaces f. Friction/heating from moving parts g. Materials with high heats of absorption h. Adiabatic compression i. Specific catalysts such as EO "rust" j. Self heating polymer k. Vent systems leading to ignition sources l. "Cool flame" potential m. Open flames, etc., outside an opened vessel	
❑ 4. Others: a. Loss of inhibitors b. Heat of adsorption c. Fire exposure d. Loss of cooling e. Accumulation of reactants	❑ 4. Others: a. Dust/air mixtures (confined)	

A2

Example Management of Change Hazard Review Form

(Courtesy Eli Lilly and Company)

Shown in Table A2.1 is a two-page form used by one company to document hazard reviews performed as part of a management of change procedure.

Table A2.1 Management of change hazard review form

Management of Change Document Number	
Management of Change Hazard Review Date	
Building / Department Affected	
Management of Change Hazard Review Team Members	
Function	**Name**
Change Proposal Owner	
Process Engineering	
Operations	
Other (specify)	
Review Scope – *Provide a description of the change being reviewed*	

Hazard Review *(Note: Document all issues discussed, even if they do not result in concerns or recommendations)*			
_____/_________ ***The current PHA Report was reviewed/referenced prior to approving this hazard review.*** Initial Date			
#	**Topics / Questions / Potential Hazards Discussed**	**Concerns / Consequences / Existing Protections**	**Recommendations**

Table A2.1 Management of change hazard review form (continued)

Other Impact Evaluation		
Area	**Impact Evaluation**	**Action**
Process Safety Information		
Safe Operating Limits		
Safety Systems		
Mechanical Integrity		
Piping Systems		
Procedures		
Training		
Contractors		
Modes of Operation		
• Normal operations		
• Start-up/restart (e.g., initial start-up, validation, after shutdown/turnaround, etc.)		
• Normal/emergency shutdown		
• Cleaning (including CIP)		
• Sampling		
• Maintenance		
• Non-routine operations • Temporary operations • Emergency operations • Filter change • Spill cleanup		
• Major, planned operations (e.g., process changeover, column packing, etc.)		

_____/__________ ***Recommendations have been added to the change management system***

Initial Date

Change Initiator: ______________________ Signature ______________ Date

Technical Reviewer: ______________________ Signature ______________ Date

Signatures imply that, to the best of your knowledge, the hazard review was thorough & technically sound

A3

Example Reactivity Checklist

(Courtesy Dow Chemical Company)

Presented in Figure A3.1 are instructions for use of a chemical reactivity checklist. The accompanying checklist is given in Table A3.1.

INSTRUCTIONS FOR USE OF REACTIVITY CHECKLIST

Note: **The Checklist should be filled out as thoroughly as possible, using the facility reactivity matrix as a reference. As much of the information as is possible - particularly that pertaining to the possible undesirable results - should be gathered prior to the team meetings, so as to most efficiently use team time in Potential Causes (scenarios) development.**

For Each Sub-System:

- Indicate (yes or no) in the first column for each undesirable result whether that result is possible for the sub-system.
- Indicate in the columns for the various modes of operation (Startup, Shutdown, Utility Failure, Normal Operation) whether each of the potential causes is applicable (N for Not, A for Applicable), and whether it is a potential Major Process Hazard (M).
- Indicate in the Comments column any relevant comments, including the Undesirable Result that is applicable.
- When the scenario listing for the study is completed, enter the scenario number in the appropriate column.

Figure A3.1 Instructions for use of example reactivity checklist

Table A3.1 Example reactivity checklist

REACTIVITY CHECKLIST FOR REVIEW *Items that must be completed to assure that reactivity considerations have been thoroughly covered. Use in conjunction with facility reactivity matrix.*							
Concern/issue	Scenarios to be evaluated					Comments	Scenario no.
A. Reactivity With Air/Oxygen	*Note:*						
Undesirable Results	*For each*						
1. Internal Explosion (>2xMAWP)	*Undesirable*						
2. Pyrophoric Materials	*Result,*						
3. Peroxide Formation	*indicate in*						
4. Lowering of SADT	*this column*						
5. Polymerization Catalysis by Air/O2	*whether (Y)*						
6. Decomposition Catalysis by Air/O2	*or not (N) it is*						
7. Cool Flame Formation Potential	*applicable for*						
8. Formation of shock-sensitive compounds	*this system*						
Potential Causes	Startup	Shutdown	Maintenance	Utility failure	Normal op'n		
1. Vacuum operation - breaking vacuum with air	*Cause Code: N=Not Applicable, A=Applicable, M=Potential MPH*						
2. Failure to purge prior to or following maintenance							
3. Improper connection of utility drop							
4. Failure of process air/O2 ratio control							
5. Introduction by contamination of inerting medium							
6. Ingress through flare/vent systems							
7. Backflow from furnace/incinerator							
8. Blowing transfer lines with air							
9. Other							
B. Reactivity With Water							
Undesirable Results							
1. Hydrate formation							
2. Exothermic reaction with water - gas liberation							
3. Exothermic reaction with water - polymer formation							
4. Exothermic reaction with water - overheating							
5. Heat of solution in water, water in material							
6. Reaction inhibition by water							
7. Formation of toxic or other hazardous materials							
Potential Causes	Startup	Shutdown	Maintenance	Utility failure	Normal op'n		
1. Heat exchanger/calandria tube leak/rupture							
2. Backflow from process scrubbers							
3. Introduction by contamination of inerting medium							
4. Improper connection from utility drop							
5. Contamination or improper transfer from distribution/upstream process							
6. Breakthrough of water from distillation A66column/stripper upset							
7. Improper/inadequate cleanup of hoses prior to transfer							
8. Improper/inadequate cleanup of lines or equipment after washing							
9. Change in composition of streams							
10. Inappropriate introduction of water during maintenance							

Table A3.1 Example reactivity checklist (continued)

REACTIVITY CHECKLIST FOR REVIEW *Items that must be completed to assure that reactivity considerations have been thoroughly covered. Use in conjunction with facility reactivity matrix.*							
Concern/issue	Scenarios to be evaluated					Comments	Scenario no.
C. Polymerization							
Undesirable Results							
1. Uncontrolled reaction leading to overheating/overpressure							
2. Instrument/equipment plugging/fouling							
(a) Partial - leading to upsets/unstable operation *Note: Consider effects of other failures during upsets*							
(b) Total - leading to process shutdown *Note: Consider effects of other failures during transition*							
3. Formation of shock sensitive compounds							
4. Other							
Potential Causes	Startup	Shutdown	Maintenance	Utility failure	Normal op'n		
1. Contamination from:							
(a) Normal transfers							
(b) Backflow from downstream processes/facilities							
(c) Vent/makeup systems							
(d) Steam, nitrogen, condensate, other utility systems							
(e) Purification system breakthrough							
2. Loss/depletion of inhibitors							
3. Introduction of wrong material in place of catalyst/inhibitor							
4. Loss of cooling							
5. Overheating (e.g. fire exposure)							
6. Accumulation of reactants due to improper feed ratio/feed sequence							
7. Accumulation of reactants due to catalyst poisoning/insufficient catalyst							
8. Accumulation of material due to improper/incomplete mixing/mass transfer							
9. Change in composition of streams							
10. Extended inactive storage at or above incubation temperature							
11. Introduction of polymerization catalytic materials during maintenance							

Table A3.1 Example reactivity checklist (continued)

REACTIVITY CHECKLIST FOR REVIEW *Items that must be completed to assure that reactivity considerations have been thoroughly covered. Use in conjunction with facility reactivity matrix.*							
Concern/issue	Scenarios to be evaluated					Comments	Scenario no.
D. Decomposition							
Undesirable Results							
1. Uncontrolled reaction leading to overheating/overpressure							
2. Formation of toxic or other hazardous materials							
3. Ignition source for other flammable release							
4. Equipment/instrument contamination or plugging, leading to upsets *Note: Consider effects of other failures during upsets*							
Potential Causes	Startup	Shutdown	Maintenance	Utility failure	Normal op'n		
1. Overheating due to:							
(a) Hot spots							
(b) Fire exposure							
(c) Loss of temperature control							
(d) Loss of cooling							
(e) Other causes							
2. Contamination with catalytic material							
3. Loss of inhibitor							
4. Loss of diluent							
5. Extended inactive storage at or above incubation temperature							
6. Mechanical shock							
7. Overpressure (lower initiation energy requirement)							
8. Introduction of wrong material in place of catalyst/inhibitor							
9. Change in composition of streams							
E. Other Unwanted Reactions *(Aldol Condensations, etc.)*							
Undesirable Results							
1. Uncontrolled reaction leading to overheating/overpressure							
2. Formation of toxic or other unwanted materials							
3. Instrument/equipment plugging/fouling							
(a) Partial - leading to upsets/unstable operation *Note: Consider effects of other failures during upsets*							
(b) Total - leading to process shutdown *Note: Consider effects of other failures during transition*							
4. Corrosion							
5. Product quality impairment							
Potential Causes	Startup	Shutdown	Maintenance	Utility failure	Normal op'n		
1. Contamination of normal feed streams							
2. Erroneous transfer from distribution/upstream facilities							
3. Contamination of steam, nitrogen, condensate, or other utility							
4. Upsets/maloperation of upstream separation equipment							
5. Backflow from flare/vent collection systems							
6. Backflow from scrubbers							
7. Unloading incorrect material							
8. Improper switchbank setup							
9. Delivery of incorrect material from stores, lab, etc.							
10. Change in composition of streams							
11. Introduction of reactive material during maintenance activity							

A4

An Inherently Safer Process Checklist

(Courtesy E.I. Du Pont de Nemours & Co., Inc.)

This checklist may be used to stimulate the thinking of inherent safety review and process hazard analysis teams, and any other individuals or groups working on process improvements. It is intended to promote "blue-sky" or "out-of the-box" thinking, and to generate ideas that might be usable in an existing facility or a "plant of the future" concept.

This checklist should not be used in a rote "yes/no" manner, nor is it necessary to answer every question. The idea is to consider what might be possible, and then determine what is feasible. The checklist should be reviewed periodically throughout the life cycle of the process. As technology changes, what was once impossible becomes possible, and what was once infeasible becomes feasible.

Users of this checklist may find it helpful to rephrase questions in order to prompt maximum creativity; for example "how might it be possible to...?" This approach can lead users to consider alternative means for reducing the hazard level inherent in the process.

The topics for this checklist have been taken from CCPS, *Guidelines for Engineering Design for Process Safety,* AIChE, New York, 1993 and Bollinger et al., *Inherently Safer Chemical Processes: A Life Cycle Approach,* AIChE, New York, 1996. It was first published in this form in Johnson et al., *Essential Practices for Managing Chemical Reactivity Hazards,* AIChE, New York, 2003. Every effort was made to ensure that this checklist is comprehensive; therefore, there may be some redundancy or overlap in questions among the different sections. It should be noted that some of the items in this checklist employ a very broad concept of inherent safety, as presented by Bollinger et al. (1996). As such, they may address inherent aspects of passive, engineered, or even administrative controls, rather than the narrower inherent safety conception of reducing the underlying process hazards that must be contained and controlled to safely operate a facility.

1 Intensification / Minimization

1.1 Do the following strategies reduce inventories of hazardous raw materials, intermediates, and/or finished products?
- Improved production scheduling
- Just-in-time deliveries
- Direct coupling of process elements
- Onsite generation and consumption

1.2 Do the following actions minimize in-process inventory?
- Eliminating or reducing the size of in-process storage vessels
- Designing processing equipment handling hazardous materials for the smallest feasible inventory
- Locating process equipment to minimize the length of hazardous material piping runs
- Reducing piping diameters

1.3 Can other types of unit operations or equipment reduce material inventories? For example:
- Wiped film stills in place of continuous still pots
- Centrifugal extractors in place of extraction columns
- Flash dryers in place of tray dryers
- Continuous reactors in place of batch
- Plug flow reactors in place of continuous-flow stirred tank reactors
- Continuous in-line mixers in place of mixing vessels

1.4 Can thermodynamic or kinetic efficiencies of reactors be improved by design upgrades (e.g., improved mixing or heat transfer) to reduce hazardous material volume?

1.5 Can equipment sets be combined (e.g., replacing reactive distillation with a separate reactor and multi-column fractionation train; installing internal reboilers or heat exchangers) to reduce overall system volume?

1.6 Can pipeline inventories be reduced by feeding hazardous materials as a gas instead of a liquid (e.g., chlorine)?

1.7 Can process conditions be changed to avoid handling flammable liquids above their flash points?

1.8 Can process conditions be changed to reduce production of hazardous wastes or by-products?

2 Substitution / Elimination

2.1 Is it possible to eliminate hazardous raw materials, process intermediates, or by-products by using an alternative process or chemistry?

2.2 Is it possible to eliminate in-process solvents by changing chemistry or processing conditions?

2.3 Is it possible to substitute less hazardous raw materials? For example:
- Noncombustible rather than flammable
- Less volatile
- Less reactive
- More stable
- Less toxic

2.4 Is it possible to use utilities with lower hazards (e.g., low-pressure steam instead of combustible heat transfer fluid)?

2.5 Is it possible to substitute less hazardous final product solvents?

2.6 For equipment containing materials that become unstable at elevated temperatures or freeze at low temperatures, is it possible to use heating and cooling media that limit the maximum and minimum temperature attainable?

3 Attenuation / Moderation

3.1 Is it possible to keep the supply pressure of raw materials lower than the design pressure of the vessels to which they are fed?

3.2 Is it possible to make reaction conditions (e.g., pressure or temperature) less severe by using a catalyst or by using a better catalyst?

3.3 Can the process be operated at less severe conditions using any other route? For example:

- Improved thermodynamic or kinetic efficiencies of reactors by design upgrades (e.g., improved mixing or heat transfer) to reduce operating temperatures and/or pressures
- Changes to the order in which raw materials are added
- Changes in phase of the reaction (e.g., liquid/liquid, gas/liquid, or gas/gas)

3.4 Is it possible to dilute hazardous raw materials to reduce the hazard potential? For example, by using the following:

- Aqueous ammonia instead of anhydrous
- Aqueous HCl instead of anhydrous
- Sulfuric acid instead of oleum
- Dilute nitric acid instead of concentrated fuming nitric acid
- Wet benzoyl peroxide instead of dry

4 Limitation of Effects

4.1 Is it possible to design and construct vessels and piping to be strong enough to withstand the largest overpressure that could be generated within the process, even if the "worst credible event" occurs (eliminating the need for complex, high-pressure interlock systems and/or extensive emergency relief systems)?

4.2 Is all equipment designed to totally contain the materials that might be present inside at ambient temperature or the maximum attainable process temperature (i.e., higher maximum allowable working temperature to accommodate loss of cooling, simplifying reliance on the proper functioning of external systems, such as refrigeration systems, to control temperature such that vapor pressure is less than equipment design pressure)?

4.3 Can passive leak-limiting technology (e.g., blowout resistant gaskets and excess flow valves) be utilized to limit potential for loss of containment?

4.4 Can process units be located to reduce or eliminate adverse effects from other adjacent hazardous installations?

4.5 Can process units be located to eliminate or minimize the following?

- Off-site impacts
- On-site impacts on employees and other plant facilities

4.6 For processes handling flammable materials, is it possible to design the facility layout to minimize the number and size of confined areas and to limit the potential for serious overpressures in the event of a loss of containment and subsequent ignition?

4.7 Can the plant be located to minimize the need for transportation of hazardous materials?

4.8 Can materials be transported in the following ways?

- In a less hazardous form
- Via a safer transport method
- Via a safer route

5 Simplification / Error Tolerance

5.1 Is it possible to separate a single, procedurally complex, multipurpose vessel into several simpler processing steps and processing vessels, thereby reducing the potential for hazardous interactions when the complexity of the number of raw materials, utilities, and auxiliary equipment is reduced for specific vessels?

5.2 Can equipment be designed so that it is difficult to create a potentially hazardous situation due to an operating or maintenance error? For example:

- Simplifying displays
- Designing temperature-limited heat transfer equipment
- Lowering corrosion potential by use of resistant materials of construction
- Lowering operating pressure to limit release rates
- Using higher processing temperatures (to eliminate cryogenic effects such as embrittlement failures)
- Using passive vs. active controls (e.g., stronger piping and vessels)
- Using buried or shielded tanks
- Using fail-safe controls if utilities are lost
- Limiting the degree of instrumentation redundancy required
- Using refrigerated storage vs. pressurized storage
- Spreading electrical feed over independent or emergency sources
- Reducing wall area to minimize corrosion/fire exposure
- Reducing the number of connections and paths
- Minimizing the number of flanges in hazardous processes
- Valving/piping/hose designed to prevent connection error
- Using fewer bends in piping
- Increasing wall strength
- Using fewer seams and joints
- Providing extra corrosion/erosion allowance
- Reducing vibration
- Using double-walled pipes, tanks, and other containers
- Minimizing use of open-ended valves
- Eliminating open-ended, quick-opening valves in hazardous service
- Improving valve seating reliability
- Eliminating unnecessary expansion joints, hoses, and rupture disks
- Eliminating unnecessary sight glasses/glass rotameters

5.3 Can procedures be designed so that it is difficult to create a potentially hazardous situation due to an operating or maintenance error? For example:

- Simplifying procedures
- Reducing excessive reliance on human action to control the process

5.4 Can equipment be eliminated or arranged to simplify material handling?

- Using gravity instead of pumps to transfer liquids
- Siting to minimize hazardous transport or transfer
- Reducing congestion (i.e., easier to access and maintain)
- Reducing knock-on effects from adjacent facilities
- Removing hazardous components early in the process rather than spreading them throughout the process
- Shortening flow paths

5.5 Can reactors be modified to eliminate auxiliary equipment (e.g., by creating a self-regulatory mechanism by using natural convection rather than forced convection for emergency cooling)?

5.6 Can distributed control system (DCS) modules be simplified or reconfigured such that failure of one module does not disable a large number of critical control loops?

Appendix B
Supplemental Questions for Hazard Identification

Included in this Appendix is an expanded list of process safety questions that can be used as a stand-alone Checklist Review or as a supplement to another technique. Note that not all questions will be applicable at each life cycle stage; e.g., some questions may be more pertinent to a design-stage hazard evaluation than for analysis of an operating facility.

- **Use these questions to help identify potential hazards.**
- **Answer each question fully, not with a simple "Yes" or "No."**
- **Some questions may not be applicable to the review of a particular process; other questions should be interpreted broadly to include similar materials or equipment in your facility.**
- **Consider the questions in terms of all operating modes (e.g., steady state, start-up, shutdown, maintenance, and upsets).**

I. Process

A. *Materials and Flowsheet*

1. What materials are hazardous (e.g., raw materials, intermediates, products, byproducts, wastes, products of unintended chemical reactions, and combustion products)? Are any prone to form vapor clouds?
   — Which ones are acutely toxic?
   — Which ones are chronically toxic, carcinogenic, mutagenic, or teratogenic?
   — Which ones are flammable?
   — Which ones are combustible?
   — Which ones are unstable, shock-sensitive, or pyrophoric?
   — Which ones have release limits specified by law or regulation?

2. What are the properties of the process materials? Consider
   — physical properties (e.g., boiling point, melting point, vapor pressure).
   — acute toxic properties and exposure limits (e.g., IDLH, LD_{50}).

- chronic toxic properties and exposure limits (e.g., TLV®, PEL).
- reactive properties (e.g., incompatible or corrosive materials, polymerization).
- combustion properties (e.g., flash point, autoignition temperature).
- environmental properties (e.g., biodegradability, aquatic toxicity, odor threshold).

3. What unwanted hazardous reactions or decompositions can develop:
   - because of improper storage?
   - because of impact or shock?
   - because of foreign materials?
   - because of abnormal process conditions (e.g., temperature, pH)?
   - because of abnormal flow rates?
   - because of missing ingredients or misproportioned reactants or catalysts?
   - because of mechanical failure (e.g., pump trip, agitator trip) or improper operation (e.g., started early, late, or out of sequence)?
   - because of sudden or gradual blockage or buildup in equipment?
   - because of overheating residual material (i.e., heels) in equipment?
   - because of a utility failure (e.g., inert gas)?

4. What data are available or should be obtained on the amount and rate of heat and gas evolution during reaction or decomposition of any materials?

5. What provisions are made for preventing runaway reactions and for quenching, short-stopping, dumping, or venting an existing runaway?

6. What provision is made for rapid passivation or disposal of reactants if required?

7. In heat-integrated units, what provisions are made to maintain temperature control when flow through one or more pathways stops?

8. Can compounds (e.g., iron sulfide, ammonium perchlorate) that are pyrophoric or sensitive to impact/shock precipitate out of the solution or form if the solution dries?

9. How are process materials stored?
   - Are flammable or toxic materials stored at temperatures above their atmospheric boiling points?
   - Are refrigerated or cryogenic storage tanks used to reduce storage pressures?
   - Are potentially explosive dusts stored in large bins?
   - Are any large inventories of flammables or toxics stored inside buildings?
   - Are inhibitors needed? How is inhibitor effectiveness maintained?

10. Is any stored material incompatible with other chemicals in the area?

11. What is done to ensure raw material identification and quality control? Are there hazards associated with contamination with common materials such as rust, air, water, oil, cleaning agents, or metals? Are there materials used that could be easily mistaken for each other?

12. What raw materials or process materials can be adversely affected by extreme weather conditions?

13. Can hazardous materials be eliminated? Have alternative processes with less toxic/reactive/flammable raw materials, intermediates, or by-products been evaluated? Can hazardous raw materials be stored in diluted form (e.g., aqueous ammonia instead of anhydrous, sulfuric acid instead of oleum)?

14. Can hazardous material inventories be reduced?
    — Can the number or size of tanks be reduced?
    — Has all processing equipment been selected and designed to minimize inventory (e.g., using wiped film stills, centrifugal extractors, flash dryers, continuous reactors, in-line mixers)?
    — Can hazardous materials (e.g., chlorine) be fed as a gas instead of a liquid?
    — Is it possible to reduce storage of hazardous intermediates by processing the materials into their final form as they are produced?

15. Can the process be performed under safer conditions?
    — Can the supply pressure of raw materials be kept below the working pressure of vessels receiving them?
    — Can reaction conditions (e.g., temperature, pressure) be made less severe by using or improving a catalyst or by increasing recycle flows to compensate for lower yields?
    — Can process steps be carried out in a series of vessels to reduce the complexity and number of feed streams, utilities, and auxiliary systems?

16. Can hazardous wastes be minimized?
    — Can waste streams be recycled?
    — Can all solvents, diluents, or "carriers" be recycled? If not, can they be minimized or eliminated?
    — Have all washing operations been optimized to reduce the volume of wastewater?
    — Can useful by-products be recovered from waste streams? Can hazardous byproducts be extracted to reduce the overall volume of hazardous waste?
    — Can hazardous wastes be segregated from non-hazardous wastes?

17. What has been done to ensure that the materials of construction are compatible with the process materials involved?

18. What changes have been made in process equipment or operating parameters since the previous safety review?

19. What changes have occurred in the composition of raw materials, intermediates, or products? How has the process been changed to accommodate these differences?

20. In view of process changes since the last hazard evaluation, how adequate is the size of:
    — other process equipment?
    — relief and flare systems?
    — vents and drains?

21. What safety margins have been narrowed by design or operating changes (e.g., to reduce cost, increase capacity, improve quality, or change products)?

22. What hazards are created by the loss of each feed, and by simultaneous loss of two or more feeds?

23. What hazards result from loss of each utility, and from simultaneous loss of two or more utilities such as:
    - electricity?
    - high, medium, or low pressure steam?
    - instrument air?
    - instrument electric power?
    - inert gas?
    - fuel gas/oil?
    - natural gas/pilot gas?
    - plant air?
    - cooling water?
    - refrigerant/brine?
    - process water?
    - deionized water?
    - ventilation?
    - process drain/sewer?

24. What are the most severe credible incidents (i.e., the worst conceivable combinations of reasonable malfunctions) that can occur?

25. What is the potential for external fire (which may create hazardous internal process conditions)?

26. How much experience do the facility and company have with the process? If limited, is there substantial industry experience? Is the company a member of industry groups that share experience with particular chemicals or processes?

27. Is the unit critical to overall facility operations on a throughput or value-added basis? Does shutdown of this unit require other units to be shut down as well?

B. Unit Siting and Layout

1. Can the unit be located to minimize the need for off-site or intra-site transportation of hazardous materials?

2. What hazards does this unit pose to the public or to workers in the control room, adjacent units, or nearby office or shop areas from:
    - toxic, corrosive, or flammable sprays, fumes, mists, or vapors?
    - thermal radiation from fires (including flares)?
    - overpressure from explosions?
    - contamination from spills or runoff?
    - noise?
    - contamination of utilities (e.g., potable water, breathing air, sewers)?
    - transport of hazardous materials from other sites?

3. What hazards do adjacent facilities (e.g., units, highways, railroads, underground pipelines) pose to personnel or equipment in the unit from:
   - toxic, corrosive, or flammable sprays, fumes, mists, or vapors?
   - overpressure from explosions?
   - thermal radiation from fires (including flares)?
   - contamination?
   - noise?
   - contamination of utilities (e.g., potable water, breathing air, sewers)?
   - impacts (e.g., airplane crashes, derailments, turbine blade fragments)
   - flooding (e.g., ruptured storage tank, plugged sewer)

4. What external forces could affect the site? Consider:
   - high winds (e.g., hurricanes, typhoons, tornadoes).
   - earth movement (e.g., earthquakes, landslides, sink holes, settling, freeze/thaw heaving, coastal/levee erosion).
   - snow/ice (e.g., heavy accumulation, falling icicles, avalanches, hail, ice glaze).
   - utility failures from outside sources.
   - releases from adjacent plants.
   - sabotage/terrorism/war.
   - airborne particulates (e.g., pollen, seeds, volcanic dust, dust storm).
   - natural fires (e.g., forest fires, grass fires, volcanism).
   - extreme temperatures (causing, for example, brittle fracture of steel).
   - flooding (e.g., hurricane surge, seiche, broken dam or levee, high waves, intense precipitation, spring thaw).
   - lightning.
   - drought (causing, for example, low water levels or poor grounding).
   - meteorite.
   - fog.

5. What provisions have been made for relieving explosions in buildings or operating areas?

6. Are there open ditches, pits, sumps, or pockets where inert, toxic, or flammable vapors could collect?

7. Should there be concrete bulkheads, barricades, or berms installed to protect adjacent personnel and equipment from explosion hazards?

8. Are operating units and the equipment within units spaced to minimize potential damage from fires or explosions in adjacent areas and to allow access for fire fighting activities? Are there safe exit routes?

9. Has equipment been adequately spaced and located to permit anticipated maintenance (e.g., pulling heat exchanger bundles, dumping catalyst, lifting with cranes)?

10. Is temporary storage provided for raw materials and for finished products at appropriate locations?

11. What expansion or modification plans are there for the facility?

12. Can the unit be built and maintained without lifting heavy items over operating equipment and piping?

13. Is there adequate access for emergency vehicles? Could access roads be blocked by trains, highway congestion, etc.?

14. Are access roads well engineered to avoid sharp curves? Are traffic signs provided?

15. Is vehicular traffic appropriately restricted from areas where pedestrians could be injured or equipment damaged?

16. Are vehicle barriers installed to prevent impact to critical equipment adjacent to high traffic areas?

II. Equipment

A. Pressure and Vacuum Relief

1. Can equipment be designed to withstand the maximum credible overpressure generated by a process upset?

2. Where are emergency relief devices needed (e.g., breather vents, relief valves, rupture disks, and liquid seals)? What is the basis for sizing these (e.g., utility failure, external fire, mispositioned valve, runaway reaction, thermal expansion, tube rupture)?

3. Is the relief system designed for two-phase flow? Should it be?

4. Is any equipment that is not protected by relief devices operating under pressure or capable of being overpressurized by a process malfunction?

5. Where are rupture disks installed in series with relief valves?
   - Is there a pressure indicator (e.g., gauge, transmitter, switch) and vent between the rupture disk and relief valve?
   - How often is the pressure indicator read? Should an automatic bleeder be installed with an excess flow check valve and pressure alarm?
   - Were the relief devices sized considering the pressure drop through the entire assembly?

6. Where rupture disks are used to vent explosive overpressures (e.g., peroxide decomposition), are they properly sized relative to vessel capacity and design?

7. Are the relief setpoints and sizes correct?
   — Is at least one relief device set at or below the design pressure of protected equipment?
   — Should multiple relief devices with staggered settings be considered to avoid chattering (particularly where the relief loads in many scenarios will be less than 25% of maximum capacity)?
   — In piping systems, does the relief setpoint allow for static head and differential pressure between the pressure source (e.g., pump) and the relief device?
   — What is maximum backpressure at the relief device? Has its capacity been corrected for this backpressure?
   — Has the relief device been resized appropriately for changes in process conditions (e.g., higher throughput, different reactants)?

8. Are the inlet piping and the outlet piping for relief devices adequate?
   — Are the lines sized for the desired flow and allowable pressure drop?
   — Are the inlet and outlet line ratings and sizes consistent with the ratings and sizes of the relief device's flanges?
   — What has been done to prevent end-of-line whipping during discharge?
   — Is the discharge piping independently supported?
   — Can the discharge piping withstand liquid slugs?
   — Have piping bends and lengths been minimized?
   — How is condensate/rain drained from the discharge piping?
   — Can steam be injected in the discharge piping to snuff fires or disperse releases? If so, is the discharge piping adequately drained and protected from freezing?
   — What prevents solids from plugging the inlet or outlet piping? Is there a purge or blowback system? Is heat tracing required? Should a rupture disk be used? Are there bird screens?
   — Are all maintenance valves car sealed or locked open? How often is this verified?

9. How are relief headers, blowdown headers, and vents kept open?
   — How often are knockout pots drained? Is there an independent high level alarm?
   — How are liquid seals kept from freezing?
   — How is condensate/ice kept from accumulating inside uninsulated headers?
   — Can autorefrigerated vapors freeze and plug the header?
   — Can heavy oils or polymers accumulate in the header?
   — Are there any low spots that could accumulate liquids?
   — Does all process discharge piping drain freely into the header, and does the header drain freely to a knockout pot or collection point?
   — Are all maintenance valves locked open and oriented so a valve stem failure will not allow the gate to fall and obstruct the piping?
   — Can the vent scrubber or adsorption bed plug?

10. Are discharges from vents, relief valves, rupture disks, and flares located to avoid hazards to equipment and personnel? Could liquids be sprayed into the air? Are vents from relief devices (e.g., between rupture disks and relief valves, between balanced bellows, and between weep holes in discharge piping) also routed to a safe location? Are flame arrestors installed?

11. Are relief devices located so that when they open, the process flow will continue cooling critical equipment (e.g., steam superheaters)?

12. What are the impacts of a flare, incinerator, or thermal oxidizer trip or flameout? What would happen if the flare gas recovery compressor tripped?

13. Are there reliable flare flameout detection devices? Is the flare equipped with a reliable ignition system?

14. What actions are required if a flare, incinerator, thermal oxidizer, or scrubber is out of service? Do procedures minimize the potential for releases until the system is returned to service?

15. Are the flare, blowdown, and off-gas systems adequately purged, sealed or otherwise protected against air intrusion? Are there suitable flame arrestors installed in the piping?

16. Will the relief devices withstand the damaging properties (e.g., corrosion, autorefrigeration, embrittlement) of the relieved material, as well as other materials that may be present in the relief header? Is the material likely to plug the internals of the relief device (e.g., balanced bellows)?

17. What provisions are there for removing, inspecting, testing, and replacing vents, vacuum breakers, relief valves, and rupture disks? Who is responsible for scheduling this work and verifying its completion?

18. What is the plant policy regarding operation with one or more disabled relief devices (e.g., inoperative or removed for testing or repair)? Is the policy followed?

19. Are the flare, blowdown, and off-gas systems capable of handling overpressure events (including loss of utilities) for the plant as it currently exists (e.g., after plant expansions and debottlenecking)? What are the worst case scenarios for the process discharging into these systems?

20. Are there separate cold and wet relief systems? Are relief valve discharges directed to the proper system?

B. Piping and Valves

1. Is the piping specification suitable for the process conditions, considering:
   - compatibility with process materials and contaminants (e.g., corrosion and erosion resistance)?
   - compatibility with cleaning materials and methods (e.g., etching, steaming, pigging)?
   - normal pressure and temperature?
   - excess pressure (e.g., thermal expansion or vaporization of trapped liquids, blocked pump discharge, pressure regulator failure)?
   - high temperature (e.g., upstream cooler bypassed)?
   - low temperature (e.g., winter weather, cryogenic service)?
   - cyclical conditions (e.g., vibration, temperature, pressure)?
   - Is the piping particularly vulnerable to external corrosion because of its design (e.g., material of construction, insulation on cold piping), location (e.g., submerged in a sump), or environment (e.g., saltwater spray)?

2. Is there any special consideration, for either normal or abnormal conditions, that could promote piping failure? For example:
   - Would flashing liquids autorefrigerate the piping below its design temperature?
   - Could accumulated water freeze in low points or in dead-end or intermittent service lines?
   - Could cryogenic liquid carry-over chill the piping below its design temperature?
   - Could heat tracing promote an exothermic reaction in the piping, cause solids to build up in the piping, or promote localized corrosion in the piping?
   - Could the pipe lining be collapsed by vacuum conditions?
   - Could a process upset cause corrosive material carry-over in the piping, or could dense corrosive materials (e.g., sulfuric acid) accumulate in valve seats, drain nipples, etc.?
   - In high temperature reducing service (e.g., hydrogen, methane, or carbon monoxide), could metal dusting cause catastrophic failure? Is the piping protected by suitable chemical addition (e.g., sulfides)?
   - Is the piping vulnerable to stress corrosion cracking (e.g., caustic in carbon steel piping, chlorides in stainless steel piping)? Should the piping be stress relieved?
   - Is the piping vulnerable to erosion? Are piping elbows and tees designed to minimize metal loss, and are they periodically inspected?
   - Could rapid valve closure or two-phase flow cause hydraulic hammer in the piping? Should valve opening/closing rates be dampened to avoid piping damage?
   - Are there flexible connections that could distort or crack?

3. Can piping sizes or lengths be reduced to minimize hazardous material inventories?

4. Have relief devices been installed in piping runs where thermal expansion of trapped fluids (e.g., chlorine) would separate flanges or damage gaskets?

5. Are piping systems provided with freeze protection, particularly cold water lines, instrument connections, and lines in dead-end service such as piping at standby pumps? Can the piping system be completely drained?

6. Were piping systems analyzed for stresses and movements due to thermal expansion and vibration? Are piping systems adequately supported and guided? Will any cast-iron valves be subjected to excessive stresses that could fracture them? Will pipe linings crack (particularly at the flange face) because of differential thermal expansion?

7. Are bellows, hoses, and other flexible piping connections really necessary? Could the piping system be redesigned to eliminate them? Are the necessary flexible connections strong enough for the service conditions?

8. What are the provisions for trapping and draining steam piping?

9. Which lines can plug? What are the hazards of plugged lines?

10. Are provisions made for flushing out all piping during start-up and shutdown? Are hoses, spools, jumpers, etc., flushed or purged before use?

11. Are the contents of all lines identified?

12. Are there manifolds on any venting or draining systems and, if so, are there any hazards associated with the manifolds?

13. Are all process piping connections to utility systems adequately protected against potentially hazardous flows?
    - Are there check valves or other devices preventing backflow into the utility supply?
    - Are there disconnects (spools, hoses, swing elbows, etc.) with suitable blinds or plugs for temporary or infrequently used utility connections?
    - Are there double blocks and bleeds for permanent utility connections?

14. Are spray guards installed on pipe flanges in areas where a spraying leak could injure operators or start fires?

15. Will the piping insulation trap leaking material and/or react exothermically with it?

16. Have plastic or plastic-lined piping systems been adequately grounded to avoid static buildup?

17. Are there remote shutoff devices on off-site pipelines that feed into the unit or storage tanks?

18. Can bypass valves (for control valves or other components) be quickly opened by operators?
    — What hazards may result if the bypass is opened (e.g., reverse flow, high or low level)?
    — What bypass valves are routinely opened to increase flow, and will properly sized control valves be installed?
    — Is the bypass piping arranged so it will not collect water and debris?
    — Is there a current log of open bypass valves kept in the control room so operators can ensure they are reclosed if necessary in an emergency?

19. How are the positions of critical valves (block valves beneath relief devices, equipment isolation valves, dike drain valves, etc.) controlled (car seals, locks, periodic checks, etc.)?

20. How are the positions of critical valves (e.g., emergency isolation valves, dump valves) indicated to operators? Is the position of all nonrising stem valves readily apparent to the operators? Do control room displays directly indicate the valve position, or do they really indicate some other parameter, such as actuator position or torque, application of power to the actuator, or initiation of a control signal to the actuator?

21. Are block valves or double block and bleed valves required:
    — because of high process temperatures?
    — because of high process pressures?
    — because the process material is likely to erode or damage valve internals?
    — because the process material is likely to collect on the valve seat?
    — for worker protection during maintenance on operating systems?

22. Are critical isolation valve actuators powerful enough to close the valves under worst case differential pressure conditions (including backflow) in the event of a rupture?

23. Are chain-operators for valves adequately supported and sized to minimize the likelihood of valve stem breakage?

24. How will control valves react to loss of control medium or signal? Do the control valves:
    — reduce heat input (cut firing, reboiling, etc.)?
    — increase heat removal (increase reflux, quench, cooling water flow, etc.)?
    — reduce pressure (open vents, reduce speed of turbines, etc.)?
    — maintain or increase furnace tube flow?
    — ensure adequate flow at compressors or pumps?
    — reduce or stop input of reactants?
    — reduce or stop makeup to a recirculating system?
    — isolate the unit?
    — avoid overpressuring of upstream or downstream equipment (e.g., by maintaining level to avoid gas blowby)?
    — avoid overcooling (below minimum desired temperature)?

25. Will control valve malfunction result in exceeding the design limits of equipment or piping?
    - Are upstream vessels between a pressure source and the control valve designed for the maximum pressure when the control valve closes?
    - Some piping's class decreases after the control valve. Is this piping suitable if the control valve is open and the downstream block closed? Is other equipment in the same circuit?
    - Is there any equipment whose material selection makes it subject to rapid deterioration or failure if any specific misoperation or failure of the control valve occurs (overheating, overcooling, rapid corrosion, etc.)?
    - Will the reactor temperature run away?
    - Is the three-way valve used in a pressure-relieving path the equivalent of a fully open port in all valve positions?

26. Is there provision in the design for a single control valve to fail:
    - in the worst possible position (usually opposite the fail-safe position)?
    - with the bypass valve open?

27. Upon a plant-wide or unit-wide loss of control medium or signal, which valves should fail to a position that is different from their normal failure positions? How were the conflicts resolved?

28. Can the safety function of each automatically controlled valve be tested while the unit is operating? Will an alarm sound if the sensing-signal-control loop fails or is deactivated? Should any bypass valves be car-sealed or locked closed?

29. Are battery limit block valves easily accessible in an emergency?

30. Are controllers and control valves readily accessible for maintenance?

C. Pumps

1. Can the pump discharge pressure exceed the design pressure of the casing?
    - Does the pump casing design pressure exceed the maximum suction pressure plus the pump shutoff pressure?
    - Is there a discharge-to-suction relief valve or minimum flow valve protecting the pump (set below the casing design pressure minus the maximum suction pressure)?
    - How would a higher density fluid affect the discharge pressure (e.g., during an upset, start-up, or shutdown)?
    - How would pump overspeed affect the discharge pressure?
    - Do any safety signals that close a pump's minimum flow bypass also shut down the pump?

2. Can the pump discharge pressure exceed the design pressure of downstream piping or equipment?

- If a downstream blockage could raise the pump suction pressure, is the downstream piping and equipment rated for the maximum suction pressure plus the pump shutoff pressure?
- If a downstream blockage would not raise pump suction pressure, is the downstream piping and equipment rated for the greater of (1) normal suction pressure plus the pump shutoff pressure or (2) maximum suction pressure plus normal pump differential pressure?

3. In parallel pump arrangements, can leakage through an idle pump's discharge check valve overpressure the suction valve, flange, and connecting piping for the idle pump?

4. Can the design temperature of the pump be exceeded?
   - What is the maximum upstream temperature?
   - Could heat removal equipment (e.g., lube oil coolers, gland oil coolers, stuffing box coolers, seal flushes) be bypassed or lose flow?
   - Could the pump run in a total recycle or blocked-in configuration?
   - Could the pump be run dry?

5. Can the pump suction be isolated from the feed source in an emergency?
   - Considering the materials, process conditions, and location, can operators safely close the isolation valve(s) during a fire or toxic release?
   - Are remotely operable valves, valve actuators, power cables, and instrument cables fireproofed?

6. Would leakage of the process fluid into the motor of a canned pump be hazardous?

D. *Compressors*

1. Can the compressor discharge pressure exceed the design pressure of the casing?
   - Does the compressor casing design pressure exceed the maximum suction pressure plus the compressor shutoff pressure? Is this true for each stage?
   - Is there a discharge-to-suction relief valve or recycle valve protecting the compressor (set below the casing design pressure minus the maximum suction pressure)?
   - How would a higher density fluid (e.g., during an upset, start-up, or shutdown) affect the discharge pressure?
   - How would compressor overspeed affect the discharge pressure?
   - Is there a relief valve for each low pressure stage capable of discharging the maximum recycle flow?
   - Do any safety signals that close a compressor's recycle valve also shut down the compressor?

2. Can the compressor discharge pressure exceed the design pressure of downstream piping or equipment?
   - If a downstream blockage could raise the compressor suction pressure, is the downstream piping and equipment rated for the maximum suction pressure plus the compressor shutoff pressure?

- If a downstream blockage would not raise compressor suction pressure, is the downstream piping and equipment rated for the greater of (1) normal suction pressure plus the compressor shutoff pressure or (2) maximum suction pressure plus normal compressor differential pressure?
- Are pulsation dampeners provided to protect against metal fatigue?

3. Is the compressor adequately protected against overpressuring of the suction piping or interstage equipment?
   - What restricts the recycle flow? Is there a tight-sealing valve in the recycle line?
   - Is there a check valve protecting the compressor and recycle line from backflow of downstream equipment or parallel compressors?
   - What pressure would result in the suction for each stage if the discharge check valve leaks when the compressor is tripped or shut down?

4. Can the design temperatures of the compressor be exceeded?
   - What is the maximum upstream temperature?
   - What is the maximum interstage temperature?
   - Could heat removal equipment (e.g., chillers, condensers, interstage coolers, lube oil coolers, cooling jackets) be bypassed, trip off, or lose its cooling media?
   - Could the compressor run in a total recycle mode?
   - Could the compressed fluid burn or exothermically decompose?

5. Are there adequate protections against upsets that could damage the compressor?
   - Are there enough suction knockout drums to protect the compressor from liquid carry-over? Will a high liquid level in the drums sound an alarm, and will high-high level trip the compressor?
   - Is the compressor suction piping heat traced?
   - Is there an automatic recycle system adequate to prevent surging?
   - Is there a check valve in the discharge of each compressor stage to protect against reverse rotation?
   - Will the compressor shut down to prevent air leakage when vacuum conditions are detected in the suction piping?
   - Will the compressor shut down when low lube oil pressure or high lube oil temperature is detected?
   - Will the compressor shut down when overspeed or insufficient load conditions are detected?

6. Can the compressor be isolated from flammable inventories in an emergency?
   - Can the compressor be shut down from the control room?
   - Can the suction, discharge, and recycle lines be remotely isolated?
   - Is there a significant inventory of flammable liquids in knockout pots before each stage, and are there remotely operable isolation valves for each stage?
   - Are remotely operable valves, valve actuators, power cables, and instrument cables fireproofed?

7. Are self-lubricated components or nonflammable synthetic lubricants used for air compressors to guard against explosion?

8. Are air compressor intakes protected against contaminants (rain, birds, flammable gases, etc.)?

9. If the compressor is in an enclosed building, are proper gas detection and ventilation safeguards installed?

E. Reactors

1. What would cause an exothermic reaction in the reactor?
   — Would quench failure or loss of external cooling cause a runaway reaction?
   — Would an excess (e.g., a double charge) or deficiency of one reactant cause a runaway reaction?
   — Would contaminants (e.g., rust, air, water, oil, cleaning agents, metals, other process materials) cause a runaway reaction?
   — Would inadequate cleaning cause a runaway reaction?
   — Would reactants added in the wrong order cause a runaway reaction?
   — Can loss of agitation in a cooled, stirred reactor lead to excessive temperature/pressure and a subsequent runaway reaction?
   — Could loss of agitation in a heated, jacketed reactor lead to localized overheating at liquid surface and a subsequent runaway reaction?
   — Could local hot spots result from partial bed obstruction?
   — Will excessive point or surface temperature lead to thermal decomposition or a runaway reaction?
   — Would delayed initiation of batch reaction during reactant addition cause a runaway reaction?
   — Could an exothermic reaction be caused by leakage of heat transfer fluid from the jacket or internal coil into the reactor?
   — Could backflow of material through a drain, vent, or relief system lead to or exacerbate a runaway reaction?
   — Will excessive preheating drive the reaction further?
   — Would a loss of purge or inerting gas cause a runaway reaction?

2. What would be the effect of an agitator
   — failing?
   — failing and later restarting?
   — being started late?
   — running too fast or too slow?
   — running in the reverse direction?

3. How is agitator motion monitored (e.g., shaft speed, motor current)?

4. Can material overcharges, solvent undercharges, overcooling, etc., lead to precipitation and loss of effective agitation?

5. Is the pressure relief for the reactor adequate?
   — What is the design basis for the relief system (e.g., cooling failure, external fire, runaway reaction)?
   — Was the potential for two-phase flow through the relief device(s) considered?
   — Is the relief device inlet protected from plugging?
   — Was the pressure drop through the reactor considered in the relief system design?
   — Could the reactor bed plug (e.g., scale, coking, catalyst attrition, structural failure) and cause overpressure in a region with no relief device?
   — Could heat transfer fluid leak into the reactor and overpressure it?
   — Could the reactor be subjected to excessive vacuum?

6. Can the design temperature of the reactor be exceeded?
   — Could the feed streams be overheated?
   — Could the reaction run away?
   — Could local hot spots develop?
   — Could the bed regeneration temperature be set too high?
   — Could uncontrolled reactions or burning occur in the bed during regeneration?
   — Could air (e.g., instrument air, plant air, regeneration air) leak into the reactor during operation?
   — Could heat transfer fluid leak into the reactor and overheat it?

7. What hazards are associated with the reactor catalyst?
   — Is the catalyst pyrophoric either before or after use?
   — Could the catalyst attack the reactor (or downstream equipment) during normal use, during an abnormal reaction, or during regeneration?
   — Is the fresh or spent catalyst toxic? Will it emit toxic gases when dumped from the reactor?

8. What hazards are associated with regenerating the catalyst or bed?
   — Is a runaway reaction possible?
   — Are regeneration feeds (e.g., air) adequately isolated during normal operation?
   — Are there interlocks to prevent simultaneous operation and regeneration?
   — How are unintended flows prevented in multiple reactor systems where one reactor is regenerated while others remain in operation?

F. *Vessels (Tanks, Drums, Towers, etc.)*

1. Are all vessels regularly inspected (e.g., x-ray, ultrasound) and pressure tested? Would the inspection method reliably detect localized damage (e.g., hydrogen blistering, fretting)? Do all pressure vessels conform to state and local requirements? Are they registered? Has the history of all vessels been completely reviewed? When were they last inspected?

2. Is the pressure relief for the vessel adequate?
   — What is the design basis for the relief system (e.g., cooling water failure, external fire, blocked flow, blowdown from upstream vessel)?

- Is a thermal expansion relief valve needed for small, liquid-filled vessels that would not otherwise require a relief valve?
- Is a vacuum relief system needed to protect the vessel during cooldown or liquid withdrawal?
- What would happen if a slug of water were fed to the vessel?

3. Can a vessel upset overpressurize downstream equipment?
   - What if the overhead pressure control valve or vent fully opens?
   - What if the liquid level were lost? Can high pressure gas blow through?
   - What if water were not separated and drained?
   - What if process material escapes through a water drawoff?

4. What hazards can occur as a result of loss of gas for purging, blanketing, or inerting?
   - How consistent is the gas supply composition?
   - How dependable are the supplies of gas, and how easily can supplies to individual units be interrupted?
   - How will a loss of the inert gas be detected?

5. What safety precautions are needed in loading liquids into, or withdrawing them from, tanks? Has the possible creation of static electricity been adequately addressed? Are diptubes used to avoid static buildup? Is all equipment properly grounded/bonded, including transport containers?

6. Can the contents of the vessel be isolated in an emergency?
   - Considering the materials, process conditions, and location, can operators safely close the isolation valves during a fire or toxic release?
   - Are there excess flow check valves or automatic isolation valves that would limit the loss of material through a downstream piping rupture?
   - Are remotely operable valves, valve actuators, power cables, and instrument cables fireproofed?
   - Can the vessel contents be pumped out or vented to a safe location?
   - Do emergency shutdowns prevent operators from emptying process materials from the unit?

7. Are all tower and drum vents and drains properly specified?
   - Are their ratings consistent with the vessel design pressure and temperature?
   - Are all drains valved and, where required, plugged, capped, or blinded?
   - Are double valves provided on regularly used drain connections for vessels? Are bleeds required?
   - Do drains on vessels that contain flashing liquids capable of autorefrigeration have double valves, with a quick-closing valve nearest the vessel?
   - Are normally closed vents plugged, capped, or blinded and, where required, also valved?
   - Is there a large vent (or vent capability) on all vessels in which human entry is planned?
   - Are all lines that could collect water adequately protected against freezing?

- — Are vents large enough for planned steamouts?
- — Are vents large enough to prevent vacuum conditions when liquids are drained from the vessel (e.g., after a washout)?

8. What vessel levels are vital for the operation of process units (e.g., levels required for pump suction pressure or surge capacity between or after process equipment)? How are these levels monitored?

9. Are the contents of all storage vessels identified?

G. *Heat Exchangers*

1. What are the consequences of a tube failure in a heat exchanger (or a heating/cooling coil failure in a vessel)?
   - — Will the fluids react, leading to high pressure, high temperature, or formation of solids?
   - — Will the fluid flash and autorefrigerate the system, possibly freezing the other fluid or embrittling the exchanger material?
   - — Will the leaking fluid cause toxic or flammable emissions in an unprotected area (e.g., at the cooling tower)?
   - — Will the leaking fluid cause corrosion, embrittlement, or other damage to equipment (including gaskets and seals) in the low pressure circuit?

2. Is the pressure relief for both sides of the heat exchanger adequate?
   - — Can the exchanger withstand exposure to the maximum pressure source upstream or downstream?
   - — What if a tube ruptures (particularly if the high-pressure side's design rating is more than 150% of the low-pressure side's rating, or if the differential pressure in a double pipe exchanger is 1000 psi or more)?
   - — What if the exchanger were exposed to an external fire?
   - — What if the cold fluid expands/vaporizes because it is blocked in?
   - — What is the pressure drop between the exchanger and the relief device protecting it?
   - — Can hot fluid (e.g., steam) condense and create a vacuum if the exchanger is blocked in?
   - — What if the fluid freezes in the exchanger?

3. Can the design temperatures of the heat exchangers be exceeded?
   - — What is the maximum upstream temperature?
   - — Could upstream heat removal equipment be bypassed, trip off, or lose its cooling medium?
   - — Could the flow of cooling medium for this exchanger be lost?
   - — Could the heating medium be too hot (e.g., loss of the steam desuperheater, hot oil temperature control failure)?
   - — Could flashing material, released by a tube failure or vent, autorefrigerate and embrittle the exchanger?
   - — Could fouling reduce the heat transfer rate below acceptable limits?

4. Will unacceptably high downstream temperatures result if the exchanger is bypassed or its cooling media is lost?
   — Will hot material cause undesirable venting from storage or rundown tanks?
   — Can personnel be burned by touching the hot piping?

5. Will unacceptably low downstream temperatures result if the exchanger is bypassed or its heating media is lost?
   — Could freezing cause plugging or damaged equipment downstream?
   — Could unvaporized gases (e.g., liquid nitrogen, LPG) flash and embrittle equipment downstream?

6. What are the consequences of low level in a boiler or reboiler? Can high pressure vapors blow through to the next vessel? Will the tubes warp or split?

7. How reliable is the cooling water supply?
   — Are motor-driven and turbine-driven pumps used?
   — Are there multiple sources of makeup water?
   — Is there any spare capacity in the cooling towers?
   — Are autostart systems regularly tested?

8. Are there adequate equipment clearances so that maintenance can be performed safely (e.g., cleaning or removal of a tube bundle)?

H. Furnaces and Boilers

1. Is the firebox protected against explosions?
   — Does the burner control system meet all applicable codes and standards (e.g., NFPA)?
   — How is the firebox purged before start-up? If steam is used, are the valves located away from the firebox? Is there a purge timer?
   — Are dedicated, positive shutoff trip valves installed in every fuel line? Must these valves be manually reset? Are bypass valves locked closed?
   — What signals will trip the furnace: low fuel pressure? high fuel pressure? loss of pilot or main flame? high stack temperature? low combustion air flow? low atomizing air/steam flow? loss of instrument air or power? low flow of water or process material?
   — How often are the furnace trips tested?
   — Are the fuel pressure sensors downstream of the fuel control valves?
   — Will air or stack dampers fail in a safe condition?
   — Can the forced draft fan overpressurize the firebox?
   — If several fireboxes share a common stack, will fuel leaking into one firebox be ignited by exhaust from the other fireboxes?
   — Could a tube failure cause an explosion?
   — Are there explosion hatches in the firebox?
   — Can flammable or combustible gases enter the firebox via the combustion air supply system?

2. Is the furnace protected against liquids in the fuel gas system?
   — Is an uninsulated fuel gas knockout drum provided for each fuel gas, pilot gas, and waste gas system?
   — Is a manual block valve accessible at least 50 feet from the furnace on each fuel line?
   — Are provisions made for draining liquids from the knockout drum (preferably to a closed system)? Does the drain need backflow protection?
   — Will the furnace trip on high level in the knockout drum?
   — Is the fuel line heat-traced/insulated from drum to burner?

3. Is the furnace protected against liquid fuel system failures?
   — Is atomizing air or steam flow monitored?
   — Is the fuel supply at higher pressure than the atomizing air or steam flow? Could a plugged burner tip cause a backflow?
   — Is the fuel supply filtered and heat traced?
   — Is a manual block valve accessible at least 50 feet from the furnace?
   — Are toe walls provided in the furnace to contain any spills?

4. Is the furnace adequately protected against tube failures?
   — Are individual pass flow controls, indications, and alarms provided?
   — Will a loss of process flow or drum level trip the furnace (but not the pilots)?
   — Are there check valves or remotely operable isolation valves in the outlet of each coil to prevent backflow in the event of a tube rupture?
   — Are there remotely operable valves (with appropriate fireproofing) in the furnace inlet lines, or are manual isolation valves located where they could be closed in the event of a fire?
   — Are relief valves provided for each coil with suitable protection against plugging (e.g., coking) the valves' inlets?
   — How would flame impingement on a tube be detected before it led to tube failure?
   — Is snuffing steam supplied to the firebox? Are the valves located where they could be opened in the event of a fire? Are there adequate traps and drains in the snuffing steam lines?

I. Instrumentation

1. Have instruments critical to process safety been identified and listed with an explanation of their function and setpoints?

2. Has overall integration and operation of the safety and control system been considered as part of the plant design?

3. What has been done to minimize response time lag in instruments directly or indirectly significant to process safety?

4. What would be the effect of a faulty sensor transmitter, indicator, alarm, or recorder? How would the failure be detected?

5. Are instruments designed to fail to the specified safe state?

6. How is the control system configured? Are there backups for all hardware components (computers, displays, input/output modules, programmable logic controllers, data highways, etc.)? How quickly can the backup be engaged? Is human action required?

7. How is programmable control software written and debugged? If there is a software error, is the backup computer also likely to fail as a result of the same error? Should extremely critical shutdowns be hardwired instead?

8. Is fault detection provided for control systems? Are there combinations of output failures that could present a process hazard?

9. Where sequence controllers are used, is there an automatic check at key steps after the controller has called for a change? Is there a check at key steps before the next sequence changes? What are the consequences of operator intervention in computer-controlled sequences?

10. Does the control system verify that operator inputs are within an acceptable range (e.g., if the operator makes a typographical error, will the control system attempt to supply 1000 lb of catalyst to a reactor that normally requires only 100 lb)?

11. What would be the consequences of a brief or extended loss of instrument power? Is there an uninterruptible power supply (UPS) for supporting the process control computer? Is it periodically tested under load? Does the UPS also support critical devices that may need to be actuated or does it only support information and alarm functions?

12. Does the operator-machine interface incorporate good human factors principles?
    - Is adequate information about normal and upset process conditions displayed in the control room?
    - Is the information displayed in ways the operators understand?
    - Is any misleading information displayed, or is any display itself misleading?
    - Is it obvious to operators when an instrument is failed or bypassed?
    - Do separate displays present information consistently?
    - What kinds of calculations must operators perform, and how are they checked?
    - Are all critical alarms immediately audible or visible to an operator? Are any alarms located in areas or buildings that are not normally staffed?
    - Are operators provided with enough information to diagnose an upset when an alarm sounds?
    - Are operators overwhelmed by the number of alarms associated with an upset or emergency? Should an alarm prioritization system be implemented? Can operators easily tell what failure/alarm started the upset (e.g., is there a first alarm or critical alarm panel)?
    - Are the displays adequately visible from all relevant working positions?
    - Do the displays provide adequate feedback on operator actions?

- — Do control panel layouts reflect the functional aspects of the process or equipment?
- — Are related displays and controls grouped together?
- — Does the control arrangement logically follow the normal sequence of operation?
- — Are all controls accessible and easy to distinguish?
- — Are the controls easy to use?
- — Do any controls violate strong populational stereotypes (e.g., color, direction of movement)?
- — Are any process variables difficult to control with existing equipment?
- — How many manual adjustments must an operator perform during normal and emergency operations?
- — When adjacent controls (e.g., valves, switches) have a similar appearance, what are the consequences if the incorrect control is used?
- — Are redundant signal or communication lines physically separated (i.e., run in separate cable trays, one run aboveground and another underground)?
- — Are signal cables shielded or segregated from power cables (i.e., to avoid electromagnetic interference and false signals)?
- — Are there control loops in the process that are not connected into the computer control system? How do operators monitor and control from the control room?

13. Are automatic controls ever used in manual mode? How do operators ensure safe operation while in manual mode?

14. What emergency valves and controls can operators not reach quickly and safely while wearing appropriate protective clothing?

15. Have procedures been established for testing and proving instrument functions and verifying their alarm setpoints are correct? How often is inspection and testing of instrumented systems performed?

16. Are the means provided for testing and maintaining primary elements of alarm and interlock instrumentation without shutting down the process?

17. Are instruments, displays, and controls promptly repaired after a malfunction? Are any instruments, displays, or controls deliberately disabled during any phase of operation? How are alarm setpoints and computer software protected from unauthorized changes?

18. What provision is made for process safety when an instrument is taken out of service for maintenance? What happens when such an instrument is not available?

19. Are instrument sensing lines adequately purged or heat traced to avoid plugging?

20. What are the effects of atmospheric humidity and temperature extremes on instrumentation? What are the effects of process emissions? Are there any sources of water (e.g., water lines, sewer lines, sprinklers, roof drains) that could drip into or spray onto sensitive control room equipment?

21. Is the system completely free of instruments containing fluids that would react with process materials?

22. What is being done to verify that instrument packages are properly installed, grounded, and designed for the environment and area electrical classification? Is instrument grounding coordinated with cathodic protection for pipes, tanks, and structures?

23. Are the instruments and controls provided on vendor-supplied equipment packages compatible and consistent with existing systems and operator experience? How are these instruments and controls integrated into the overall system?

J. Electrical Power

1. What is the area electrical classification?
   — What process characteristics affect the classification, group, and division?
   — Are the hardware (e.g., motors, forklifts, vent fans, radios) and protective techniques consistent with the area electrical classification?
   — Was all equipment tested and approved by an independent laboratory (e.g., Underwriters Laboratories or Factory Mutual), or is additional testing required?
   — Are any new protective techniques being employed?

2. Is all auxiliary electrical gear (e.g., transformers, breakers) located in safe areas (e.g., from hazardous materials and flooding)?

3. Are electrical interlocks and shutdown devices made fail-safe?
   — What is the purpose of each interlock and shutdown?
   — Can the interlock and shutdown logic be simplified?
   — How is continued use of protective devices ensured?
   — How often are the interlocks and shutdowns tested under load?

4. How completely does the electrical system parallel the process?
   — What faults in one part of the plant will affect operation of other independent parts of the plant?
   — How are the plant's instrument and control power supplies protected from faults or other voltage disturbances?
   — Are primary and spare equipment powered from independent buses?
   — Is there an emergency power supply for critical loads?

5. Is the electrical system simple in schematic and physical layout so that it can be operated in a straightforward manner?

6. Are the electrical system instruments arranged so that equipment operation can be monitored?

7. What are the overload and short circuit protective devices?
   — Are they located in circuits for optimum isolation of faults?

- — Will they act quickly enough?
- — What is the interrupting capacity?
- — How are they coordinated?
- — Are they tested under load? How often?
- — Are they sensitive to voltage or frequency variations?

8. Can operators safely open or reset breakers in an emergency?

9. What bonding and grounding is provided?
   - — Does it protect against static buildup?
   - — Does it provide lightning protection?
   - — Does it provide for personnel protection from power system faults?

10. Are trucks and railcars properly grounded during loading/unloading operations?

11. What electrical equipment can be taken out of service for preventive maintenance without interrupting production? Can the equipment be safely locked out? How?

12. Are conduits sealed against flammable vapors?

K. Miscellaneous

1. Are special seals, packing, or other closures necessary for severe service conditions (e.g., toxic, corrosive, high/low temperature, high pressure, vacuum)?

2. Do major pieces of rotating equipment have adequate equipment integrity shutdowns to minimize major damage and long-term outages (e.g., lube oil shutdowns)?

3. Is the equipment's vibration signature routinely monitored to detect incipient failures? How is excessive vibration detected? Will excessive vibration trip large rotating equipment such as
   - — turbines?
   - — motors?
   - — compressors?
   - — pumps?
   - — cooling tower fans?
   - — blowers?

4. What is the separation of critical and operating speeds? Will the equipment trip on overspeed? Could overspeed or imbalance cause the equipment to disintegrate?

5. Are all turbine overspeed trips set below the maximum speed of the driven equipment?

6. Are there provisions for operation or safe shutdown during power failures?

7. Are check valves fast-acting enough to prevent reverse flow and reverse rotation of pumps, compressors, and drivers?

8. What procedure exists for ensuring an adequate liquid level or flow in any liquid flushed, cooled, or lubricated seals?

9. Are there full-flow filters in lube oil systems?

10. Are there provisions for trapping and draining steam turbine inlet and exhaust lines? Are there separate visible-flow drain lines from all steam turbine points?

11. Are adequate service factors on gears in shock services provided?

12. Are the mechanical loads imposed on equipment acceptable considering
    — thermal expansion?
    — piping weight?
    — overfilling the vessel?
    — high winds?
    — snow, ice, and water accumulation?

13. Are the foundations, supporting structures, and anchor points adequate for
    — vessel(s) completely filled with water (or process material)?
    — high winds?
    — ground movement?
    — snow/ice/water accumulation?
    — anticipated floor loading?
    — relief device discharges (thrust or reactive loads)?

14. In cases where glass or other fragile material is used, can durable materials be substituted? If not, is the fragile material adequately protected to minimize breakage? What is the hazard resulting from breakage?

15. Are sight glasses provided only where positively needed? On pressure vessels, do sight glasses have the capability to withstand the maximum pressure? Are they equipped with excess flow valves? Are they frequently inspected for cracks/damage?

16. What provisions have been made for dissipation of static electricity to avoid sparking? Will currents be induced in large rotating equipment?

17. How are the piping and equipment protected from corrosion?
    — Are corrosion inhibitors used?
    — Are the pipes and vessels lined?
    — Is there a cathodic protection system?
    — Are corrosion-resistant materials used?
    — Is the exterior painted or coated?

18. What could cause a catastrophic failure of the piping or equipment (e.g., hydrogen cracking, thermal shock, external impact)?

19. Are there suitable barricades between process equipment and adjacent roadways? Are overhead pipe racks protected from crane impacts?

20. Does all equipment comply with applicable laws and regulations, codes and standards, and company guidelines?

21. What tests will be performed to detect specification errors, manufacturing defects, transportation damage, construction damage, or improper installation before the equipment is put into service? What ongoing tests, inspections, and maintenance are performed to ensure long-term reliability and integrity of the equipment?

III. Operations

1. What human errors may have catastrophic consequences? Have critical jobs and tasks been identified? Have the mental and physical aspects of such jobs been analyzed for both routine and emergency activities? What has been done to reduce the likelihood and/or consequences of potential human errors in the performance of these jobs?

2. Is a complete, current set of procedures for normal operations, start-ups, shutdowns, upsets, and emergencies available for operators to use? How are specific, up-to-date procedures maintained? Do the operators themselves help review and revise the procedures? How often? Are known errors allowed to remain uncorrected?

3. What process equipment or parameters have been changed? Have the operating procedures been appropriately revised and have operators been trained in the new procedures?

4. Are procedures written so workers can understand them, considering their education, background, experience, native language, etc.? Is a step-by-step format used? Are diagrams, photographs, drawings, etc., used to clarify the written text? Are cautions and warnings clearly stated in prominent locations? Does procedure nomenclature match equipment labels? Are there too many abbreviations and references to other procedures?

5. How are new operating personnel trained on initial operations, and how are experienced operating personnel kept up to date? Is there regular training on emergency procedures, including drills on simulated emergencies?

6. How do workers demonstrate their knowledge before being allowed to work independently? Is there a testing and verification system?

7. Are checklists used for critical procedures? Is only one action specified per numbered step? Are any instructions embedded in explanatory notes? Are the steps in the correct sequence? Do steps requiring control actions also specify the expected system response?

8. Do operator practices always comply with written procedures? How are differences detected and resolved? Who can authorize changes and deviations from the written

procedures? Does such authorization include a review of the safety implications of the change or deviation?

9. How thorough is the operators' knowledge of the process chemistry and potential undesired reactions?

10. Do the procedures specify safe operating limits for all materials and operations? What process variables do, or could, approach those limits? How quickly could safety limits be exceeded? Can operators detect and respond to upsets before safety limits are exceeded, or are automatic systems provided?

11. What procedures or operations must be monitored by process engineers or other technically trained personnel? Is this requirement documented?

12. Is all important equipment (vessels, pipes, valves, instruments, controls, etc.) clearly and unambiguously labeled with name, number, and contents? Does the labeling program include components (e.g., small valves) that are mentioned in the procedures even if they are not assigned an equipment number? Are the labels accurate? Who is responsible for maintaining and updating the labels?

13. What special clean-up, purging, or draining requirements are there before start-up? How are these requirements checked?

14. How are utility system failures handled?
    - Is there a plant-wide response procedure?
    - Are load-shedding priorities defined?
    - Are there backup electrical supplies (e.g., diesel generators)?
    - Can the steam system operate without electrical power (i.e., with steam-driven fans and feedwater pumps)?
    - Is there at least one boiler that can start without steam (e.g., with motor-driven fans and feedwater pumps)?

15. Is the process difficult to control (e.g., limited time to respond to upset conditions)? Are operators overwhelmed by low-priority alarms during an upset?

16. Have there been "near miss" incidents that could have been much more serious, given other operating situations or operator responses?

17. Is equipment left unattended under automatic control? If so, what is the strategy for responding to alarm conditions?

18. Should television cameras be installed
    - to watch loading/unloading racks?
    - to watch flare tips?
    - to watch for process material releases?
    - to watch for intruders?

19. What loading and unloading operations are performed?
    - What procedures control these operations?
    - Who performs these operations?
    - How is training/familiarization conducted for company and noncompany personnel involved in these operations?
    - How is surveillance or supervision maintained?
    - How are hookups performed? Are there any physical means to prevent reversed connections or connections to the wrong tank?
    - How is the transport container grounded/bonded? Is the electrical continuity verified?
    - How is the raw material or product composition verified?
    - Is the composition verified before any material transfer takes place?

20. Are adequate communications provided to operate the facility safely (telephones, radios, signals, alarms)?

21. Are shift rotation schedules set to minimize the disruption of workers' circadian rhythms? How are problems with worker fatigue resolved? What is the maximum allowable overtime for a worker, and is the limit enforced? Is there a plan for rotating workers during extended emergencies?

22. Are there enough operators on each shift to perform the required routine and emergency tasks?

IV. Maintenance

1. Are written procedures available and followed for:
    - hot work?
    - hot taps and stopples (including metal inspection before welding)?
    - opening process lines?
    - confined space or vessel entry?
    - work in an inert atmosphere?
    - lockout/tagout?
    - work on energized electrical equipment?
    - blinding before maintenance or vessel entry?
    - pressure testing with compressible gases?
    - use of supplied-air respiratory equipment?
    - removal of relief devices from operating equipment?
    - digging and power excavation?
    - cranes and heavy lifts?
    - contractor work?
    - entry into operating units?

2. What procedures govern crane/heavy equipment usage in an operating unit?
    - Is operator certification required?
    - Are equipment/cable inspections and certifications current?
    - How are underground voids or piping positioned before a heavy lift is performed?

3. Is it necessary to shut down the process completely to safely repair a piece of equipment? Are there provisions for blanking off all lines into equipment that people may enter? Are other precautions necessary to protect operators, mechanics, and service personnel?

4. How often is the process equipment cleaned? What chemicals and maintenance equipment are used? Are nozzles and manholes sized and located for safe cleanout, maintenance access, and emergency removal of people from vessels?

5. What is the preventive maintenance schedule, and is it adequate to ensure the reliability of safety-critical equipment and instrumentation?
   — Is vibration monitoring needed?
   — Do valves, agitators, etc., require regular greasing?
   — Must seal oil and lube oil levels be monitored?
   — Must lubricants be changed periodically?
   — Must oil mist systems be checked for water, low spots, mist generator failure, etc.?

6. What process hazards are introduced by routine maintenance procedures?

7. Do platforms provide adequate clearance for safe maintenance of equipment?

8. Consider the consequences of a breakdown of each piece of equipment during operation. Can it be safely bypassed, isolated, drained, cleaned/purged, and repaired? How is overpressure protection provided while the equipment is isolated?

9. What provisions are made for spare machines or spare parts for critical machines? Are there important pieces of individual equipment that are not spared and/or would require a long time to replace (e.g., compressors, reactors, heaters, specialty vessels)?

10. Is material control maintained for material and supplies to be used in the units (e.g., weld rod, piping and fittings, gaskets, rupture disks)?

11. Are the right tools available and used when needed? Are special tools required to perform any tasks safely or efficiently? What steps are taken to identify and provide special tools?

12. What kind of special housekeeping is required? Will accumulation of small spills cause slippery floors, or powder accumulation possibly cause a dust explosion?

13. What hazards do adjacent units pose to maintenance workers? Consider:
    — normal exhausts and vents
    — emergency relief and blowdown
    — spills and unintended releases
    — fires and explosions.

V. Personnel Safety

A. Building and Structures

1. What standards are being followed in the design of stairways, platforms, ramps, and fixed ladders? Are they well lit?

2. Are sufficient general exit and escape routes available from operating areas, shops, laboratories, and offices? Are the exits appropriately marked? Are alternate means of escape from roofs provided? Is protection provided to persons using the escape routes?

3. Are doors and windows hung to avoid projecting into or blocking walkways and exits?

4. Is structural steel grounded?

5. Where operations are potentially hazardous from the standpoints of fire and explosion, are controls housed in separate structures? If not, are control room windows kept to a minimum and glazed with laminated safety glass? Is the control room structure blast resistant?

6. Does the control room provide a safe haven during incidents, protecting operators from potential fires, explosions, and toxic releases? What is the design basis for the protection? What are the evacuation plans? If a shelter-in-place strategy is used, are there enough SCBAs for control room personnel and others who may come there in an emergency?

B. Operating Areas

1. What fire and explosion hazards are workers exposed to, and how are the hazards mitigated? Are there:
   — flammable conditions in process equipment?
   — combustible materials near hot process equipment?
   — spills/releases of flammables or combustibles?
   — accumulation of flammables or combustibles (e.g., dusts, oily sumps)?
   — cleaning solvents?
   — strong oxidizers (e.g., peroxides, oxygen gas)?
   — ignition sources (e.g., open flames, welding, resistance heaters, static)?

2. How is high pressure vented from the area?

3. Has a safe storage and dispensing location for flammable liquid drums been provided?

4. What chemical hazards are workers exposed to, and how are they mitigated? (Consider raw materials, intermediates, products, by-products, wastes, unintended reactions, and combustion off-gases.) Are there:
   — asphyxiants?
   — irritants?
   — poisons?
   — carcinogens?
   — mutagens?
   — teratogens?

5. Where may workers be exposed to chemical hazards? Are special protective measures (e.g., special ventilation) required? Consider:
   — collecting samples?
   — gauging tanks, vessels, or reservoirs?
   — charging raw materials?
   — withdrawing or packaging products?
   — loading/unloading trucks, railcars, or drums?
   — cleaning filters or strainers?
   — purging/draining process materials from lines and vessels?
   — draining/venting wastes?

6. Have workers been notified of the hazards, and are material safety data sheets available? Are appropriate warning signs and labels posted? Are medical personnel aware of the hazards and trained/equipped to render appropriate treatment?

7. Can the process be better designed to minimize or eliminate exposure to toxic substances?

8. Is adequate general and local ventilation furnished for hazardous fumes, vapors, dust, and excessive heat? How was the adequacy of ventilation determined for the current activities? Are air intakes well clear of sources of harmful contaminants?

9. Are there any confined or partially confined areas (e.g., instrument cabinets, analyzer buildings, tank pits) where inert gas leaks could collect and asphyxiate workers?

10. Are all utility connections (e.g., steam, water, air, nitrogen) clearly and unambiguously labeled? If a color-coding scheme is used, are all pipes the proper color?

11. Will personnel require medical surveillance or air monitoring for radiation, biological, or chemical contaminants? (One time only or continuous?)

12. Is personal protective equipment required, such as:
    — head protection (bumps, falling objects, etc.)?
    — eye protection (particulates, fragments, liquid splashes, strong light, etc.)?
    — ear protection (noise)?
    — face protection (liquid splashes, ultraviolet exposure, etc.)?
    — respiratory protection (dusts, mists, vapors, inert gases, etc.)?
    — skin/body protection (liquid splashes, vapors, burns, contamination, etc.)?

— hand protection (cuts, burns, liquids, etc.)?
— wrist protection (repetitive motions)?
— back protection (heavy lifting)?
— toe protection (trips, falling objects, etc.)?

13. Is appropriate personal protective equipment available and located accessibly for
— normal operations?
— process upsets?
— minor spills?
— major spills and fires?

14. Are emergency showers and eyebaths provided? In cold climates, is tempered water supplied or is the shower enclosed so workers will not suffer exposure in cold weather? Is water flow alarmed in the control room?

15. What first aid and medical treatment are required for unusual exposure? Have personnel who may be involved (coworkers, emergency response personnel, medical personnel, etc.) been notified of any special hazards or precautions?

16. Can workers carry hazardous substances home on contaminated clothing?

17. What pressure hazards are workers exposed to, and how are they mitigated? Are there:
— compressed air tools?
— high pressure gas or steam leaks?
— discharges from vents or relief devices?
— blowing particulates?
— hydraulic hammers?
— container or equipment ruptures (e.g., unvented gear boxes, dust collectors, high pressure hoses)?
— vacuums (e.g., compressor suction, blower inlet, vacuum hose)?

18. Are vents located so that discharges, including liquids, do not endanger personnel, public, or property? Are all vents above the highest liquid level possible?

19. What temperature hazards are workers exposed to, and how are they mitigated? Are there:
— hot surfaces (including surfaces that would be hot only in unusual circumstances such as a cooler being bypassed)?
— hot exhaust gases?
— steam/condensate blowdown?
— cold flashing liquids or vapors?
— refrigerated or cryogenic surfaces?
— extreme ambient temperatures (outdoors or indoors)?
— heavy or nonporous protective clothes?

20. What mechanical hazards are workers exposed to, and how are they mitigated? Are there:
    - sharp edges or points?
    - obstacles likely to cause head injury or tripping?
    - slippery surfaces?
    - heavy weights to be lifted?
    - falling or toppling objects?
    - unguarded (e.g., rail-less, cageless) or unstable platforms/ladders?
    - ejected parts or fragments?
    - unguarded moving equipment (pulleys, belts, gears, augers, pistons, etc.)?
    - unguarded pinch points/nips?
    - unexpected movements of unsecured objects or ruptured hoses?

21. Are emergency stop switches and/or cables provided for all equipment? Does the equipment stop quickly enough?

22. Are steam, water, air, electrical, and other utility outlets arranged to keep aisles and operating floor areas clear of hoses and cables? Are there any temporary or permanent process interconnections blocking walkways?

23. Are free-swinging hoists avoided? Are hoists equipped with safety hooks and limit switches, if motorized? Do all cranes, hoists, monorails, hooks, jacks, and slings conform to applicable design standards and guidelines?

24. Are elevators equipped with shaftway door interlocks and car gate contacts? Are there safety astragals on doors that could pinch workers as they close?

25. Is there an alarm system for medical emergencies? Are emergency communication devices (and instructions) readily available in areas where workers may need to summon help (e.g., elevators, loading docks)?

26. Is every effort being made to handle materials mechanically rather than manually?

27. What vibration hazards are workers exposed to, and how are they mitigated? Are there:
    - vibrating tools or material handling equipment?
    - structural vibrations?
    - sonic flow vibrations?
    - high levels of noise?

28. What electrical hazards are workers exposed to, and how are these hazards mitigated? Do they include:
    - shock?
    - burn?
    - arcing/electrical explosion?
    - unexpected energization?

29. Are positive disconnects and interlocks being installed for lockout of all energy sources?

30. What radiation hazards are workers exposed to, and how are they mitigated? Do they include:
    - ionizing radiation?
    - ultraviolet light?
    - high intensity visible light?
    - infrared radiation?
    - microwave radiation?
    - laser beams?
    - intense magnetic fields?

31. Are there at least two exits from hazardous work areas?

32. How good is the lighting system?
    - Adequate for safe normal operation?
    - Adequate for routine maintenance?
    - Adequate for shutdown during a power failure?
    - Adequate for escape lighting during a fire?

C. Yard

1. Are material loading/unloading operations continuously monitored by an operator (in the yard or via closed circuit television)?

2. Is yard lighting adequate?

3. Are roadways laid out with consideration for the safe movement of pedestrians, vehicles, and emergency equipment?

4. Are flammable liquid tank car and tank truck loading and unloading docks bonded or grounded?

5. Are safe means provided on loading platforms for access to work areas of tank cars and trucks? Are counterweight cables checked periodically?

6. Are employees who work atop railroad cars and trucks protected against falls?

7. Is safe access provided for employees who work atop storage tanks?

8. Are railroad car puller control stations fully protected against broken cable whiplash? What will protect the operator from being caught between a cable or rope and the capstan or cable drum?

VI. Fire Protection

1. What combustible mixtures can occur within equipment:
   — because of normal process conditions?
   — because of abnormal process conditions?
   — because of a loss or contamination of gas for purging, blanketing, or inerting?
   — because of moving liquids into and out of vessels (e.g., tank breathing)?
   — because of dust?
   — because of improper start-up, shutdown, or restoration after maintenance?
   — because dissolved or chemically bound oxygen was released and accumulated?
   — because of condensation in the ducts?

2. What is the approximate inventory of flammable liquids in the equipment? Are inventory amounts kept to a minimum?

3. How have major storage tanks or vessels been located to minimize the hazard to process equipment if the tanks catch fire or rupture? Are liquid-filled tanks near the ground?

4. What combustible materials are present? How are they protected from fire, sparks, and excessive heat?

5. Are fire walls, partitions, or barricades provided to separate high-value property, high-hazard operations, and units important for production continuity? Do fire doors have fusible link closures?

6. Can all process lines and utilities (especially those containing fuels or high pressure steam) be isolated at the unit battery limits?

7. Are there ignition sources present? Mechanical spark sources? Are worker smoking areas clearly defined and enforced?

8. Is insulation provided on all hot equipment and piping that could ignite a spill of any process material?

9. Is odorant added to all flammable gases used in enclosed areas (e.g., control rooms, kitchens, camps, boiler rooms)?

10. Are sheltered or enclosed areas (e.g., pumphouses, compressor buildings, boiler rooms) adequately ventilated to prevent accumulation of flammable gases? Are vents properly located at high and/or low points, considering the density of the gases involved?

11. In confined areas, how is open-fired equipment prevented from igniting flammable releases?

12. Are tanks, buildings, and structures adequately protected against lightning?

13. Are there flame and detonation arresters where appropriate (e.g., tank vents)? Are they properly specified for the actual service conditions? When were they last tested or inspected?

14. What protection has been provided for dust hazards? Is explosion suppression equipment needed to stop an explosion once started? Are there blast gates in the ducts?

15. How are fires or potential fires detected (e.g., smoke detectors, heat detectors, gas detectors, water flow sensors)? Have suitable locations been selected for fire detectors and alarms (pull boxes and sirens)? Can personnel identify the type of alarm and the location of the fire?

16. Are fire fighting techniques defined for all materials? Is the technique usable in the work area? Is the preferred fire extinguishing method readily available in the area?

17. Are there any extinguishing media that are prohibited (because they are not effective, they react with some other chemical present in the area, or they are harmful to equipment)? Are any prohibited extinguishers available in the area? If water is prohibited, are there warning signs in the area?

18. Is there adequate fire fighting equipment?
    - What firewater hydrants serve the area? Are there hose standpipes inside buildings?
    - What fixed or portable water cannons or monitor nozzles are provided for coverage of manufacturing facilities or storage facilities in open areas (not within buildings)?
    - What automatic sprinklers are provided in buildings with combustible construction or contents? Is this adequate for high-piled storage areas?
    - What total flooding or local-application fire suppression systems (CO2, Halon®, etc.) have been provided?
    - What type, size, location, and number of fire extinguishers are provided?
    - What flammable liquid storage tank protection (e.g., foam, deluge) has been provided?
    - Is equipment containing volatile flammable materials (e.g., spheres) or materials above their autoignition temperature (e.g., hot pumps) protected by deluge systems? Do the deluge systems adequately protect small-diameter piping attached to vessels (particularly spheres and bullets)?
    - Is sprinkler protection provided for fin-fan coolers?
    - Is snuffing steam provided for all fired equipment?
    - Is inert gas or steam provided for all combustible reactor or absorber beds (e.g., activated carbon beds)?
    - Are there mobile equipment and trained crews that can respond quickly?
    - Are hydrocarbon drainage systems equipped with explosion traps and vents?

19. What procedures are followed in the event of a fire?
    - To what extent should operators, maintenance workers, or contractors attempt to fight fires?
    - Have all fire fighters been trained?

- — Who decides when to call the fire brigade?
- — Who decides when to call outside fire brigades?
- — Where is the emergency command center, and how is it staffed?
- — When were these procedures last practiced?

20. What are the capabilities of the fire brigade?
    - — How is the fire brigade assembled during the day shift? off-shifts?
    - — What training does the fire brigade receive? Does it include first aid?
    - — What procedures do fire fighters follow when entering a unit?
    - — What protective equipment is available to the fire fighters? Are enough SCBAs available? Will bunker gear withstand exposure to process chemicals?
    - — What fire fighting equipment is available in the facility? from mutual aid groups? from the community?

21. What is the capability of firewater supplies?
    - — What is the maximum firewater demand?
    - — How long will supplies meet the maximum demand?
    - — Are any alternate supplies available?
    - — Are there redundant firewater pumps with diverse drivers (electric, steam, diesel)?
    - — Are there contaminants (e.g., mud, shells, gravel) in the firewater supply that could damage fire fighting equipment? How often is equipment flushed out?

22. Have the underground fire mains been extended or looped to supply additional sprinkler systems, hydrants, and monitor nozzles? Are there any dead ends? What sectional control valves have been provided?

23. Are important fire protection resources (e.g., fire hall, firewater pumps) located where they can be threatened by fires or explosions in the facility?

24. How is process equipment protected from external fire?

25. Is load-bearing structural steel, which is exposed to potential flammable liquid or gas fires, fireproofed to a sufficient height above a fire-sustaining surface to protect it? Are cable trays similarly protected?

26. Are critical isolation valves fire-safe, and will their actuators withstand fire exposure?

27. Has adequate drainage been provided to carry spilled flammable liquids and water used for fire fighting away from buildings, storage tanks, and process equipment? Are drain valves outside any dikes? Can the drains and dikes accommodate the water used during fire fighting? Will burning materials float into adjacent areas?

28. Is the control room adequately protected against external fires or explosions? Do any glass windows face process areas where explosions might occur?

29. Are fire protection systems periodically tested? Is there a program to ensure that fire protection systems are in service? Does the program provide priority maintenance for equipment found out of service?

30. Are there strong administrative controls requiring permits and/or notification before fire protection equipment can be taken out of service or used for normal operation (e.g., auxiliary cooling) or maintenance (e.g., equipment flushing)?

VII. Environmental Protection

1. Are there any chemicals handled that are particularly sensitive from an environmental standpoint? (carcinogens, volatile toxics, odorants)

2. Have all effluent streams been defined? Are they hazardous? What is their disposition? Are scrubbers required? Have permit requirements been addressed? What has been done to minimize effluents and wastes? Will any hazardous materials such as heavy metals reach the waste treatment plant?

3. Does surface water runoff require any special treatment? Is surface drainage adequate? Can it be protected (e.g., with sandbags) from process material spills?

4. How are effluents monitored (e.g., sampled) for unacceptable emissions? What is the lag time between measurement and alarm or notification? Do emission points include:
   — stacks and vents?
   — ventilation exhausts?
   — surface water runoff?
   — discharges to city sewers?
   — discharges to surface water bodies?
   — discharges or seepage to groundwater?

5. What precautions are necessary to meet environmental requirements and protect human health? Are there specific environmental restrictions that will limit operations?

6. Will maintenance work require special precautions to prevent odor problems, air pollution, or sewer contamination?

7. Is the sampling system arranged so any initial blowdown is vented to a closed system instead of to the atmosphere or sewer?

8. What are the hazards of sewered materials during normal and abnormal operation? Consider:
   — runaway reactions?
   — flammable concentrations, either from the sewered material or from reactions (e.g., hydrogen evolution) in the sewer?

— toxic fumes?
— environmental contamination?
— cross-contamination of process and sanitary sewers?

9. What is the potential for releases in the process area, and where would they go? What hazards would result from these releases? Are any special precautions necessary for leak-prone equipment (e.g., bellows, rotating seals)?

10. What prevents or limits spills during loading/unloading operations?
    — Is there remote shutdown/isolation capability?
    — Are there excess flow check valves or automatic shutdowns?
    — Are the trucks/railcars chocked?
    — Are railcars protected against collision or inadvertent movement?
    — Are hoses inspected/pressure tested/replaced regularly?
    — Are there high level and/or pressure alarms on storage tanks (particularly remote tanks)?

11. Are storage areas diked? Are the dikes large enough? Are any dikes damaged or breached? Are proper drainage programs implemented to ensure the integrity of the dikes when required? What would happen if the dike overflowed (e.g., because of fire fighting activities)?

12. Are there toxic gas monitors and alarms in process and material storage areas? How often are they tested?

13. What procedures are followed in the event of a release?
    — To what extent should operators, maintenance workers, or contractors attempt to contain and clean up releases?
    — Have the people who will clean up releases been trained?
    — Who decides when to call the spill response team?
    — Who decides when to call outside emergency response teams?
    — Who notifies corporate management and public authorities?
    — Who decides to evacuate the unit, facility, or community?
    — Where is the emergency command center, and how is it staffed?
    — When were these procedures last practiced?

14. Are there adequate, reliable means of reporting emergencies to a response team and to applicable government officials or agencies?

15. Are there adequate, reliable means of sounding an evacuation alarm to all building or area occupants?

16. Is there a written evacuation plan for the unit, facility, and community?
    — Are the process operations shut down, or can they be left on automatic control?
    — Are assembly points, evacuation routes, and alternates clearly marked?
    — Are emergency control centers established?

— Are there spill containment procedures?
— Are there re-entry and cleanup procedures?
— Has the plan been coordinated with local authorities?
— Has the plan been tested and appropriately revised?

17. Are up-to-date emergency shutdown and evacuation plans posted? Are they effectively communicated to transient workers (e.g., outside contractors)?

18. What are the nearest and/or largest onsite and offsite populations? How far away are they? Are there any locations that present special evacuation problems (e.g., schools, hospitals, nursing homes, large population centers)?

19. Are containment and clean-up techniques defined for all materials? Is the technique usable in the work area? Are appropriate protective equipment and clean-up supplies on hand in readily accessible locations? Are different procedures or supplies required to handle products of undesired reactions?

20. Are there any suppression, absorption, or cleaning media that are prohibited (because they are not effective, they react with some other chemical present in the area, or they are harmful to equipment)? Are any media of this type available in the area? If water is prohibited, are there warning signs in the area?

21. What are the capabilities of the spill response team?
— How is the spill response team assembled during the day shift? off-shifts?
— What procedures do emergency personnel follow when entering a unit?
— What protective equipment is available to the emergency personnel? Are enough SCBAs available? Will protective gear withstand exposure to process chemicals?
— What release suppression, collection, and cleanup equipment is available in the facility? from mutual aid groups? from the community?

22. Can wastes be safely handled? Can the material be decontaminated, recycled, or destroyed? Have arrangements for disposal been completed?

23. What means is provided for disposal of off-specification products or aborted batches?

24. Are empty containers for packaged raw materials and intermediates systematically recycled or disposed of by acceptable methods?

VIII. Management and Policy Issues

1. Is upper management's commitment to employee health and safety clear? What policy statements communicate this commitment to employees? Do workers understand these policies, and are they convinced of upper management's sincerity?

2. Do supervisors and workers believe that safety has higher (or at least equal) status with other business objectives in the organization? How does the company promote a "safety first" approach?

3. Have supervisors and workers been specifically told to err on the safe side whenever they perceive a conflict between safety and production? Will such decisions be supported throughout the management chain?

4. Is there a policy that clearly establishes which individuals have the authority to stop work if safety requirements are not met?

5. Is management of worker health and safety an essential part of a manager's daily activities? How are managers held accountable for their health and safety record, and how do the rewards and penalties compare to those for production performance?

6. Is health and safety regularly discussed in management meetings at all levels? Do such discussions involve more than a review of injury statistics? What actions are taken if an injury occurs? Are near misses discussed, and is any action taken to prevent recurrence?

7. Are there clear procedures during emergencies for communications between workers and emergency response personnel, plant management, corporate management, and public authorities? Are they regularly practiced?

8. Is the mutual aid network documented by formal agreements?

9. Are the responsibilities for utility system maintenance and operation clearly defined throughout the plant? Are interfaces between different organizations recognized?

10. Are workers encouraged to ask supervisors for assistance? Do workers know when to seek assistance? Are workers penalized for "unnecessary" shutdowns when they truly believe there is an emergency?

11. Are workers encouraged to discuss potential human errors and near misses with their supervisors? Are such worker disclosures treated as evidence of worker incompetence, as unwarranted criticism of management, or as valuable lessons to be shared and acted upon? What criteria and procedures exist for reporting and investigating incidents and near misses? Are they followed consistently? Do the investigations go into enough depth to identify the root causes of worker errors? How are the human factors engineering deficiencies identified during the investigation of an incident corrected at (1) the site of the original incident, (2) similar sites at the same facility, and (3) similar sites at other facilities?

12. Is there a written training policy applicable to all workers?
    - What safety objectives are established, and how is attainment of such objectives monitored?
    - Are training records kept?

- — How are retraining needs identified?
- — How are workers trained on new processes, equipment, and procedures?
- — What training is given to workers changing jobs or taking additional responsibilities?
- — What training is given to new workers?
- — How is training effectiveness assessed?
- — What training is required before a worker can "step up" to substitute for an absent foreman or supervisor?

13. Are there adequate controls on contractor personnel? Do they have to meet the same safety standards required of company personnel? Are there different requirements for long-term and short-term contractors?

14. Does company policy require that all safety-related equipment (alarms, shutdowns, relief devices, trips, deluges, etc.) be tested periodically? What failures are tolerated until the next planned shutdown?

15. What is the company policy for designing and operating facilities in different jurisdictions (e.g., are pressure vessels designed and maintained to code standards, whether or not the state requires it)? Are the design and operating practices in this facility consistent with those in other facilities?

16. Are there adequate controls on design changes? Are changes coordinated with operations so procedures and training materials can be updated? Are field changes by operations or maintenance personnel handled in the same way as engineering changes?

17. Are engineering drawings or models up to date, including those related to environmental management permits?

18. What administrative control is necessary to ensure replacement of proper materials during construction/modification/maintenance to avoid excessive corrosion and to avoid producing hazardous compounds and reactants?

19. What is the company policy toward compliance with process safety guidelines published by industry or trade groups such as the Chemical Manufacturers Association, the American Petroleum Institute, or the Chlorine Institute? Have they been followed in this design?

20. Is there an audit program that regularly reviews safety compliance? Do workers participate on the audit teams? Who sees and responds to audit reports?

21. Are there programs for identifying and helping workers with substance abuse or mental health problems? What counseling, support, and professional advice is available to workers during periods of ill health or stress? What is the company policy on reassigning or terminating workers who are unable/unfit to perform their jobs?

Appendix **C**

Symbols and Abbreviations for Example Problem Drawings

Table C.1 illustrates the abbreviations and Figure C.1 illustrates the symbols used for process instrumentation and equipment in the example problem drawings.

Table C.1 Abbreviations used in example problem drawings

CWS	Cooling Water Supply	PSL	Pressure Switch Low
CWR	Cooling Water Return	PSLL	Pressure Switch Low-Low
FAL	Flow Alarm Low	PSV	Pressure Safety Valve
FC	Fail Closed	PT	Pressure Transmitter
FI	Flow Indicator	TAH	Temperature Alarm High
FO	Fail Open	TAHH	Temperature Alarm High-High
FT	Flow Transmitter	TAL	Temperature Alarm Low
LAL	Level Alarm Low	TIC	Temperature Indicating Controller
LRC	Level Recording Controller	TI	Temperature Indicator
LT	Level Transmitter	TT	Temperature Transmitter
PAH	Pressure Alarm High	S/D	Shutdown
PAL	Pressure Alarm Low	UVL	Ultraviolet Light Detector
PALL	Pressure Alarm Low-Low	XAL	Analyzer
PI	Pressure Indicator		

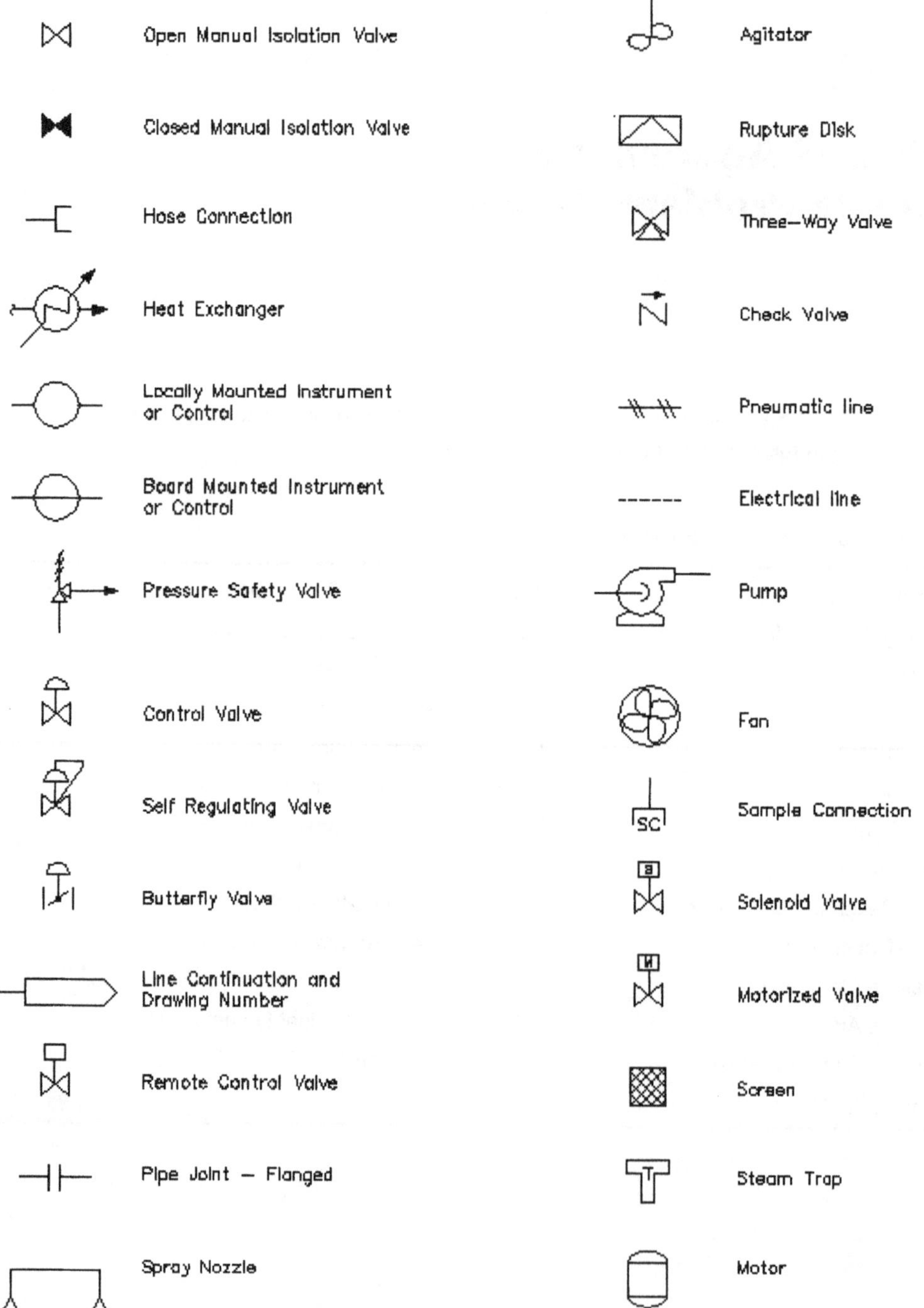

Figure C.1 Symbols used in example problem drawings

Appendix D
Software Aids

Table D.1 lists hazard evaluation software aids identified as available at the time of publication. Fault Tree Analysis, Event Tree Analysis and consequence analysis software are not included in this tabulation. Users must determine for themselves what is available to meet a specific need.

Table D.1 Hazard evaluation software aids

Product(s)	Source	Hazard evaluation methods supported*	Platform
*e*PHA™	Unwin Company Columbus, Ohio, USA	HAZOP, What-If, What-If/Checklist, FMEA, Hazard Identification	Microsoft Excel®
HazardReview *LEADER*™	ABS Consulting Houston, Texas, USA	HAZOP, What-If, Checklist, FMEA	Microsoft Access®
HAZOP Manager	Lihou Technical & Software Services, Birmingham, UK	HAZOP, FMEA, Hazard Identification	Proprietary database
HAZOP+; Reliability Workbench	Isograph Inc. Irvine, California, USA	HAZOP, What-If, FMEA	Custom Windows®-based
HAZOPtimizer™	ioMosaic Corporation Salem, New Hampshire, USA	HAZOP, FMEA, What-If, Checklist	Microsoft Excel
PHA Manager, LOPA Manager	Berwanger, Inc. (Siemens) Houston, Texas, USA	PHAs	Relational database
PHAPlus™	Risk Management Professionals, Inc., Mission Viejo, California, USA	HAZOP, What-If/Checklist	Microsoft Access
FMEA-Pro® PHA-Pro®	Dyadem International Ltd., Richmond Hill, Ontario, Canada	HAZOP, What-If, What-If/Checklist, FMEA, PreHA	Proprietary database
PHAWorks®	PrimaTech Inc. Columbus, Ohio, USA	HAZOP, What-if, What-If/Checklist, FMEA, PreHA	Stand-alone Windows-based

* Some of these products also allow documentation of order-of-magnitude quantitative risk analysis methods or combined techniques such as LOPA, HAZOP/LOPA and/or SIL determination

Appendix E

Chemical Compatibility Chart

The purpose of the Compatibility Chart[1] (Figure E.1) is to show chemical combinations known or believed to be dangerously reactive in the case of unintentional mixing, to indicate whether bulk chemicals can be shipped in adjacent cargo tanker holds. The chart provides a broad grouping of chemicals with an extensive variety of possible binary combinations.

Although one group can be considered dangerously reactive with another group under normal (ambient) conditions where an "X" appears on the chart, there are wide variations in the reaction rates of individual chemicals within the broad groupings. Some individual materials in one group will react violently with some of the materials in another group; others will react slowly, or not at all. Also, even though two groups are generally compatible, there may be specific chemical combinations (e.g., trichloroethylene [36] and caustics [5]) that are incompatible, and combinations may be reactive with one another under non-ambient or impure conditions such as at elevated temperatures or with a catalytic contaminant present. ***The Chart should, therefore, not be used as an infallible guide. It is only an aid in the safe storage and handling of chemicals, with the recommendation that proper safeguards be taken to avoid unintentional mixing of binary mixtures for which an "X" appears on the chart.***

Chemical compatibility must always be reviewed by a qualified chemist or chemical engineer before storage or operations. Readers desiring to use this chart should obtain the CHRIS database with its definitions of reactivity groups and lists of exceptions to the general chart. Documented exceptions to Figure E.1 are listed at the end of this Appendix.

The following procedure explains how the chart can be used in determining compatibility information:

1. Determine the chemical group of a particular chemical.
2. Enter the chart with the chemical group. Proceed across the page. An "X" indicates a reactivity group that forms an unsafe combination with the chemical in question.

For example, crotonaldehyde belongs to Group 19 (aldehydes). The Compatibility Chart shows that chemicals in this group should be segregated from sulfuric and nitric acids, caustics, ammonia, and all types of amines (aliphatic, alkanol, and aromatic). As explained in the notes in CHRIS, crotonaldehyde is also incompatible with Group 1, non-oxidizing mineral acids, even though Group 19 materials are generally compatible with Group 1 materials.

A computerized Chemical Reactivity Worksheet, that employs chemical reactivity groups similarly to the Compatibility Chart, is available from the U.S. National Oceanic and Atmospheric Administration (NOAA).[2] Additional information on identifying chemical incompatibility hazards can be found in CCPS' Concept Book on managing chemical reactivity hazards[3] and in the ASTM standard E 2012 on preparing chemical compatibility charts.[4]

REACTIVE GROUPS \ CARGO GROUPS	1. NON-OXIDIZING MINERAL ACIDS	2. SULFURIC ACID	3. NITRIC ACID	4. ORGANIC ACIDS	5. CAUSTICS	6. AMMONIA	7. ALIPHATIC AMINES	8. ALKANOLAMINES	9. AROMATIC AMINES	10. AMIDES	11. ORGANIC ANHYDRIDES	12. ISOCYANATES	13. VINYL ACETATE	14. ACRYLATES	15. SUBSTITUTED ALLYLS	16. ALKYLENE OXIDES	17. EPICHLOROHYDRIN	18. KETONES	19. ALDEHYDES	20. ALCOHOLS, GLYCOLS	21. PHENOLS, CRESOLS	22. CAPROLACTAM SOLUTION	30. OLEFINS	31. PARAFFINS	32. AROMATIC HYDROCARBONS	33. MISCELLANEOUS HYDROCARBON MIXTURES	34. ESTERS	35. VINYL HALIDES	36. HALOGENATED HYDROCARBONS	37. NITRILES	38. CARBON DISULFIDE	39. SULFOLANE	40. GLYCOL ETHERS	41. ETHERS	42. NITROCOMPOUNDS	43. MISCELLANEOUS WATER SOLUTIONS
1. NON-OXIDIZING MINERAL ACIDS		X			X	X	X	X	X	X	X	X	X			X	X																			
2. SULFURIC ACID	X		X	X	X	X	X	X	X	X	X	X	X	X	X	X	X	X	X	X	X	X	X				X			X			X	X		X
3. NITRIC ACID		X			X	X	X	X	X	X	X	X	X	X	X	X	X	X	X	X	X		X		X	X	X	X						X		
4. ORGANIC ACIDS		X			X	X	X	X				X				X	X																			
5. CAUSTICS	X	X	X	X							X	X				X	X		X	X	X	X													X	
6. AMMONIA	X	X	X	X						X	X	X	X			X	X		X																X	
7. ALIPHATIC AMINES	X	X	X	X							X	X	X	X	X	X	X	X	X	X	X	X									X				X	
8. ALKANOLAMINES	X	X	X	X							X	X	X	X	X	X	X		X												X				X	
9. AROMATIC AMINES	X	X	X								X	X							X																X	
10. AMIDES	X	X	X			X						X									X															
11. ORGANIC ANHYDRIDES	X	X	X		X	X	X	X	X																											
12. ISOCYANATES	X	X	X	X	X	X	X	X	X	X										X		X											X			X
13. VINYL ACETATE	X	X	X			X	X	X																												
14. ACRYLATES		X	X				X	X																												
15. SUBSTITUTED ALLYLS		X	X				X	X																												
16. ALKYLENE OXIDES	X	X	X	X	X	X	X	X																												
17. EPICHLOROHYDRINS	X	X	X	X	X	X	X	X																												
18. KETONES		X	X				X																													
19. ALDEHYDES		X	X		X	X	X	X	X																											
20. ALCOHOLS, GLYCOLS		X	X		X		X					X																								
21. PHENOLS, CRESOLS		X	X		X		X			X																										
22. CAPROLACTAM SOLUTION		X			X		X					X																X								

Figure E.1 Cargo compatibility chart from *CHRIS Manual* ("X" indicates incompatible groups; see text)

Compatibility Chart Exceptions: Nonreactive Combinations

The binary combinations listed in Table E.1 have been tested as prescribed in Appendix III of Reference 1 and found not to be dangerously reactive. These combinations are exceptions to the Compatibility Chart (Figure E.1) and may be stowed in adjacent tanks.

Table E.1 Not dangerously reactive exceptions

Member of reactive group	Compatible with
Acetone (18)	Diethylenetriamine (7)
Acetone cyanohydrin (0)	Acetic acid (4)
Acrylonitrile (15)	Triethanolamine (8)
1,3-Butylene glycol (20)	Morpholine (7)
1,4-Butylene glycol (20)	Ethylamine (7) Triethanolamine (8)
Gamma-Butyrolactone(0)	N-Methyl-2-pyrrolidone (9)
Caustic potash, 50% or less (5)	Isobutyl alcohol (20) Ethyl alcohol (20) Ethylene glycol (20) Isopropyl alcohol (20) Methyl alcohol (20) iso-Octyl alcohol (20)
Caustic soda, 50% or less (5)	Butyl alcohol (20) tert-Butyl alcohol, Methanol mixtures Decyl alcohol (20) Iso-Decyl alcohol Diacetone alcohol (20) Diethylene glycol (40) Ethyl alcohol (40%, whiskey) (20) Ethylene glycol (20) Ethylene glycol, Diethylene glycol mixture (20) Ethyl hexanol (Octyl alcohol) (20) Methyl alcohol (20) Nonyl alcohol (20) iso-Nonyl alcohol (20) Propyl alcohol (20) Propylene glycol (20) Sodium chlorate (0) iso-Tridecanol (20)
Dodecyl and Tetradecylamine mixture (7)	Tall oil, fatty acid (34)
Ethylenediamine (7)	Butyl alcohol (20) tert-Butyl alcohol (20) Butylene glycol (20) Creosote (21) Diethylene glycol (40) Ethyl alcohol (20) Ethylene glycol (20) Ethyl hexanol (20) Glycerine (20) Isononyl alcohol (20) Isophorone (18) Methyl butyl ketone (18) Methyl isobutyl ketone (18) Methyl ethyl ketone (18) Propyl alcohol (20) Propylene glycol (20)
Oleum (0)	Hexane (31) Dichloromethane (36) Perchloroethylene (36)
1,2-Propylene glycol (20)	Diethylenetriamine (7) Polyethylene polyamines (7) Triethylenetetramine (7)
Sodium dichromate, 70% (0)	Methyl alcohol (20)
Sodium hydrosulfide solution (5)	Iso-Propyl alcohol (20)
Sulfuric acid (2)	Coconut oil (34) Coconut oil acid (34) Palm oil (34) Tallow (34)
Sulfuric acid, 98% or less (2)	Choice white grease tallow (34)

Compatibility Chart Exceptions: Reactive Combinations

The binary combinations listed below have been determined to be dangerously reactive, based on either data obtained in the literature or on laboratory testing that has been carried out in accordance with procedures prescribed in Appendix III of Reference 1. These combinations are exceptions to the Compatibility Chart (Figure E.1) and may not be stowed in adjacent cargo tanks.

- Acetone cyanohydrin (0) is not compatible with Groups 1-12, 16, 17 and 22.
- Acrolein (19) is not compatible with Group 1, Non-Oxidizing Mineral Acids.
- Acrylic acid (4) is not compatible with Group 9, Aromatic Amines.
- Acrylonitrile (15) is not compatible with Group 5 (Caustics)
- Alkylbenzenesulfonic acid (0) is not compatible with Groups 1-3, 5-9, 15, 16, 18, 19, 30, 34, 37, and strong oxidizers.
- Allyl alcohol (15) is not compatible with Group 12, Isocyanates.
- Alkyl (C7-C9) nitrates (34) is not compatible with Group 1, Non-oxidizing Mineral Acids.
- Aluminum sulfate solution (43) is not compatible with Groups 5-11.
- Ammonium bisulfite solution (43) is not compatible with Groups 1, 3, 4, and 5.
- Benzenesulfonyl chloride (0) is not compatible with Groups 5-7, and 43.
- 1,4-Butylene glycol (20) is not compatible with Groups 1-9.
- gamma-Butyrolactone (0) is not compatible with Groups 1-9.
- Caustic soda solution, 50% or less (5) is not compatible with 1,4-Butylene glycol (20).
- Crotonaldehyde (19) is not compatible with Group 1, Non-Oxidizing Mineral Acids.
- Cyclohexanone, Cyclohexanol mixture (18) is not compatible with Group 12, Isocyanates.
- 2,4-Dichlorophenoxyacetic acid, Triisopropanolamine salt solution (43) is not compatible with Group 3, Nitric acid.
- 2,4-Dichlorophenoxyacetic acid, Dimethylamine salt solution (0) is not compatible with Groups 1-5, 11, 12, and 16.
- Dimethyl hydrogen phosphite (34) is not compatible with Groups 1 and 4.
- Dimethyl naphthalene sulfonic acid, sodium salt solution (34) is not compatible with Group 12,
- Formaldehyde, and strong oxidizing agents.
- Dodecylbenzenesulfonic acid (0) is not compatible with oxidizing agents and Groups 1, 2, 3, 5, 6, 7, 8, 9, 15, 16, 18, 19, 30, 34, and 37.
- Ethylenediamine (7) is not compatible with Ethylene dichloride (36).
- Ethylene dichloride (36) is not compatible with Ethylenediamine (7).

- Ethylidene norbonene (30) is not compatible with Groups 1-3 and 5-8.
- 2-Ethyl-3-propylacrolein (19) is not compatible with Group 1, Non-Oxidizing Mineral Acids.
- Ferric hydroxyethyylethylenediamine triacetic acid, Sodium salt solution (43) is not compatible with Group 3, Nitric acid.
- Fish oil (34) is not compatible with Sulfuric acid (2).
- Formaldehyde (over 50%) in Methyl alcohol (over 30%) (19) is not compatible with Group 12, Isocyanates.
- Formic acid (4) is not compatible with Furfural alcohol (20).
- Furfuryl alcohol (20) is not compatible with Group 1, Non-Oxidizing Mineral Acids and Formic acid (4).
- 2-Hydroxyethyl acrylate is not compatible with Groups 2, 3, 5-8 and 12.
- Isophorone (18) is not compatible with Group 8, Alkanolamines.
- Magnesium chloride solution (0) is not compatible with Groups 2, 3, 5, 6 and 12.
- Mesityl oxide (18) is not compatible with Group 8, Alkanolamines.
- Methacrylonitrile (15) is not compatible with Group 5 (Caustics).
- Methyl tert-butyl ether (41) is not compatible with Group 1, Non-oxidizing Mineral Acids.
- Naphtha, cracking fraction (33) is not compatible with strong acids, caustics or oxidizing agents.
- o-Nitrophenol (0) is not compatible with Groups 2, 3, and 5-10.
- Octyl nitrates (all isomers) *see* Alkyl (C7-C9) nitrates.
- Oleum (0) is not compatible with Sulfuric acid (2) and 1,1,1-Trichloroethane (36).
- Phthalate based polyester polyol (0) is not compatible with group 2, 3, 5, 7 and 12.
- Pentene, Miscellaneous hydrocarbon mixtures (30) are not compatible with strong acids or oxidizing agents.
- Polyglycerine, Sodium salts solution (20) is not compatible with Groups 1, 4, 11, 16, 17, 19, 21, and 22.
- Sodium acetate, Glycol, Water mixture (1% or less Sodium hydroxide) (34) is not compatible with Group 12 (Isocyanates).
- Sodium chlorate solution (50% or less) (0) is not compatible with Groups 1-3, 5, 7, 8, 10, 12, 13, 17, and 20.
- Sodium dichromate solution (70% or less) (0) is not compatible with Groups 1-3, 5, 7, 8, 10, 12, 13, 17, and 20.
- Sodium dimethyl naphthalene sulfonate solution (34) is not compatible with Group 12, Formaldehyde and strong oxidizing agents.

- Sodium hydrogen sulfide, Sodium carbonate solution (0) is not compatible with Groups 6 (Ammonia) and 7 (Aliphatic amines).
- Sodium hydrosulfide (5) is not compatible with Groups 6 (Ammonia) and 7 (Aliphatic amines).
- Sodium hydrosulfide, Ammonium sulfide solution (5) is not compatible with Groups 6 (Ammonia) and 7 (Aliphatic amines).
- Sodium polyacrylate solution (43) is not compatible with Group 3, Nitric Acid.
- Sodium salt of Ferric hydroxyethylethylenediamine triacetic acid solution (43) is not compatible with Group 3, Nitric acid.
- Sodium silicate solution (43) is not compatible with Group 3, Nitric acid.
- Sodium sulfide, hydrosulfide solution (0) is not compatible with Groups 6 (Ammonia) and 7 (Aliphatic amines).
- Sodium thiocyanate (56% or less) (0) is not compatible with Groups 1-4.
- Sulfonated polyacrylate solution (43) is not compatible with Group 5 (Caustics).
- Sulfuric acid (2) is not compatible with Fish oil (34), or Oleum (0).
- Tallow fatty acid (34) is not compatible with Group 5, Caustics.
- 1,1,1-Trichloroethane (36) is not compatible with Oleum (0).
- Trichlorethylene (36) is not compatible with Group 5, Caustics.
- Triethyl phosphite (34) is not compatible with Groups 1 and 4.
- Trimethyl phosphite (34) is not compatible with Groups 1 and 4.
- 1,3,5-Trioxane (41) is not compatible with Group 1 (Non-oxidizing mineral acids) and Group 4 (Organic acids).

Appendix E References

1. *Chemical Hazards Response Information System (CHRIS) Manual,* U.S. Coast Guard, 400 Seventh St. SW, Washington, DC 20590., www.chrismanual.com, updated September 2001.
2. NOAA Chemical Reactivity Worksheet, Version 1.9, U.S. National Oceanic and Atmospheric Administration, http://response.restoration.noaa.gov/chemaids/react.html.
3. R. Johnson, S. Rudy and S. Unwin, *Essential Practices For Managing Chemical Reactivity Hazards,* American Institute of Chemical Engineers, New York, 2003.
4. ASTM E 2012-00, Standard Guide for the Preparation of a Binary Chemical Compatibility Chart, ASTM International, West Conshohocken, Pennsylvania, 2000.

Appendix **F**

Organizations Offering Process Safety Enhancement Resources

Table F.1 lists some professional and industry organizations that offer process safety enhancement resources. This cannot purport to be a complete listing; however, it may point the user to some helpful starting points for meeting a particular need. Manufacturing companies and chemical user groups should be contacted for information regarding the safe storage and handling of particular chemicals.

Some consulting companies also provide resources that may prove useful. CCPS' Professional Services Directory (www.aiche.org/ccps) lists what various CCPS member companies profess to offer in the way of process safety services.

Table F.1 Professional and industry organizations offering process safety enhancement resources

Organization	Selected examples of programs offered
Air & Waste Management Association Pittsburgh, Pennsylvania (412) 232-3444 awma.org	• Exchange of technical and managerial information about air pollution control and waste management • Books, journals, videotapes
American Chemical Society Washington, DC (202) 872-4600 chemistry.org	• Health & safety and chemical properties referral services • Chemical safety manual for small businesses • Hazard Communication Standard information • Laboratory safety and design information • Laboratory safety short courses
American Chemistry Council Arlington, Virginia (703) 741-5000 americanchemistry.org	• Management guidelines • Responsible Care® • National Chemical Referral and Informational Center (including CHEMTREC) • Community Awareness and Emergency Response (CAER)
American Industrial Hygiene Association Akron, Ohio (216) 873-2442 aiha.org	• Emergency Response Planning Guidelines • Workplace Environmental Exposure Levels • Hygiene Guides (Toxic Properties Surveys) • Professional Development Seminars

Table F.1 Professional and industry organizations offering process safety enhancement resources

Organization	Selected examples of programs offered
American Institute of Chemical Engineers New York, New York (212) 591-8100 aiche.org, sache.org	• Center for Chemical Process Safety (CCPS) concept and guidelines books, *Process Safety Beacon,* Safety Alerts, Conferences • Design Institute for Emergency Relief Systems • Design Institute for Physical Property Data • Loss Prevention Symposia, Process Plant Safety Symposia, Ammonia Safety Symposia • Continuing education short courses • Center for Waste Reduction Technology • Safety and Chemical Engineering Education undergraduate teaching resources
American National Standards Institute New York, New York (212) 642-4900 ansi.org	• Consensus standards on various subjects • Special publications and handbooks • Conference and seminars on emerging technologies and standards development • *ANSI Newsletter*
American Petroleum Institute Washington, DC (202) 682-8000 api.org	• Process Hazards Management Task Force • Process hazards and process safety seminars • Technical standards and recommended practices (e.g., fire protection, facility maintenance) • Equipment Inspection Guides • Operator and maintenance training • Recommended Practices
American Society of Mechanical Engineers New York, New York (212) 705-7722 asme.org	• Boiler and Pressure Vessel Code • National Board (repair of pressure vessels and safety valves) • Non-destructive testing • Professional development courses • System of accreditation for manufacturers of equipment • Technical publications • Industry Advisory Board
American Society of Safety Engineers Des Plaines, Illinois (708) 692-4121 asse.org	• Seminars and workshops • Continuing education courses in system safety and incident prevention • Audio-visual programs
ASTM International Philadelphia, Pennsylvania (215) 299-5400 astm.org	• Standards development • CHETAH™ program for estimating physical properties • Standard technology training courses • Continuing technical education • Publication Information Center

Table F.1 Professional and industry organizations offering process safety enhancement resources

Organization	Selected examples of programs offered
Center for Chemical Process Safety	*See* American Institute of Chemical Engineers
Chemical Industries Association Ltd. London, United Kingdom 020 7834 3399 cia.org.uk	• Chemical Industry Safety & Health Council • Hydrogen Fluoride Producers and Users Sector Group • Codes of practice for chemicals with major hazards: ammonia, chlorine, ethylene dichloride, ethylene oxide, hydrogen chloride (anhydrous), phosgene
Chlorine Institute Arlington, Virginia (703) 741-5760 chlorineinstitute.org	• Publications on enhancing safety in manufacturing, shipping, handling and storage of chlorine • Guidelines for chlorine storage and handling
Compressed Gas Association Chantilly, Virginia (703) 788-2700 cganet.com	• Guidelines for compressed gas storage and handling • Safety posters • *Handbook of Compressed Gases* • Training, seminars, and videos
Dangerous Goods Advisory Council Washington, DC (202) 728-1460 dgac.org	• Publications of standards • Monthly newsletter • Educational services: video conferences, self-study courses • Basic and advanced courses on: — Transportation of hazardous materials and waste — FAA certification course on air transportation of dangerous goods (FAA-ICAO approved)
European Chemical Industry Council Brussels, Belgium +32 2 676 72 29 cefic.org	• Online glossary, resources, and links • European chemical industry information • Responsible Care®
Federal Emergency Management Agency Washington, DC (202) 646-3923 fema.gov	• Handbook of chemical hazard analysis procedures • Emergency management activities • Disaster assistance coordination • Public awareness and education programs • Emergency Management Institute
The Institution of Chemical Engineers Rugby, United Kingdom (0788) 578214 icheme.org, epsc.org	• *Loss Prevention Bulletin* (case histories) • Information exchange • Training modules • Conferences on major accident prevention • European Process Safety Centre (epsc.org)
IEEE New York, New York (212) 419-7900 ieee.org	• *IEEE Transactions on Reliability* special issues: Chemical Process Reliability, Safety, and Risk Management • Journals and conference publications • Technical Activities Guide

Table F.1 Professional and industry organizations offering process safety enhancement resources

Organization	Selected examples of programs offered
International Institute of Ammonia Refrigeration Arlington, Virginia (703) 312-4200 iiar.org	• Ammonia codes and standards • Bulletins, guidelines • RMP & PSM Compliance Library
Instrumentation, Systems and Automation Society Research Triangle Park, North Carolina (919) 549-8411 isa.org	• Standards and practices related to instrumentation and electrical systems • Consensus practice related to safety instrumented systems • Certification programs • Technical training • Books, publications • Conferences
Mary Kay O'Connor Process Safety Center Texas A&M University College Station, Texas (979) 845-3489 process-safety.tamu.edu	• Process safety research • Annual process safety symposium • Continuing education courses • Safety alerts
NACE International Houston, Texas (281) 228-6200 nace.org	• Guidelines for protecting piping and equipment from corrosion
National Association of Manufacturers Washington, DC (202) 637-3000 nam.org	• Case histories of incident causes and prevention • Computer analysis of accidents (NAM Safe System) • Process Hazard Task Force • Hazard Training Task Force • Risk Management Committee
National Fire Protection Association Quincy, Massachusetts (617) 770-3000 nfpa.org	• Consensus standards related to fire and explosion prevention • Fire safety seminars • Training materials
National Institute for Chemical Studies Charleston, West Virginia (304) 346-6264 www.nicsinfo.org	• Chronic health effects study • Voluntary reduction of routine emissions • Community safety assessment program • Emergency response database • Citizens guide for environmental issues
National Safety Council Itasca, Illinois (630) 285-1121 nsc.org	• Accident Prevention Manual for Industrial Operations • Safety Training Institute • Newsletters • Injury data • Safety Video Programs

Table F.1 Professional and industry organizations offering process safety enhancement resources

Organization	Selected examples of programs offered
National Oceanic & Atmospheric Administration, U.S. Dept. of Commerce Washington, DC (202) 482-6090 noaa.gov	• CAMEO chemical database • ALOHA dispersion model • Chemical Reactivity Worksheet
Organisation for Economic Co-operation and Development Paris, France +33 1.45.24.82.00 oecd.org	• *OECD Guiding Principles for Chemical Accident Prevention, Preparedness and Response*; other guidance documents • Workshops • International Directory of Emergency Response Centres for Chemical Accidents
Semiconductor Equipment and Materials International San Jose, California (408) 943-6900 www.semi.org	• Semiconductor manufacturing safety guidelines • Fire safety evaluation checklist for semiconductor equipment using hazardous production materials
Society for Risk Analysis McLean, Virginia (703) 790-1745 sra.org	• Conferences and workshops on risk analysis • Journals (*Risk Analysis; Journal of Risk Research*)
The Society of the Plastics Industry Washington, DC (202) 974-5200 plasticsindustry.org	• Publications on plastics safety, incident prevention, and health hazards • Seminars and workshops • Organic Peroxides Producers Safety Division
Synthetic Organic Chemical Manufacturers Association Washington, DC (202) 721-4100 socma.com	• Worker training and certification • ChemStewards® program
System Safety Society Unionville, Virginia (540) 854-8630 system-safety.org	• *Journal of System Safety* • Standards for system safety • Educational programs • International System Safety Conference • Programs to broaden the application of system safety
Technical Association of the Pulp and Paper Industry Atlanta, Georgia (770) 446-1400 tappi.org	• Publications (*TAPPI Journal*, conference proceedings, technical information) • Self-study courses • Videotapes • Human Resource Development Committee • Conferences, seminars and short courses
The World Bank Washington, DC (202) 473-1000 worldbank.org	• *Techniques for Assessing Industrial Hazards: A Manual* (ISBN 0-8213-0779-7)

Table F.1 Professional and industry organizations offering process safety enhancement resources

Organization	Selected examples of programs offered
US Chemical Safety and Hazard Investigation Board Washington, DC (202) 261-7600 csb.gov	• Incident investigation reports and digests • Hazard investigations • Videos • Safety bulletins
US Coast Guard 2100 Second St SW, Washington, DC 20593 (202) 426-9568 uscg.mil; chrismanual.com	• *Chemical Hazard Response Information System* (CHRIS) • Maritime Transportation Security Act
US Department of Labor, Occupational Safety & Health Administration Washington, DC 20210 (800) 488-7087 osha.gov	• OSHA alliances and cooperative programs • Consultation services for employers • Safety and Health Information Bulletins (SHIBs) • Voluntary Protection Programs
US Environmental Protection Agency Washington, DC (202) 260-2090 epa.gov	• Risk Management Program (RMP) guidance • *Hazardous Materials Emergency Planning Guide*, NRT-1 • Safety Alerts

Selected Bibliography

General references pertaining to hazard evaluation procedures are given in this Selected Bibliography. See references at the end of each chapter and methodology section for topic-specific literature.

Center for Chemical Process Safety, *Guidelines for Risk Based Process Safety,* ISBN 978-0-470-16569-0, American Institute of Chemical Engineers, New York, 2007.

Center for Chemical Process Safety, *Guidelines for Safe and Reliable Instrumented Protective Systems,* ISBN 978-0-471-97940-1, American Institute of Chemical Engineers, New York, 2007.

F. Crawley, M. Preston and B. Tyler, *HAZOP: Guide to Best Practice,* ISBN 0-85295-427-1, Institution of Chemical Engineers, Rugby, UK, 2000.

D. A. Crowl and J. F. Louvar, *Chemical Process Safety: Fundamentals with Applications, 2nd Ed.,* Prentice Hall, Upper Saddle River, New Jersey, 2002.

H. R. Greenberg and J. J. Cramer, eds., *Risk Assessment and Risk Management for the Chemical Process Industry*, ISBN 0-471-28882-9, John Wiley & Sons, New York, 1991.

M. G. Gressel and J. H. Gideon, "An Overview of Process Hazard Evaluation Techniques," *American Industrial Hygiene Association Journal,* Vol. 52, No. 4, April 1991.

V. L. Grose, *Managing Risk — Systematic Loss Prevention for Executives,* Prentice Hall, Englewood Cliffs, NJ, 1987.

N. Hyatt, *Guidelines for Process Hazards Analysis (PHA, HAZOP), Hazards Identification, and Risk Analysis,* ISBN 978-0849319099, Dyadem Press, Richmond Hill, Ontario, 2003.

S. A. Lapp, "The Major Risk Index System," *Plant/Operations Progress*, Vol. 9, No. 3, July 1990.

F. P. Lees, "The Hazard Warning Structure of Major Hazards," *Transactions of the Institution of Chemical Engineers*, Vol. 60, No. 211, London, 1982.

S. Mannan, ed., *Lees' Loss Prevention in the Process Industries, 3rd Ed.,* Elsevier Butterworth-Heinemann, ISBN 0-7506-7555-1, Oxford, UK, 2005.

P. F. McGrath, "Using Qualitative Methods to Manage Risk," *Reliability Engineering and System Safety*, Vol. 29, 1990.

Organisation for Economic Co-operation and Development, *OECD Guiding Principles for Chemical Accident Prevention, Preparedness and Response, 2nd Ed.,* Paris, 2003.

R. H. Perry and D. W. Green, *Perry's Chemical Engineer's Handbook, 8th Ed.*, McGraw-Hill, New York, 2007.

F. Redmill and T. Anderson (eds.), *Developments in Risk-Based Approaches to Safety: Proceedings of the 14th Safety-critical Systems Symposium,* ISBN 978-1-84628-333-8, Springer-Verlag, London, 2006.

B. Skelton, *Process Safety Analysis: An Introduction*, Institution of Chemical Engineers, ISBN 0-88415-666-4, Rugby, England, 1997.

U.S. Department of Defense, Military Standard System Safety Program Requirements, MIL-STD-882D, Washington, DC, 2000.

U.S. Department of Energy, "DOE Handbook: Chemical Process Hazard Analysis," DOE-HDBK-1100-2004, Washington, DC, August 2004. Available from National Technical Information Service, Springfield, Virginia 22161, and from www.energy.gov.

U.S. Department of Energy, "Example Process Hazard Analysis of a Department of Energy Water Chlorination Process," DOE/EH-0340, Washington, DC, September 1993. Available from National Technical Information Service, Springfield, Virginia 22161, and from www.energy.gov.

Index

A

B

C

D

E

F

G

H

Q

R

S

T

V

W